FORTSCHRITTE DER CHEMIE ORGANISCHER NATURSTOFFE

PROGRESS IN THE CHEMISTRY OF ORGANIC NATURAL PRODUCTS

BEGRÜNDET VON · FOUNDED BY

L. ZECHMEISTER

HERAUSGEGEBEN VON · EDITED BY

W. HERZ
TALLAHASSEE, FLA.

H. GRISEBACH
FREIBURG i. BR.

G. W. KIRBY
GLASGOW

VOL. 36

VERFASSER · AUTHORS

C. W. J. CHANG · I. FLAMENT · J. A. MATSON · T. NISHIDA
G. OHLOFF · F. W. WEHRLI · A. J. WEINHEIMER

1979

SPRINGER-VERLAG WIEN GMBH

Mit 11 Abbildungen. With 11 Figures

Originally published by Springer-Verlag Wien-New York in 1979
Softcover reprint of the hardcover 1st edition 1979

Library of Congress Catalog Card Number AC 39-1015

ISSN 0071-7886

ISBN 978-3-7091-3267-8 ISBN 978-3-7091-3265-4 (eBook)
DOI 10.1007/978-3-7091-3265-4

Inhaltsverzeichnis. Contents

Mitarbeiterverzeichnis. List of Contributors

Chang, Assoc. Prof. Dr. C. W. J., Faculty of Chemistry, University of West Florida, Pensacola, FL 32504, U.S.A.

Flament, Dr. I., Laboratoire de Recherches, Firmenich SA, Case Postale 239, CH-1211 Genève 8, Switzerland.

Matson, Dr. J. A., Research Associate, Department of Medicinal Chemistry and Pharmacognosy, University of Houston, Houston, TX 77004, U.S.A.

Nishida, Dr. T., Swedish Tobacco Company, Research Department, P.O.B. 17007, S-10462 Stockholm 17, Sweden.

Ohloff, Dr. G., Laboratoire de Recherches, Firmenich SA, Case Postale 239, CH-1211 Genève 8, Switzerland.

Wehrli, Dr. F. W., Varian AG, Steinhauserstrasse, CH-6300 Zug, Switzerland.

Weinheimer, Prof. Dr. A. J., Department of Medicinal Chemistry and Pharmacognosy, University of Houston, Houston, TX 77004, U.S.A.

The Use of Carbon-13 Nuclear Magnetic Resonance Spectroscopy in Natural Products Chemistry

By F. W. WEHRLI, Varian AG, NMR Applications Laboratory, Zug, Switzerland, and T. NISHIDA, Swedish Tobacco Company, Research Department, Stockholm, Sweden

With 10 Figures

Contents

1. Introduction

Since the late 1950's when proton n.m.r. spectroscopy was first used in organic natural products studies the technique has increasingly contributed to the rapid advancement of this important area of chemistry. Although the potential utility of ^{13}C n.m.r. was recognized very early, essentially no application of ^{13}C n.m.r. appeared in the literature prior to 1966 and 95% of the existing data are less than five years old. The initially slow growth had its cause in inadequate instrumentation, insufficient sensitivity being the main obstacle. This situation drastically changed with the advent and commercial availability of broadband excitation and Fourier transform methods, giving natural-abundance ^{13}C n.m.r. and its numerous chemical applications a tremendous impetus. Today ^{13}C spectra can be recorded on sample quantities down to the submilligram level, which until recently even withstood proton n.m.r.

Paralleling the development of experimental techniques considerable progress has also been made on an understanding of spectral parameters, in particular their stereochemical implications on natural products. Although the large majority of data present up to now deals with known structures, an adequate basis now exists which allows the chemist to use the technique for tackling *real* problems on *unknown* molecules.

A few years ago HIGHET and SOKOLOSKI (*1*) in this series reviewed newer n.m.r. methods in structure investigation of natural products by devoting half of their review to describe the principle of pulsed Fourier transform n.m.r. spectroscopy, general features and basic experimental techniques related to ^{13}C n.m.r. spectroscopy. Several introductory review articles have also appeared recently to demonstrate the usefulness of ^{13}C n.m.r. for structure analysis of natural products (*2*, *3*) and in biosynthetic studies (*4*—*9*).

In order to avoid unnecessary duplication and since comprehensive coverage would present an almost impossible task, the authors of the present review have chosen to cover mainly the 1974—1977 literature and to quote earlier primary literature only in exceptional cases. For topics of more general nature the reader will be referred to the pertinent specialist reports.

2. ^{13}C NMR Spectral Assignments

Carbon-13 chemical shifts are readily obtained from the proton noise-decoupled (PND) spectrum in which non-equivalent carbons resonate as separate single lines. Although proton-coupled ^{13}C spectra contain much more information, they cannot, in general, be analyzed completely because of severe overlap of spin multiplets as they occur in the spectra of complex organic molecules such as natural products. Moreover, recording of such spectra demands up to two orders of magnitude more spectrometer time. Fortunately, some of the inherent coupling information can also be gained from the residual splitting patterns observed in the single-frequency off-resonance decoupled (SFORD) spectra. A prerequisite for any application of ^{13}C n. m. r. spectroscopy is the proper and unambiguous assignment of the resonances in the molecule under investigation. This requirement is a stringent one and uncertainities which exist in the assignment must be emphasized and clearly mentioned to avoid incorrect assignments in future studies. In this section several recently developed assignment techniques are reviewed. For a more detailed discussion of this subject the reader is referred to Chapter 3 in Reference (*10*).

2.1. Single Frequency Decoupling

When the proton decoupler frequency is at exact resonance of some protons, the carbon atom directly bonded to them will appear as a singlet (selective proton decoupling). Off-resonance irradiation, on the other hand, causes compressed multiplets, characterized by a residual splitting J , while from the multiplicity the number of attached protons can be derived. The residual coupling J^r can be correlated with the decoupler offset frequency ΔF (in Hz) and the decoupler power $\gamma H_2/2\pi$ (in Hz)

$$J^r = {}^1J_{CH}\,\Delta F/(\gamma H_2/2\pi). \quad (1)$$

This approximation is valid if $\gamma H_2/2\pi \gg \Delta F$ and ${}^1J_{CH}$.

Incremental variation of the proton decoupler frequency provides a set of spectra containing differently spaced residual multiplets. The effective ^{13}C resonance frequencies are then linearly related to the decoupler frequency. In this manner a series of straight lines is obtained which intersect at the frequency which corresponds to the proton resonance frequency. These experiments therefore afford a correlation between ^{13}C and ^1H chemical shifts (*11*).

However, off-resonance patterns of CH_2's and CH's in complex organic molecules are not always clearly resolved multiplets. Instead rather complicated multiline patterns are often obtained.

Recently deviations from simple first-order multiplets were systematically studied and the utilization of off-resonance pattern recognition applied to signal assignment and structure elucidation (*12*).

Basically two situations have been encountered, both of which lead to deviations from simple first-order off-resonance patterns. The first case has its cause in chemical shift non-equivalence of the appended protons. Provided that ${}^1J_{CH_A}$ and ${}^1J_{CH_B}$ in the methylene moiety CH_AH_B are equal, a 1 : 2 : 1 triplet is obtained in the proton coupled ^{13}C spectrum. However, in order to afford this pattern in SFORD spectra, a further condition has to be satisfied, i.e. $\delta H_A = \delta H_B$. For strongly anisochronous methylene protons double doublets are therefore anticipated instead of triplets. This is exemplified by a SFORD spectrum shown in Fig. 1 for the exocyclic olefinic methylene carbon belonging to the α,β-unsaturated

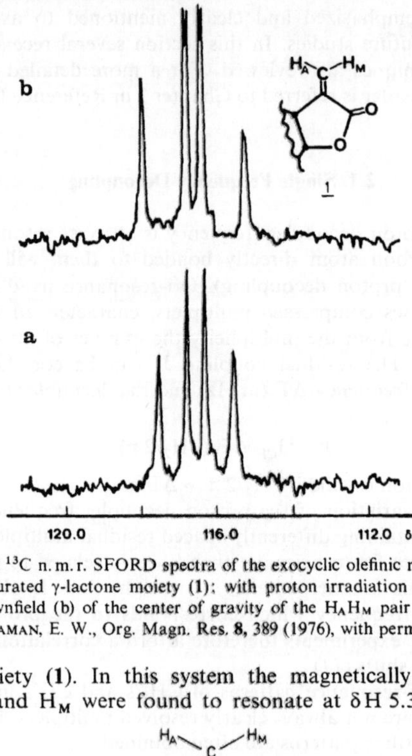

Fig. 1. 25.2 MHz ^{13}C n.m.r. SFORD spectra of the exocyclic olefinic methylene resonance of the α,β-unsaturated γ-lactone moiety (**1**); with proton irradiation 1.7 ppm upfield (a) and 2.3 ppm downfield (b) of the center of gravity of the H_AH_M pair of doublets. (From HAGAMAN, E. W., Org. Magn. Res. **8**, 389 (1976), with permission)

γ-lactone moiety (**1**). In this system the magnetically non-equivalent protons H_A and H_M were found to resonate at δH 5.34 and δH 6.11,

(**1**)

respectively. Since the difference $J^{r}_{CH_A} - J^{r}_{CH_B}$ is independent of ΔF, which is readily verified on the basis of eq. (**1**), the two central transitions remain invariant. A further source of geminal non-equivalence are carbons in conformationally rigid systems whose SFORD spectra therefore often give rise to non-trivial multiplets.

The second cause of complex off-resonance multiplets is the intervention of second-order coupling (*13—15*). The simplest spin system which may provide second-order multiplets is the AA'X or ABX system in which the two protons are located on adjacent carbons:

$$H_A - {}^{13}C_1 - C_2 - H_B .$$

Although, in principle, such a spin system is always second order, the second-order effects usually vanish in the proton-coupled spectra since ${}^1J_{C(1)H(A)} \gg {}^3J_{H(A)H(B)}$. However, high power proton irradiation reduces ${}^1J_{CH}$ to ${}^1J^{r}_{CH}$ which may now become comparable in magnitude to ${}^3J_{HH}$

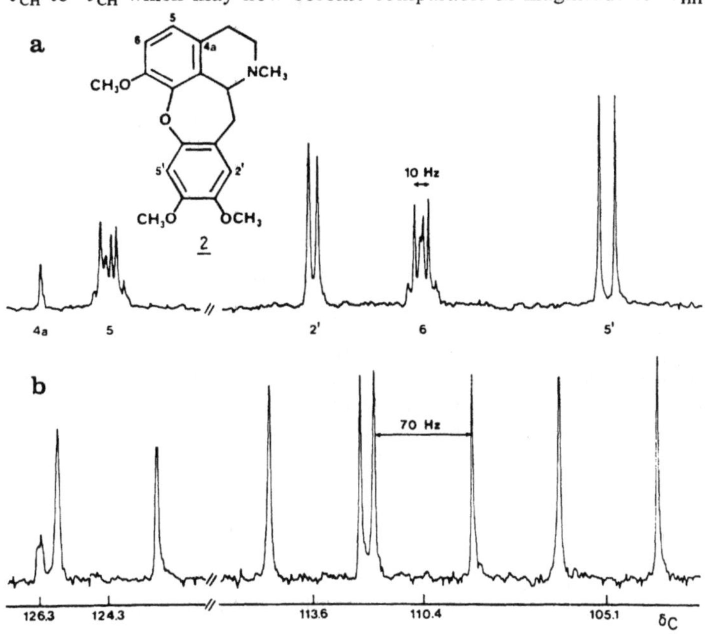

Fig. 2. 25.2 MHz ^{13}C n.m.r. SFORD spectra of the aromatic CH region in the alkaloid cularine (**2**). Traces (a) and (b) have been recorded at a different decoupler power level corresponding to a residual splitting of the methine resonance at 110.4 ppm of 10 Hz (a) and 70 Hz (b). (From HAGAMAN, E. W., Org. Magn. Res. **8**, 389 (1976), with permission)

(*13*). For tetrahedral carbons with $^1J_{CH} \sim 100$ to 200 Hz second-order splittings will thus become most pronounced when $^1J^r_{CH} \sim 0.1\,^1J_{CH}\;^3J_{HH}$. As has been demonstrated (*12*) the second-order effects therefore are largely under experimental control through variation of either the decoupler offset or the decoupler field strength. How this may be exploited for signal assignment is illustrated by the partial ^{13}C SFORD spectra (Fig. 2) displaying some of the aromatic resonances in the alkaloid cularine (**2**). In trace (a) two of the proton-bearing carbons (C-5 and C-6) exhibit second-order multiplets whereas the isolated carbons C-2′ and C-5′ give simple doublets. By contrast all multiplets in trace (b) appear as first-order doublets which is due to the lower decoupler power level used.

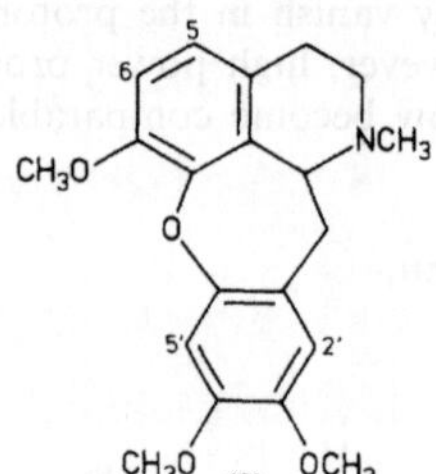

(**2**)

2.2. Proton-Coupled Spectra

Although they are potentially useful diagnostic aids, ^{13}C-^1H spin-spin coupling constants are difficult to obtain because of frequent overlap of resonances and the complexity of the spin systems. ERNST and co-workers (*16*) have recently outlined the possibility of obtaining subspectra of individual carbons within a molecule as a special application of two-dimensional spectroscopy. However, the method requires very large storage capacities for storing the two-dimensional data arrays if sufficient resolution is to be achieved.

An alternative to the previously mentioned technique consists in selectively exciting the resonance of interest. FREEMAN *et al.* (*17*) have proposed applying a train of short radiofrequency pulses at such rate that the magnetization of the nucleus of interest performs an *integral* number of precessions between two pulses. Such a condition can be met for only *one* particular chemically shifted nucleus whose magnetization experiences a cumulative tipping whilst the remainder of signals is annihilated. If, in addition the decoupler is turned off during data acquisition, a proton-coupled single-carbon spectrum is obtained. This technique was applied in order to obtain proton-coupled spectra of C_α and C_β in menthone (**3**) (*18*). The relative stabilities of rotamers (**3a**), (**3b**), and (**3c**) in menthone and analogs have been the subject of considerable debate.

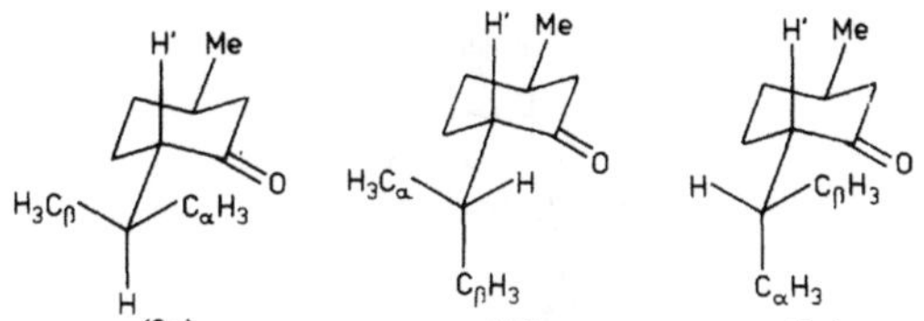

(**3a**) (**3b**) (**3c**)

Since the vicinal coupling constant $^3J_{CCCH}$ is well known to be stereospecific, the knowledge of the coupling constant between the isopropyl carbon and H′ was expected to settle the controversy. From the methyl

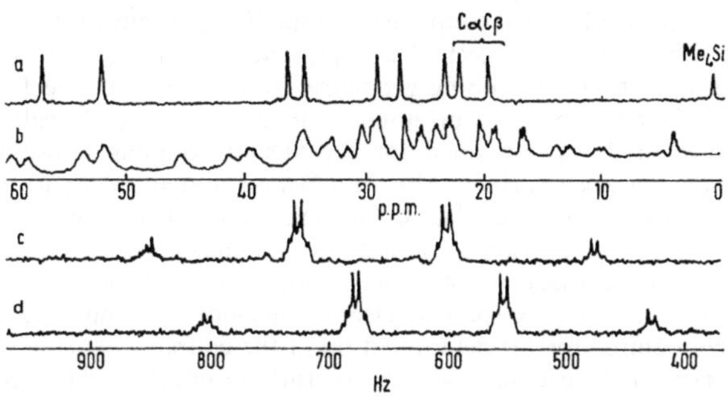

Fig. 3. 20 MHz ^{13}C n.m.r. spectra and subspectra of menthone (**3**), (a) PND; (b) proton-coupled; (c), (d) proton-coupled subspectra of the isopropyl methyls C_α and C_β (assignments are uncertain). (From FREEMAN, R., G. A. MORRIS, and M. J. T. ROBINSON, J. C. S. Chem. Commun. 754 (1976), with permission)

carbon subspectra in Fig. 3c and d, which are indistinguishable as far as their fine structure is concerned, $^3J_{CCCH} = 2.5$ Hz was obtained which, is only compatible with rotamer (**3 a**) in which both methyls are *gauche* relative to H′. It is obvious from the proton-coupled overall spectrum in Fig. 3 b that this information could not have been obtained due to crowding of lines.

The result obtained with menthone could further be corroborated on the basis of menthol (**4**) for which the stable rotamer is the one indicated. In this case coupling constants of 5 Hz ($^3J^{trans}_{CCCH}$) and 2.5 Hz ($^3J^{cis}_{CCCH}$) could be extracted.

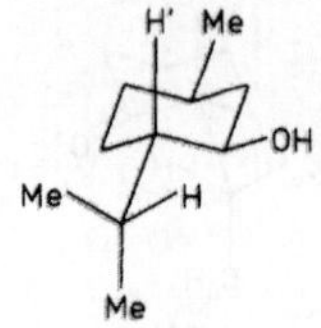

(**4**)

This technique certainly has further potentially interesting applications. Only minor software and hardware modifications are necessary to provide this capability on existing spectrometers.

The three-bond vicinal coupling constant $^3J_{CCCH}$ in substituted alkenes has recently been the subject of a comprehensive study, in view of its possible utilization for probing geometric isomerism (*19*). The J^{trans}/J^{cis} ratio was found to be most pronounced in α,β-unsaturated acids where an average of 2 was obtained for the vicinal coupling constants involving the carboxyl carbon. Also in α,β-unsaturated aldehydes $J^{trans}_{CHO,H}$ is consistently greater than $J^{cis}_{CHO,H}$. However, substituent effects are considerable and a comparison with model compounds may be necessary for unambiguous identification of E or Z configuration if only one isomer is available. In complex spectra selective decoupling techniques turned out to be required for the determination of the coupling constants. Care is necessary in these cases since the perturbing effect of the decoupler field causes a reduction of J. The latter effect in turn may be used for signal assignment (*20*) since the percentage reduction is related to the proton chemical shift.

Three-bond vicinal coupling constants were determined in connection with the study of morellin (**5**) an antibiotic isolated from *Garcinia morella*, isomorellin (**6**), moreollin (**7**) and isomoreollin (**8**).

Assignment of stereochemistry in this case could be made on the basis of the relative chemical shifts observed for the aldehydic and methyl carbons, respectively (steric compression shifts) as indicated in the formulae. That the aldehydic carbon is more highly shielded in morellin (189.4 ppm) is consistent with the Z configuration whereas the lower shielding (194.5 ppm) in isomorellin points to the E configuration.

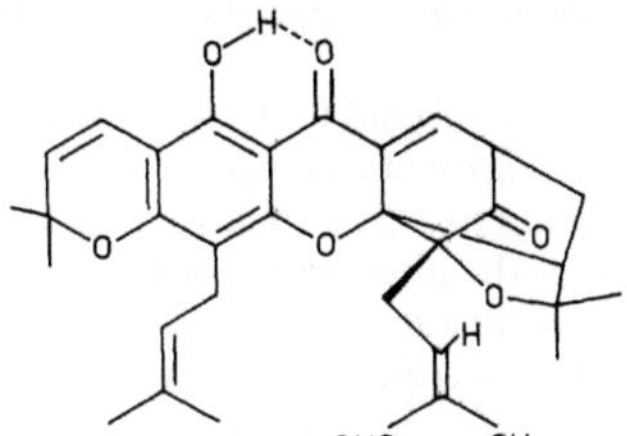

(5)

δ_{CHO} = 189.4 ppm
$^3J_{CH}$ = 10.9 Hz
δ_{CH_3} = 16.4 ppm
$^3J_{CH}$ = 6.5 Hz

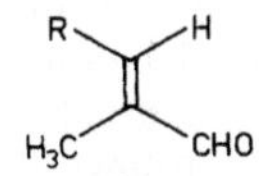

(6)

δ_{CHO} = 194.5 ppm
$^3J_{CH}$ = 9.7 Hz
δ_{CH_3} = 8.5 ppm
$^3J_{CH}$ = 7.5 Hz

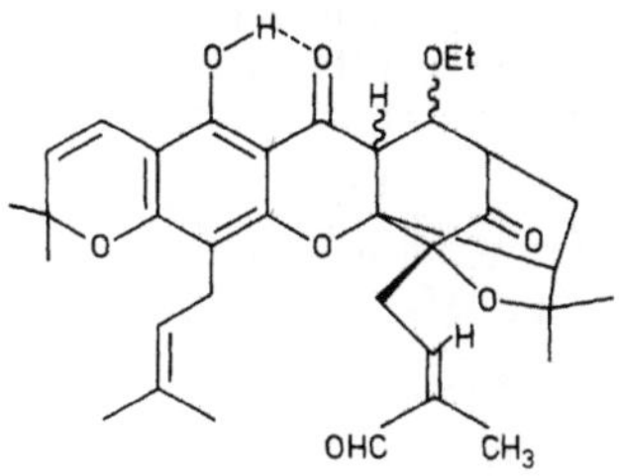

(7)

δ_{CHO} = 191.4 ppm
$^3J_{CH}$ = 11.0 Hz
δ_{CH_3} = 16.5 ppm

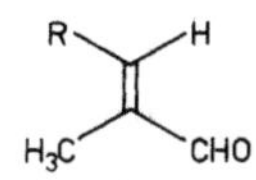

(8)

δ_{CHO} = 195.2 ppm
$^3J_{CH}$ = 9.7 Hz
δ_{CH_3} = 9.3 ppm

Analogous arguments apply to the methyl shielding. This assignment is in full agreement with the observed three-bond coupling constants $^3J_{CHO,H}$ of 10.9 Hz (*trans*) and 9.7 Hz (*cis*). The values are also in ac-

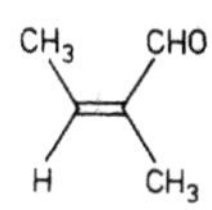

$^3J_{CHO,H}$ = 11.0 Hz

(9)

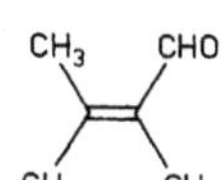

$^3J_{CHO,H}$ = 9.7 Hz

(10)

cordance with those found in the model compounds angelaldehyde (**9**) and tiglaldehyde (**10**).

^{13}C-^{1}H spin-spin coupling has been found to be a valuable tool for the assignment of signals in substituted aromatic compounds where chemical shift arguments usually are inadequate. GÜNTHER and co-workers (*21*) pointed out a distinct difference in the appearance of the multiplets of α and β carbons in symmetrically disubstituted benzenes. Whereas the α carbon always gives rise to broad non-resolved patterns, a neat doublet of doublets is obtained for the β carbon. This has been substantiated with computed spectra. The differing appearance of the two types of multiplets is suggested as a "fingerprint" for establishing assignments.

In carbocyclic aromatics the (transoid) three-bond coupling constant prevails over the smaller two-bond and four-bond couplings. In benzene, for example, $^{2}J_{CCH} = +1.0$ Hz, $^{3}J_{CCCH} = +7.4$ Hz, $^{4}J_{CCCCH} = -1.1$ Hz (*22*). Introduction of electronegative substituents usually results in an increase of $^{2}J_{CCH}$. In heteroaromatic systems $^{2}J_{CCH}$ can also become large (*23*).

Assignment of the aromatic carbons in (**11**) isolated from *Pimpinella anisum* L. (*24*) was accomplished on the basis of its single resonance ^{13}C

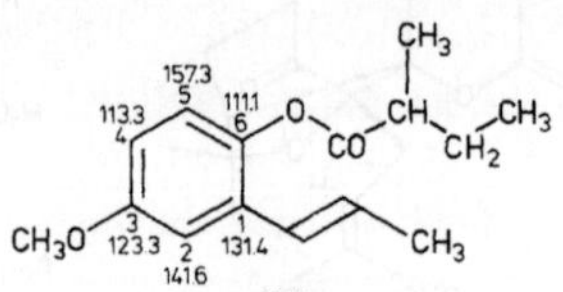

(**11**)

spectrum in which C-4 appears as a doublet of doublets (δ_C 113.3, $^{1}J_{CH} = 162$ Hz, $^{3}J_{CCCH} = 6.4$ Hz). The smaller of the two splittings corresponds to the coupling constant between C-4 and H-2.

Long range ^{13}C-^{1}H coupling constants have also been found to be valuable parameters in the interpretation of the spectra of flavones (*25, 26*) and phenol derivatives (*27*). The intramolecularly hydrogen-bonded phenolic proton couples to C-6 of 2-acylphenols, such as salicylaldehyde (**12**) and 4′,5,7-trihydroxyflavanone (**13**). In the latter compound the

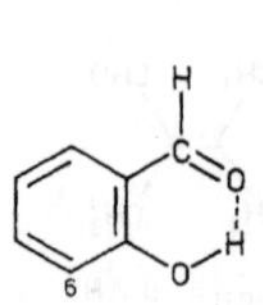

(**12**)

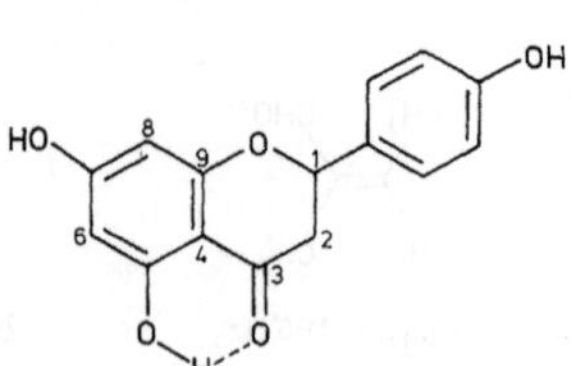

(**13**)

following coupling constants between the chelated hydroxyl proton and aromatic carbons (*25*) could be extracted: $^{2}J_{HOC(5)} = 5.0 \pm 0.5$ Hz, $^{3}J_{HOCC(4)} \approx {}^{3}J_{HOCC(6)} = 4.5 \pm 1$ Hz and $^{4}J_{HOCCC(7)} = 1.5 \pm 0.5$ Hz. These coupling constants were readily identified by comparing the spectrum with that of the deutero isotopomer obtained upon deuteron exchange.

The availability of ^{13}C-^{1}H spin-spin coupling constants may still not provide adequate arguments for unequivocal assignment. Some time ago an elegant double resonance experiment, based on the generalized nuclear Overhauser experiment, was suggested (*28*, *29*). The method uses weak selective pulses for inverting the spin populations on the energy levels of a particular proton transition. This induces large intensity changes for the lines corresponding to the connected ^{13}C transitions. If the inverting proton π pulse is short enough, a full inversion of populations ensues (selective population inversion SPI) whereas only a partial transfer of population occurs for relatively long pulses (selective population transfer SPT) due to partial relaxation during the pulse. This distinction is immaterial, however, if the technique is solely used for assignment purposes. Requisite for its successful use is that the ^{1}H pulse be sufficiently selective so as to excite only *one* proton transition.

Very recently PACHLER and co-workers (*30*) have performed SPI experiments in connection with the full assignment of the ^{13}C spectrum of oxaline (**14**), a novel alkaloid from *penicillum oxalicum*. Inversion of the

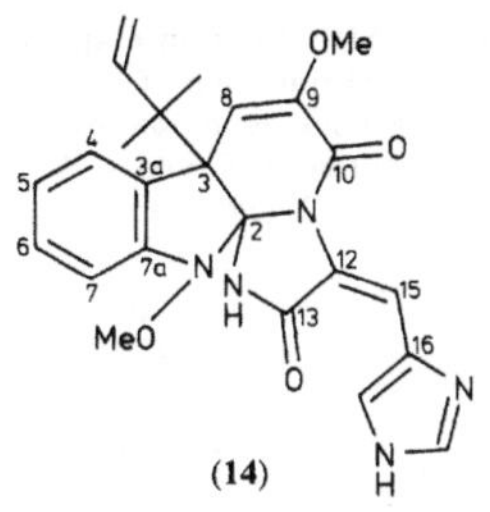

(**14**)

high-field C-8 proton transition, for example, affects the resonances at δ 157.3 and δ 101.6 thereby permitting assignment of the former to C-10 and the latter to C-2 (Fig. 4 b). Analogously, inversion of the low-field C-15 proton transition led to enhancement of the carbon signals pertaining to C-13 (Fig. 4 c). The same technique has been used for the assignment of quaternary carbon resonances in aflatoxin B_1 (**15**) (*31*). Here, the SPI experiments permitted identification of the quaternary carbon resonances C-7 and C-11 which are difficult to designate otherwise.

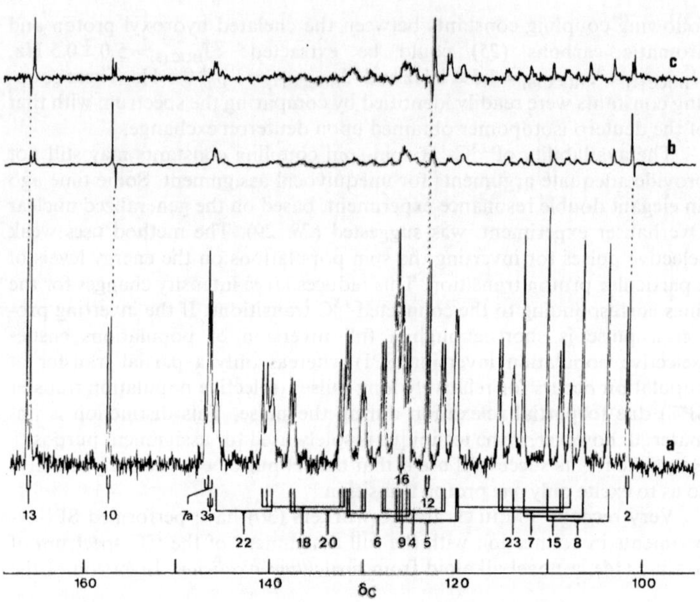

Fig. 4. Low-field region of the 25.2 MHz ^{13}C n.m.r. spectra of oxaline (**14**); (a) proton-coupled; (b) SPI experiment, high-field transition of C-8 proton selectively inverted using $T_{\pi} = 0.08$ sec; (c) same as (b) but π pulse acting on low-field transition of C-15 proton. (From NAGEL, D. W., K. G. R. PACHLER, P. S. STEYN, R. VLEGGAAR, and P. L. WESSELS, Tetrahedron **32**, 2625 (1976), with permission)

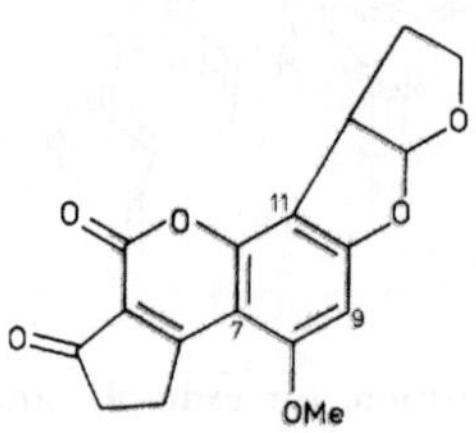

(15)

2.3. Isotopic Substitution

Specific ^{13}C enrichment at a specific carbon site in the molecule not only identifies the labelled carbon unambiguously but provides useful information for designation of more remote carbons due to long range

^{13}C-^{13}C couplings. This powerful technique recently demonstrated that incorrect assignments were made earlier for some monosaccharides including glucose (*32*). Apart from the large one-bond coupling ($^{1}J_{C(1)C(2)} \sim 46$ Hz) C-3 is found to be coupled to C-1 in the β anomer ($^{2}J_{C(1)C(3)} \sim 4$ Hz) but not in the α anomer. By contrast C-5 couples to C-1 in the α anomer ($^{2}J_{C(1)C(5)} \sim 2$ Hz). This behaviour has been interpreted in terms of a Karplus-type relationship where the angle is defined by the relative orientation of C-3 or C-5 and the electronegative oxygen substituent at C-1. Apart from $^{2}J_{CCC}$, $^{3}J_{CCCC}$ is well known to obey a dihedral angle relationship (*10*).

Substitution of protons by deuterons provides useful complementary information in the ^{13}C spectral assignment procedure. The signals due to CD (or CD_2, CD_3) usually do not appear in the PND spectra. This may be ascribed to the concomitant partial loss of the NOE, signal splittings due to CD coupling and comparatively long T_1's. Adjacent carbons are shifted ca. 0.1 ppm upfield (per deuteron) which is a consequence of the deuteron-induced isotope effect. Very small but measurable isotope shifts (upfield) are also detectable on the carbons three bonds removed from deuterium. The effects are best observed by recording the spectrum of a 1 : 1 mixture of the protio and deutero isotopomers.

The technique was found to aid assignments in steroids (*33*), peptides (*34*), carbohydrates (*35*) and terpenoids (*36*). In order to differentiate the signals arising from ring A and ring B, respectively, 5,6,7,8-tetradeuteroflavone and 2′,3′,4′,5′,6′-pentadeuteroflavone were synthesized (*37*) and their spectra compared.

Intramolecular hydrogen-bonded hydroxyl protons undergo exchange at a rate which may be sufficiently slow so that the OH and OD iso-

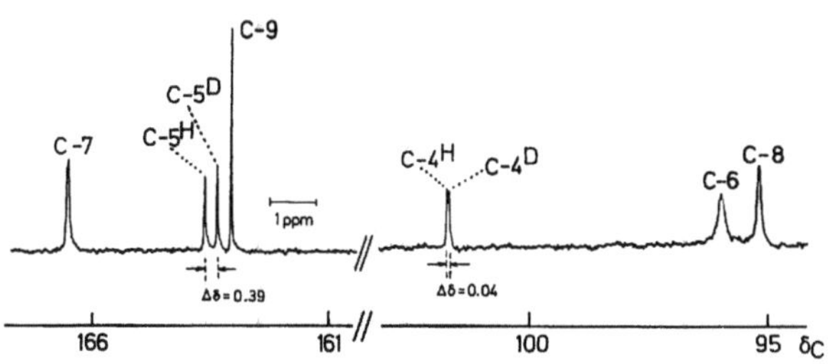

Fig. 5. Low-field region of the 25.2 MHz PND ^{13}C n.m.r. spectrum of naringenin (**13**) in DMSO-d_6. The spectrum was recorded from a mixture of isotopomers obtained after exchanging 50% of the labile protons by deuterons (*38*)

topometers co-exist. In the case of naringenin (**13**), for example, separate signals were obtained for some of the carbons from a sample in which 50% of the labile protons had been exchanged by deuterons (*38*). Fig. 5 which displays the low-field region of the spectrum recorded in DMSO-d_6 clearly shows separate resonances for C-5 and C-4. The unusually large geminal isotope shift for C-4 (0.39 ppm) is likely to be a so-called secondary isotope effect having its cause in a change in the structure of the hydrogen bond. The C-7 resonance, although it is not split like the one for C-4, is significantly broadened, which is accounted for by a vicinal isotope shift.

An analogous situation is encountered in naphtoquinones (*39*) possessing hydroxyl groups in peri position. In penicillin methyl ester (**16**)

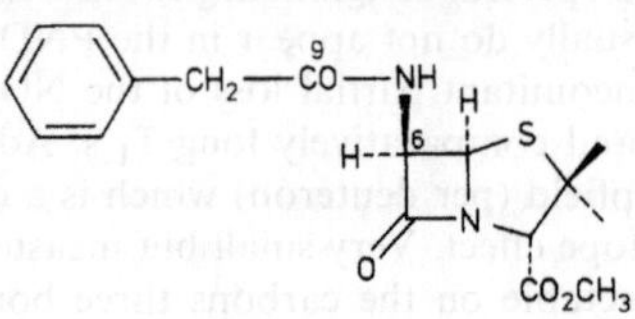

(**16**)

addition of CD_3OD was found to induce a 0.1 ppm upfield shift of C-6 and C-10 which is attributed to a geminal isotope effect brought about by replacement of NH by ND (*40*). Similar isotope shifts were observed in the spectrum of totarol (**17**), recorded in DMSO-d_6 (*41*). The acidic

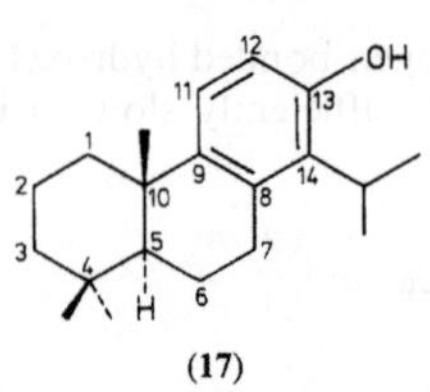

(**17**)

proton (or deuteron, respectively) is strongly associated with the solvent. The three resonances which shifted upfield upon replacement of OH by OD were designated to C-12, C-13 and C-14, thus correcting an earlier erroneous assignment (*42*).

2.4. Lanthanide Shift Reagents

Effects of paramagnetic rare earth ions binding to polar groups of the molecule, and resulting in chemical shift dispersion, have been widely used for over a decade. The total isotopic shift induced by a lanthanide ion arises from unpaired electron spins and can be ascribed to two different effects. One is a *through-bond scalar effect* also designated *Fermi contact shift*, the other is a *through-space effect* and hence *dipolar* in nature. This latter is also termed *pseudo-contact* shift. It is of primary interest in n.m.r. applications of organic compounds because the induced shifts contain information on molecular geometry. Except for the carbons nearest to the site of complexation the incremental shifts can generally be rationalized in terms of a dipolar model based on the McConnell-Robertson equation

$$\Delta H_i/H = K\, r_i^{-3}\, (3 \cos^2 \chi_i - 1). \qquad (2)$$

In eq. 2, which assumes axial symmetry of the complex, $\Delta H_i/H$ is the relative shift of nucleus i, r_i represents its distance to the paramagnetic centre and χ_i expresses the angle between r_i and the symmetry axis. Apart from the geometrical dependences the actual shifts are also related to the lanthanide : substrate molar ratio. The reagents most commonly used are of type Ln $(dpm)_3$ where the acronym "dpm" stands for "dipivaloylmethanato". Experimentally it was found (*43*) that Yb $(dpm)_3$ induces the largest shifts (ca. 2.5 times larger than for the more commonly used Eu $(dpm)_3$) and minimal contact contributions.

By fitting dipolar shifts to eq. 2 the topology of a molecule can be established. Separation of the dipolar contribution from the total observed shift may for example be accomplished by using predicted pseudo-contact shifts which are obtained from the space coordinates furnished by the analysis of the lanthanide-induced proton shifts (*44*). The latter are known to arise mainly from pseudo-contact interaction. A further small contribution to the observed shift is purely diamagnetic in origin and simply results from the inductive effect brought about by complexation. Yb $(dpm)_3$-induced "pure" pseudo-contact shifts were for example obtained from the ^{13}C spectra of cholestan-3β-ol (*45*). Excellent agreement with computed shifts confirmed the assignment of all 27 carbons.

Although the angular term of eq. 2 cannot be ignored, in principle, χ_i varies only slightly for the majority of carbons in the molecule and the shifts thus are nearly proportional to $1/r^3$. Even if it is used in this somewhat simplistic manner, the method can serve as an assignment tool. This is illustrated in Fig. 6 which shows a correlation stick diagram for manool (**18**) with and without shift reagent added (*46*). The spectra permit distinction between C-11 and C-2, for example. Whereas the remote backbone carbon is virtually unaffected by the shift reagent, C-11 close to the site of complexation experiences a 1.5 ppm downfield shift.

The effects exerted by a lanthanide reagent can be exploited in yet another manner. The concomitant dispersion of the proton spectrum may

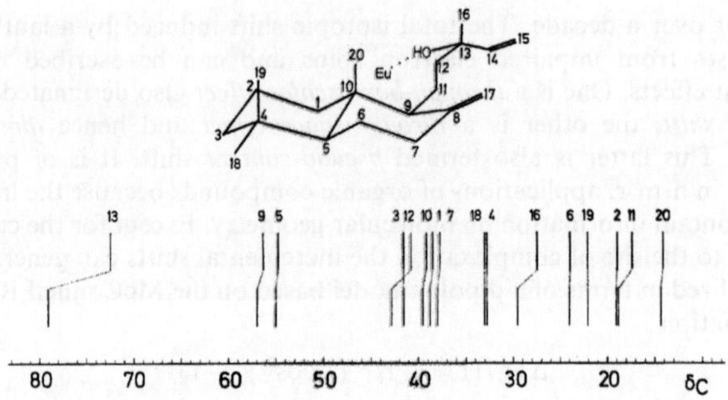

Fig. 6. Stick-diagram of the high-field region of the ^{13}C n.m.r. spectrum of manool (**18**) with and without shift reagent added (1 : 2 europium/substrate ratio)

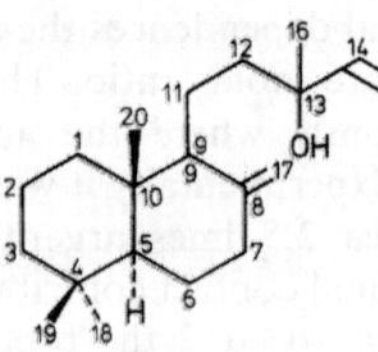

(**18**)

permit selective irradiation, thus providing correlation of proton and ^{13}C shifts, which might not be possible on the basis of the diamagnetic spectrum.

2.5. Spin-Lattice Relaxation

Pulse Fourier transform ^{13}C n.m.r. spectroscopy provides easy access to a further spectral parameter: the spin-lattice relaxation time T_1. For details regarding theory, experimental techniques and applications the reader is referred to the pertinent reviews (*47*, *48*, *49*). Experimentally it has been shown that in organic molecules of intermediate size (MW ~ 250—1000) both proton-bearing and quaternary carbons are relaxed almost exclusively by the *^1H-^{13}C dipole-dipole mechanism*. If the molecule tumbles at a rate much greater than the Larmor resonance frequency (so-called *motional narrowing* limit) the relaxation rate $1/T_{1C}$, may be expressed as

$$1/T_{1C} = \hbar^2 \gamma_C^2 \gamma_H^2 \Sigma_i r_{CH}^{-6} \tau_C . \qquad (3)$$

In eq. (3) γ_C and γ_H represent the magnetogyric ratios of ^{13}C and ^{1}H, respectively, r_{CH_i} is the internuclear distance between the carbon in question and proton i, and τ_C stands for the reorientational correlation time which is a measure for the molecular rate of reorientation (typically 10^{-12} to 10^{-10} sec for natural products molecules). A further condition that has to be satisfied for the validity of eq. 3 demands that the molecule reorient in an isotropic fashion, which is strictly true only for spherical molecules. Nevertheless, relaxation of a large percentage of organic molecules can be described by this simple relationship. Because of the sixth power C-H distance dependence the protons to be considered in the relaxation of protonated carbons are confined to those directly attached to the carbon in question so that $1/T_{1C}$ becomes proportional to the number N of bonded hydrogens. Hence a methylene carbon is predicted to relax twice as fast as its methine counterpart or alternatively NT_1 should be equal for all skeletal carbons. In particular, in situations where crowding of lines prevents an unequivocal differentation of multiplicities on the basis of the SFORD spectra, relaxation measurements may offer an attractive alternative for signal assignment. For this purpose a quantitative evaluation of T_1 data is often unnecessary and one single inversion-recovery spectrum can provide the desired information. In a $180°$-τ-$90°$ experiment the carbon signals appear inverted if $\tau/T_1 \ll 1$ but positive for $\tau/T_1 \gg 1$. By an appropriate choice of the pulse interval a spectrum may be obtained in which the methylene resonances appear positive whereas the more slowly relaxing methines give rise to inverted lines. However this holds only true if the C-H bond vectors of the carbons considered reorient at the same rate. Internal motion such as it occurs for methyls, or segmental motion of flexible sidechains, according to eq. 3 is accompanied by a lengthening of T_1 since the effective correlation time (also denoted τ_{eff}) shortens relative to τ_C.

An example of an inversion-recovery T_1 trace is provided in Fig. 7 for cholestane. For this spectrum $\tau = 0.5$ sec had been chosen. It is evident that the resonances due to the methylenes of the steroidal backbone are all positive and of approximately equal height (dashed line) while the corresponding methines appear slightly negative (dotted line). By contrast the methyls C-18, 19, 21, 26 and 27, which experience internal rotation and therefore possess longer T_1's, are almost unrelaxed, giving rise to strongly negative excursions. A deviating behaviour is also observed for the sidechain carbon resonances C-23, 24 and 25 whose intensities reflect increased motion.

If backbone carbons of the same multiplicity exhibit unequal T_1's this is an indication of anisotropic overall motion. In the androstane and cholestane series, for example, C-3 was found to relax consistently more rapidly than the remainder of backbone methylenes (*50*). It is well known

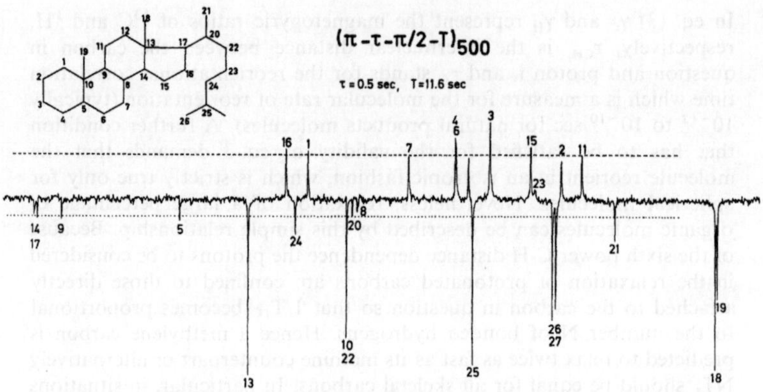

Fig. 7. 25.2 MHz inversion-recovery ^{13}C spectrum of cholestane (ca. 1-molar in $CDCl_3$) under conditions as indicated. It may be noted that backbone methine carbons appear slightly negative (dotted line) whereas corresponding methylene carbons are positive (dashed line). (From WEHRLI, F. W., and T. WIRTHLIN, Interpretation of Carbon-13 NMR Spectra, p. 253, Heyden, London, 1976, with permission)

that carbons whose C-H bond vectors are collinear with the principal axis for reorientation relax faster (*51*). Since this is also the case in steroids which preferentially reorient around the steroidal axis, slightly enhanced relaxation for C-3 results. This behaviour is also borne out by the spectrum in Fig. 7 which exhibits increased intensity of the C-3 resonance.

Anisotropic motion, if quantitative information is to be obtained, requires a more rigorous treatment in terms of a rotational diffusion model (*48*, *49*). A least-squares treatment of rotational diffusion has recently been presented (*52*). This permits extraction of the diffusion coefficients using experimental T_1's and atomic coordinates. Interesting conclusions could be drawn from such an analysis regarding motion and conformation of the visual chromophores, the retinals (*53*, *54*). In all *trans*-retinal (**19**) motion around the olefinic carbon chain was found to be ten times faster than that around the perpendicular axis. This ratio reduces to approximately 5 and 2 in 13-*cis*-retinal (**20**) and 11-*cis*-retinal (**21**), respectively. For species 19 the best fit was obtained for a dihedral angle of 60° between the six-membered ring and the olefinic sidechain.

Methyl carbons are often difficult to designate since they usually resonate in a narrow spectral region. Their relaxation rates are normally governed by internal motion. A freely spinning methyl carbon is known to relax up to 9 times more slowly than in the absence of internal motion

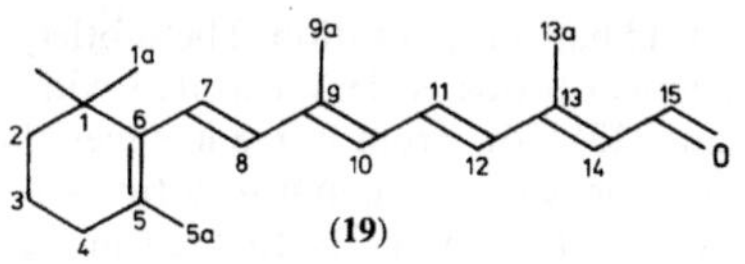

(19)

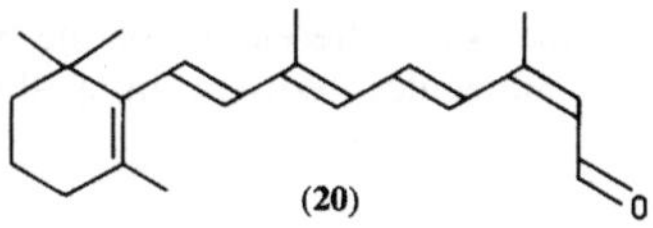

(20)

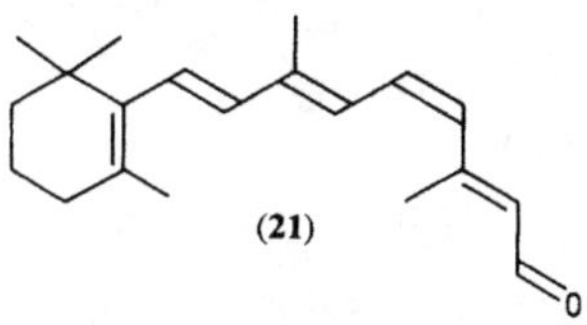

(21)

(*55*). The ratio $3\,T_1(CH_3)/T_1(CH)$, where $T_1(CH)$ represents the relaxation time of a backbone methine carbon of the same molecule, qualitatively reflects the degree of internal motion. This latter is of course related to the energy barrier to internal motion. In steroids, e.g. it was found that as the number of 1,3-diaxial interactions, experienced by the angular methyls C-18 and C-19, increases, their barrier to rotation is lowered and hence T_1 is lengthened (*50*). In ring A-substituted androstanes C-19 consistently relaxes more slowly than C-18, which is due to the former possessing a larger number of 1,3 interactions with axial protons. The same phenomenon was observed in labdane diterpenoids (*46*) in which the sterically crowded axial methyls at position 4 and 10 relax more slowly than those at position 8 and 13 which possess a smaller number of 1,3-diaxial interactions. Shorter relaxation times are also found for the equatorial methyl carbon at position 4 whose rotation is essentially

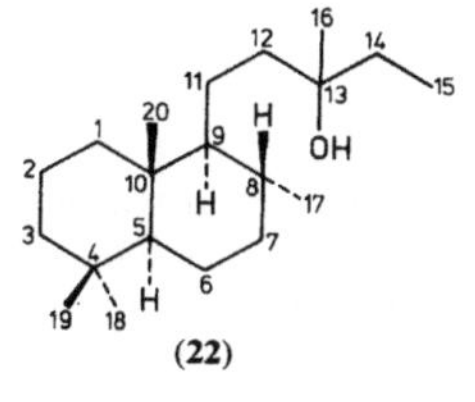

(22)

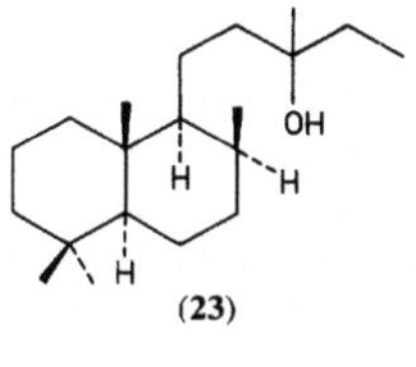

(23)

governed by a three-fold barrier to rotation. The subtlety of spin relaxation times as a stereochemical probe is demonstrated with 8α- and 8β-tetrahydromanool (**22**) and (**23**) (*56*). From Fig. 8 it is seen that in the β epimer (**23**) NT_1 of C-20 is approximately equal to 9 times NT_1(C-8), showing this methyl to behave as a free rotor. By contrast in the α epimer (**22**) the steric interaction exterted by the axial C-17 is removed thus increasing the barrier to rotation of C-20 with a concomitant decrease in T_1(C-20) (5.03 vs. 2.98 sec). From the NT_1 differentials of the backbone carbons it is inferred that both molecules reorient in a slightly anisotropic fashion.

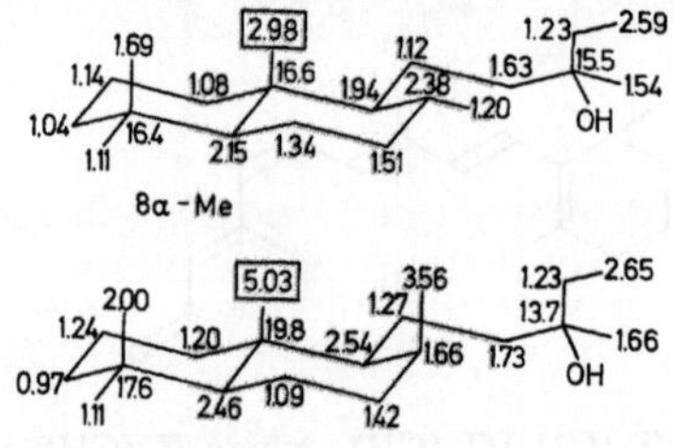

Fig. 8. T_1 relaxation map of 8α- and 8β-tetrahydromanool (**22, 23**) (*56*). The data were obtained at 25.2 MHz, ca. 0.5-molar in $CDCl_3$

Interesting conclusions were recently drawn from T_1 data regarding the conformation of neuraminic acids (*57*) (**24**). The observation of

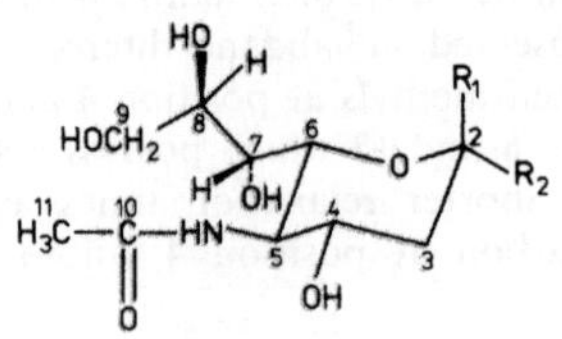

(**24**)

nearly identical NT_1 values for ring carbons C-3, -4, -5, -6 and C-7 and C-8 of the side chain points to a conformation in which the side chain carbons C-7 and C-8 reorient in an isotropic fashion, jointly with the ring. However, C-9 provides evidence for internal motion. These observations are in agreement with the conformational model (**25**).

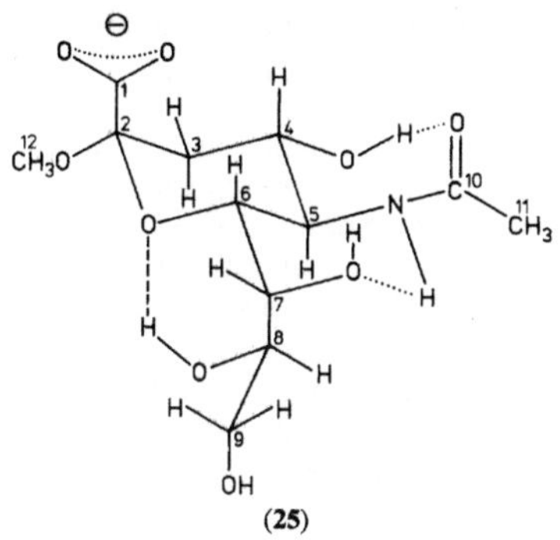

(25)

Quaternary carbons which are particularly difficult to designate, may often be identified on the basis of their relative relaxation times. As stated earlier these carbons in the large majority of organic molecules of the type discussed in this review, are predominantly relaxed by the ^{13}C-^{1}H dipole mechanism with their relaxation rate given by eq. 3. In the sum $\Sigma_i\ r_{CH}^{-6}$ in eq. 3 the carbon-proton internuclear distances involving protons in geminal and vicinal position are to be considered. Because of the inverse sixth power dependence of r_{CH} it is primarily the number of protons attached to the α carbons which governs the relaxation rate of such carbons. This approach has been chosen to assign the

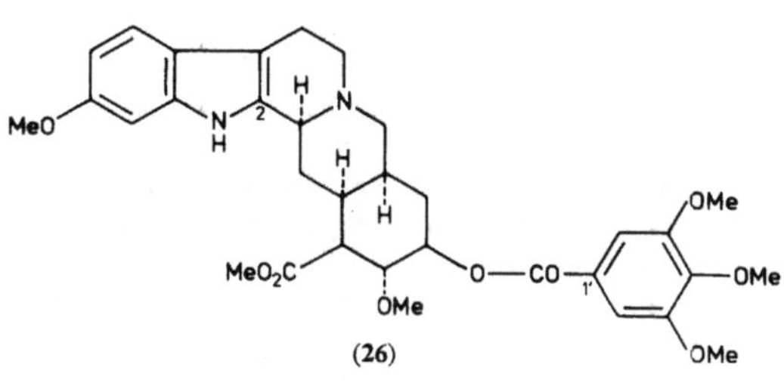

(26)

quaternary carbon resonances in some alkaloids (*58*). In resperpine (**26**), for example, all 11 quaternary carbons could be assigned with the aid of combined relaxation and chemical shift arguments. The shortest T_1's were found for the quarternary carbons at the ring B/C junction (1.6 sec) while the largest value was ascribed to the carbon in the trimethoxy-benzoyl moiety with no protons in geminal position (12.8 sec). T_1 data also supported the assignment of quaternary carbons in the retinals (*54*) (**19**—**21**).

Because of the smaller magnetic moment of the deuteron

$$(\gamma\,(^1H)/\gamma\,(^2H) \approx 6.5)$$

substitution of a proton by a deuteron lengthens T_1 of a geminal quaternary carbon. In thiamine hydrochloride (*59*) C-2′ and C-4′ have essentially equal T_1's in H_2O. Solution of the compound in D_2O results in pentadeuteration (**27**). Due to deuteration of the amine group T_1(C-4′) is found to increase which allows its distinction from C-2′.

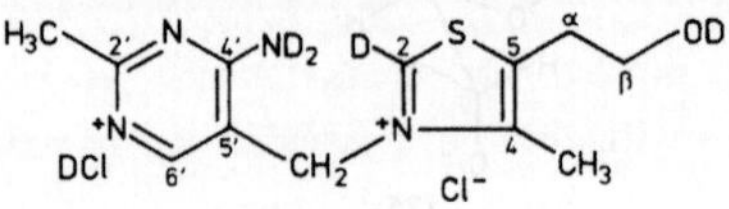

(**27**)

A combination of specific deuteration and spin-lattice relaxation measurements was also used to assign the quaternary carbon resonances in a number of indolo [2, 3-a] quinolizidines (**28**) (*60*). Dideuteration at

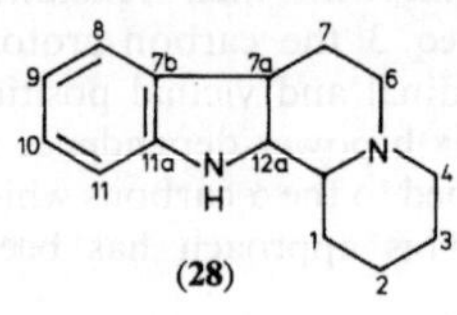

(**28**)

C-7 resulted in a dramatic increase of T_1 (C-7 a) and T_1 (C-7 b) but also C-11 a and C-12 a were significantly affected by the deuterium substitution.

A rather unusual mechanism was recently detected to invoke relaxation of some nonprotonated carbons in nucleotides (*61*). In such molecules quaternary carbons bonded to nitrogen, but having no hydrogens two bonds removed, are substantially relaxed by a ^{14}N-^{13}C dipolar mechanism, in spite of the weakness of the magnetic moment of ^{14}N. This is possible thanks to the inefficacy of the competing ^1H-^{13}C dipolar mechanism.

In addition to intrinsic relaxation mechanisms, relaxation may artificially be enhanced by the addition of paramagnetic relaxation reagents (PARR). Due to the much larger magnetic moment of the electron a powerful electron-nuclear mechanism may be generated in this manner. Most suitable for structural investigations are PARR's which bind to polar sites of the molecule in analogy to the shift reagents. Among these Gd $(dpm)_3$ appears to be particularly promising (*62*). Since the nature of the interaction is largely dipolar, a $1/r^6$ dependence of the relaxation rate is predicted thus providing geometrical information.

The properties of a variety of PARR's have recently been examined (*63*). At doping concentrations up to 5×10^{-2} M it appears that the limiting relaxation rate depends only on the PARR concentration and on the substrate-PARR complex stability constant.

3. ^{13}C Spectral Data of Natural Products

^{13}C n.m.r. spectroscopy may be used in natural products chemistry in a variety of ways and at various stages of the structure characterization process. One of the most trivial applications aims at checking the identity of two compounds. This is achieved by simply comparing the spectrum of a known compound with that of the sample under investigation.

Thanks to the sensitivity of the carbon-13 chemical shift the method is also ideally suited for the detection of stereoisomers or other closely related structures, a problem often encountered during isolation and purification of natural products. If one deals with unknown samples, ^{13}C n.m.r. furnishes key information such as the total carbon number, the number of CH_3, CH_2, CH and quaternary carbons, the number of unsaturations and finally, the number and nature of oxygen and nitrogen-containing functions. These data can be gathered without a great deal of spectroscopic expertise and without having recourse to one of the more elaborate techniques discussed in Section 2.

The ^{13}C spectral information in conjunction with complementary techniques such as ^{1}H n.m.r., i.r., m.s. etc. usually provides a set of possible structures. Arguments in favour of one of the postulated structures may then be obtained by comparing the ^{13}C shift data with those of structurally related compounds. It is for this reason that availability of a large body of spectral data greatly facilitates analysis. Progress has recently also been made with regard to an understanding of substituent effects which may in addition give access to the stereostructure of the compounds.

Because of the mass of spectral data in the literature a full account of the procedures employed for assigning individual carbons would unduly enlarge the scope of the present review. Likewise the authors felt unable to evaluate critically all chemical shifts reported. However,spectral features of particular interest and novelty will be outlined and some erroneous assignments will be corrected in the following section. Chemical shifts are given in ppm relative to tetramethylsilane and, unless otherwise noted, the data were obtained in deuterochloroform. Ambiguous assignments are marked with an asterisk in the formulae.

3.1. Terpenoids and Steroids

For reasons of biogenetic kinship and also because of the close spectroscopic and structural similarity the authors decided to treat isoprenoids and steroids jointly.

Monoterpenes

A comprehensive ^{13}C study of nearly one hundred monoterpenes, which includes assignment of all carbon atoms, has been reported recently (*64*). Together with some additional data for bicyclic monoterpenes (*65*) the chemical shifts in these systems are given in formulae (**29**—**130**). The usefullness of such a compilation of reference data for correlation purposes

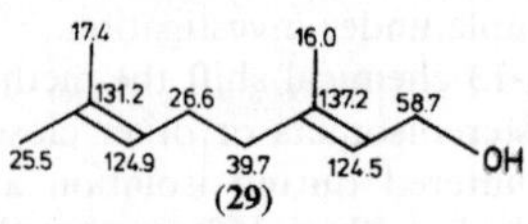

(29)

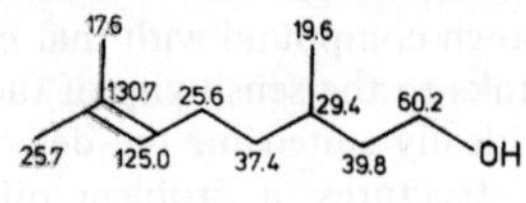

(30)

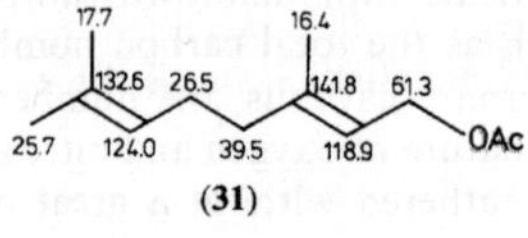

(31)

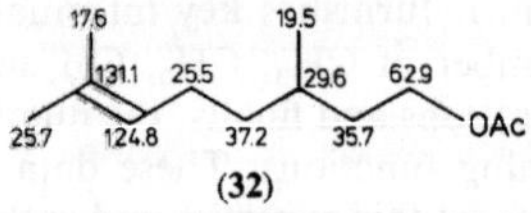

(32)

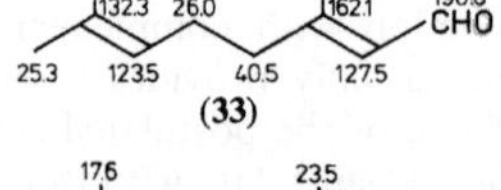

(33)

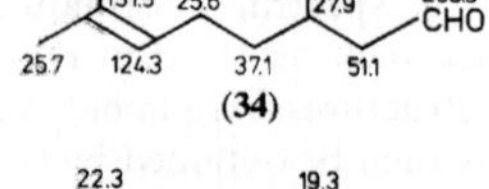

(34)

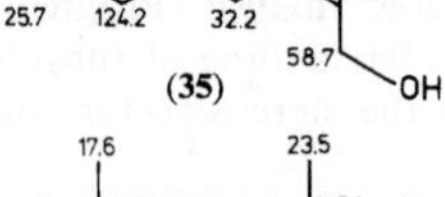

(35)

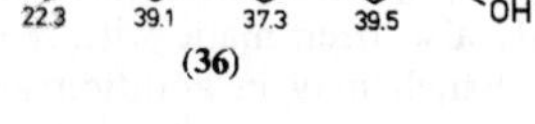

(36)

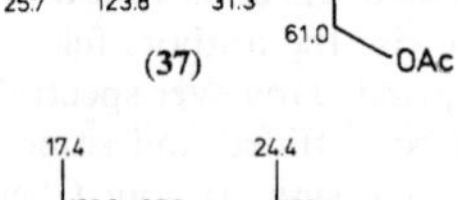

(37)

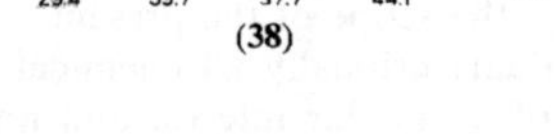

(38)

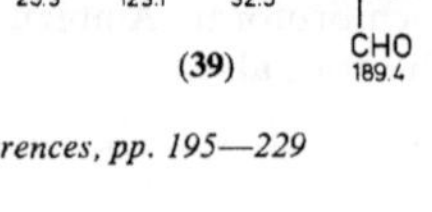

(39)

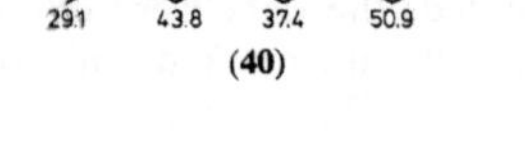

(40)

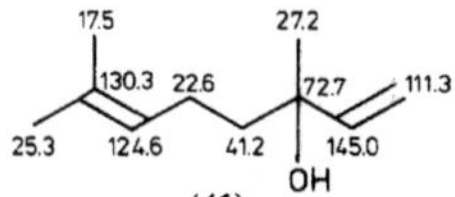

(41)

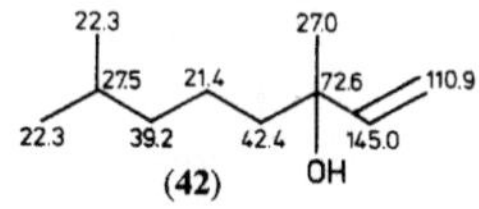

(42)

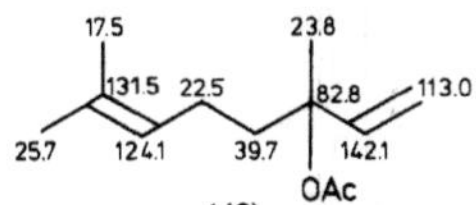

(43)

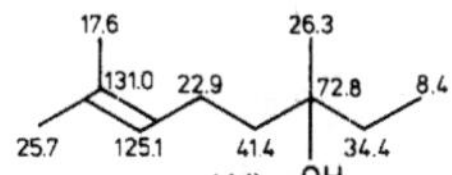

(44)

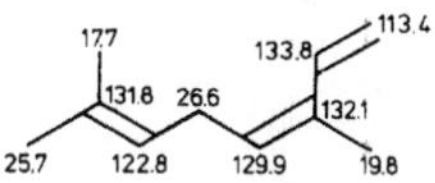

(45)

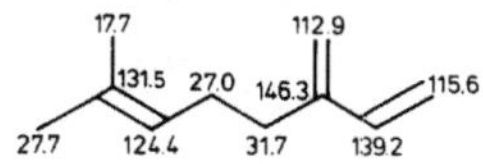

(46)

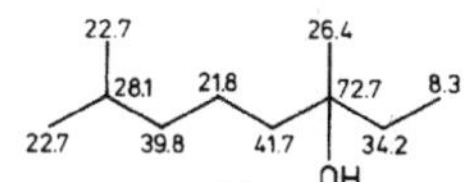

(47)

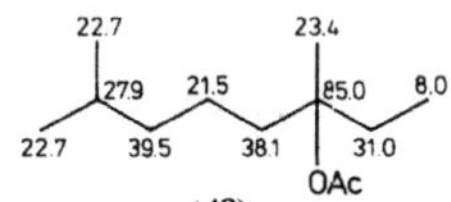

(48)

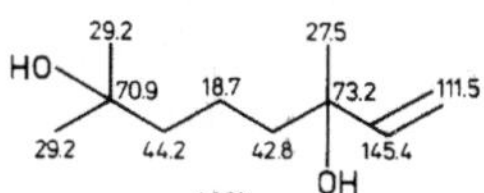

(49)

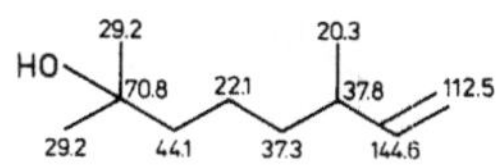

(50)

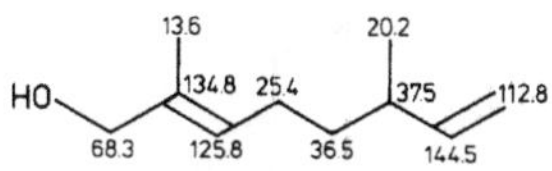

(51)

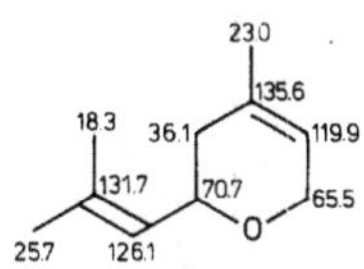

(52)

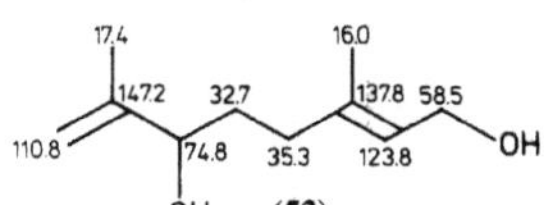

(53)

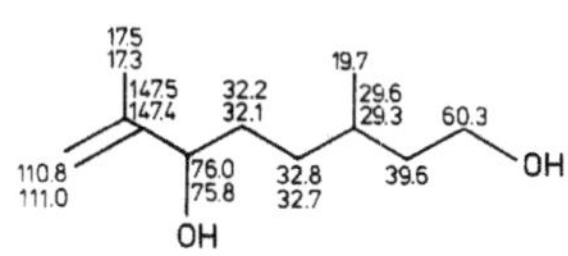

(54) (diastereomers)

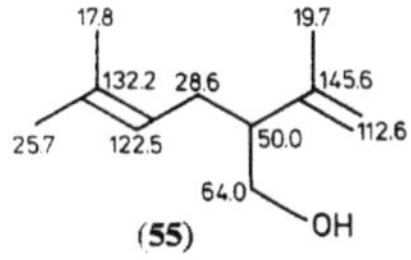

(55)

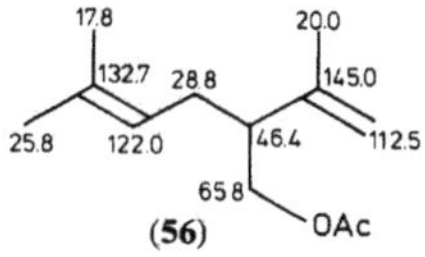

(56)

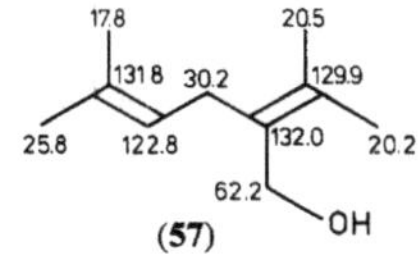

(57)

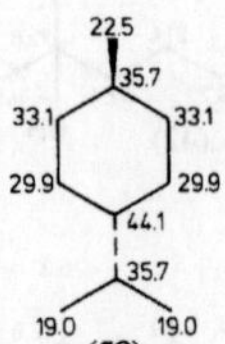

(58)

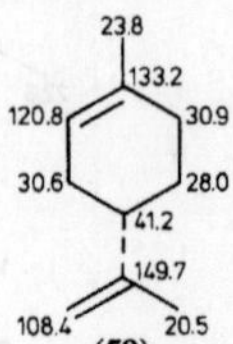

(59)

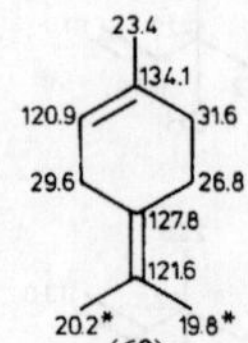

(60)

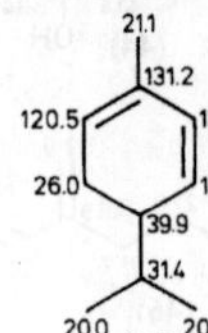

(61)

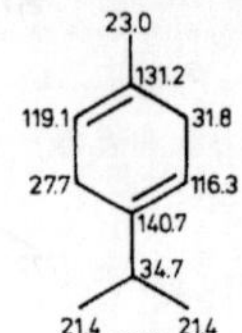

(62)

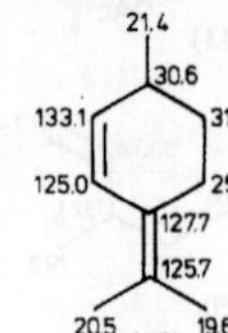

(63)

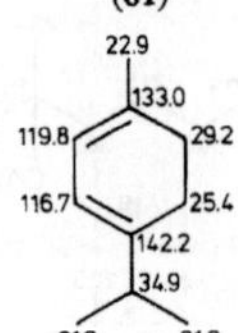

(64)

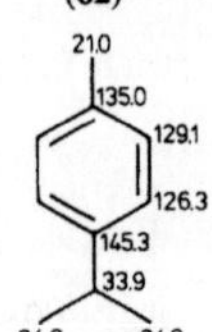

(65)

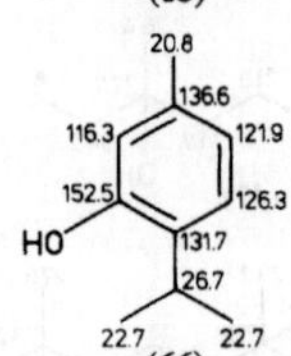

(66)

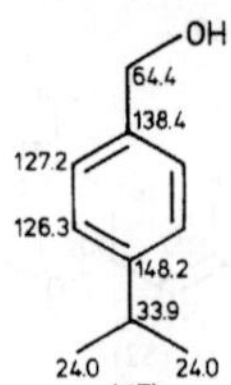

(67)

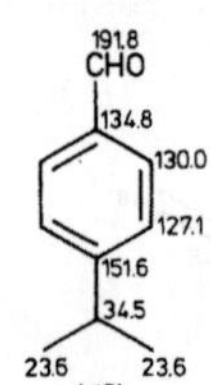

(68)

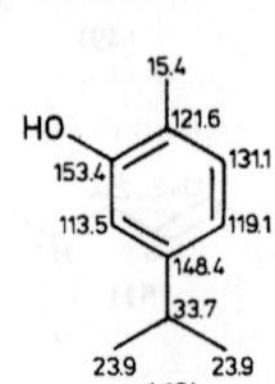

(69)

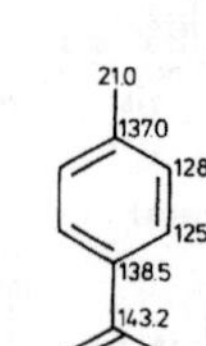

(70)

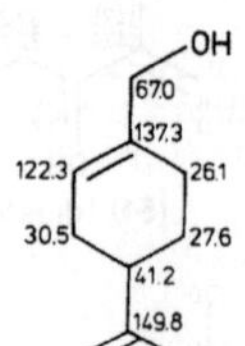

(71)

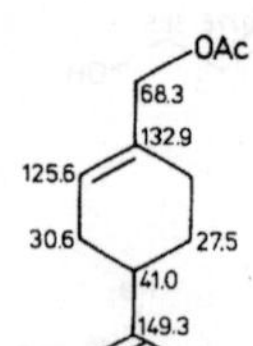

(72)

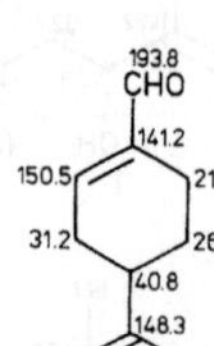

(73)

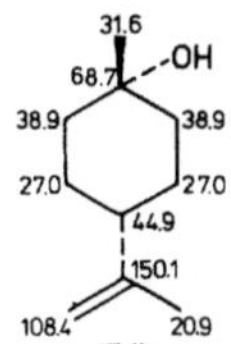

(74)

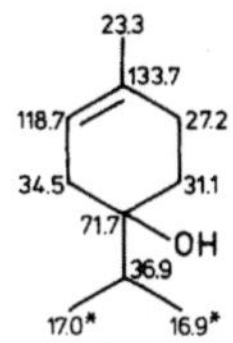

(75)

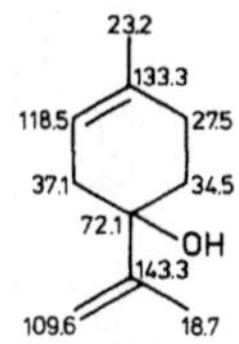

(76)

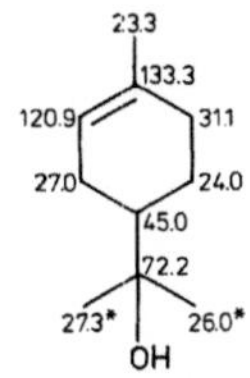

(77)

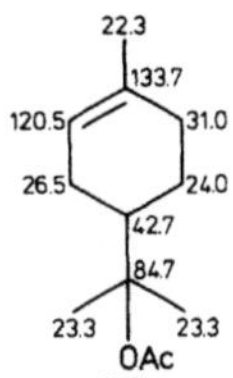

(78)

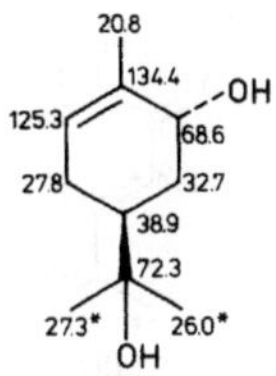

(79)

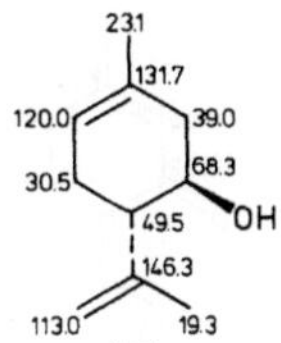

(80)

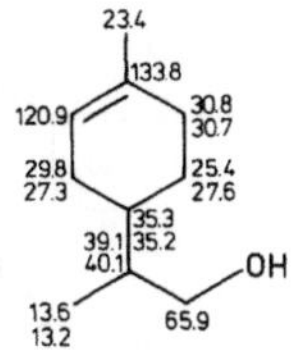

(81) (diastereomers)

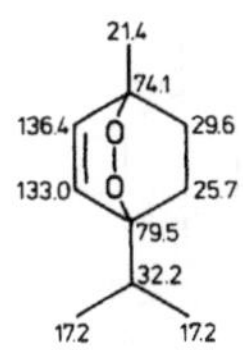

(82)

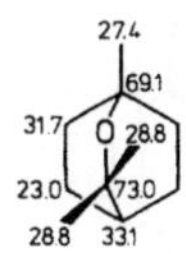

(83)

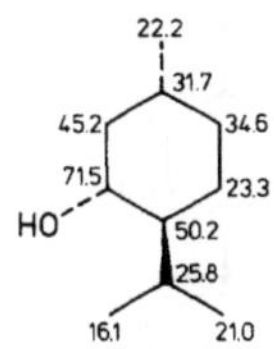

(84)

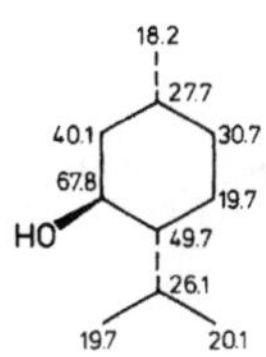

(85)

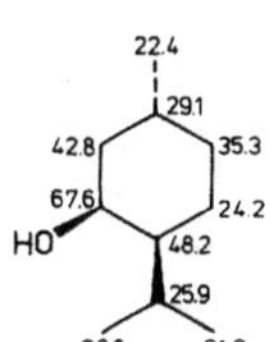

(86)

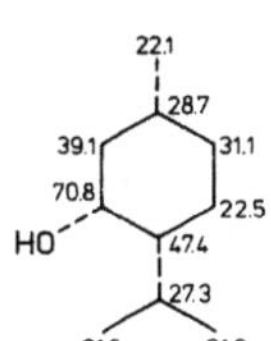

(87)

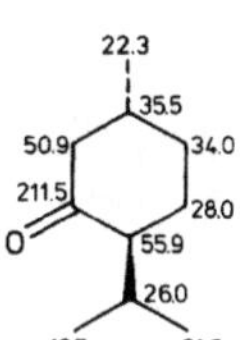

(88)

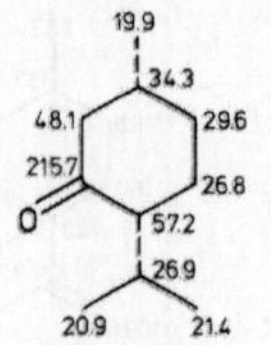

(89)

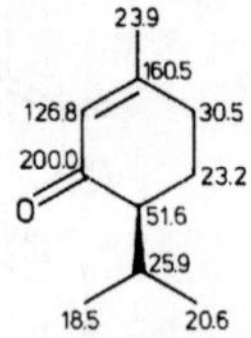

(90)

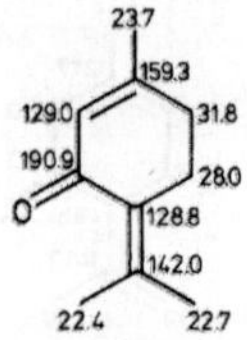

(91)

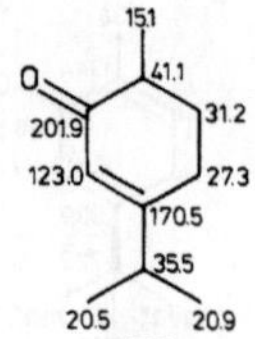

(92)

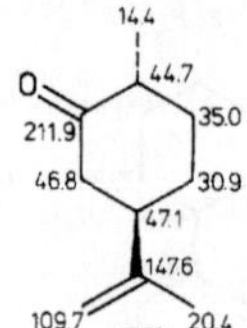

(93)

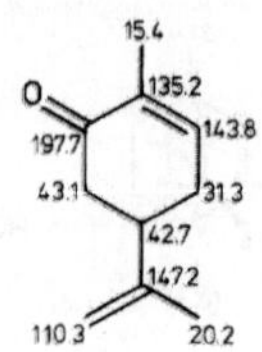

(94)

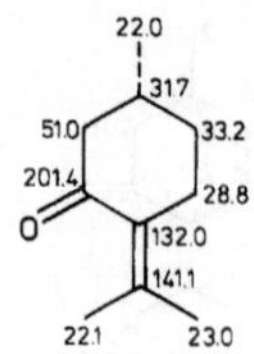

(95)

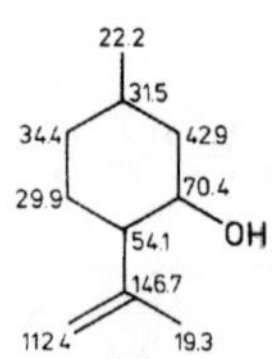

(96)

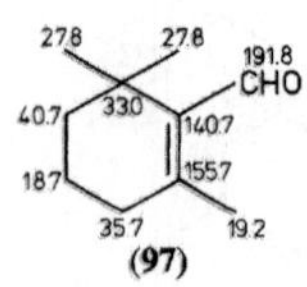

(97)

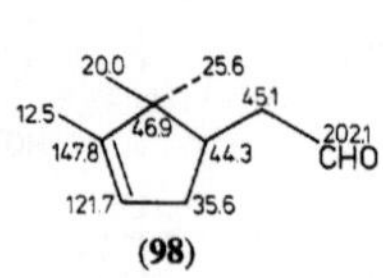

(98)

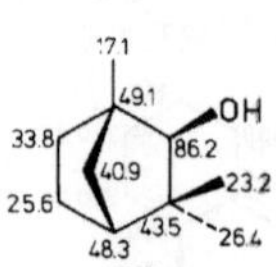

(99)

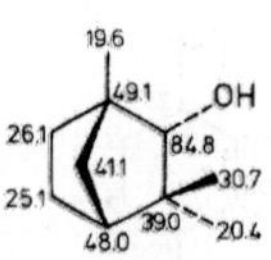

(100)

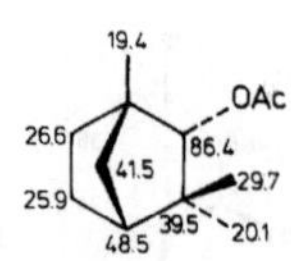

(101)

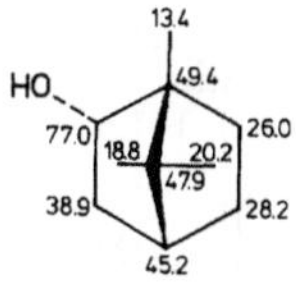

(102)

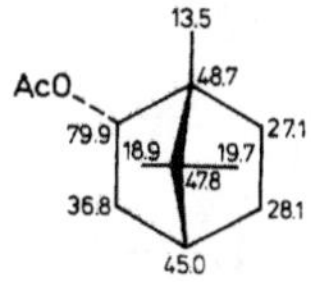

(103)

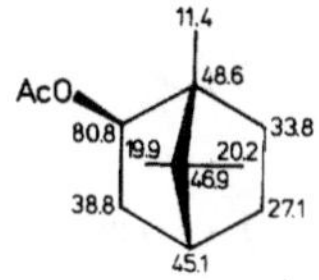

(104)

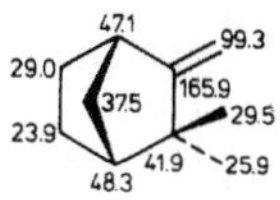

(105)

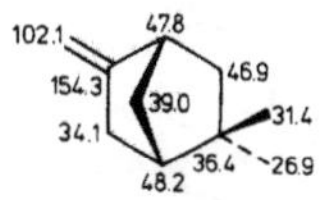

(106)

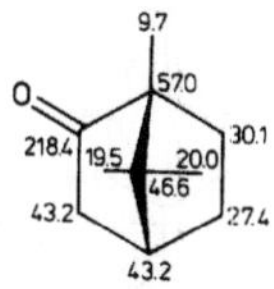

(107)

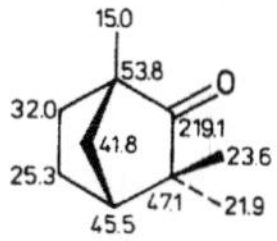

(108)

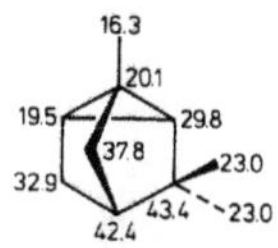

(109)

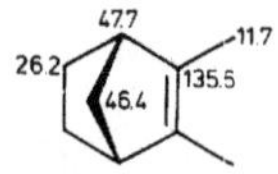

(110)

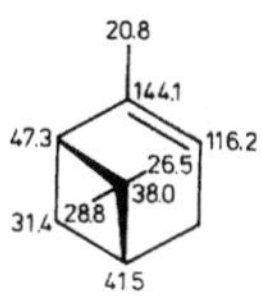

(111)

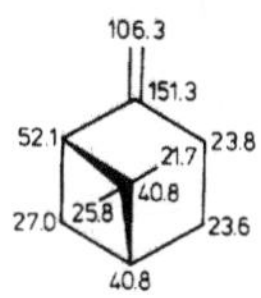

(112)

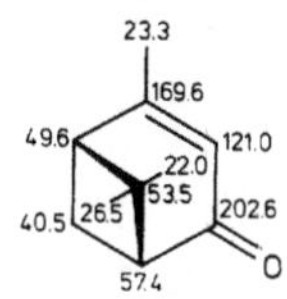

(113)

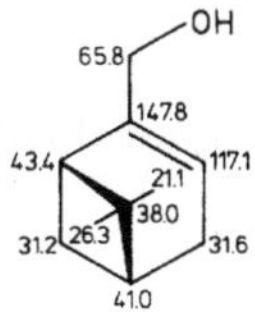

(114)

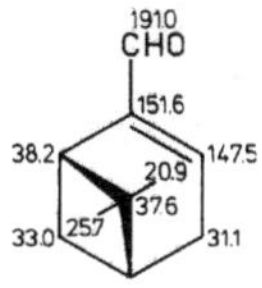

(115)

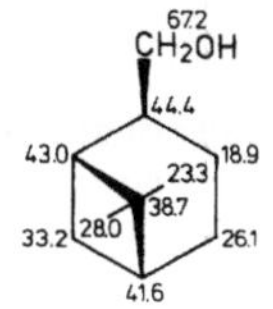

(116)

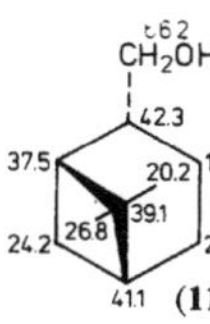

(117)

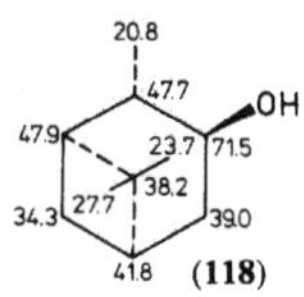

(118)

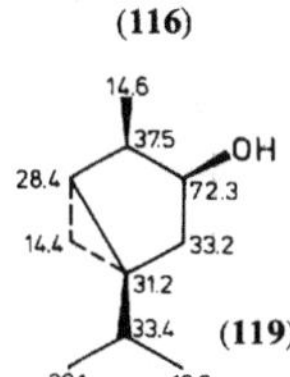

(119)

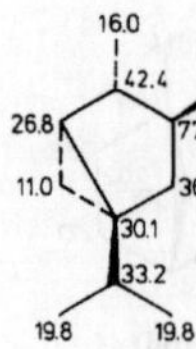

(120)

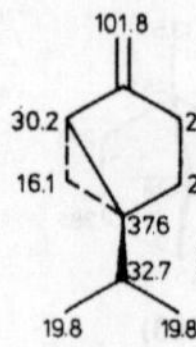

(121)

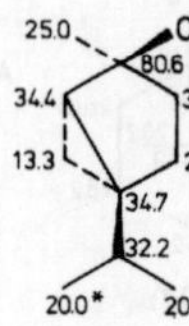

(122)

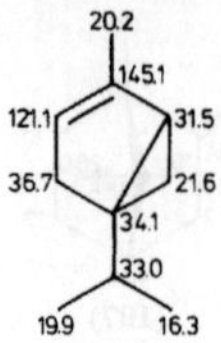

(123)

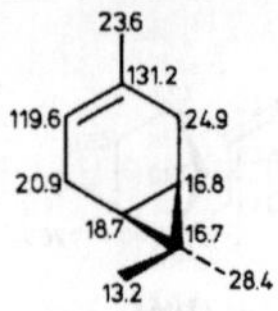

(124)

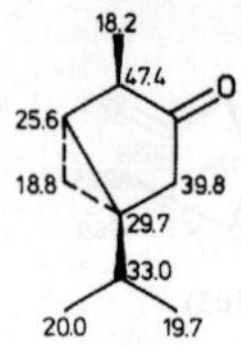

(125)

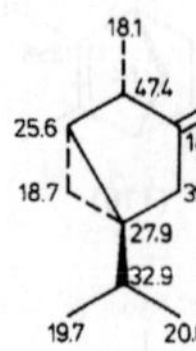

(126)

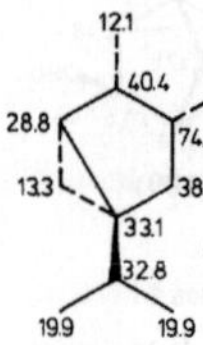

(127)

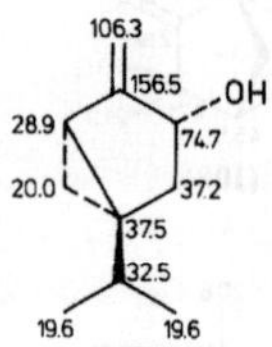

(128)

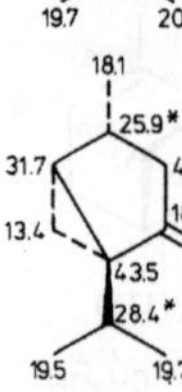

(129)

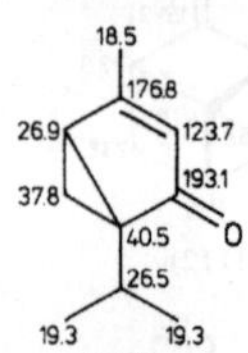

(130)

is obvious. Details of the procedure have been reviewed elsewhere (*66*). The technique may be exemplified with cannabidiol (**131**) (*67*). The monoterpene moiety in (**131**) is very similar to limonene (**59**) and assignments

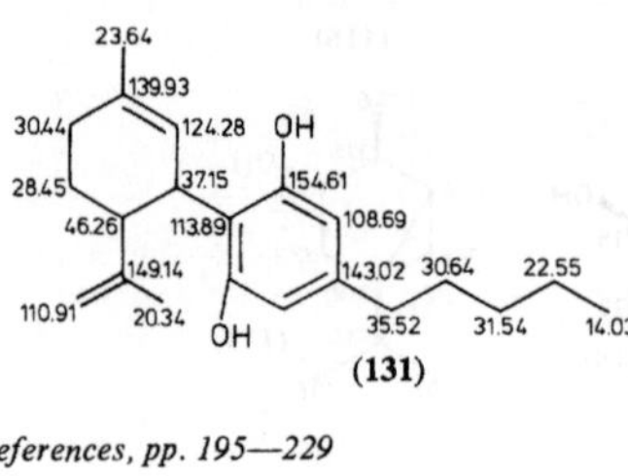

(131)

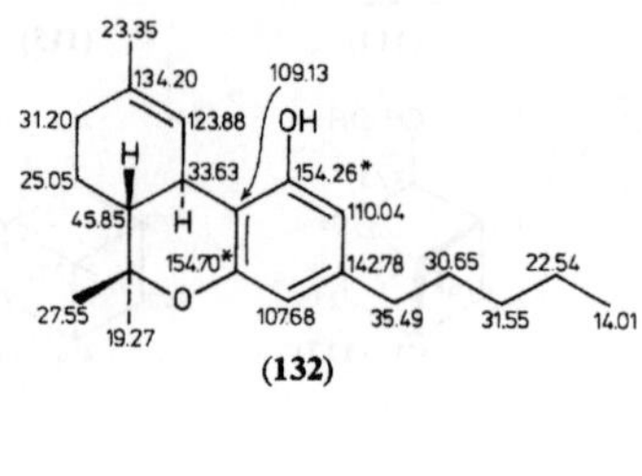

(132)

of this part of the molecule thus are straightforward on the basis of shift comparison, further aided by the multiplicities in the SFORD spectrum. The sidechain and aromatic carbons are readily identified by comparing the chemical shifts with those in Δ^1-tetrahydrocannabinol (**132**) (*68*). It is noteworthy that the PND spectrum of (**131**) exhibits the usual sharp peaks except for those pertaining to C-2′/C-6′ and C-3′/C-5′, which are broad at ambient temperature indicating restricted rotation of the phenyl ring around the C-3—C-1′ bond axis. Restricted rotation of the phenyl moiety had been suggested earlier from its ^{1}H n.m.r. spectrum (*69*).

The ^{13}C chemical shift data of the linalyl oxides (**133**) and (**134**) (*70*) and their dehydration products (**135**) and (**136**) served as a basis for analysis of the natural sesquiterpene (+) davanone (**137**) (*71*).

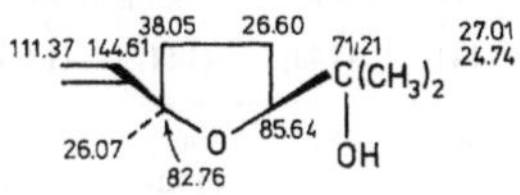

(**133**)

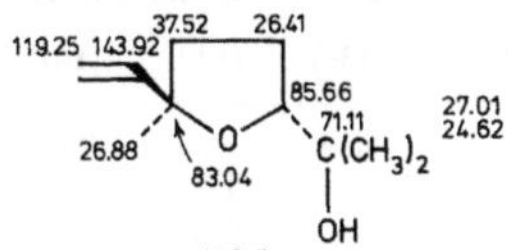

(**134**)

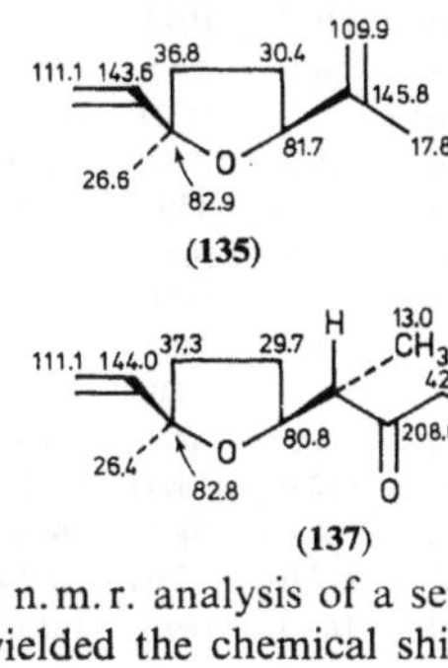

(**135**)

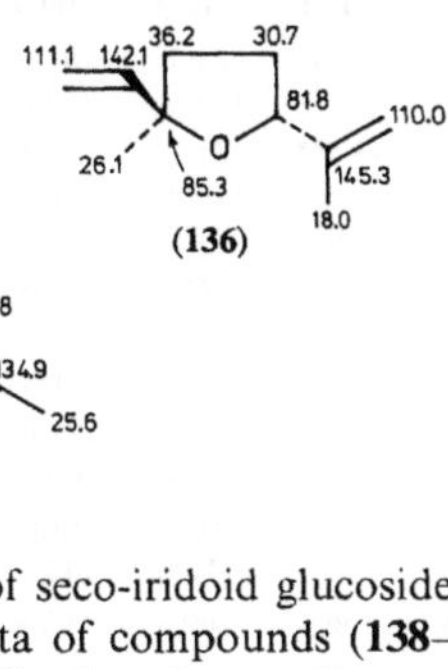

(**136**)

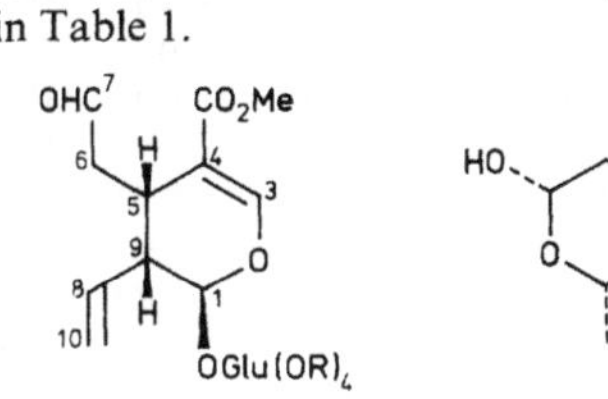

(**137**)

A ^{13}C n.m.r. analysis of a series of seco-iridoid glucosides and their acetates yielded the chemical shift data of compounds (**138**—**146**) (*72*). Chemical shifts and one-bond ^{13}C—^{1}H spin-spin coupling constants are compiled in Table 1.

(**138**) R = H
(**139**) R = Ac

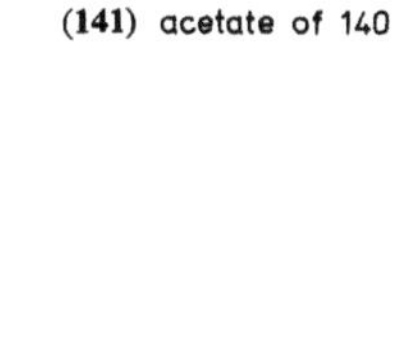

(**140**)
(**141**) acetate of 140

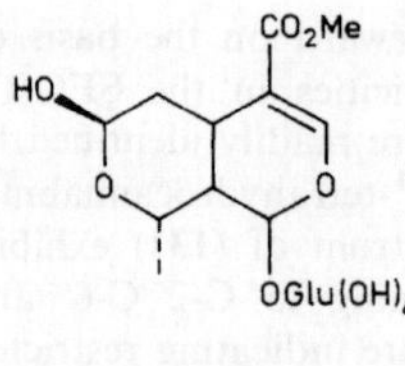

(142)

(143) acetate of 142

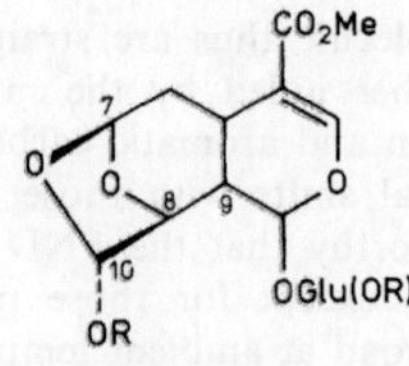

(144) R = H

(145) R = Ac

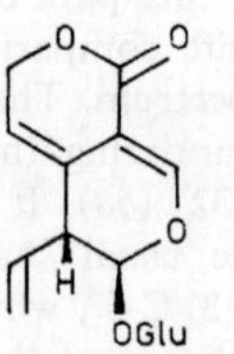

(146)

Table 1. *^{13}C Chemical Shifts and Coupling Constants $^{1}J_{CH}$* for Seco-Iridoid Glucosides and Acetates* (**138—146**) (*72*)

	Secologanin		α-Morroniside		β-Morroniside		Secogalioside		Gentio-picroside
Carbon	(**138**)	(**139**)	(**140**)	(**141**)	(**142**)	(**143**)	(**144**)	(**145**)	(**146**)
1	97.6	95.4	96.2	94.5	96.2	94.2	96.3	95.7	98.5
	(174)		(170x)	(173)	(170x)	(175)	(174x)	(168x)	(177)
3	154.0	150.7	154.9	151.9	154.9	151.9	151.1	152.2	150.4
	(194)		(192)	(192)	(192)	(192)	(192)	(192)	(196)
4	109.6	109.1	110.2	109.9	110.9	110.7	109.8	109.7	104.4
5	27.5	25.1	31.1	30.0	26.8	26.1	25.2	24.6	125.2
	(134x)		(130x)	(130)	(130x)		(132)	(132x)	
6	44.6	43.5**	36.1	32.9	33.4	31.4	34.6	34.4	117.8
			(130x)	(130x)	(130x)	(130x)	(130x)	(130x)	(169)
7	206.8	199.7	95.9	93.5	91.6	91.1	103.3	103.8	71.2
	(177)		(160x)	(160x)	(170)	(172)	(174)	(174)	(154)
8	133.8	131.8	73.8**	73.3	65.9	67.1	78.9	77.2	133.9
	(156)			(145x)			(160)	(162x)	(161)
9	44.6	43.1**	38.7	38.8	39.3	39.2	37.2	36.8	45.4
	(132x)		(130)	(130x)	(130)	(129)	(130x)	(132x)	(133)
10	121.6	120.6	19.6	18.8	19.6	18.8	96.7	95.1	119.5
	(161)		(127)	(127)	(127)	(127)	(176x)	(178)	(160)
C=O	169.8	166.1	169.8	165.8	169.8	165.9	169.5	166.6	167.5
OMe	52.6	51.1	52.6	51.2	52.6	51.1	52.7	51.3	—
	(147)		(147)	(146)	(147)	(146)	(147)	(147)	
1′	99.6	95.4	99.5	96.5	99.5	96.6	100.4	98.4	99.6
	(161)		(162)	(163)	(162)	(163)	(162)	(162)	(162)
2′	73.5	70.4	73.6**	70.8	73.6	70.9	73.6	71.0	73.3
3′	76.6**	72.0	76.8**	72.4	76.8**	72.5	76.7**	72.7	76.5**
4′	70.5	67.9	70.5	68.3	70.5	68.4	70.4	68.4	70.3
5′	77.2**	72.0	77.1**	71.8	77.1**	72.0	77.1**	72.1	77.1**
6′	61.6	61.4	61.6	61.6	61.6	61.5	61.6	61.9	61.6

* In Hz; ±2 Hz: if the value is followed by an "x": ±5 Hz.

** Assignments may have to be reversed.

An unusual seco-iridoid monoterpenoid named xylomollin (**147**) was recently isolated from *Xylocarpus molluscensis* (*73*). Its unique structure accomodated an acetal as well as a hemiacetal function. Xylomollin is formally derived from secologanin (**138**).

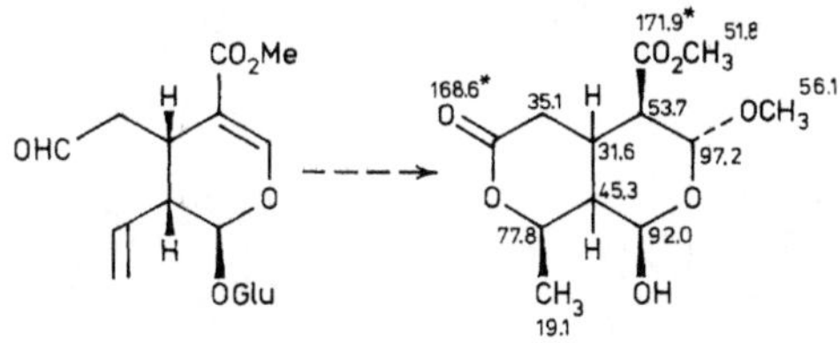

(**138**) (**147**) in Pyr-d_5

Sesquiterpenes

The ^{13}C n.m.r. spectra of an unusual *cis*-eudesmansesquiterpene alcohol, 10-epijunenol (**148**), which was isolated from galbanum resin, provided distinct evidence for a *cis* ring fusion. This permitted structures (**148a**) and (**148b**) to be taken into consideration. The high-field shift

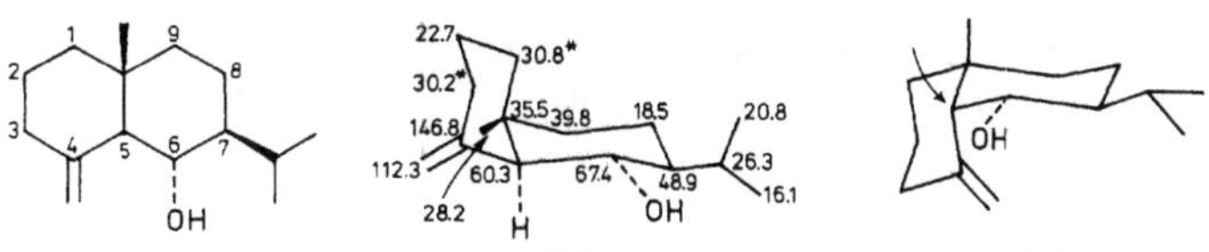

(**148**) (**148a**) (**148b**)

of C-1 can be rationalized in terms of a 1,3-diaxial interaction involving C-1 and the axial hydrogens of C-6 and C-8. The chemical shifts of C-7 (close to C-4 of menthol (**84**) and the methylene carbon adjacent to the double bond both support structure (**148a**) (*74*).

Structures of the plant growth inhibitor, cuauthemone (**149**) and its diester (**150**) were elucidated with the aid of their ^{13}C SFORD and inversion-recovery spectra. The latter technique turned out to be particularly beneficial for studying the congested highfield region of the spectra (*75*).

In a study of stress metabolites in potatoes, two vetispirane derivatives (**151**) and (**152**) could be isolated from potato tubers (*76*). Their ^{13}C n.m.r. spectra supported the structure proposed. The signals assignable to the spiro carbons resonating at 50.2 and 52.8 ppm were of particular help.

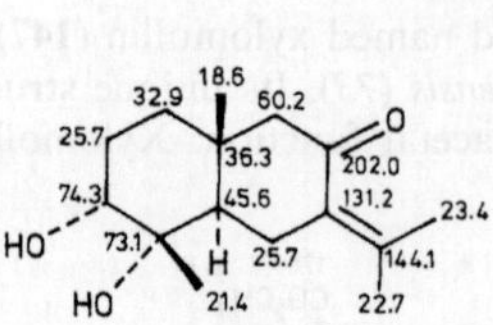

(149)

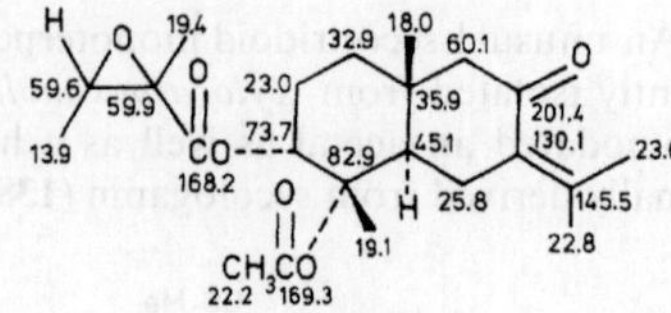

(150)

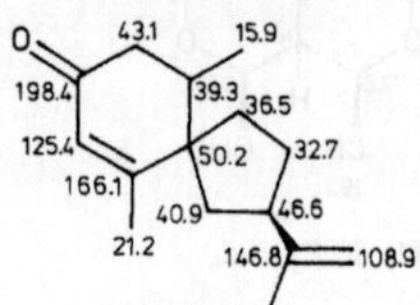

(151)

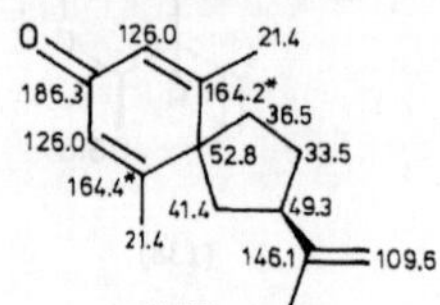

(152)

In the course of a structural study of some stress metabolites the acyclic (**153**—**155**) and cyclic (**156**—**158**) sesquiterpenes could be isolated

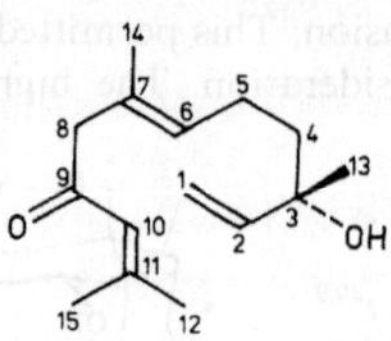

(153)

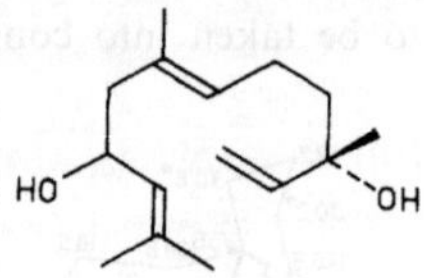

(154)

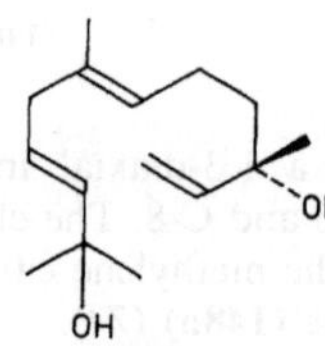

(155)

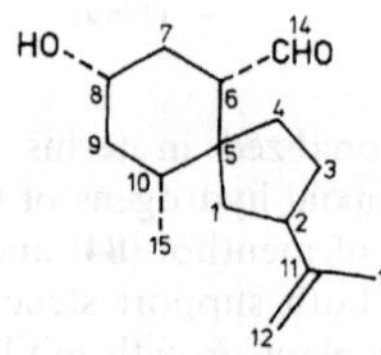

(156)

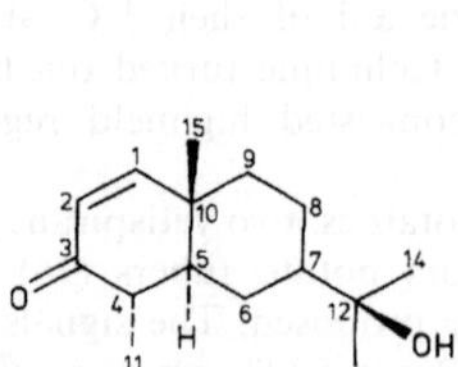

(157)

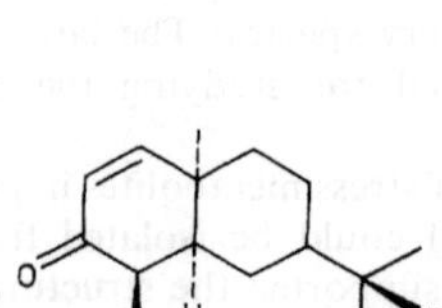

(158)

from egg plant (*Selanum melongena*) inoculated with fungi. Their structures were determined by means of joint ^{1}H and ^{13}C n.m.r. spectroscopy (*77*). Chemical shifts are listed in Table 2.

Table 2. ^{13}C *Chemical Shift Data for Compounds* (**153—158**) (*77*)

Carbon	(**153**)	(**154**)$^{+}$	(**155**)	(**156**)	(**157**)	(**158**)
1	111.6	111.8 (111.7)	111.7	33.3	160.9	156.7
2	144.8	144.9 (145.1)	144.9	41.0	125.9	128.5
3	73.2	73.5 (73.4)	73.4	32.6	204.0	201.9
4	41.8	41.8	42.1	25.9	49.1*	48.6*
5	23.0	23.0 (22.9)	22.9	46.9	45.3*	48.5*
6	122.8	127.7	125.0*	58.4	26.1	24.5
7	129.6	134.8	134.1	41.8	44.5*	42.1*
8	55.2	48.1	42.4	69.3	22.4	23.9
9	198.9	66.0	125.2*	40.3	40.0	41.0
10	129.2	128.5 (128.4)	139.4	47.4	36.2	38.2
11	155.4	131.8 (131.9)	70.6	147.3	13.4	12.0
12	27.6	25.7	29.9	108.8	72.6	72.7
13	27.9	28.0 (28.2)	27.9	21.2	27.2*	27.1
14	16.4	16.2	16.2	204.9	27.6*	27.1
15	20.7	18.2	29.9	16.4	20.8	27.1

$^{+}$ Shielding values in parenthesis refer to second diastereomer.
* Ambiguous assignments.

The structure of a new sesquiterpene furan-3-aldehyde (**159**) from *Lactarius vellereus* and *L. pergamenus* has been elucidated with the aid of several spectroscopic techniques including ^{13}C n.m.r. (*78*).

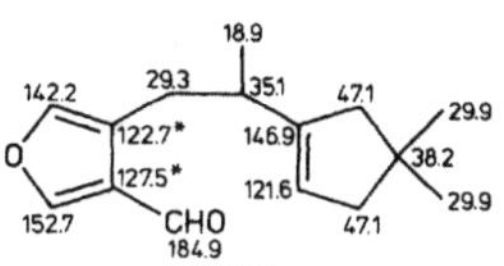

(**159**)

Epoxydecompostin (**160**) from *Lepidospartum squamatum* gray was characterized on the basis of its ^{13}C spectrum (*79*).

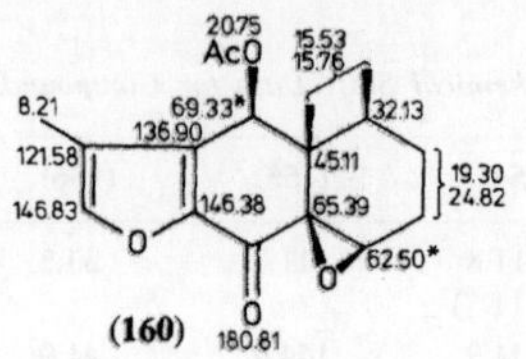

(**160**)

^{13}C n.m.r. spectral data have been reported in connection with the study of a new class of sesquiterpenes (**161**—**163**), named capnellanes, which were found in certain soft corals (*80*). Their chemical shifts are compiled in Table 3.

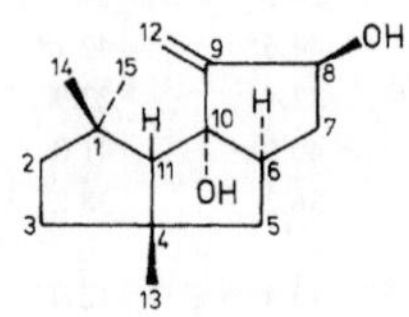

(**161**)

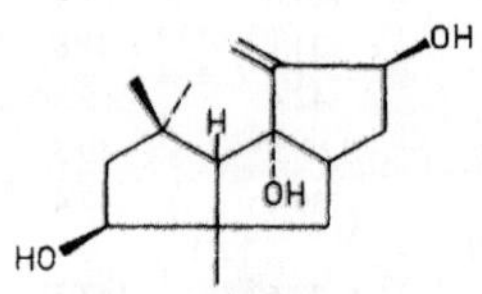

(**162**)

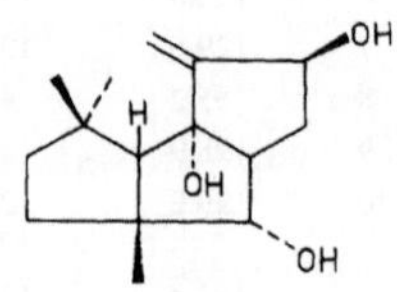

(**163**)

Table 3. ^{13}C *Chemical Shifts and Multiplicity Data for Compounds* (**161**—**163**)

Carbon	(**161**)	(**162**)	(**163**)
1	43.3	38.6	43.7
2	42.7	51.7	42.4
3	41.4	81.4	32.9
4	49.3	52.6	53.2
5	45.6	45.3	82.8
6	48.7	49.8	56.1
7	36.8	38.1	34.6
8	72.4	73.8	72.4
9	160.3	161.5	159.9
10	88.8	89.8	86.0
11	64.6	65.5	64.0
12	107.5	109.1	108.4
13	31.5*	25.0*	30.8*
14	30.3*	32.9	31.4*
15	32.2	26.1*	24.1

* Ambiguous assignments.

Structure (**164**) was suggested for PR toxin, a mycotoxin from *Penicillium roqueforti*, on the basis of its spectral characteristics (*81*). Spectral data including ^{13}C n.m.r. have been used for elucidating the structure of the fungal metabolite LL-N313J (**165**) (*82*).

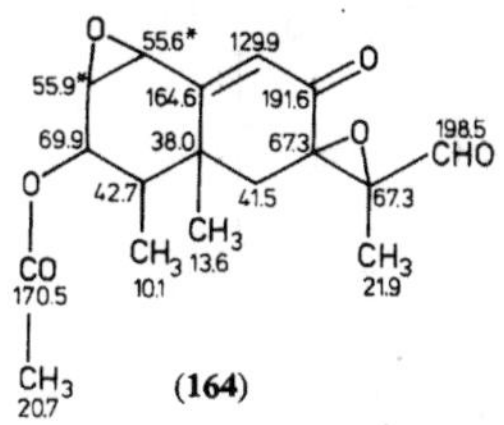

(**164**)

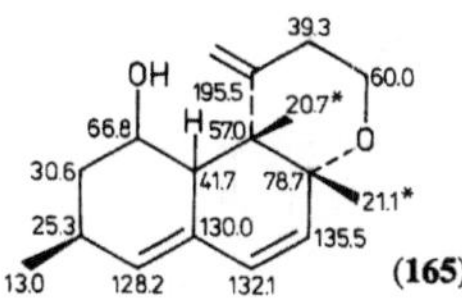

(**165**)

A total assignment has recently been given for a host of trichotecane derivatives (**166**—**176**) (*83*). Signal designation was accomplished by means of selective decoupling and chemical shift comparison with model compounds. Chemical shifts and multiplicity data for verrucarol and

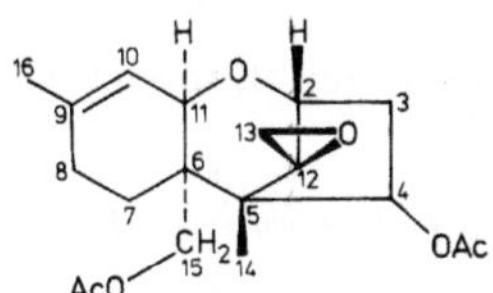

(**166**)

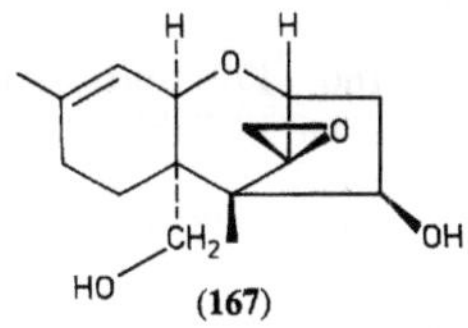

(**167**)

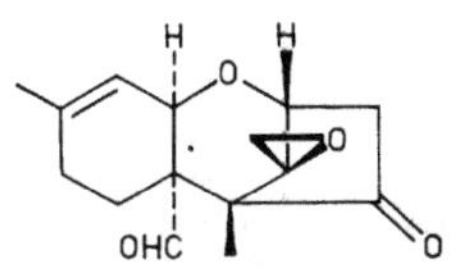

(**168**)

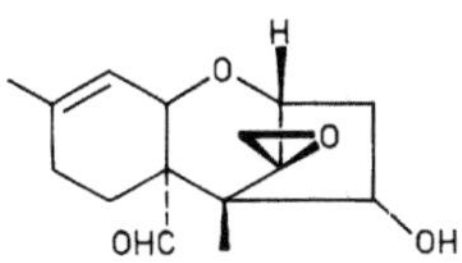

(**169**)

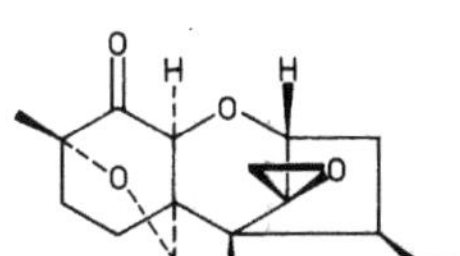

(**170**)

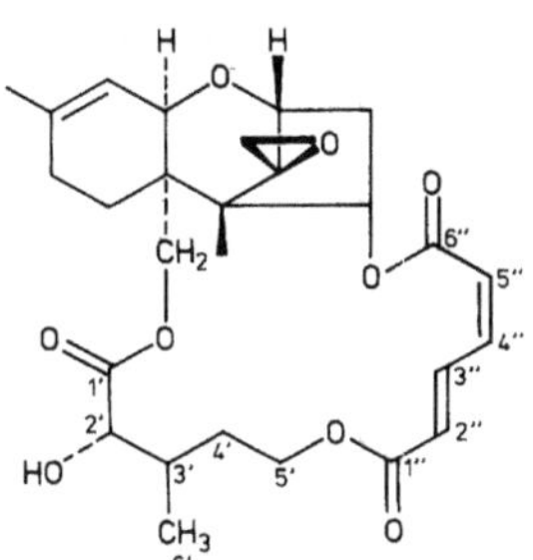

(**171**)

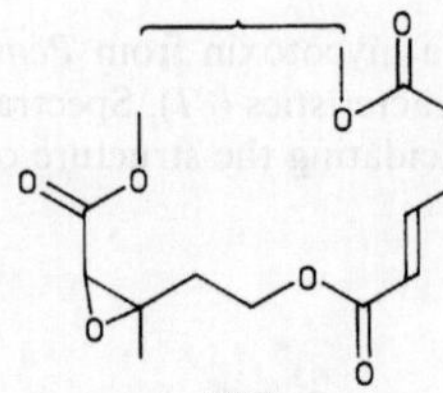

(172)

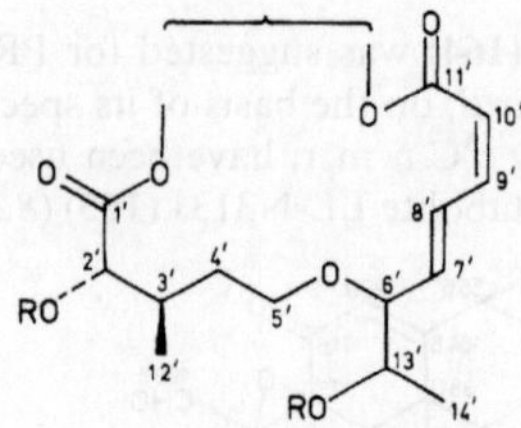

(173) R = H
(174) R = Ac

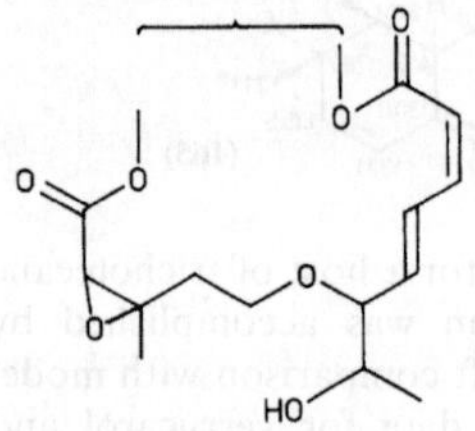

(175)

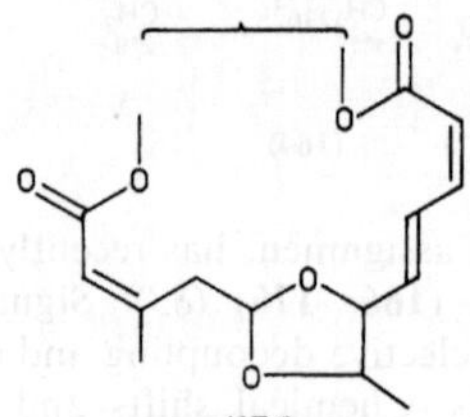

(176)

its derivatives (**166—170**) and verrucarin A and B, roridin A, D and H (**171—176**) are found in Tables 4 and 5, resp.

Table 4. ^{13}*C Chemical Shift and Multiplicity Data for Compounds* (**166—170**) (*83*)$^{+}$

Carbon	(166)	(167)	(168)	(169)	(170)
2	78.4d	78.5d	75.8d	78.7d	79.3*
3	36.0t	39.5t	41.6t	39.7t	39.3t
4	74.7d	74.0d	212.7s	73.4d	72.8d*
5	48.2s	48.7s	53.4s*	48.8s	46.1s**
6	42.7s	43.7s	54.7s*	54.0s	44.7s**
7	20.7t	20.9t	18.6t	20.5t	21.2t
8	27.4t	28.0t	27.2t	27.8t	31.5t
9	139.6s	140.4s	142.2s	140.3s	76.4s
10	118.2d	118.7d	117.7d	118.8d	208.5s
11	66.2d	66.4d	65.4d	65.3d	75.1d*
12	64.7s	65.6s	64.1s	64.8s	64.8s
13	47.1t	47.4t	49.3t	46.9t	46.5t
14	5.8q	6.8q	5.5q	6.5q	6.3q
15	63.0t	62.1t	199.4d	202.3d	66.2t
16	22.4q	22.9q	23.1q	23.1q	17.9q
CH_3(Ac)	20.2q				
C=O(Ac)	169.6s				
C=O(Ac)	189.8s				

*, ** Ambiguous assignments.

$^{+}$ The symbols following the shift value indicate the signal multiplicity (s = singlet, d = doublet, t = triplet, q = quartet).

Table 5. ^{13}C *Chemical Shift and Multiplicity Data for Compounds* **(171—176)** (*83*)

Carbon	**(171)**	**(172)**		**(173)**	**(174)**	**(175)**	**(176)**
2	78.6d	78.7d		78.8d	78.8d[b]	78.8d	79.0d
3	34.6t	34.8t		34.6t	34.6t	34.9t	34.8t
4	75.3d	75.4d		74.2d	74.3d	74.3d	74.0d
5	49.2s	49.0s		49.1s	49.1s	49.0s	48.9s
6	43.9s	43.6s		43.6s	43.7s	43.1s	43.2s
7	19.7t	19.8t		20.0t	20.7t	20.4t	20.5t
8	27.2t	27.4t		27.5t	27.2t	27.4t	27.6t
9	140.7s	140.4s		140.4s	140.7s	140.1s	139.9s
10	117.7d	118.0d		118.2d	119.9d[c]	118.4d[c]	118.6d[c]
11	66.6d	66.9d		66.9d	67.0d	66.9d	67.6d
12	64.9s	64.9s		64.9s	64.9s	65.1s	65.3s
13	47.4t	47.5t		47.4t	47.4t	47.4t	47.3t
14	7.0q	7.3q		7.2q	6.9q	6.8q	7.0q
15	63.1t	63.6t		64.2t	63.3t	64.3t	63.0t
16	22.9q	22.8q		22.9q	22.9q	22.9q	22.9q
1′	174.3s	167.4s		174.5s	[f]	167.8s	166.0s
2′	73.8d	58.0d		75.3d	76.6d	57.9d	119.0d[c]
3′	32.9d	61.1s		36.7d	32.9d	62.9s	154.4s
4′	31.9t	36.9t		33.0t	33.7t	39.4t	47.7t
5′	60.8t	60.4t		69.5t	68.2t	67.3t	100.8d
6′	9.8q	15.8q		83.7d	79.1d[b]	85.3d	81.9d[e]
1″	165.8s[b]	165.8s[b]	C(7′)	139.0d	137.6d	138.1d	134.6d
2″	127.2d[a]	127.2d[a]	C(8′)	126.0d	126.9d	126.2d	126.2d
3″	138.6d	138.0d	C(9′)	143.6d	142.8d	142.9d	142.4d
4″	138.6d	138.0d	C(10′)	117.2d	117.6d[c]	117.8d[c]	118.9d[c]
5″	125.5d[a]	125.6d[a]	C(11′)	166.3s	166.3s	166.1s	166.0s
6″	165.1s[b]	164.9s[b]	C(12′)	14.4q	14.6q[d]	17.2q[d]	18.2q[d]
			C(13′)	70.4d	70.5d	70.5d	76.8d[e]
			C(14′)	18.0q	14.1q[d]	17.9q[d]	16.3q[d]
			CH_3(Ac)		20.1q		

[a, b, c, d, e] Ambiguous assignments.
[f] C-1′ and two acetyl carbonyl signals occur between C 168.9 and 170.3.

Further members of the same family of terpenoids studied by ^{13}C n.m.r. comprise species **(177—184)** (*84*). Chemical shift and multiplicity data are given in Table 6.

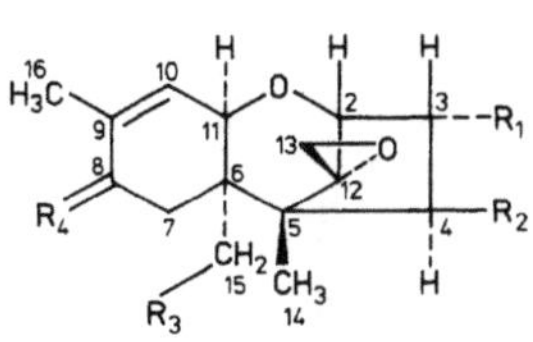

(177) $R_1 = OAc$; $R_2 = OAc$; $R_3 = OAc$; $R_4 = \alpha\text{-}O\text{-}COCH_2CH(CH_3)_2$
(178) $R_1 = OH$; $R_2 = OAc$; $R_3 = OAc$; $R_4 = \alpha\text{-}O\text{-}COCH_2CH(CH_3)_2$
(179) $R_1 = OH$; $R_2 = OH$; $R_3 = OAc$; $R_4 = \alpha\text{-}O\text{-}COCH_2CH(CH_3)_2$
(180) $R_1 = OH$; $R_2 = OH$; $R_3 = OH$; $R_4 = \alpha\text{-}O\text{-}COCH_2CH(CH_3)_2$
(181) $R_1 = OH$; $R_2 = OH$; $R_3 = OH$; $R_4 = \alpha\text{-}OH$
(182) $R_1 = OH$; $R_2 = OAc$; $R_3 = OAc$; $R_4 = H_2$
(183) $R_1 = H$; $R_2 = OCOCH{=}CHCH_3$; $R_3 = H$; $R_4 = O$
(184) $R_1 = H$; $R_2 = OCOCH{=}CHCH_3$; $R_3 = H$; $R_4 = \alpha\text{-}H,\beta\text{-}7,8\text{-}epoxy$

Table 6. ^{13}C *Chemical Shifts and Signal Multiplicities in Trichothecenes* (**177**—**184**) (*84*)

Compound	Carbon							
	2	3[a]	4[a]	5	6	7	8	9
Acyl T-2 toxin (**177**)	76.2(d)	78.7(d)	78.2(d)	47.7(s)	42.0(s)	26.7(t)	67.2(d)	135.0(s)
T-2 toxin (**178**)	78.7(d)	76.0(d)	82.4(d)	48.2(s)	42.2(s)	26.7(t)	67.4(d)	134.6(s)
HT-2 toxin (**179**)	78.7(d)	79.1(d)	79.8(d)	47.7(s)	41.8(s)	26.7(t)	68.9(d)	134.2(s)
T-2 triol (**180**)	78.7(d)	79.3(d)	79.3(d)	47.5(s)	43.1(s)	27.4(t)	68.1(d)	133.7(s)
T-2 tetraol (**181**)	78.5(d)	79.1(d)	80.0(d)	48.0(s)	45.1(s)	28.5(t)	64.8(d)	138.1(s)
Diacetoxyscirpenole (**182**)	79.3(d)	76.7(d)	83.1(d)	49.1(s)	44.0(s)	20.6(t)	27.4(t)	140.6(s)
Trichothecin (**183**)	73.4(d)	35.8(t)	78.7(d)	48.2(s)	42.6(s)	41.3(t)	175.3(s)	138.8(s)
Crotocin (**184**)	74.3(d)	36.2(t)	78.5(d)	47.7(s)	41.3(s)	58.1(d)	50.4(d)	137.2(s)
	10	11	12	13	14	15	16	
Acyl T-2 toxin (**177**)	124.0(d)	66.3(d)	63.6(s)	41.4(t)	6.1(q)	63.6(t)	19.4(q)	
T-2 toxin (**178**)	124.2(d)	66.3(d)	64.1(s)	46.2(t)	6.4(q)	63.9(t)	19.4(q)	
HT-2 toxin (**179**)	124.6(d)	67.4(d)	64.3(s)	45.3(t)	6.6(q)	63.9(t)	19.4(q)	
T-2 triol (**180**)	125.1(d)	66.3(d)	64.5(s)	45.3(t)	6.4(q)	61.7(t)	19.4(q)	
T-2 tetraol (**181**)	121.6(d)	67.6(d)	64.3(s)	45.1(t)	6.3(q)	61.0(t)	19.9(q)	
Diacetoxyscirpenole (**182**)	119.8(d)	67.4(d)	64.5(s)	46.4(t)	6.4(q)	63.4(t)	22.5(q)	
Trichothecin (**183**)	137.2(d)	69.2(d)	65.4(s)	46.4(t)	4.9(q)	14.4(q)	17.7(q)	
Crotocin (**184**)	122.9(d)	69.4(d)	65.9(s)	46.8(t)	6.2(q)	15.9(q)	21.0(q)	

[a] The C-3 and C-4 assignments in (**177**), (**179**), and (**181**) could be reversed.

In an investigation of aromatic sesquiterpenoids from *Emmotum nitens* ^{13}C n. m. r. spectroscopy provided corroborative evidence for structures (**185**) and (**186**) (*85*).

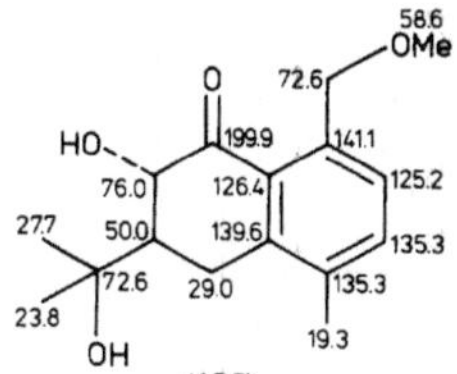

(**185**)

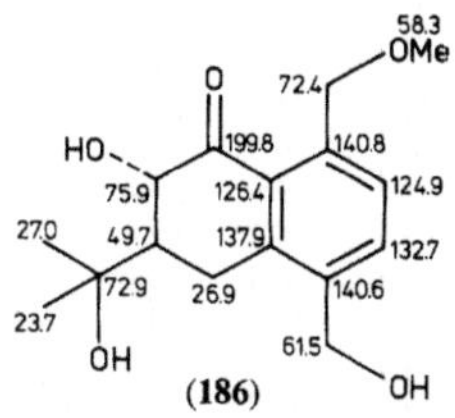

(**186**)

The ^{13}C spectra of two dihydroagarofurans (**187**) and (**188**) were found to be consistent with the proposed structure and stereochemistry (*86*).

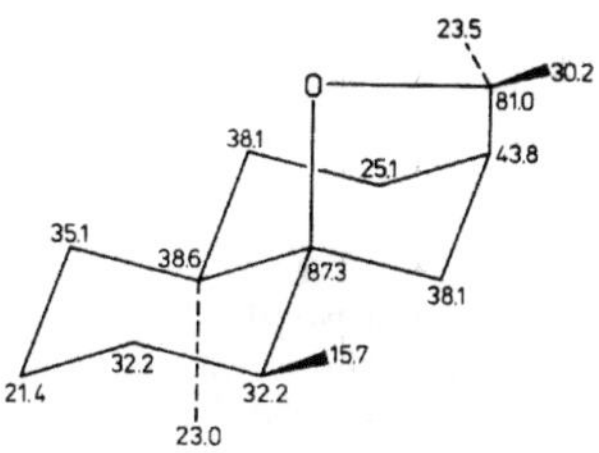

(**187**)

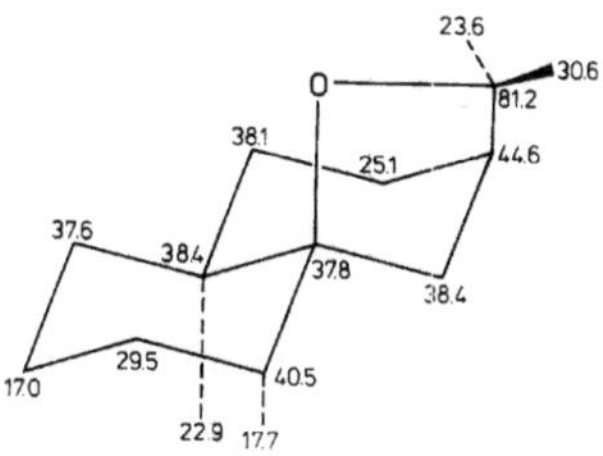

(**188**)

Structure and stereochemistry of some unusual polyalcohols isolated from *Celastrus paniculatus* Willd have recently been reported (*87*). In this connection the ^{13}C spectra of compounds (**189—193**) were examined. Chemical shifts are presented in Table 7. Since one of the minor constituents occurred as a small impurity, its spectrum was obtained by subtraction of the spectrum of the pure compound.

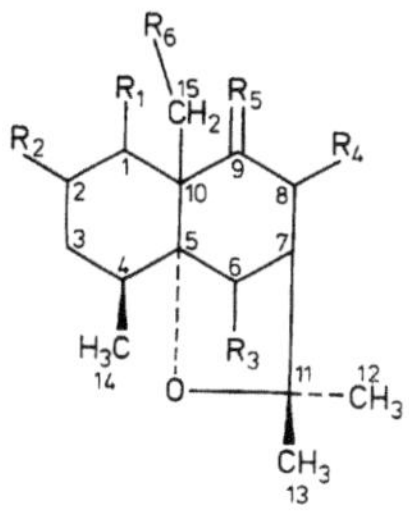

(**189—193**)

	R_1	R_2	R_3	R_4	R_5	R_6
(**189a**)	β-OH	H	H	α-OH	H,α-OH	OH
(**189b**)	β-OAc	H	H	α-OAc	H,α-OH	OAc
(**189c**)	β-OAc	H	H	α-OAc	H,α-OAc	OAc
(**189d**)	β-OH	H	H	α-OH	O	OH
(**189e**)	β-OAc	H	H	α-OAc	O	OAc
(**190**)	β-OH	H	α-OH	β-OH	H,α-OH	OH
(**191**)	β-OH	H	α-OH	α-OH	H,β-OH	H
(**192a**)	β-OH	H	α-OH	α-OH	H,α-OH	OH
(**192b**)	β-OAc	H	α-OAc	α-OAc	H,α-OH	OAc
(**193**)	β-OH	α-OH	H	β-OH	H,α-OH	H

Table 7. ^{13}*C Chemical Shifts on Poly Alcohols* (**189—193**) (*87*)*

Carbon	(189) a	(189) b	(189) c	(189) d	(189) e	(190)	(191)	(192) a	(192) b	(193)
1	69.7	71.8	71.8	71.6	71.2	71.0	76.8	69.7	71.6	80.0
2	25.7	22.2	22.6	23.5	22.4	26.0	25.3	26.0	22.6	73.4**
3	27.4	26.7	26.5	26.6	25.8	27.6	27.2	27.6	26.6	36.3
4	40.0	39.3	39.6	39.4	39.4	33.5	33.7	33.6	33.6	40.9
5	87.4	88.6	86.3	89.4	89.0	91.3	92.0	91.3	90.8	88.0
6	35.6	36.1	36.4	36.6	37.2	74.8	74.6	74.1	76.1	30.3
7	50.5	48.2	48.1	50.4	48.7	58.1	56.4	57.8	53.5	51.1
8	71.7	74.6	72.2	77.5	77.1	74.8	73.2	71.4	73.5	73.0**
9	69.8	69.0	68.2	199.8	71.4	79.3	70.0	69.5	66.1	
10	51.9	51.1	50.9	64.3	60.9	53.1	45.7	52.8	51.9	47.9
11	80.9	82.9	81.9	83.4	83.7	80.2	81.5	80.9	83.2	80.1
12	25.7	24.9	24.5	24.9	24.7	26.4	26.1	26.0	26.6	24.5
13	31.8	30.4	31.2	30.4	30.0	31.5	31.8	31.9	30.9	30.8
14	17.2	16.7	17.1	15.0	15.0	17.6	17.5	17.4	16.4	17.8
15	61.3	63.2	63.7	62.3	61.5	62.9	10.9	63.8	63.8	18.6

* In DMSO-d$_6$ except (**189b**) and (**192b**).
** Ambiguous assignments.

In the sesquiterpene lactone tenulin (**194**) the absolute configuration of all except two (C-11 and C-16) asymmetric carbons was established with the aid of a variety of spectroscopic techniques (*88*). The ^{13}C PND spectrum of (**194**) and (**195**) afforded two sets of lines each indicating

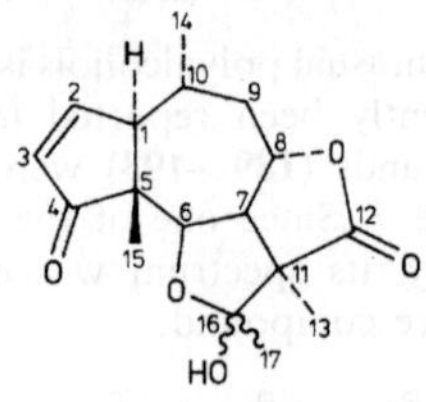

C-17 α - Me
C-17 β - Me

(**194**)

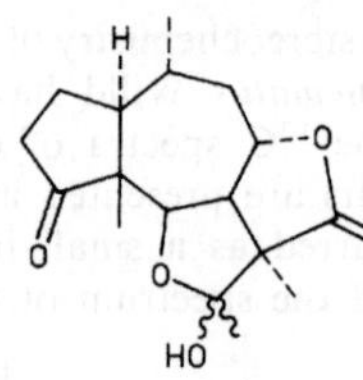

C-17 α - Me
C-17 β - Me

(**195**)

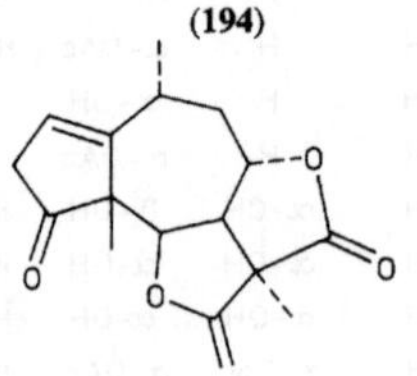

(**196**)

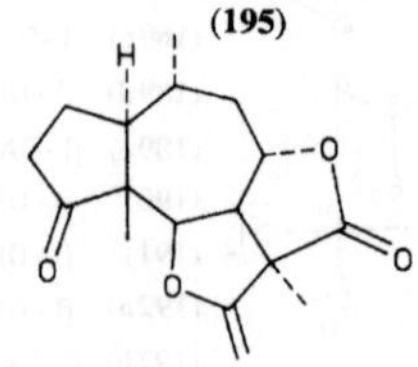

(**197**)

Table 8. ^{13}C *Chemical Shifts in Tenulin* (**184**) *and its Derivatives* (**195—201**) (*86*)

Carbon	(194) a	(194) b	(195) a	(195) b	(196)	(197)	(198)	(199)	(200)	(201)
1	54.3	54.7	48.3	48.6	150.1	48.4	52.7	47.3	173.2	47.8
2	162.6	162.3	24.7	24.2	122.7	24.4	161.1	24.3	138.4	23.0
3	130.4	130.0	34.6	34.5	39.7	34.4	130.3	35.1	207.0	35.8
4	212.7	212.3	221.6	221.2	217.6	220.3	210.6	219.8	37.2	209.7
5	56.3	55.3	54.5	53.4	53.4	53.4[a]	54.5	54.3	44.0	64.0
6	77.4	74.0	78.4	74.8	78.6	49.1	66.0	67.5	66.0	201.6
7	63.3	63.8	61.7	62.0	60.2	59.6	55.1	53.6	58.4	61.8
8	76.5	76.0	76.1	75.6	75.5	75.3	75.8	75.9	75.8	76.5
9	42.9	42.5	42.4	42.0	41.3	41.8	44.7	44.9	38.1	44.9
10	28.4	28.0	30.8	30.3	291.8	30.3	27.2	29.5	31.3	29.2
11	58.8	58.8	58.5	58.3	56.5	54.4*	37.0	37.0	41.2	36.8
12	176.4	176.42	175.6	177.6	175.4	75.4	177.4	177.6	178.0	177.2
13	18.3*	19.0*	18.0*	18.5*	21.2*	21.3*	20.6*	20.6*	20.6	20.1
14	19.6*	20.4*	13.7	13.8	16.2	13.7	19.7	14.8	8.12*	14.8
15	19.9*	20.4*	19.8*	19.6*	19.2*	19.7*	13.9	13.8	12.0*	18.7
16	108.4	105.2	107.6	104.4	161.4	162.0	169.2	169.2		
17	24.7	27.3	24.6	26.8	83.7	82.7	20.0*	19.8*		

* Ambiguous assignments.

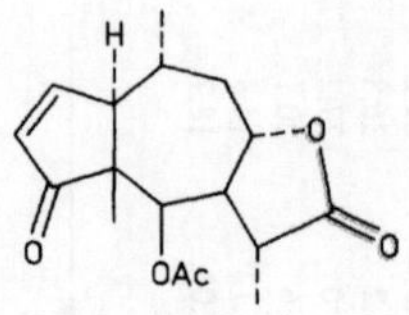

(198)

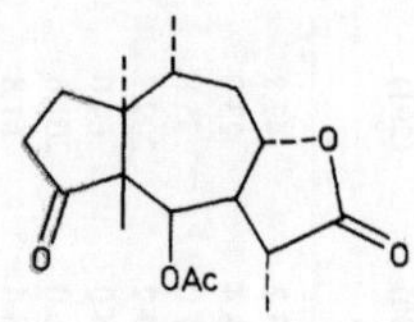

(199)

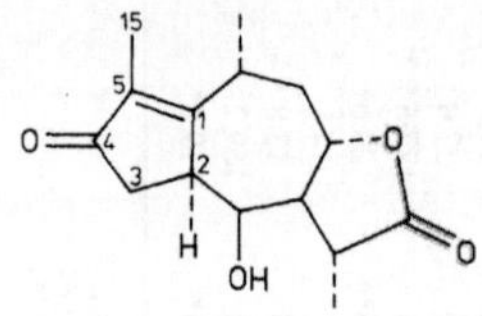

(200)

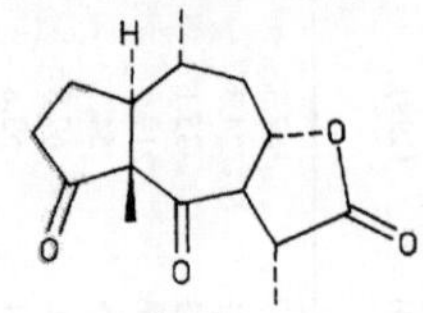

(201)

the presence of stereoisomers. In contrast, **(196)** and **(197)**, formed by dehydration of **(194)** and **(195)**, resp., turned out to be spectroscopically homogeneous. The chemical shift differences for the two sets of signals in compounds **(194)** and **(195)** was greatest for C-16 and neighbouring carbons C-8, C-11 and C-12. From this it was inferred that naturally occurring tenulin consisted of a mixture of C-16 epimers (*88*). Chemical shifts of tenulin and its derivatives are collated in Table 8.

In order to establish a data basis for future structural studies and for shedding some light on the stereochemical dependence of ^{13}C chemical shifts, the spectra of santonin derivatives **(202—205)** were recorded and

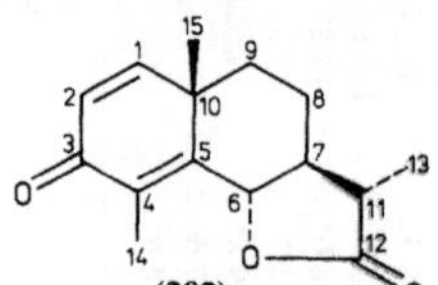

(202)

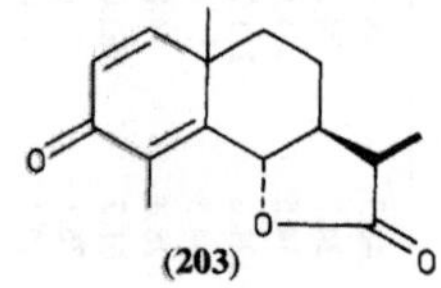

(203)

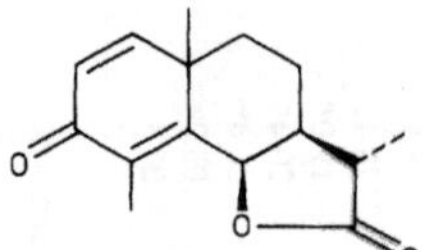

(204)

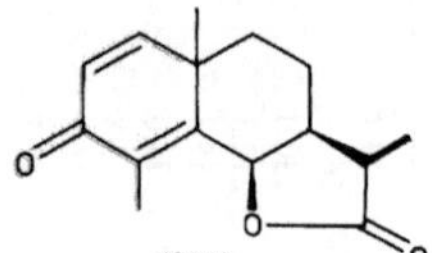

(205)

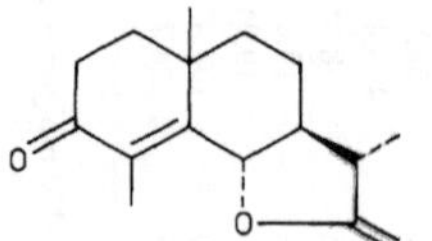

(206)

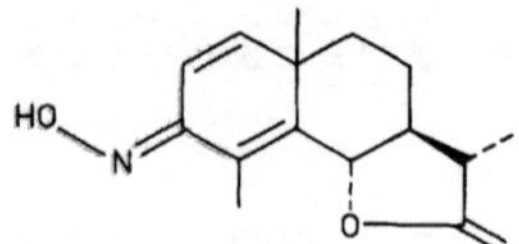

(207)

References, pp. 195—229

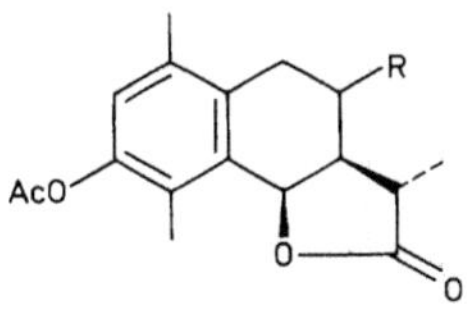

(208) R = H
(210) R = OAc

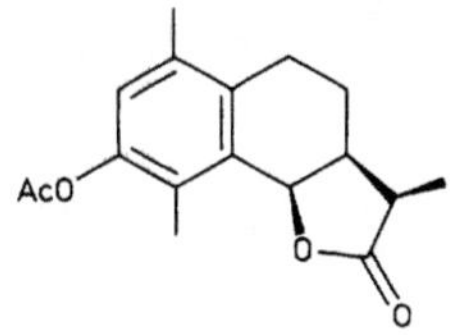

(209)

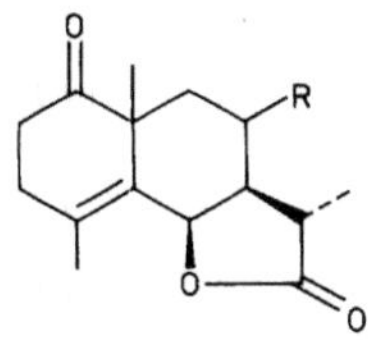

(211) R = OH
(212) R = H

chemical shifts assigned (*89*). The results indicated that the ^{13}C method may provide a simple method for determining the stereochemistry of the lactone ring fusion and the configuration of the methyl group at C-11. This work was later extended to a number of related eudesmanolides (**206**—**212**) (*90*). Chemical shifts in santonin derivatives (**202**—**212**) are provided in Table 9.

The ^{13}C n.m.r. spectra of the eudesmanolides granilin (**213**) and ivasperin (**214**) isolated from *Ambrosia polystachya* DC have been reported along with those of other naturally occurring ring-A hydroxylated iso-alantolactones (**215**—**218**) (*91*). The data are found in Table 10.

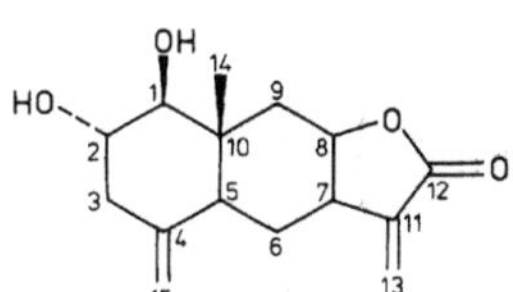

(214)

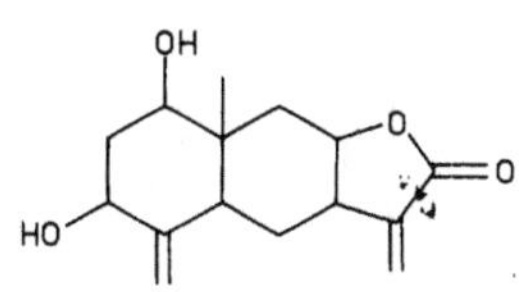

(213)

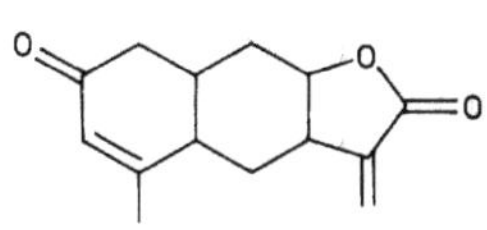

(215)

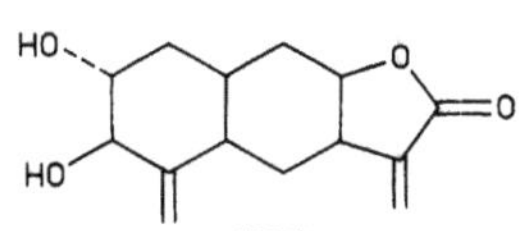

(216)

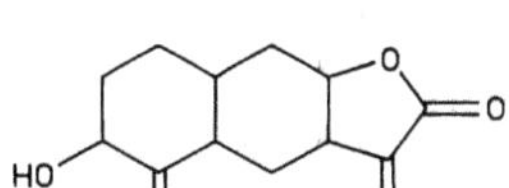

(217)

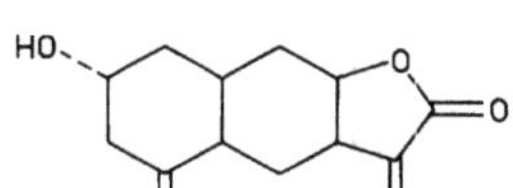

(218)

Table 9. ^{13}C *Chemical Shifts in Eudesmanolides* (**202—212**) (*90*)

Carbon	(**202**)	(**203**)	(**204**)	(**205**)	(**206**)	(**207**)	(**208**)	(**209**)	(**210**)	(**211**)	(**212**)
1	155.1	155.0	157.5	157.6	42.1	145.6	**	**	**	214.3	213.9
2	125.9	126.0	126.1	126.1	33.9	112.5	124.3	124.4	124.0	29.6	30.8*
3	186.0	186.1	186.2	186.3	198.5	150.7	148.1	148.1	148.4	35.0	35.5
4	128.4	128.8	137.8	137.6	128.5	122.5	**	**	**	139.4	139.3
5	151.5	151.9	148.0	149.3	152.9	159.5	**	**	**	127.0	128.1
6	81.5	80.8	76.5	76.9	82.1	82.3	75.6	75.5	79.6	77.5	76.6
7	54.0	49.5	43.8	41.8	53.4	53.7	42.1	40.0	53.0	51.0	42.8
8	23.8	20.3	23.4	18.3	23.7	23.8	23.8*	19.7	70.7	67.3	23.6
9	39.3	38.2	34.9	34.7	38.7*	38.5	24.9*	25.9	35.5	39.2	31.1*
10	41.7	41.8	39.4	39.5	38.5*	41.1	**	**	**	46.7	46.1
11	41.2	38.2	44.2	41.3	41.4	41.4	40.8	40.7	40.5	41.5	44.4
12	177.4	178.3	179.6	179.0	177.6	178.0	179.4	178.7	178.3	180.4	180.0
13	12.5	9.9	14.9	9.6	12.5	12.5*	14.6	9.5	14.1*	15.6	15.2
14	10.9	11.0	11.0	11.0	11.3	12.2*	12.1	11.9	13.6*	19.8	19.5
15	25.3	25.2	25.2	24.9	23.4	26.0	20.7	20.8	20.7	25.1	24.4

* Ambiguous assignments.

** Signal obscured by C_6F_6 resonance.

Table 10. ^{13}C *Shieldings in Sesquiterpenes* (**213**—**218**) (*91*)

Carbon	(**213**)**	(**214**)	(**215**)	(**216**)**	(**217**)	(**218**)
1	73.4	85.2	54.2	49.2	40.0	51.0
2	38.8	72.2	196.9	73.2	32.1	66.9
3	74.2	44.3	127.7	79.5*	73.0	46.3
4	149.2	147.5	159.6	150.1	151.0	146.1
5	43.2	45.6	44.9	45.2	44.6	45.6
6	25.9	28.3	26.9	28.6	27.5	27.3
7	39.4	41.4	40.8	41.2	40.3	40.5
8	76.7	79.0	75.6	78.5*	76.7	76.6
9	33.0	39.3	40.6	41.6	41.0	41.2
10	39.1	39.5	35.7	34.5	34.0	33.9
11	141.4	144.3	141.6	143.7	142.0	142.1
12	169.6	172.9	169.6	172.3	170.5	170.3
13	119.0	121.3	120.6	120.9	120.3	120.1
14	16.8	13.3	18.3	19.0	17.7	18.7
15	108.8	109.9	21.5	106.1	103.5	108.9

* Ambiguous assignment.
** In CH_3OH.

In the furanosesquiterpenes (**219**—**226**) a distinction between quaternary carbons C-7 and C-11 was made on the basis of the residual long-range coupling involving proton C(12)-H. The spectrum of (**224**) was found to remain invariant within the range of temperatures from −90° C to +100° C. This implies that the 10-membered ring probably exists as one single rigid conformer in solution (*92*).

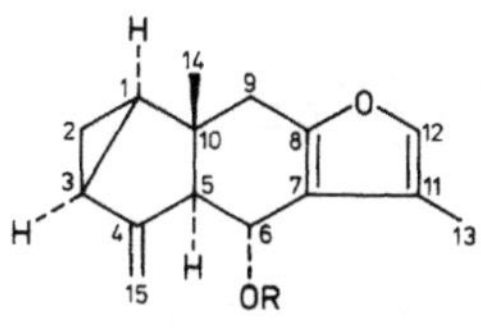

(**219**) R=H
(**220**) R=Ac

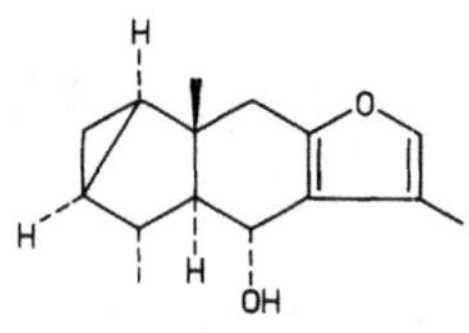

(**221**)

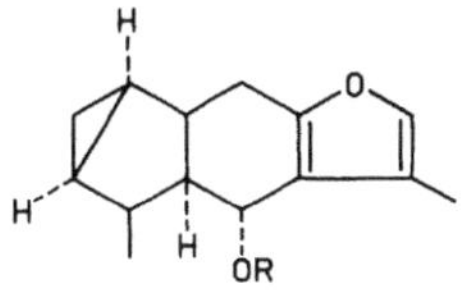

(**222**) R=H
(**223**) R=Ac

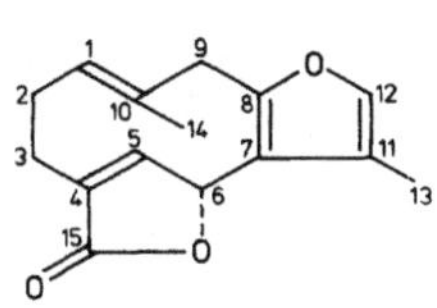

(**224**)

Table 11. ^{13}C *Chemical Shifts of Furanosesquiterpenes* (**219**—**226**) (*92*)

Carbon	(**219**)	(**220**)	(**221**)	(**222**)	(**223**)	(**224**)[a]	(**225**)[a]	(**226**)
1	27.4	27.2	28.3	29.1	29.1	130.3 (2.5)	130.0 (1.6)	143.6
2	16.6	16.7	16.7	10.9	11.0	25.8 (2.8)	23.0 (1.8)	114.8
3	23.1	23.1	26.8	24.5	24.6	26.8 (4.5)	26.6 (2.9)	123.7
4	150.0	149.0	39.5	31.9	32.0	135.1 (2.6)	61.3 (6.2)	136.6
5	68.4	64.8	71.1	66.1	62.9	151.7 (4.9)	65.6 (4.6)	46.8
6	65.1	65.9	67.6	64.2	65.9	74.2 (4.2)	73.2 (2.7)	72.2
7	120.9	117.7	121.4	121.3	118.3	115.4 (2.3)	113.7 (1.2)	113.6
8	152.5	153.7	152.9	153.1	154.3	152.8 (2.2)	153.2 (1.1)	152.6
9	38.4	38.3	39.1	40.8	40.9	40.7 (1.8)	40.4 (0.9)	34.9
10	41.3	41.6	43.6	43.6	44.0	130.8 (4.0)	131.6 (1.4)	40.9
11	120.1	119.2	119.8	120.1	119.2	122.2 (1.3)	122.2 (0.6)	119.8
12	138.5	138.8	138.4	138.2	138.6	137.2 (1.0)	137.1 (0.4)	138.6
13	9.1	8.7	9.2	9.1	8.6	8.4 (0.4)	8.3 (0.2)	8.2
14	18.2	18.3	20.4	19.8	20.0	15.7 (2.9)	15.8 (1.4)	18.7
15	108.0	107.9	17.7	13.6	13.9	173.4 (8.7)	171.5 (5.0)	170.9
Carbonyl	—	171.0	—	—	170.7	—	—	—
Methyl	—	21.1	—	—	21.0	—	—	—

[a] Values in parentheses are $Eu(FOD)_3$-induced shifts [$Eu(FOD)_3$: substrate $\simeq 1:2$].

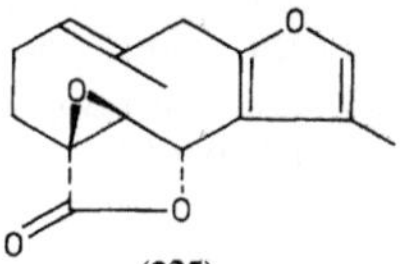

(225)

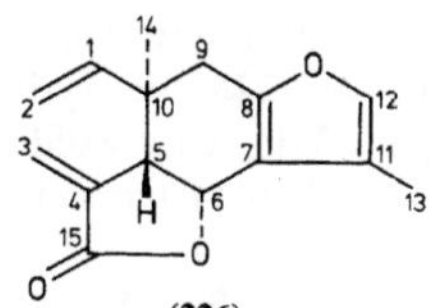

(226)

Chemical shifts of sesquiterpenes (**219**—**226**) are given in Table 11. A conformational analysis has been carried out on laurenobiolide (**227**) (*93*). Whereas the ^{13}C spectrum at room temperature displayed only one set of signals, four separate sets of lines could be discerned below $-15°$ C corresponding to an intensity ratio of ca. 5 : 4 : 3 : 1. This suggested that each of the four sets of peaks originates from one of the four major low-energy conformers (**227A**—**227D**). Conformer (**227A**) and (**227B**) were identified by 100 MHz ^1H n.m.r. spectroscopy. Evidence for (**227C**) and (**227D**) was provided by high field proton spectra.

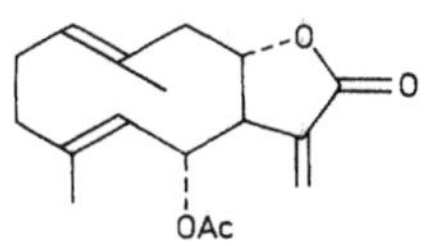

(227)

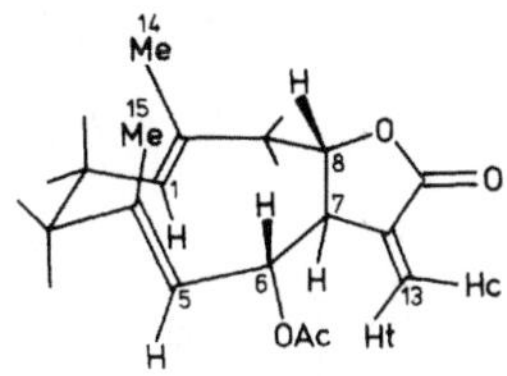

(227A)

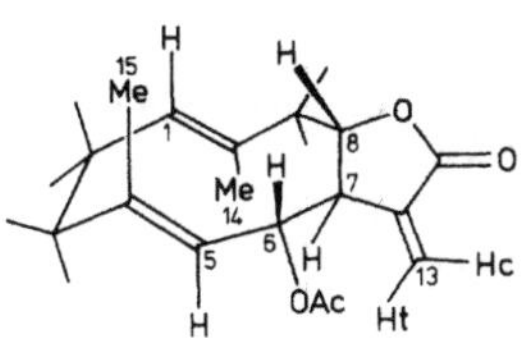

(227B)

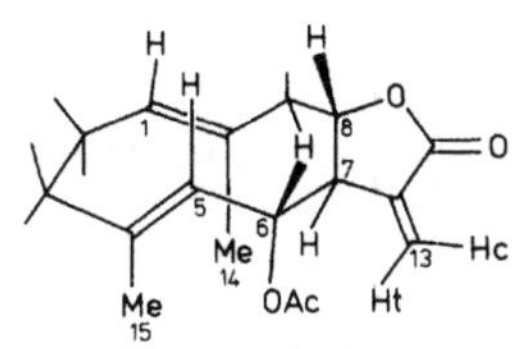

(227C)

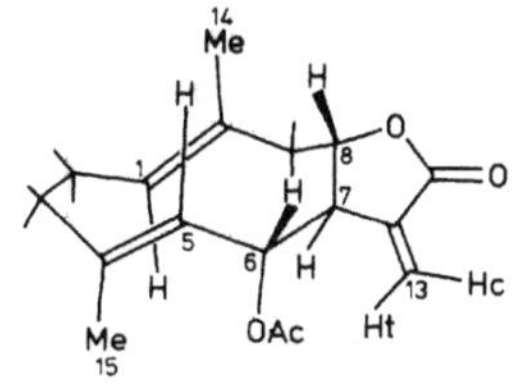

(227D)

^{13}C n.m.r. spectroscopy was also utilized in an analysis of germacranolides tifruticin (**228**), deoxytifruticin (**229**) and tirotundin (**230**) (*94*) and eupassopin (**231**), eupassofilin (**232**) and eupassopilin (**233**) (*95*). Chemical shifts in these systems are found in Table 12.

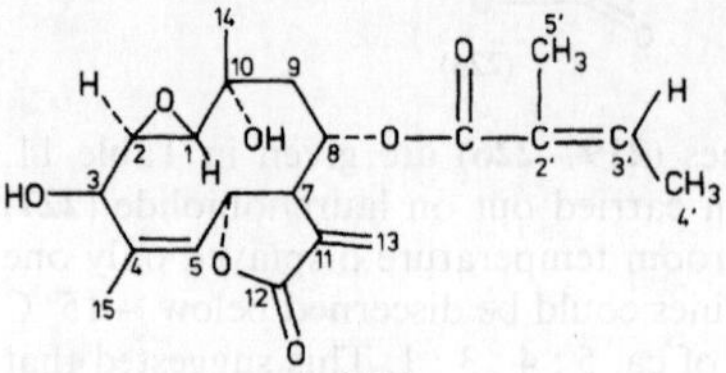

(228)

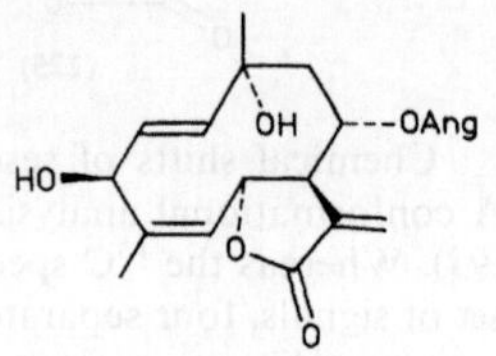

(229)

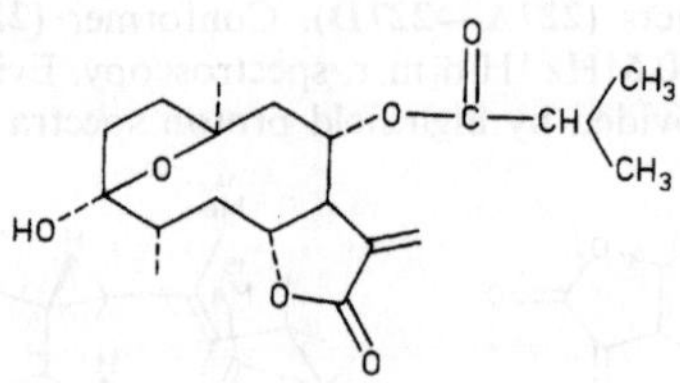

(230)

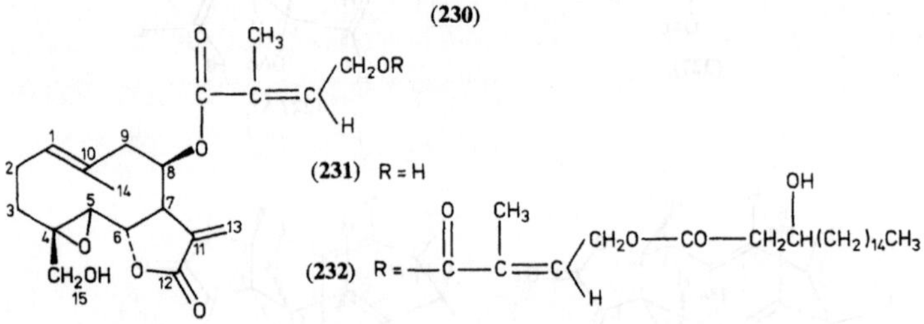

(231) R = H

(232) R =

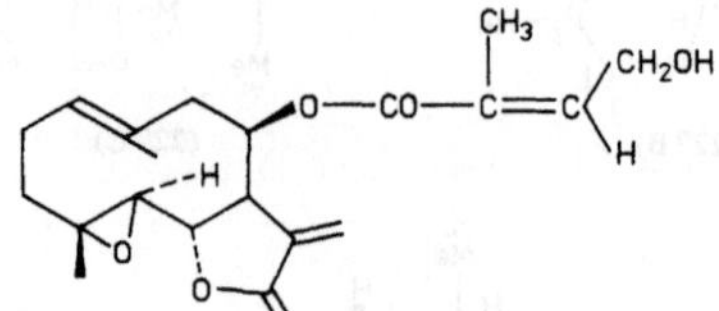

(233)

Complete assignments have been reported for the ^{13}C spectra of melampodin B (**234**) (*96*) and melampodin (**235**) (*97*). In the latter case an unambiguous assignment was attained with the aid of selective decoupling techniques and chemical shift arguments (steric effects). Distinction,

Table 12. ^{13}C *Chemical Shifts in Germacranolides* (**228**—**333**) (*94, 95*)

Carbon	(**228**)*	(**229**)*	(**230**)*	(**231**)	(**232**)	(**233**)
1	60.0	131.9	38.9	128.8	128.8	128.5
2	56.9	130.5	38.4	31.8	32.0	35.6
3	69.9	72.9	108.8	23.9	23.9	23.9
4	143.8	146.8	43.4	64.5	64.4	61.8
5	126.8	125.9	38.0	66.8	67.0	66.2
6	74.4	73.9	81.3	75.0	74.8	75.6
7	51.2	51.3	47.9	79.8	49.8	41.4
8	67.1	68.9	69.8	74.0	74.2	74.0
9	41.3	43.8	42.4	44.0	44.1	43.5
10	67.4	70.8	80.0	132.2	132.1	131.2
11	134.9	135.5	137.2	127.0	121.3	126.7
12	167.3	168.0	169.4	169.0	168.5	168.5
13	122.9	122.6	121.4	122.8	122.7	122.3
14	27.8	29.5	26.9	12.8	12.9	12.4
15	25.4	25.1	**	60.8	60.8	16.9
1′	169.3	169.4	176.1	166.2	165.7	165.9
2′	127.1	127.1	34.1	136.4	136.3	136.0
3′	140.1	140.7	**	142.9	137.3	142.6
4′	20.5	20.4	**	59.6	61.2	59.1
5′	15.9	15.9	—	19.6	19.9	11.4

* Tentative assignments based purely on shift comparison with literature data.
** No attempt was made to assign the methyl resonances at 18.6, 18.7 and 19.1.

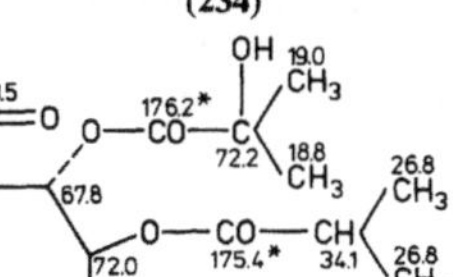

(**234**)

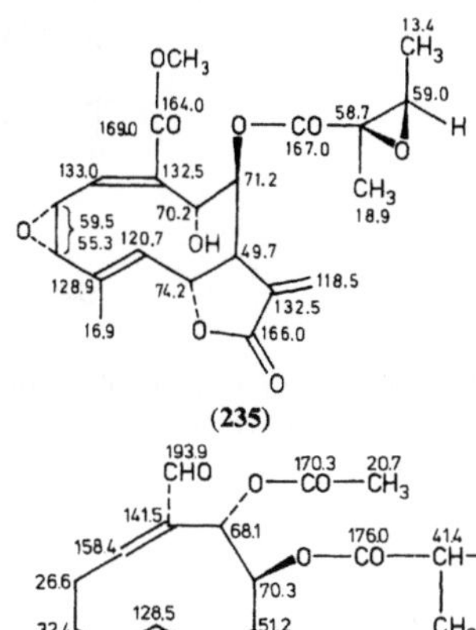

(**235**)

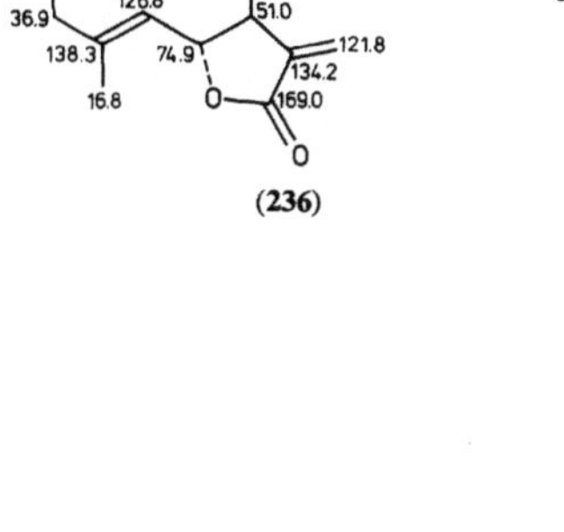

(**236**)

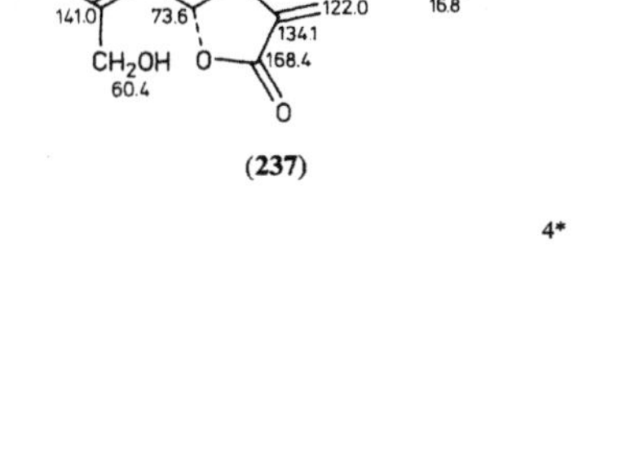

(**237**)

for example, could be made between C-2 and C-3 (59.5 and 55.3 ppm, resp.) by considering the steric congestion between of C (3)-H and the C-4 methyl group, which results in an upfield shift of C-3. Two more melampodins were subsequently analyzed and their structures determined, *i.e.* acanthospermal A (**236**) and B (**237**) (*98*).

Chemical and spectroscopic evidence led to structure (**238**) for pycnolide, a compound isolated from *Liatris pycnostachya* (*99*). Chemical

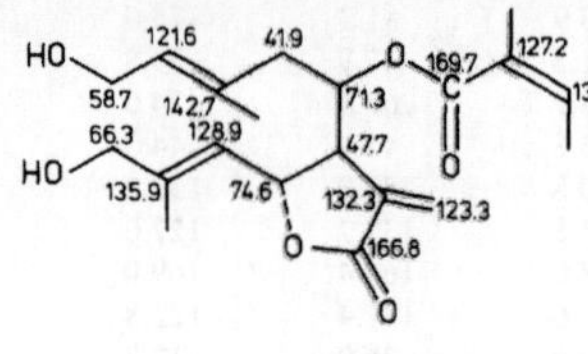

(238)

shifts but no assignments were reported for vanillosmin (**239**), (*100*) isolated from *Vanillosmopsis erythropappa*. This compound was found to be identical with eremanthin on the basis of their ^{13}C spectra.

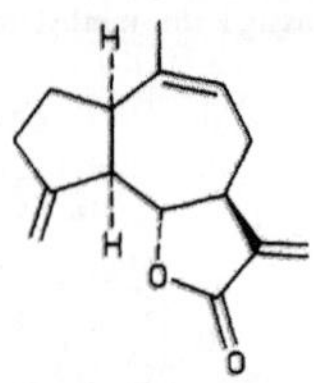

(239)

The ^{13}C spectra of a host of synthetic cyclopropane-containing terpenoids (**240**—**252**) have been analyzed. A systematic evaluation of the available chemical shift data permitted the establishment of the stereo-

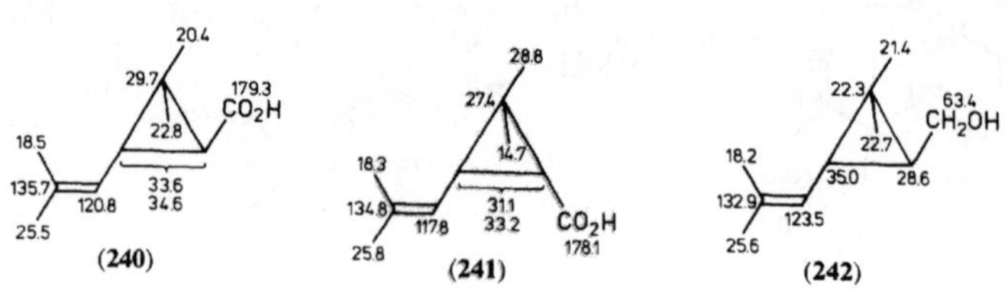

(240) **(241)** **(242)**

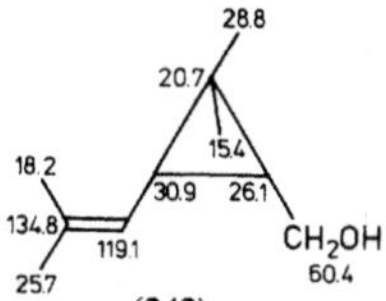

(243)

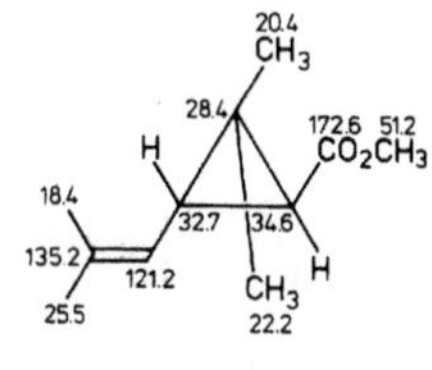

(244)

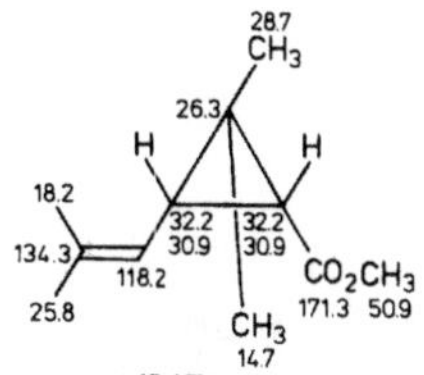

(245)

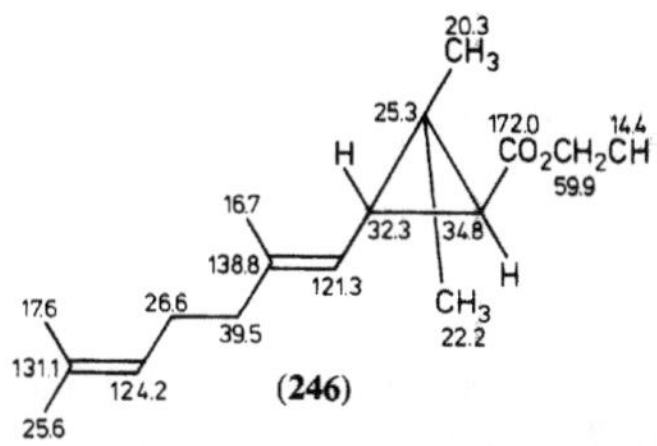

(246)

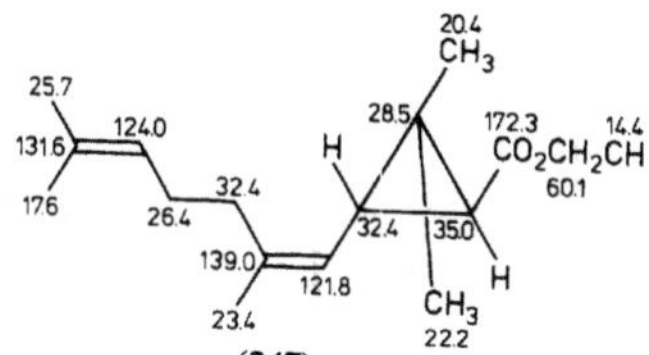

(247)

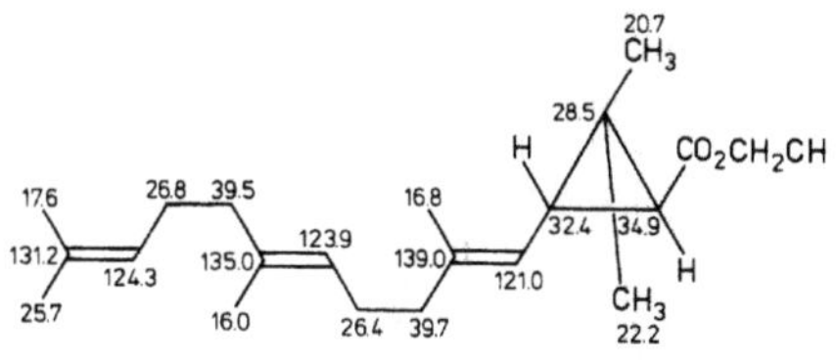

(248)

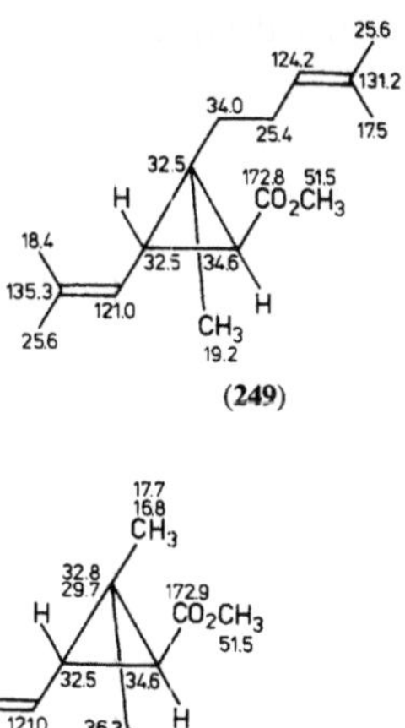

(249)

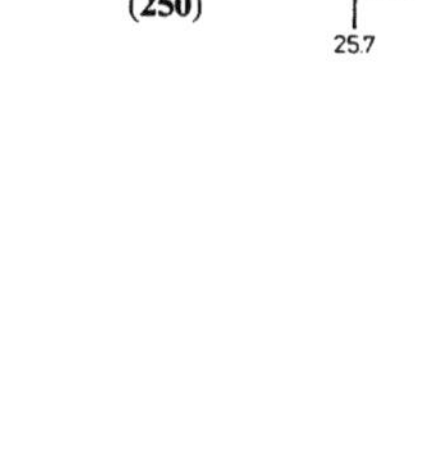

(250)

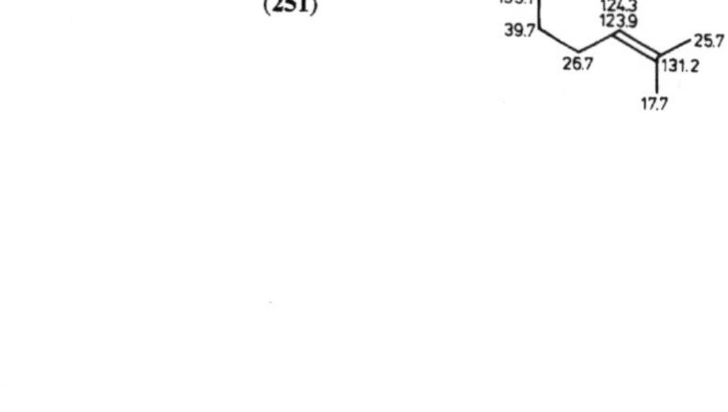

(251)

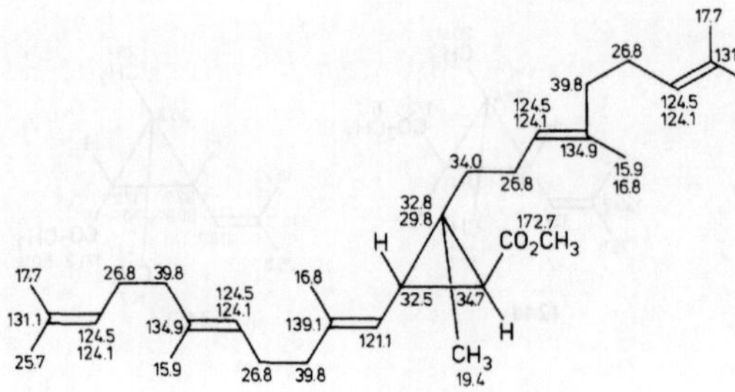

(252)

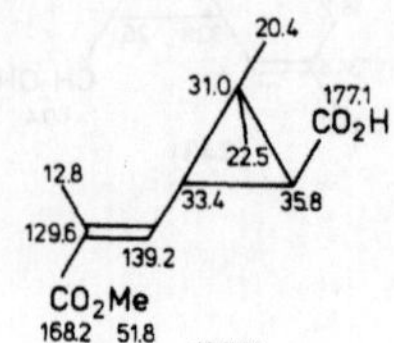

(253)

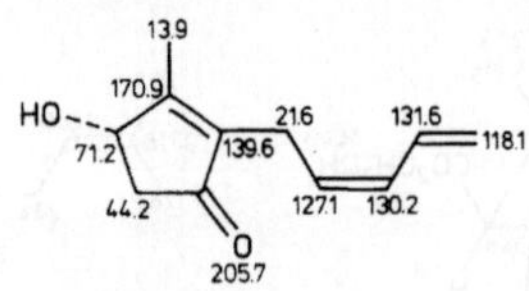

(254)

chemistry at the cyclopropane moiety and the trisubstituted double bond (*101*). Chemical shifts in (**240**), (**253**) and (**254**) served as a basis for interpreting the spectra of some natural pyrethrins (**255 a—f**) (*102*).

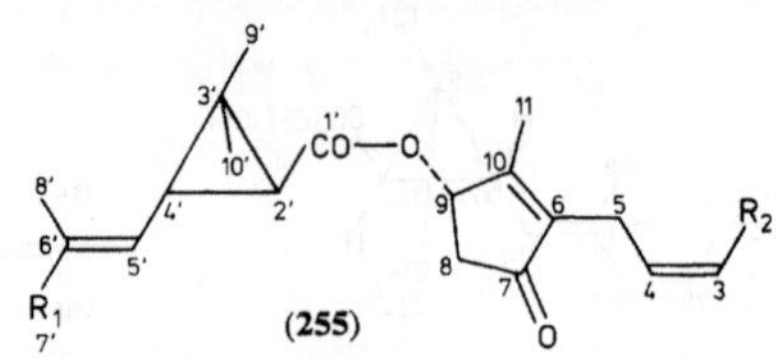

(255)

$R_1 = Me$; $R_2 = Me$
$R_1 = Me$; $R_2 = Et$
$R_1 = Me$; $R_2 = -CH=CH_2$
$R_1 = CO_2Me$; $R_2 = H$
$R_1 = CO_2Me$; $R_2 = Et$
$R_1 = CO_2Me$; $R_2 = -CH=CH_2$

^{13}C chemical shift data were reported in connection with the structure of vulgarone (**256**), a sesquiterpene ketone with a new carbon skeleton. Shieldings in this system were correlated with those of the monoterpene chrysanthenone (**257**) (*103*). A revised structure has recently been proposed for prenylsolanone (**258**) (*104*) on the basis of its ^{13}C spectrum. The absolute configuration of this compound was also determined.

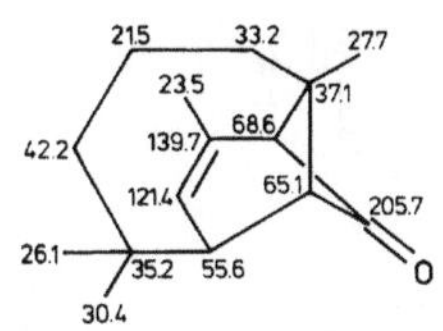

(256)

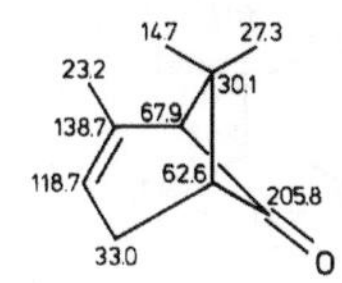

(257)

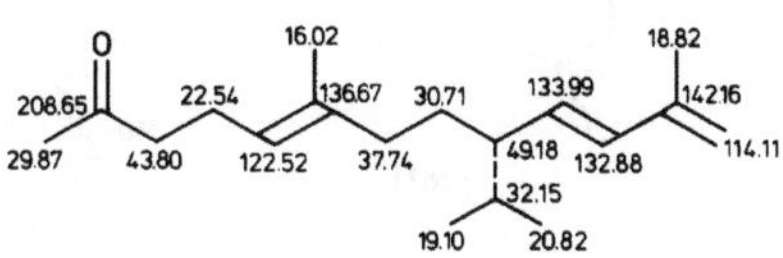

(258)

Diterpenes

A ^{13}C spectroscopic study involving a number of various labdanoids resulted in the total assignment of their spectra (**259**—**270**) (*105*—*107*). Introduction of a keto group at C-3 in (**264**) has been found to induce a marked effect on the shieldings of the vicinal geminal methyl carbons, *i.e.* −6.8 ppm at C-18 (equatorial) and −0.5 ppm at C-19 (axial). This is of considerable diagnostic value in the structure determination of terpenoids (*105*). The observed shieldings for C-1, C-8 and C-20 in (**259**) shed some light on the conformation of the sidechain with respect to the C-11—C-12 bond orientation. From the data it is inferred that rotamer (**259 a**) prevails (*106*).

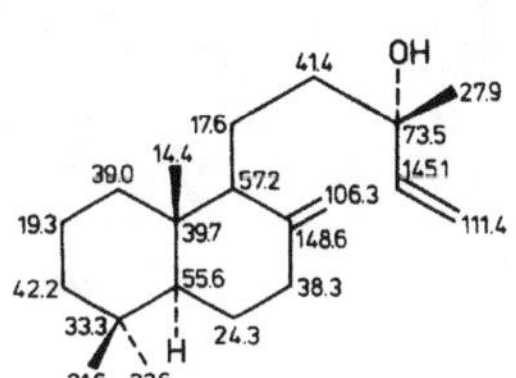

(259)

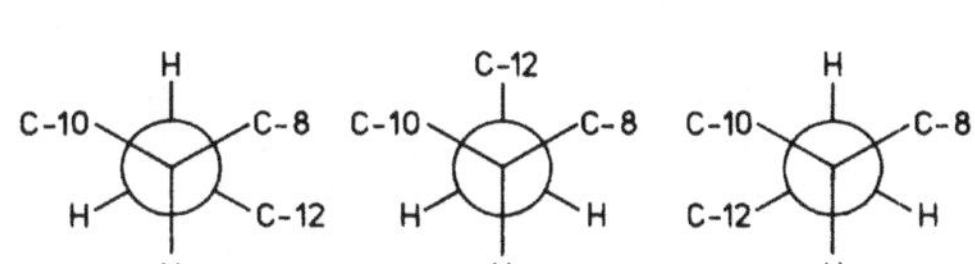

a b c

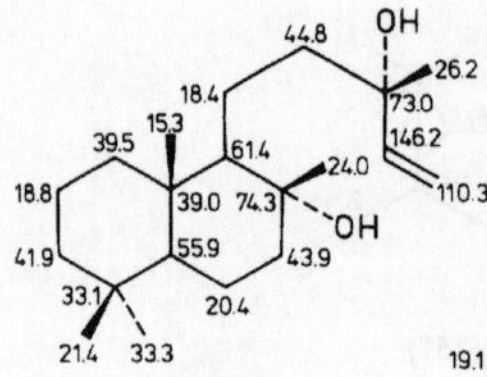

(260)

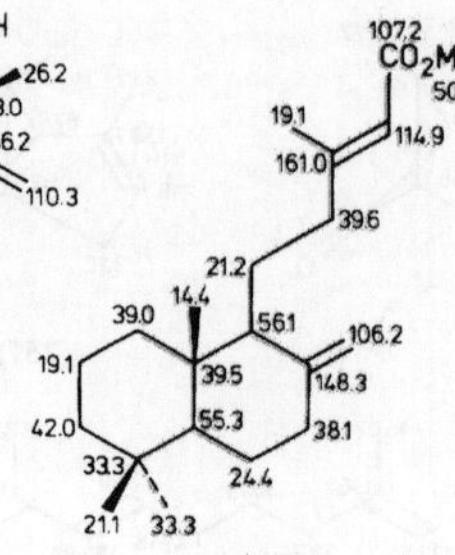

(261)

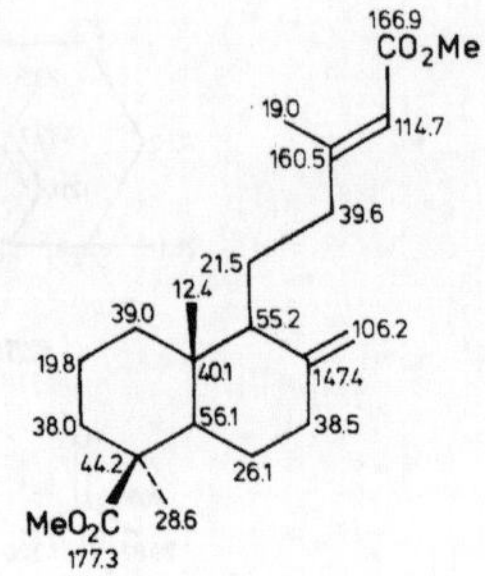

OMe : 50.9; 50.5

(262)

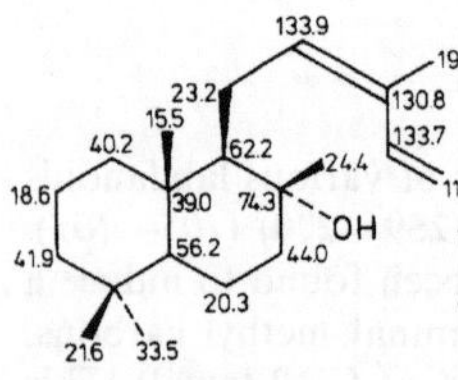

(263)

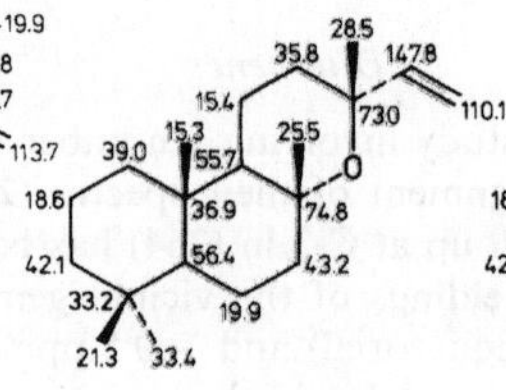

(264)

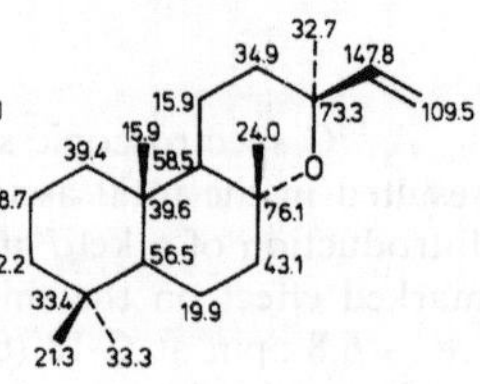

(265)

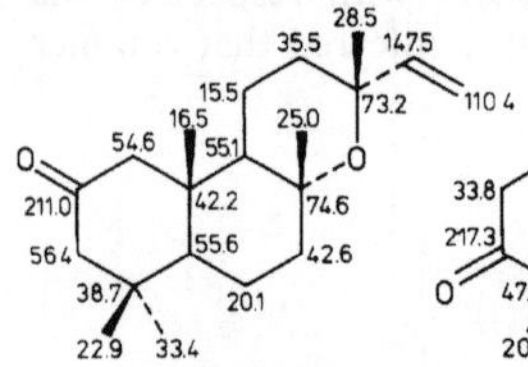

(266)

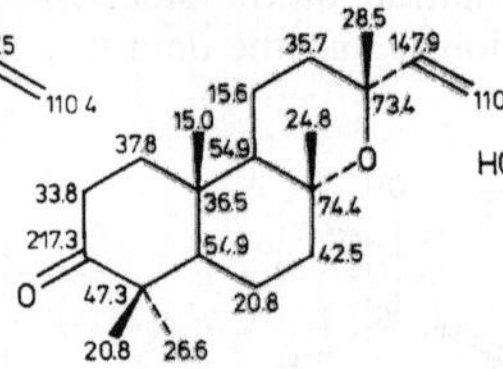

(267)

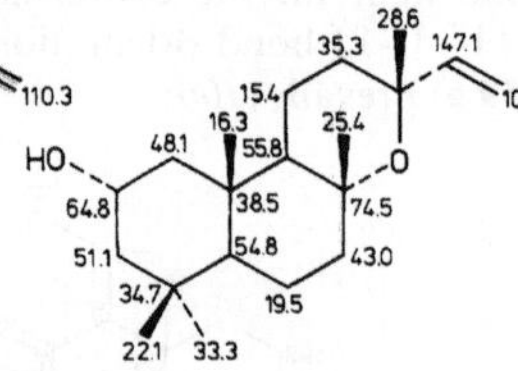

(268)

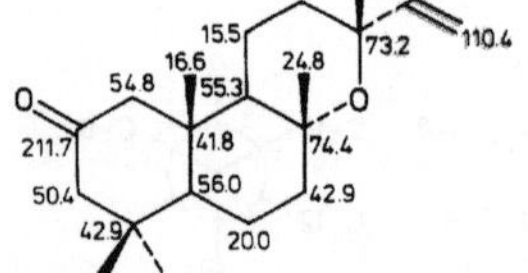

(269)

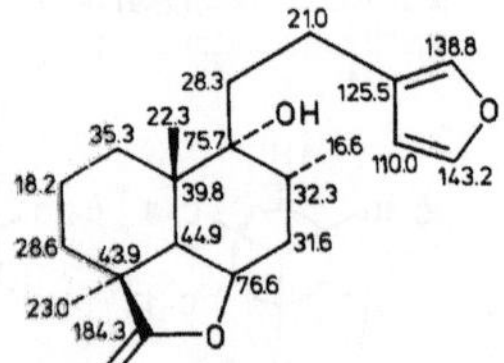

(270)

The structure of ballotinone has been shown to be consistent with formula (**271**), which is substantiated by its ^{13}C and proton spectra (*108*). ^{13}C shielding data for ballotinone and related compounds are indicated in formulae (**271**—**274**).

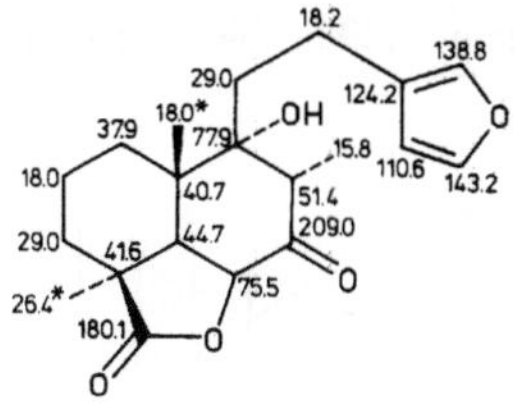

(**271**)

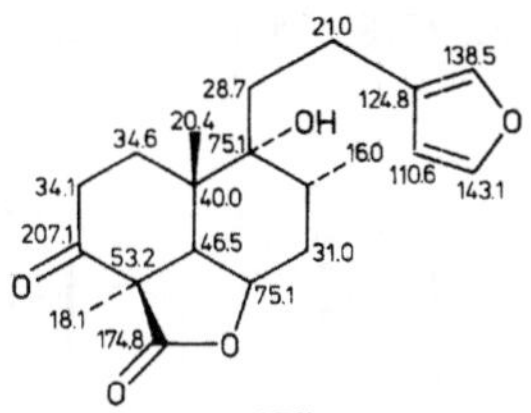

(**272**)

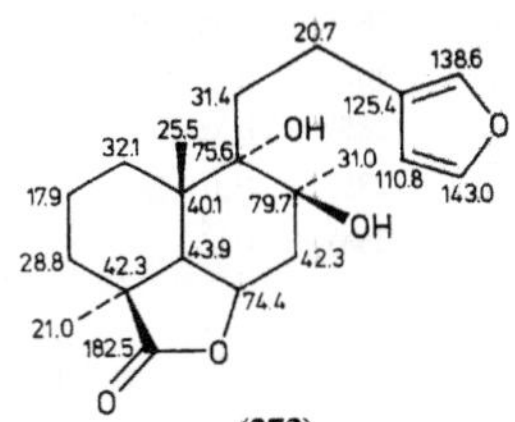

(**273**)

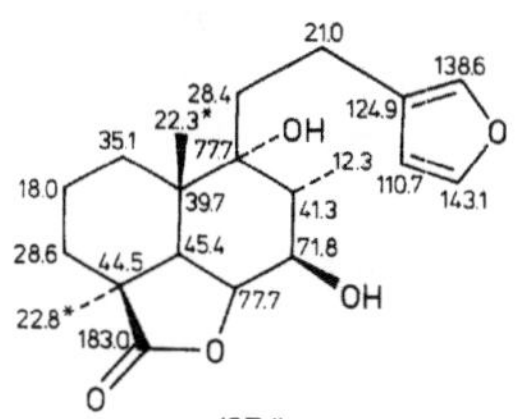

(**274**)

Combined ^{13}C and proton n. m. r. elucidated the structure of the novel furanoid diterpene potamogetonin (**275**). Chemical shift data were also provided for the related compounds sciadin (**276**) and nepetaefuran (**277**) (*109*).

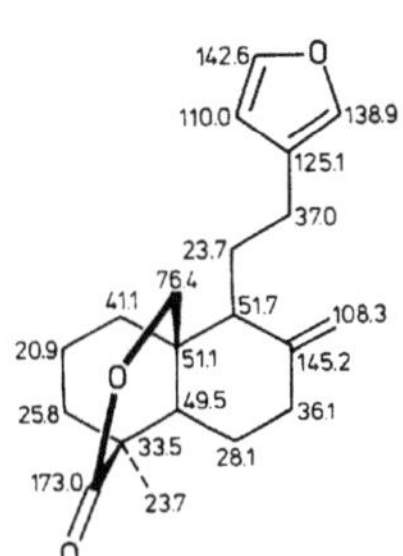

(**275**)

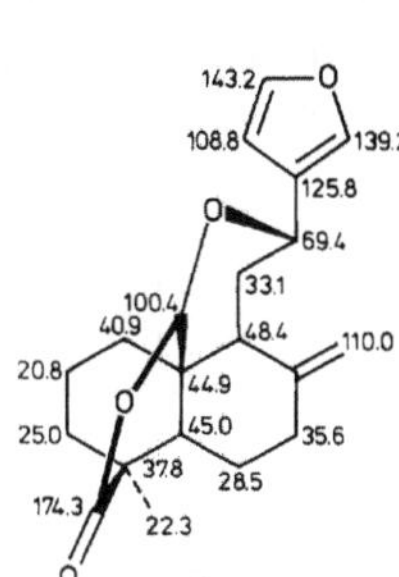

(**276**)

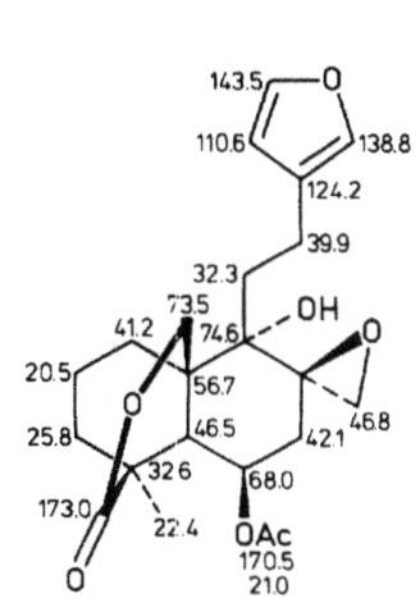

(**277**) C-8, C-19: not observed

Systematic studies of substituent effects have been carried out on a series of podocarpanes, podocarpenes (*36*) and podocarpatrienes (*41*). Unequivocal assignments could be made with the aid of joint application of a variety of techniques including specific deuteration, shift reagents and spin-lattice relaxation measurements. ^{13}C chemical shift data for podocarpanes (**278**—**289**) and podocarpenes (**290**—**300**) are summarized

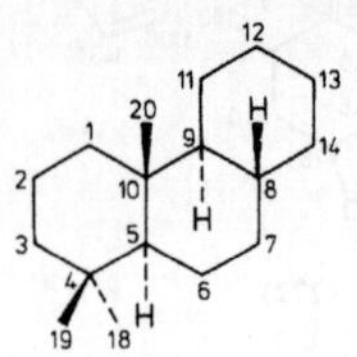

(278)

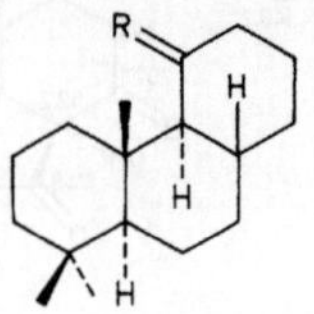

(279) R=H, α-OH
(280) R=H, β-OH
(287) R=O

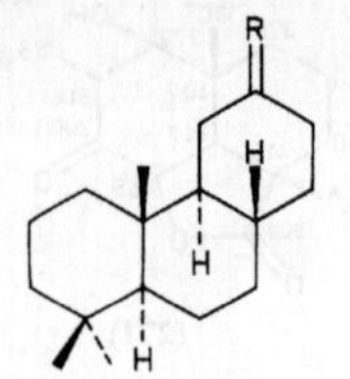

(281) R=H, α-OH
(282) R=H, β-OH
(285) R=H, α-OAc
(286) R=H, β-OAc
(288) R=O

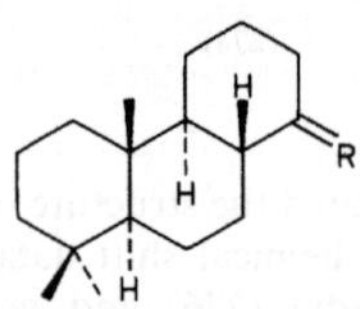

(283) R=H, α-OH
(284) R=H, β-OH
(289) R=O

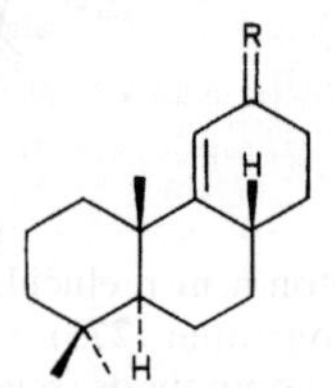

(290) R=H_2
(291) R=H, α-OH
(292) R=H, β-OH
(293) R=H, α-OAc
(294) R=H, β-OAc
(295) R=O

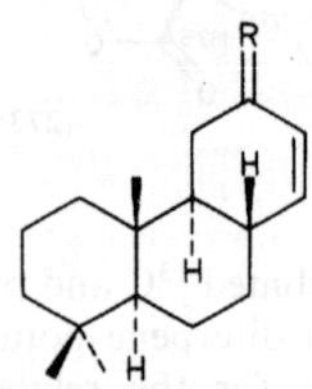

(296) R=H, α-OH
(297) R=H, β-OH
(298) R=H, β-OAc
(299) R=O

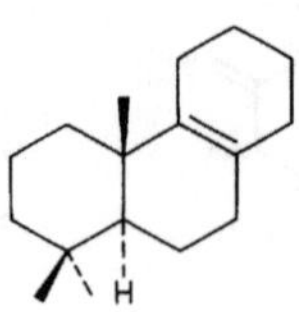

(300)

Table 13. ^{13}C *Chemical Shifts in Podocarpanes* (**278**—**289**) *and Podocarpenes* (**290**—**300**) (*36*)

Compound	Carbon 1	2	3	4	5	6	7	8	9	10	11	12	13	14	18	19	20
Podocarpane (**278**)	39.1	19.0	42.4	33.3	55.5	21.8	35.9	36.8	56.3	36.9	25.1	27.1	26.4	35.5	33.6	22.0	14.3
Podocarpan-11-ol (**279**)	41.7	19.1	42.0	33.6	55.3	21.9	35.7	36.4	62.0	39.0	71.9	38.4	23.9	34.9	33.9	22.1	15.0
Podocarpan-11-ol (**280**)	39.6	18.7	42.2	33.2	55.7	21.7	36.0	30.1	58.9	37.5	66.4	35.4	19.1	35.6	34.0	22.1	16.7
Podocarpan-12-ol (**281**)	38.8	18.9	42.2	33.3	55.4	21.7	35.4	36.4	48.5	36.4	31.5	67.0	32.3	28.6	33.6	21.7	14.1
Podocarpan-12-ol (**282**)	39.0	18.9	42.2	33.2	55.2	21.6	34.9	35.6	53.9	36.7	34.6	71.4	35.2	32.9	33.6	21.9	14.3
Podocarpan-14-ol (**283**)	39.3	18.9	42.2	33.3	54.9	21.5	30.6	40.7	48.2	36.7	24.8	20.0	33.6	71.0	33.6	22.0	14.1
Podocarpan-14-ol (**284**)	39.4	19.0	42.2	33.2	55.1	21.1	31.0	44.1	53.9	36.7	24.4	24.0	35.4	76.8	33.6	21.9	14.3
Podocarpan-12-yl acetate (8) (**285**)	38.9	18.8	42.2	33.3	55.4	21.6	35.3	36.1	49.4	36.4	29.2[a]	70.7	29.0[a]	28.8	33.6	21.9	14.0
Podocarpan-12-yl acetate (9) (**286**)	39.0	18.8	42.1	33.2	55.1	21.6	34.7	35.7	53.6	36.8	30.9[a]	74.4	31.5[a]	32.4	33.6	21.9	14.2
Podocarpan-11-one (**287**)	38.4	18.7	42.3	33.3	54.9	21.2	35.8	40.0	67.4	36.7	211.8	44.3	26.7	34.5	33.7	22.0	14.6
Podocarpan-12-one (**288**)	38.7	18.7	42.0	33.2	54.8	21.4	34.3[a]	35.6	56.1	37.3	41.0	213.4	41.1	34.2[a]	33.6	21.9	13.8
Podocarpan-14-one (**289**)	39.4	19.0	42.0	33.3	54.6	20.6	26.3[a]	49.7	75.3	37.6	24.4	26.5[a]	41.8	213.7	33.5	21.9	13.9
Podocarp-9(11)-one (**290**)	37.6	19.2	42.5	33.8	54.0	21.2	36.4	33.4	151.5	39.6	114.3	26.0	22.2	32.0	33.4	21.9	21.1
Podocarp-9(11)-en-12-ol (**291**)	37.3	19.0	42.3	33.8	53.5	22.0	35.7	33.3	155.2	39.6	117.4	66.2	29.7	26.6	33.3	21.9	21.0
Podocarp-9(11)-en-12-ol (**292**)	37.4	19.0	42.3	33.7	53.8	21.9	36.2	33.4	153.1	39.3	119.3	67.4	31.1	28.6	33.3	21.9	19.0
Podocarp-9(11)-en-12-yl acetate (**293**)	37.3	19.0	42.2	33.8	53.3	21.9	35.5	33.4	157.4	39.8	113.4	69.5	26.8[a]	26.6[a]	33.3	21.9	21.1
Podocarp-9(11)-en-12-yl acetate (**294**)	37.3	19.0	42.2	33.8	53.8	21.9	35.9	33.3	155.7	39.7	114.8	70.8	27.0	27.9	33.3	21.9	20.9
Podocarp-9(11)-en-12-one (**295**)	36.7	18.7	41.7	33.9	53.1	21.4	35.2	34.2	175.9	40.9	119.5	200.4	35.8	29.4	33.3	22.0	21.1
Podocarp-13-en-12-ol (**296**)	38.4	18.8	42.3	33.3	55.7	21.9	33.7	36.5	47.4	36.2	30.6	65.0	126.7	137.8	33.5	21.7	14.6
Podocarp-13-en-12-ol (**297**)	38.4	18.8	42.3	33.3	55.4	21.9	33.9	36.3	52.6	36.4	32.3	69.6	129.9	135.0	33.4	21.6	14.3
Podocarp-13-en-12-yl acetate (**298**)	38.4	18.8	42.2	33.3	55.3	21.9	33.8	36.2	52.3	36.5	28.1	72.5	125.6	137.0	33.4	21.6	14.2
Podocarp-13-en-12-one (**299**)	38.0	18.6	42.0	33.2	54.8	21.7	32.9	36.6	54.0	36.6	37.8	201.1	128.1	155.9	33.4	21.7	14.4
Podocarp-8-ene (**300**)	36.5	19.0[a]	42.0	33.3	51.9	19.1[a]	32.6	138.4	125.9	37.8	23.0[a]	23.9[a]	23.6[a]	30.8	33.3	21.7	19.3

[a] Assignments may have to be interchanged.

in Table 13. The effects of substituents on the shieldings of α, β, γ and δ carbons are discussed. Particularly noteworthy are the δ effects observed. Introduction of an equatorial hydroxyl group at C-11 in (**278**) is found to invoke a 2.6 ppm downfield shift for C-1. This is in agreement with earlier findings in steroids where *syn* axial interactions of a hydroxyl group with a methyl carbon produce a pronounced deshielding of the latter. *Syn* axial methyl-methyl interactions, on the other hand, are a common occurrence in terpenoids. While on the basis of a simple chemical shift estimate δ_c 11.8 is predicted for C-20 of (**278**), a value of 14.3 ppm was observed, thus emphasizing the importance of the δ effect exerted by the axial methyl C-19.

Very recently an analysis of the ^{13}C spectra of a number of different tricarbocyclic ring-C aromatic diterpenoids (**301—320**) (*3*, *41*) and (**321—326**) (*3*) has been presented. Chemical shifts for compounds (**301—320**)

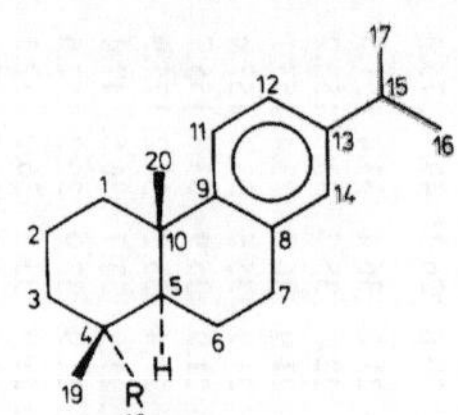

(**301**) R = CH_3
(**308**) R = CH_2OH
(**311**) R = CHO
(**314**) R = $COOCH_3$

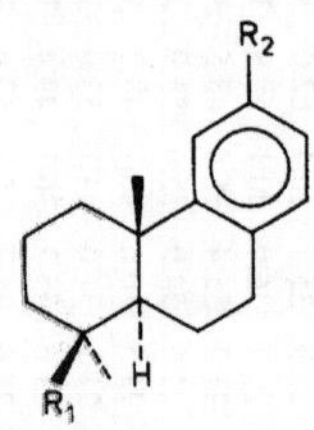

(**302**) R_1=CH_3 ; R_2=OH
(**303**) R_1=CH_3 ; R_2=OCH_3
(**309**) R_1=CH_2OH ; R_2=OCH_3
(**312**) R_1=CHO ; R_2=OCH_3
(**315**) R_1=$COOCH_3$; R_2=OCH_3
(**319**) R_1=$COOCH_3$; R_2=H

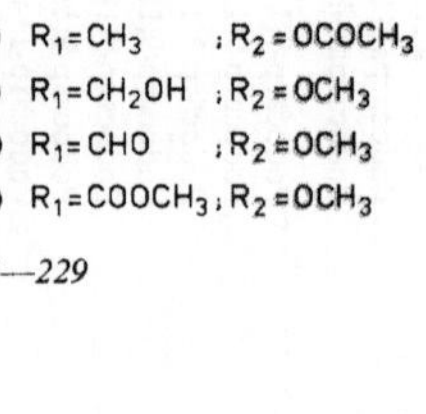

(**304**) R_1=CH_3 ; R_2=$OCOCH_3$
(**310**) R_1=CH_2OH ; R_2=OCH_3
(**313**) R_1=CHO ; R_2=OCH_3
(**316**) R_1=$COOCH_3$; R_2=OCH_3

(**305**)

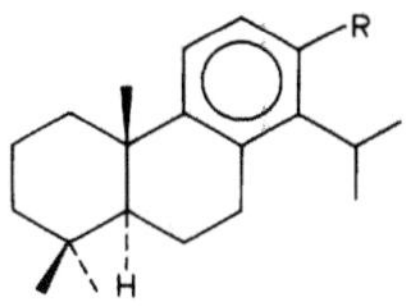

(306) R = OH

(307) R = OCOCH$_3$

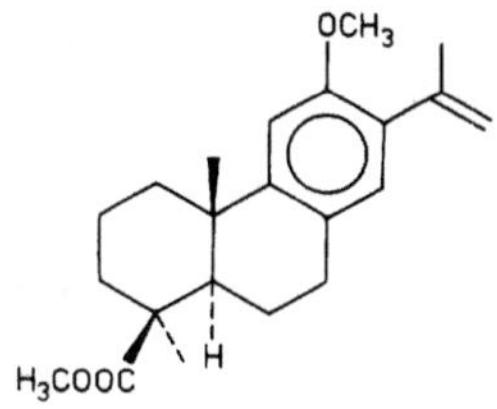

(317)

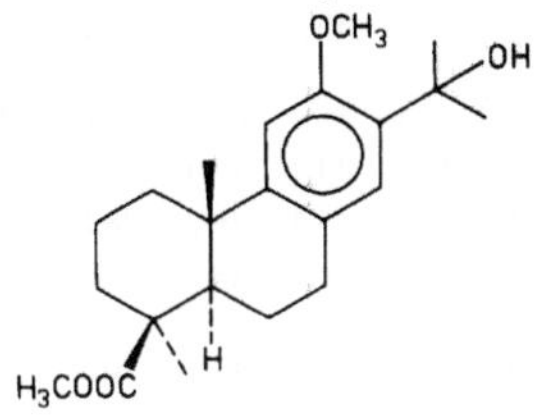

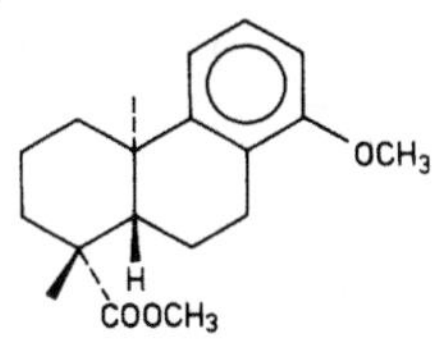

(318) (320)

are compiled in Table 14, those for compounds (**321**—**324**) are indicated in the formulae. The data are found to be sufficiently characteristic to allow stereochemical conclusions. For example, when the values for compounds having an axial carbinol group (**309**, **310**) are compared with those of dehydroabienol (**308**), the carbinol carbon is less shielded in the latter by ca. 7 ppm. Conversely C-5 is less shielded by ca. 7 ppm in the compounds possessing an axial carbinol function.

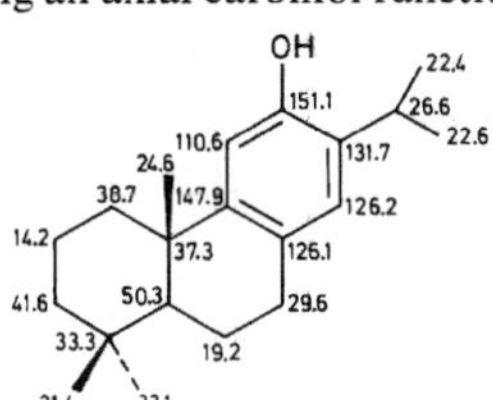

(321)

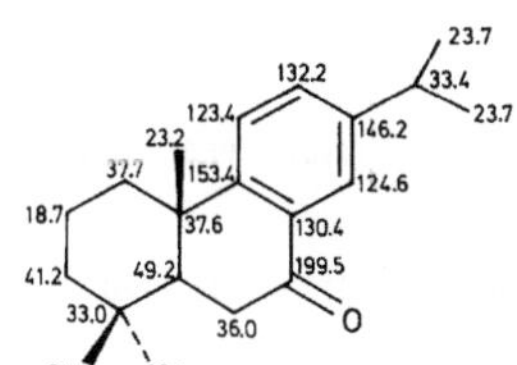

(322)

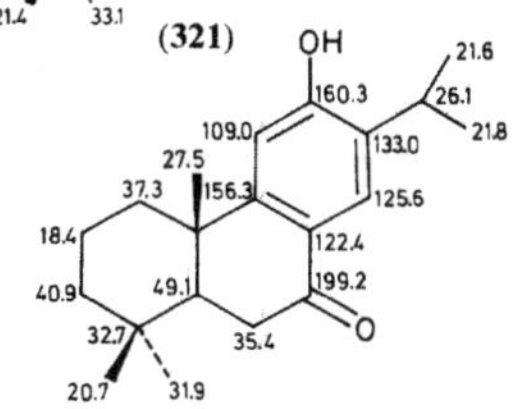

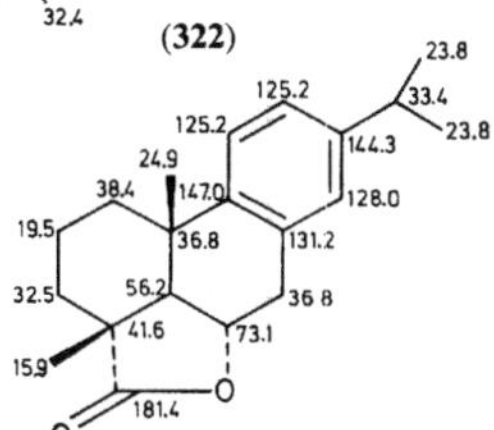

(323) (324)

Table 14. ^{13}C *Chemical Shifts in*

Compound		Carbon 1	2	3	4	5	6	7	8	9	10
Dehydroabietane	**(301)**	38.9	19.3	41.8	33.4	50.4	19.1	30.5	134.8	147.6	37.5
12-Hydroxy-8,11,13-podocarpatriene	**(302)**	38.7	19.3	41.7	33.4	50.2	19.1	29.6	127.5	151.7	37.9
12-Methoxy-8,11,13-podocarpatriene	**(303)**	38.9	19.3	41.7	33.5	50.3	19.2	29.6	127.5	151.4	38.0
Ferruginol acetate	**(304)**	38.8	19.2	41.7	33.3	50.1	19.1	30.0	133.1	148.8	37.6
13-*O*-Methylsempervirol	**(305)**	39.0	19.4	41.8	33.4	50.8	19.2	30.6	142.1	134.3	37.4
Totarol	**(306)**	39.6	19.5	41.6	33.2	49.2	19.4	28.8	133.9	143.1	37.7
Totarol acetate	**(307)**	39.4	19.4	41.5	33.3	49.3	19.3	28.6	134.1	148.0	38.1
Dehydroabietol	**(308)**	38.5	18.7[a]	35.1	37.9[b]	44.0	18.9[a]	30.1	134.8	147.3	37.4[b]
12-*O*-Methylpodocarpol	**(309)**	38.9	19.0	35.1	38.7	51.2	19.2	30.1	127.1	151.0	37.9
12-Methoxy-8,11,13-abietatrien-19-ol	**(310)**	39.0	19.0	35.2	38.8	51.4	19.4	30.4	126.5	147.6	37.8
Dehydroabietal	**(311)**	37.8	17.8	32.1	49.7	42.8	21.3	29.7	134.4	146.2	36.2
12-*O*-Methylpodocarpal	**(312)**	38.4	19.0	33.9	48.6	51.9	19.2	30.4	126.8	148.4	38.3
12-Methoxy-3,11,13-abietatrien-19-al	**(313)**	38.5	19.1	33.9	48.7	52.1	19.3	30.6	126.3	145.5	38.1
Methyl dehydroabietate	**(314)**	38.0	18.6	36.7	47.7	44.9	21.7	30.0	134.7	146.9	37.0
Methyl 12-*O*-methylpodocarpate	**(315)**	39.4	20.0	37.6	44.0	52.8	21.1	31.2	127.6	149.3	38.6
Methyl 12-methoxy-8,11,13-abietatrien-19-oate	**(316)**	39.6	20.1	37.8	44.1	53.0	21.3	31.4	127.1	146.0	38.5
Methyl 12-methoxy-8,11,13,15-abietatetraen-19-oate	**(317)**	39.5	20.0	37.7	44.0	52.9	21.2	31.2	130.4[a]	148.0	38.7
Methyl 12-methoxy-15-hydroxy-8,11,13-abietatrien-19-oate	**(318)**	39.5	20.0	37.7	44.0	52.8	21.2	31.4	127.4	147.5	38.5
Methyl 8,11,13-podocarpatrien-19-oate	**(319)**	39.4	20.0	37.7	44.0	52.9	21.0	32.1	135.4	148.1	38.5
Methyl *ent*-14-methoxy-8,11,13-podocarpatrien-19-oate	**(320)**	39.6	20.1	37.7	44.0	52.4	20.4	25.8	124.6	149.5	38.5

[a] Assignment may be interchanged.
[b] Assignment may be interchanged.

Aromatic Diterpenoids (**301—320**) (*41*)

Carbon													
11	12	13	14	15	16	17	18	19	20	OCH_3	CO	CH_3	OCH_3
124.2	123.8	145.3	126.8	33.5	24.0	24.0	33.3	21.6	24.9				
112.7[a]	153.2	111.1[a]	130.0				33.3	21.6	24.7				
110.1[a]	157.7	110.7[a]	129.7				33.3	21.7	24.7	55.2			
118.0	146.2	136.6	126.9	27.2	23.0	23.1	33.4	21.6	24.8		170.0	21.0	
122.1	133.2	154.4	110.2	26.9	22.8	23.0	33.4	21.6	24.9	55.2			
122.9	114.4	152.0	131.0	27.2	20.4	20.4	33.2	21.6	25.2				
123.2	120.9	147.0	136.3	27.0	21.0	20.9	33.2	21.6	25.0		169.8	21.6	
124.2	123.8	145.5	126.8	33.4	24.0	24.0	72.2	17.4	25.3				
110.2[a]	157.7	110.9[a]	129.8				26.8	65.1	25.6	55.2			
106.7	155.1	134.4	126.2	26.4	22.7	22.9	26.8	65.4	25.7	55.6			
124.1	124.0	145.9	127.0	33.4	23.9	23.9	205.8	14.0	25.1				
111.3[a]	157.9	110.7[a]	129.9				24.2	205.6	24.1	55.2			
107.0	155.3	134.8	126.6	26.5	22.7	22.8	24.2	205.7	24.0	55.4			
124.1	123.9	145.7	126.9	33.5	24.0	24.0	179.1	16.5	25.1				51.9
111.2	157.8	111.2	129.8				28.5	177.8	22.9	55.1			51.2
107.5	155.2	134.6	126.5	26.5	22.7	22.9	28.6	178.0	22.9	55.5			51.2
108.2	155.0	127.3[a]	129.7	143.9	114.7	23.3	28.5	177.9	22.9	55.7			51.2
108.3	155.1	133.5	126.3	72.2	29.9	29.8	28.5	177.7	22.8	55.3			51.1
125.4[a]	125.8[a]	125.6[a]	129.1				28.5	177.9	23.0				51.2
177.7	126.1	106.4	157.0				28.5	177.9	22.7				51.2

Table 15. ^{13}C *Chemical Shifts for Kanranoid Diterpenes* **(332—349)** (*116*)

Compound	Carbon atom 1	2	3	4	5	6	7	8	9	10	11	12	13	14	15	16	17	18	19	20	OCH₃
(332)	41.3	18.7	42.0	33.3	56.1	20.3	40.4	44.2	56.1	39.3	18.1	33.3	44.2	39.9	49.2	156.0	102.8	33.7	21.7	17.6	
(333)	41.0	18.5	41.9	33.2	56.1	19.2	40.3	42.5	55.0	39.4	18.5	29.7	47.9	37.5	55.2	222.5		33.6	21.7	18.0	
(334)	39.0*	33.9	217.1	47.0	53.6	20.3	39.9*	42.2	54.2	38.5	19.0	29.3	47.6	36.9	54.6	221.5		27.3	21.0	18.1	
(335)	41.0	18.7	37.6	43.7	53.9	20.7	40.6	42.4	56.7	39.7	18.7	29.5	47.7	37.3	54.9	222.7		28.9	184.2	16.1	
(336)	38.9	19.6	38.9	46.1	49.0*	70.9	79.0	46.3	51.4*	38.7	18.1	33.4	43.8	42.5	45.4	154.3	103.3	28.9	181.4	18.5	52.8
(337)	39.0	19.6	39.0	45.3	50.1*	67.9	80.1	45.1	52.7*	38.8	18.1	33.2	43.8	42.6	45.8	153.9	103.4	29.1	181.0	18.4	52.7[a]
(338)	40.6	19.1	38.0	43.7	53.8	21.8	39.2	41.6	56.9	39.2	20.4	41.3	80.1	46.9	47.4	155.9	102.9	28.6	177.9	15.2	51.1
(339)	40.8	19.1	38.9	43.7	47.0	21.4	37.7	46.0	56.5	39.4	18.0	33.3	40.6	36.4	81.6	153.7	106.1	28.9	184.6	15.9	
(340)	42.0	18.6	42.0	33.2	56.2	20.4	40.3	45.3	56.8	39.3	18.0	26.9	49.0	37.7	58.0	79.4	24.5	33.5	21.6	18.0	[b]
(341)	49.0	63.8	37.3	43.8	49.0	25.4	34.8	47.4	51.2	40.3	18.2	32.4	42.1	36.2	82.4	159.4	108.5		175.6	16.4	52.8
(342)	48.2	65.8	39.1	43.9	48.7	22.7	33.7	47.3	51.4	39.8	18.2	32.3	42.0	36.5	82.6	159.6	108.7	175.8		16.5	51.4
(343)	56.1	208.5	42.8	45.3	48.3	24.9	34.3	47.4	51.7	42.0	18.1	32.1	42.0	35.6	82.2	159.6	108.7		173.0	17.0	52.1
(344)	55.2	208.9	43.6	45.1	47.7	23.0	33.2	47.7	50.4	42.0	18.1	32.1	42.0	35.8	82.2	159.3	108.0	174.9		16.9	52.0
(345)	55.3	207.8	43.0	45.3	47.5	24.1	32.8	51.7	50.2	43.0	18.1	31.7	37.9	35.9	209.6	148.9	115.1		175.4	16.7	51.7
(346)	48.3	64.0	37.4	43.7	48.3	24.4	33.2	52.2	50.8	40.8	18.2	32.0	38.0	36.5	210.3	149.1	114.9		175.3	16.2	51.4
(347)	47.6	63.8	42.6	58.6	53.1	23.1	34.8	48.7	51.0	40.0	18.3	32.5	42.2	36.3	82.4	159.5	108.7	172.7	172.7	17.0	52.1
																					52.8
(348)	49.9	202.2	53.4	42.3	48.2	24.2	34.1	47.3	51.6	43.3	18.0	32.0	41.9	35.6	82.0	159.2	108.8		171.5	16.7	52.1
(349)	45.0	67.7	33.5	43.3	48.6	24.0	34.8	44.5	51.5	40.3	18.8	29.4	46.8	33.5	80.3	216.8			174.3	16.8	52.7[c]

* Assignments may be interchanged. [a] Acetate 21.2 and 169.8. [b] Acetate 171.4 and 21.4. [c] Acetates 20.6, 21.3, 169.9, and170.1.

Structural studies of hallol (**325**) (*110*), momilactone A (**326**) and B (**327**) (*111*), (−)-thermarol (**328**) (*112*), stemolide (**329**) (*113*), ajugarin (**330**) (*114*) and floribundic acid (**331**) (*115*) employed at least in part ^{13}C n. m. r. spectroscopy.

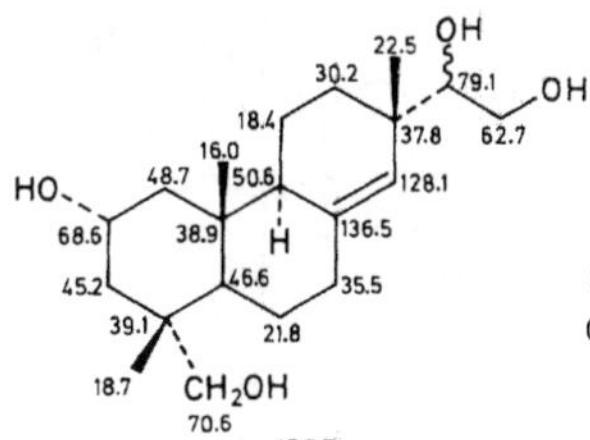

(**325**)

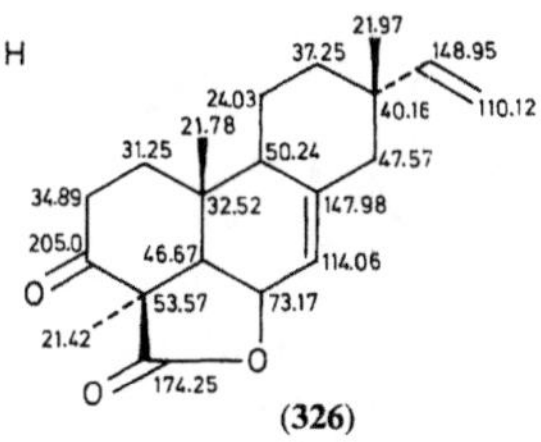

(**326**)

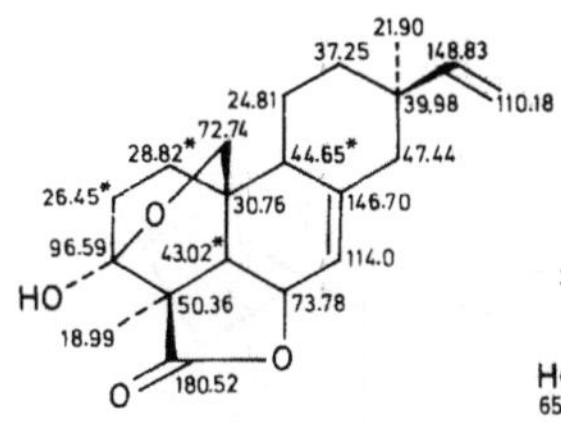

(**327**)

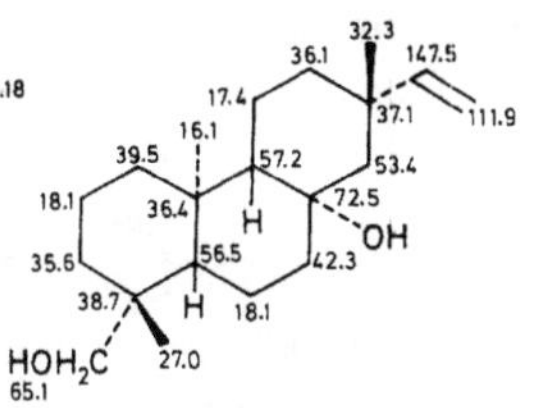

(**328**)

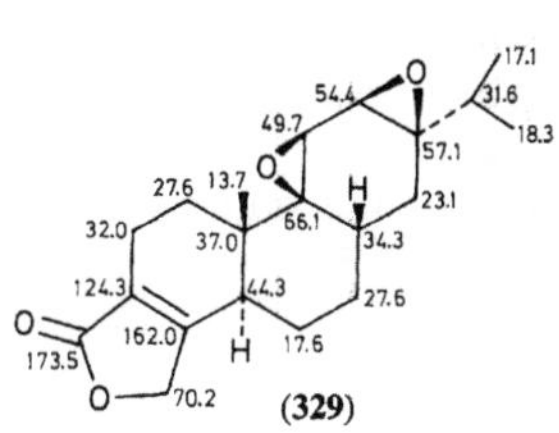

(**329**)

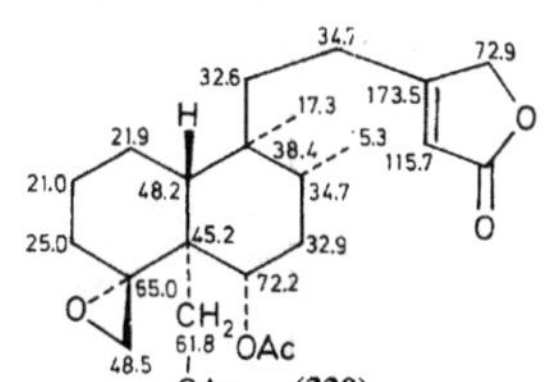

(**330**)

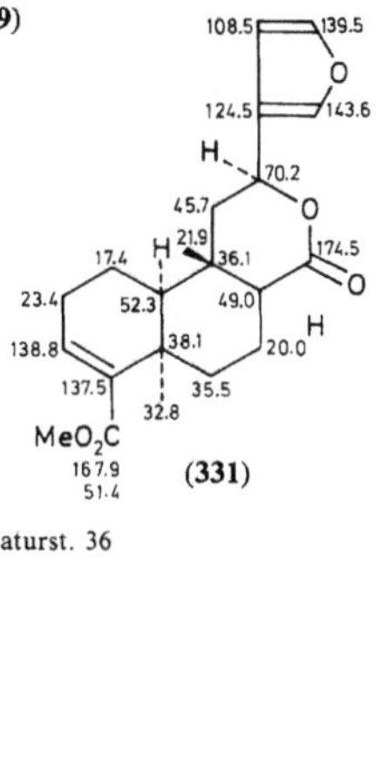

(**331**)

In Table 15 ^{13}C chemical shifts determined in a study of kauranoid diterpenes are collated (*116*). Utilization of ^{13}C shift data for the determination of the site of hydroxylation in systems (**332**—**349**) was proposed.

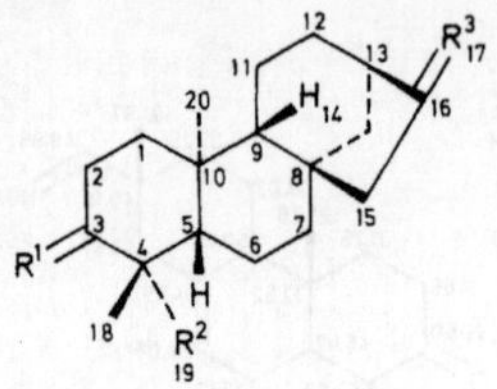

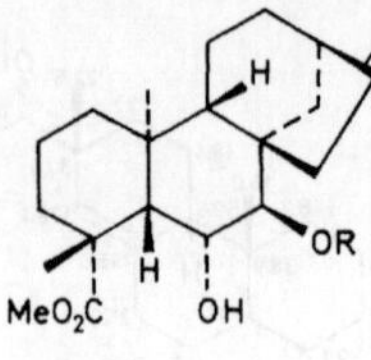

(**332**) $R^1 = H_2, R^2 = Me, R^3 = CH_2$
(**333**) $R^1 = H_2, R^2 = Me, R^3 = O$
(**334**) $R^1 = O, R^2 = Me, R^3 = O$
(**335**) $R^1 = H_2, R^2 = CO_2H, R^3 = O$
(**340**) $R^1 = H_2, R^2 = Me, R^3 = \alpha$-OH, β-Me

(**336**) R = H
(**337**) R = Ac

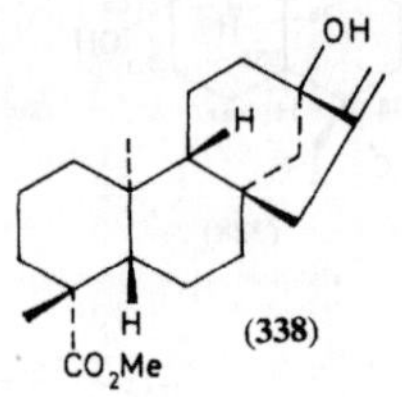

(**338**)

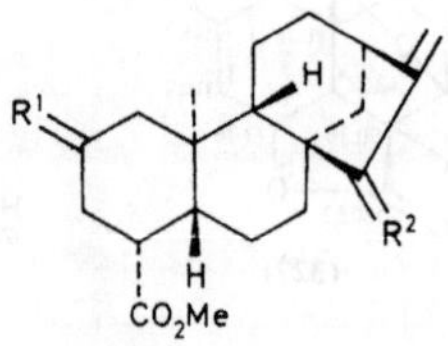

(**341**) $R^1 = H, \beta$-OH, $R^2 = \alpha$-OH, H
(**343**) $R^1 = O, R^2 = \alpha$-OH, H
(**345**) $R^1 = R^2 = O$
(**346**) $R^1 = H, \beta$-OH, $R^2 = O$

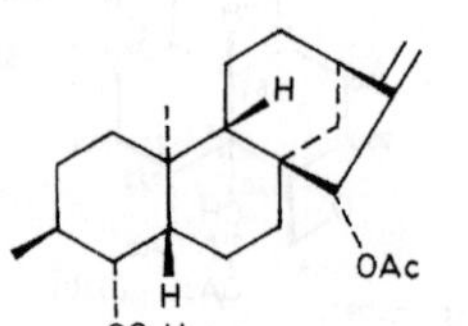

(**339**)

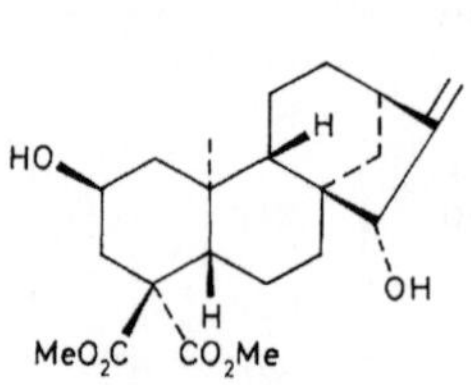

(**347**)

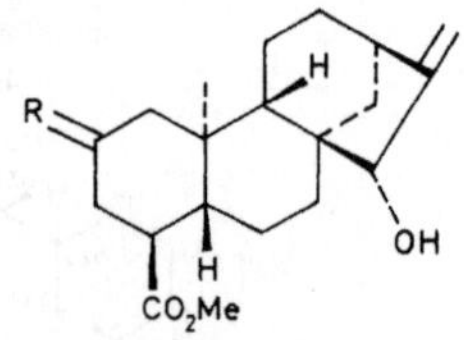

(**342**) R = H, β-OH
(**344**) R = O

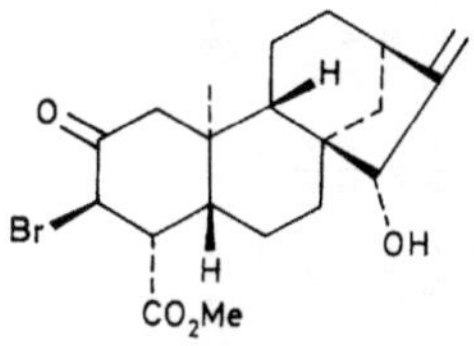

(348)

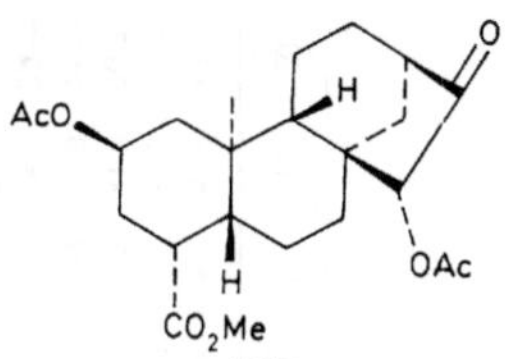

(349)

The data served as a basis for an investigation of the microbiological hydroxylation of hydroxy- and dihydroxykaurenolides by *Rhizopus arrhizus* (*117*). The results indicate that (**350**) affords (**357**) and (**359**). Similarly (**351**) was found to be converted to its 11 α-hydroxy derivative

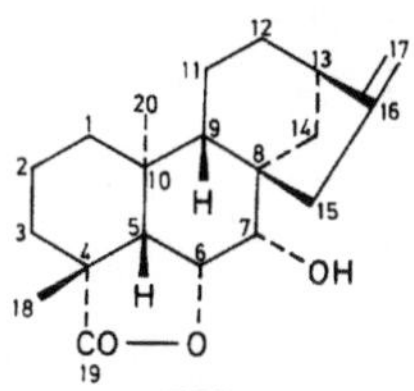

(**350**)

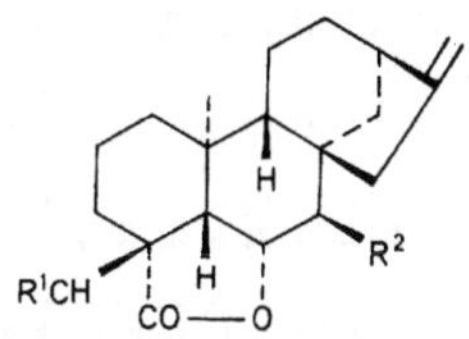

(**351**) $R^1 = H, R^2 = OH$

(**352**) $R^1 = H, R^2 = OAc$

(**353**) $R^1 = R^2 = OAc$

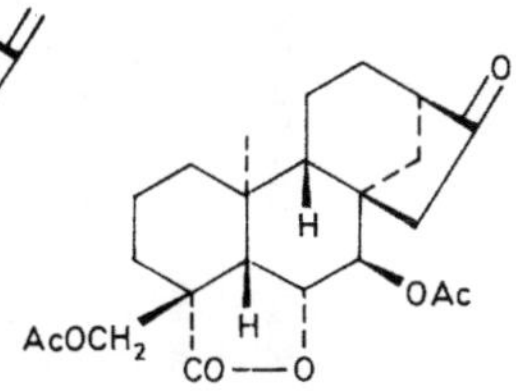

(**354**)

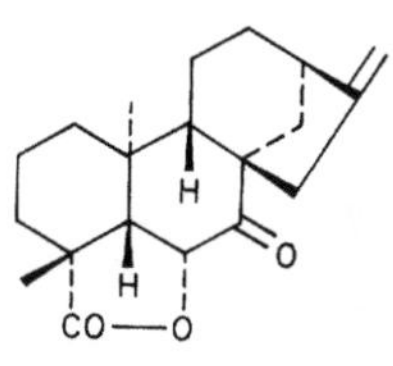

(**355**)

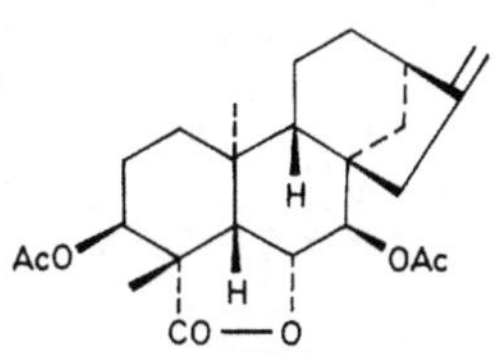

(**356**)

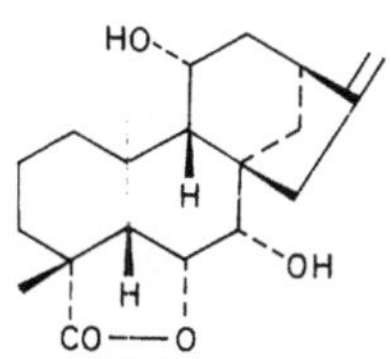

(**357**)

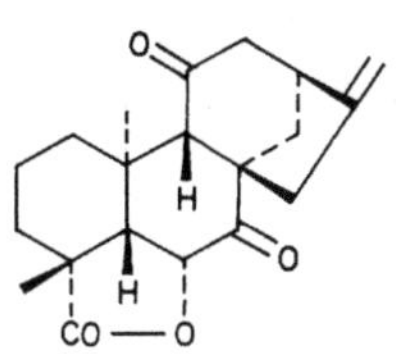

(**358**)

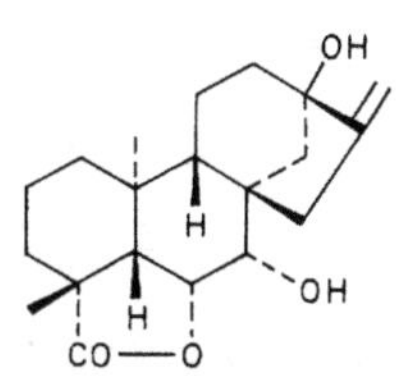

(**359**)

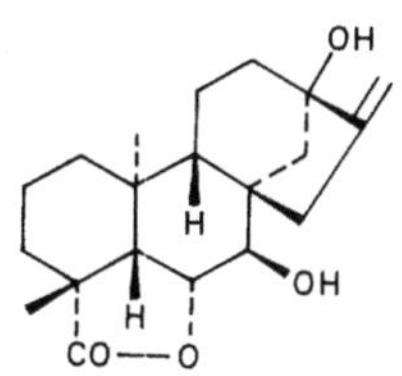

(**360**)

Table 16. ^{13}C *Chemical Shifts for Kaurenolides* (**350—360**) (*117*)

Compound	Carbon atom 1	2	3	4	5	6	7	8	9	10	11	12	13	14	15	16	17	18	19	20
(**350**)	38.2	17.3	28.15	42.1	51.2	75.7	71.6	45.4	53.7	33.5	17.8	33.2	37.3	34.2	50.6	158.5	106.8	25.1	181.9	19.1
(**351**)	37.6	17.35*	28.1	42.1	51.75	83.7	71.6	45.1	55.4	38.2	17.3*	32.6	37.4	34.3	42.1	159.4	106.9	25.5	182.5	20.5
(**352**)[a]	37.3	17.35	28.15	41.2	51.75	80.15	73.2	44.3	55.5	34.1	17.1	32.5	37.3	34.2	43.2	158.2	107.7	25.7	181.4	20.3
(**353**)[b]	36.3	16.6	23.05	45.8	48.05	81.05	72.9	44.2	55.8	34.2	17.1	32.5	37.4	33.9	43.0	158.0	107.75	67.95	178.9	21.0*
(**354**)[c]	36.0	16.4	23.0	45.8	47.9	80.05	72.8	42.8	56.5	34.3	17.7	26.2	42.8	31.7	49.8	221.3		67.6	178.4	20.9*
(**355**)	34.4	16.9	28.8	41.4	52.5	76.3	208.9	53.6	57.0	35.3	17.8	31.7	37.3	31.8	47.9	156.65	108.0	26.3	179.8	51.6
(**356**)[d]	34.9	23.8	68.4	44.7	51.5	80.2	72.8	44.3	55.7	34.0	17.1	32.3	37.2	34.0	43.1	157.9	107.8	21.4	179.0	19.1
(**357**)	38.8	17.3	27.8	41.6	50.7	75.6	71.05	45.8	57.1	34.5	65.2	45.0	38.0	36.2	50.5	157.6	107.9	25.5	181.8	21.2
(**358**)	37.8	17.65	28.3	41.4	51.6	75.8	207.1	54.6	68.1	35.3	205.8	49.0	36.8	33.9	51.2	152.6	110.7	26.15	179.5	15.8
(**359**)	37.1	17.35	28.3	42.8	51.5	76.1	72.25	43.7	52.2	34.2	19.4	39.6	78.4	41.4	48.05	157.3	106.1	24.2	181.9	20.6
(**360**)	37.4	17.2	27.9	42.8	51.9	83.1	71.5	42.7	55.2	34.2	18.5	39.1	77.7	40.5	42.2	159.7	107.1	25.5	181.8	21.2

[a] Acetate 20.9 and 169.95. [b] Acetates 21.05 (2), 170.0 and 170.05. [c] Acetates 20.95 (2) and 170.3 (2). [d] Acetates 20.9 and 21.05, 169.6 and 169.8. * Assignment may be interchanged.

and (**360**). Chemical shifts for hydroxykaurenolides (**350**—**360**) are collected in Table 16. In extension of this work it was shown that (**361**) was transformed into triols (**363**) and (**364**) (*118*).

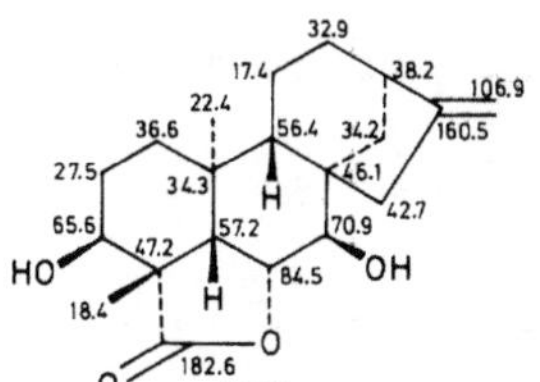

(**361**)

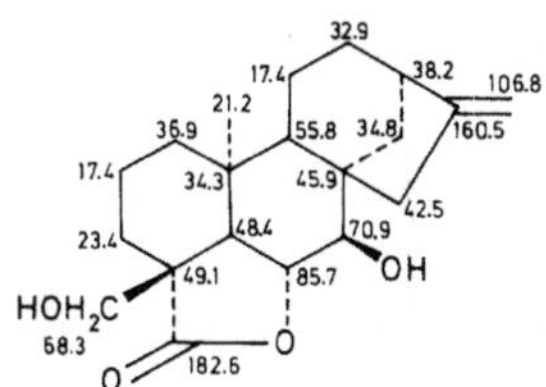

(**362**)

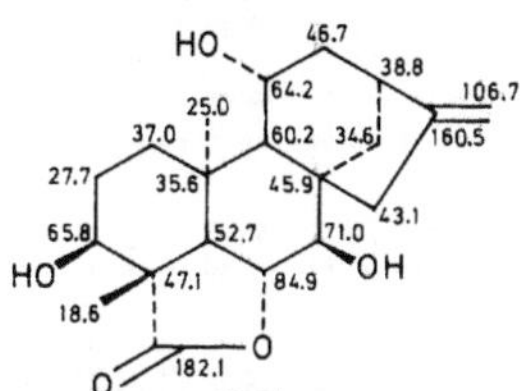

(**363**)

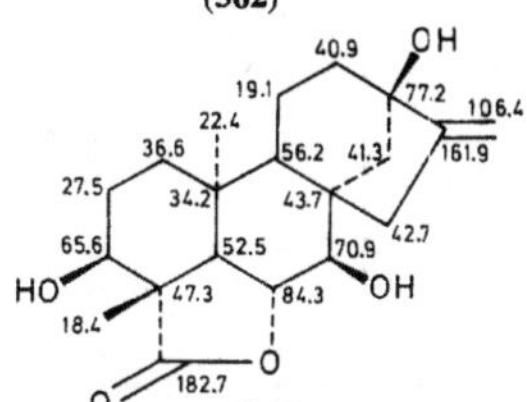

(**364**)

Four new hydroxylated ent-kauran-19-oic acids were isolated from *Eupatorium album* L. and their structures determined. ^{13}C chemical shifts are given for one representative (**365**) (*119*). A revised structure

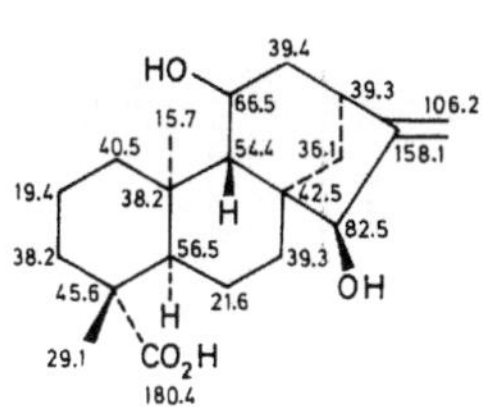

(**365**)

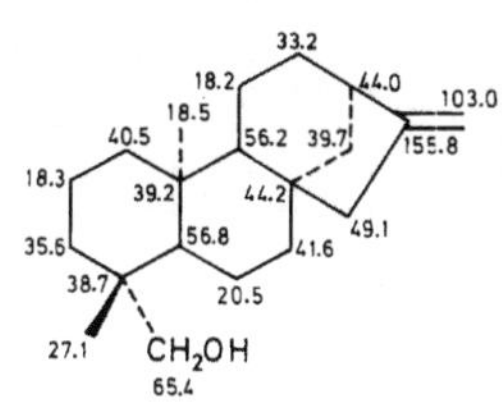

(**366**)

was proposed for *ent*-16-kauren-19-ol (**366**), extracted from coffee (*120*). Some kauren-type diterpenes were also examined in conjunction with a study of diterpene glucosides (*121*). Chemical shifts for terpenoids (**367**—**377**) are presented in Table 17.

Table 17. ^{13}C *Chemical Shifts for Kaurene-Type Diterpenes* (**367**—**377**) (*121*)

Carbon	(367)	(368)	(369)	(370)	(371)	(372)	(373)	(374)	(375)	(376)	(377)
1	40.5 (40.5)	41.1	40.8 (40.9)	41.1	40.0 (39.9)	40.9 (40.9)	40.9 (40.7)	40.2	40.9	41.1	41.0 (40.9)
2	18.9[b] (18.7)[b]	19.8	19.5 (19.2)	19.8	19.3 (18.8)	19.6 (19.2)	19.6 (19.2)	19.6	19.9	19.8	19.4[b] (19.2)[b]
3	42.2 (42.1)	38.6	38.2 (38.1)	38.6	38.3 (38.0)	38.3 (38.1)	38.3 (38.1)	38.5	38.5	38.7	38.3 (38.2)
4	33.3 (33.3)	43.8	43.9 (43.9)	43.9	43.9 (43.7)	44.0 (43.9)	44.0 (43.9)	43.9	44.0	43.9	44.0 (43.9)
5	56.2[c] (56.3)[c]	57.1	56.9 (57.2)	57.1	56.2 (56.1)	57.1 (57.1)	56.7 (56.4)	56.4	56.7	57.0	56.8 (56.9)
6	20.4 (20.3)	22.5	22.2 (21.9)	22.6	20.6 (20.0)	21.7 (21.1)	22.2 (21.6)	20.9	22.3	22.9	21.3 (21.0)
7	41.3 (41.3)	41.5	41.4 (41.4)	41.8	30.6[b] (36.4)[b]	36.6[b] (36.3)[b]	36.8 (36.3)	37.3	36.5	42.7	44.0 (43.9)
8	44.3 (44.3)	44.4	44.4 (44.3)	41.8	52.5 (52.3)	48.2 (47.8)	46.3 (45.7)	51.0	45.4	44.9	49.5 (49.2)
9	56.1[c] (56.1)[c]	55.2	55.1 (55.2)	54.3	51.8 (51.5)	54.0 (53.5)	46.0 (45.5)	63.3	55.3	56.3	48.2 (48.1)
10	39.5 (39.4)	39.9	39.6 (39.5)	39.8	40.3 (40.6)	39.9 (39.7)	39.4 (39.1)	39.3	38.3	40.0	39.7 (39.6)
11	18.3[b] (18.2)[b]	18.6	18.6 (18.4)	20.8	18.7 (18.2)	18.7 (18.3)	18.3 (18.3)	65.3	65.7	18.9	19.2[b] (19.0)[b]
12	33.5 (33.3)	33.3	33.3 (33.2)	40.7	32.4 (32.1)	33.1 (32.6)	33.9 (33.1)	41.6	42.6	26.8	25.1 (24.9)
13	44.3 (44.0)	44.2	44.2 (43.9)	79.8	38.3 (38.0)	42.8 (42.3)	40.7 (40.1)	37.8	39.9	45.8	45.0 (44.8)
14	40.0 (39.9)	39.9	39.8 (39.7)	47.4	34.1[b] (33.6)[b]	36.2[b] (35.3)[b]	39.4 (38.9)	34.9	39.4	37.8	39.7 (39.6)
15	49.9 (49.2)	49.2	49.1 (49.0)	48.1	209.6 (210.3)	82.6 (82.7)	82.2 (82.5)	208.8	82.9	53.9	135.6 (135.1)
16	155.8 (156.0)	155.7	155.7 (155.8)	157.6	150.2 (149.4)	161.1 (160.3)	159.6 (158.4)	151.9	159.4	81.6	142.2 (142.5)
17	103.4 (102.8)	103.5	103.6 (103.1)	102.9	113.9 (114.2)	107.8 (108.2)	104.4 (104.7)	110.8	105.6	66.4	15.4 (15.2)
18	33.7 (33.7)	29.3	28.6 (28.8)	29.3	28.6 (28.5)	28.7 (28.8)	28.7 (28.7)	29.3	29.3	29.3	28.7 (28.7)
19	21.2 (21.7)	179.9	177.5 (178.0)	180.0	177.5 (177.7)	177.7 (178.0)	177.7 (178.0)	179.9	180.0	180.1	177.6 (178.0)
20	17.7 (17.6)	16.0	15.5 (15.4)	15.9	15.5 (15.3)	15.9 (15.7)	15.8 (15.5)	16.0	15.9	15.4	15.4 (15.2)
COOMe			51.1 (51.1)		51.2 (51.0)	51.1 (51.2)	51.1 (51.1)				51.1 (51.1)

[b], [c] Assignments may be interchanged.

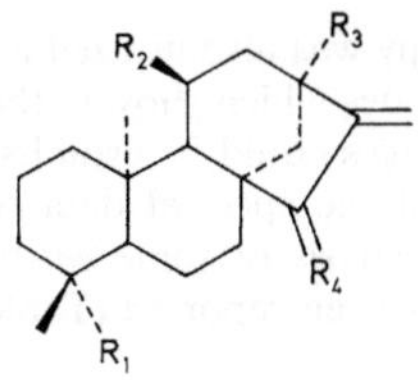

(367) R_1 = CH_3 , R_2 = H , R_3 = H , R_4 = H_2
(368) R_1 = CO_2H , R_2 = H , R_3 = H , R_4 = H_2
(369) R_1 = CO_2Me , R_2 = H , R_3 = H , R_4 = H_2
(370) R_1 = CO_2H , R_2 = H , R_3 = OH , R_4 = H_2
(371) R_1 = CO_2Me , R_2 = H , R_3 = H , R_4 = O
(372) R_1 = CO_2Me , R_2 = H , R_3 = H , R_4 = $<^{H}_{OH}$
(373) R_1 = CO_2Me , R_2 = H , R_3 = H , R_4 = $<^{OH}_{H}$
(374) R_1 = CO_2H , R_2 = OH , R_3 = H , R_4 = O
(375) R_1 = CO_2H , R_2 = OH , R_3 = H , R_4 = $<^{OH}_{H}$

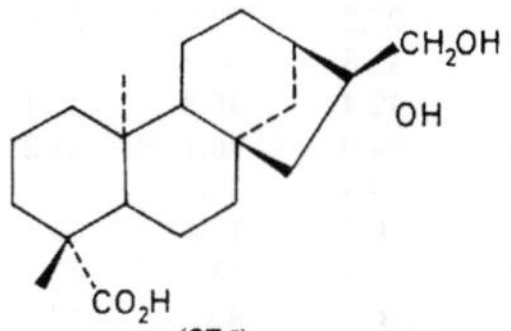

(376)

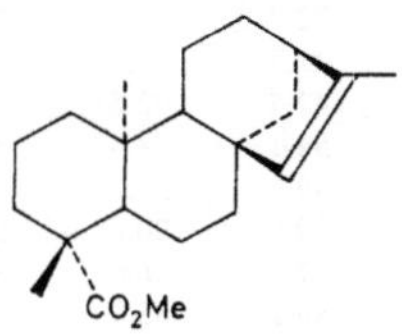

(377)

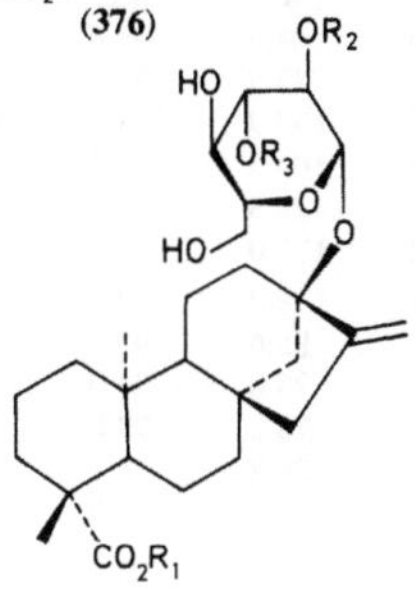

	R_1	R_2	R_3
(378)	H	H	H
(379)	H	Glc	H
(380)	Glc	Glc	H
(381)	H	Glc	Glc
(382)	Glc	Glc	Glc

Glc = β - glucopyranosyl

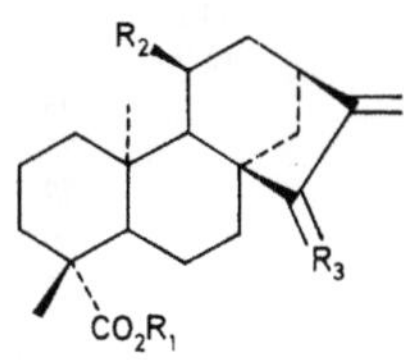

	R_1	R_2	R_3
(383)	Glc	H	$<^{OH}_{H}$
(384)	Glc	OH	$<^{OH}_{H}$
(385)	Glc	OH	=O
(386)	H	H	$<^{OH}_{H}$

^{13}C n.m.r. spectroscopy was also utilized in a structural investigation of *Stevia* diterpene glucosides which showed that location and configuration of the glucoside linkages could be established in this way. Table 18 provides a compilation of the spectral data for glucosides (**378**—**386**). A complete ^{13}C n.m.r. analysis of some *ent*-beyerane and *ent*-beyerene derivatives (**387**—**391**) has been reported (Table 19) (*122*). Based upon a

Table 18. ^{13}C *Chemical Shifts for Glucosides* (**378**—**385**) (*121*)

Carbon	(**378**)	(**379**)	(**380**)	(**381**)	(**382**)	(**383**)	(**384**)	(**385**)
1	41.0	40.9	40.7	41.0	40.9	40.8	40.5	39.9
2	19.7	19.7	19.2	19.8	19.4	19.6	19.5	19.2
3	38.4	38.5	38.1	38.7	38.3	38.4	38.3	38.1
4	43.8	43.8	43.9	43.8	44.0	44.2	44.1	44.0
5	56.9	56.9	57.3	57.0	57.4	57.2	57.1	56.8
6	22.5	22.4	22.0	22.5	22.2	22.0	21.8	20.5
7	41.6	41.5	41.5	41.8	41.8	36.7	36.3	37.0
8	42.1	42.4	42.5	42.2	42.4	46.3	45.4	50.8
9	54.1	54.1	53.8	54.1	54.0	46.1	55.2	63.1
10	39.7	39.6	39.7	39.8	39.8	39.6	38.3	39.2
11	20.6	20.5	20.6	20.6	20.7	18.4	65.7	65.3
12	38.4	37.2	36.6	37.8	37.3	33.8	42.5	41.4
13	86.4	86.1	85.9	86.8	86.6	40.7	39.9	37.7
14	44.6	44.7	44.3	44.5	44.5	39.6	39.9	34.8
15	48.2	47.9	47.5	47.9	47.9	82.3	82.8	208.8
16	153.7	154.0	154.3	153.7	153.9	159.7	159.4	151.8
17	104.9	104.9	104.5	104.6	104.5	104.3	105.5	110.7
18	29.2	29.2	28.2	29.3	28.3	28.6	28.6	28.5
19	180.0	180.1	177.0	180.2	177.0	176.9	176.8	176.6
20	15.7	15.9	15.4	16.1	15.5	16.1	15.7	15.8
G*-1			95.6		95.6	95.7	95.7	95.7
2			73.8		73.8	74.0	73.9	73.9
3			79.0		78.8	79.0	79.2	79.2
4			70.8		70.8	71.1	71.1	71.0
5			79.0		78.8	79.0	79.2	79.0
6			61.9		62.3	62.1	62.1	62.1
G1-1	99.4	97.6	97.7	97.8	97.9			
2	75.3	84.1	84.3	80.5	80.7			
3	(77.9)	(77.6)	77.9	88.0	87.8			
4	71.5	71.1	(71.3)	69.8	70.5			
5	(78.5)	(77.8)	77.9	(78.4)	(78.3)			
6	62.5	62.3	62.5	(62.3)	(62.3)			
G2-1		106.2	106.5	(104.4)	(104.5)			
G3-1				(104.6)	104.5			

* Values in brackets are tentative assignments.

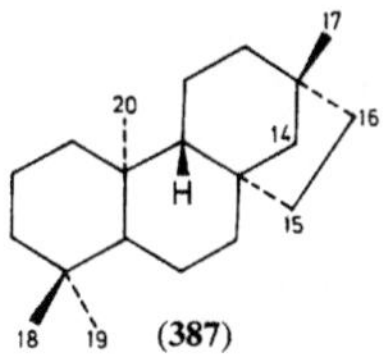

(387)

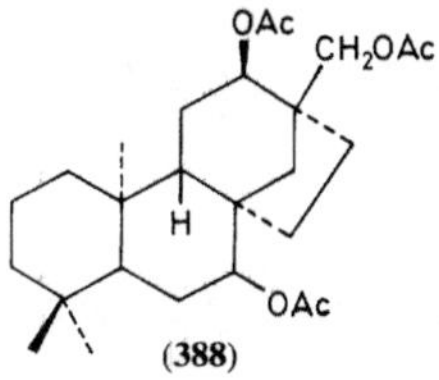

(388)

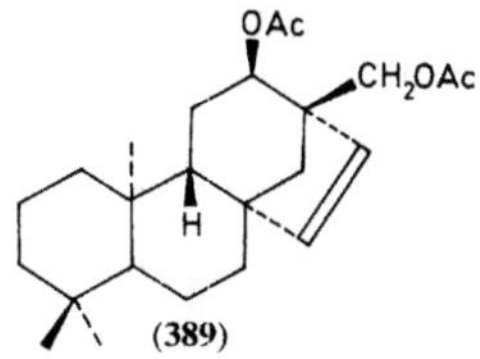

(389)

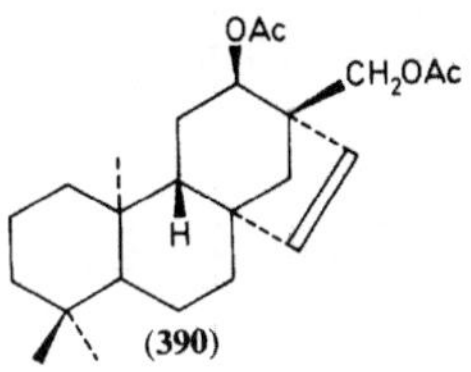

(390)

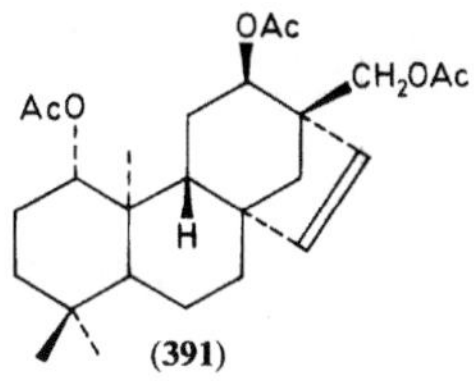

(391)

Table 19. ^{13}C *Chemical Shifts of Ent-Beyerenes and Ent-Beyerenes* (**387**—**391**) (*122*)

Carbon	(**387**)	(**388**)	(**389**)	(**390**)	(**391**)
1	39.7	39.3	38.2	38.6	81.5
2	18.5	18.2	18.4	18.3	24.9
3	42.0	41.8*	41.9	41.9	39.2
4	33.1	32.5	33.1	32.5	32.9
5	56.5	47.7	55.8	47.8	54.9
6	20.4*	25.0**	20.0	25.2*	19.7
7	41.2**	78.2	36.4	75.2	36.6
8	44.9	47.4***	48.3	52.0**	48.6
9	56.9	48.5	49.3	44.9	49.2
10	37.6	37.3	36.9	36.9	41.6
11	20.2*	25.1**	26.4	26.1*	28.7
12	40.0**	72.4	68.9	68.4	68.8
13	39.2	46.3***	50.8	51.1**	50.5
14	57.7	41.6*	49.2	44.9	48.9
15	33.6	32.2	139.3	136.6	138.8
16	37.6	30.4	131.4	133.1	131.9
17	27.1	67.5	65.5	65.2	65.3
18	33.7	33.3	33.5	33.2	32.9
19	21.9	21.6	21.8	21.6	21.5
20	15.1	14.7	14.2	13.7	10.5

*, **, *** Assignments may have to be interchanged.

Table 20. ^{13}C *Chemical Shifts for Gibberellins* **(392—401)** *(123) and* **(402—412)** *(124)*

Carbon	(392)	(393)	(394)	(395)	(396)	(397)	(398)	(399)	(400)	(401)
1	30.7	30.4	27.4	27.1	30.6	25.5	132.1	132.4	129.1	129.2
2	19.4	19.2	28.1	28.1	34.9	27.7	133.2	132.9	134.2	134.2
3	34.3	34.2	70.2	69.9	199.8	71.3	69.9	69.8	70.2	70.3
4	49.1	49.0	54.6	54.7	53.1	53.1	53.4	53.5	52.1	52.2
5	58.0	57.8	51.6	51.1	56.7	53.0	52.2	52.2	53.4	53.4
6	51.8	52.9	51.1	52.7	52.8	50.2	50.5	50.2	50.5	50.4
7	178.0	172.5	173.4	172.6	173.6	172.5	175.8	173.1	172.3	172.3
8	51.3	49.7	51.6	49.8	51.3	51.0	52.2	52.4	50.5	51.0
9	53.9	54.1	53.7	53.9	53.4	52.5	51.7	51.6	50.9	51.0
10	93.6	92.9	94.2	93.9	92.5	92.8	90.9	91.1	90.2	90.0
11	16.1	16.7	16.1	16.8	16.9	17.0	16.0	15.9	16.9	16.9
12	31.4	24.9	31.5	25.0	24.9	36.3	31.4	31.4	38.3	36.4
13	38.8	44.7	38.9	44.8	44.5	84.3	38.6	38.6	78.0	84.1
14	36.8	34.5	36.8	34.6	34.6	40.2	36.5	36.4	43.1	39.8
15	44.3	50.4	44.5	50.5	50.2	42.2	44.4	44.5	44.7	42.7
16	156.5	220.1	156.8	220.6	219.3	153.4	156.4	156.5	156.7	153.4
17	107.5		107.5			107.9	107.8	107.7	107.6	108.2
18	17.2	17.1	14.7	14.6	10.3	14.5	14.6	14.4	14.3	14.3
19	179.6	178.8	178.6	178.6	171.5	176.3	178.5	178.9	177.0	176.9

Table 20 *(continued)*

Carbon	**(402)**	**(403)**	**(404)**	**(405)**	**(406)**	**(407)**	**(408)**
1	30.8	65.9	133.0	149.3	26.4	71.6	111.7
2	35.2	45.1	133.5	129.1	28.6	45.7	33.1
3	201.9	200.9	69.9	193.1	70.9	203.8	70.5
4	63.7	64.1	54.1	65.5	54.0	61.3	49.2
5	57.5	53.0*	53.4	62.4	51.0*	55.2*	46.9*
6	51.9	51.4	51.4	51.4	53.1	52.9	50.6*
7	173.0	173.0	173.3	172.4	174.5	173.5	175.5
8	51.9	53.8	51.0	50.4	51.6	51.3	50.7
9	52.8	47.0*	51.4	52.1*	54.0*	50.7*	47.1*
10	93.8	95.4	91.4	90.8	93.4	94.4	141.6
11	18.2	17.8	17.6	17.5	19.9	21.9	19.4
12	39.5	39.6	39.6	39.3	36.0	35.9	38.9
13	78.1	78.2	78.2	78.1	49.3	51.1	79.3
14	46.1	46.2	45.1	45.1	47.8**	47.8**	49.5
15	43.2	43.2	43.6	43.2	50.7**	50.3**	39.7
16	157.8	157.8	158.6	158.1	219.0	218.4	156.1
17	10.7	10.4	14.8	11.7	14.8	10.5	21.6
18	174.8	173.6	173.5	174.2	179.0	174.9	177.0
OMe	52.8	52.6	52.4	52.6	52.4	52.6	51/8/52.0

*, ** Assignments may have to be interchanged.

total assignment of various gibberellins (**392**—**401**) (*123*) and (**402**—**408**) (*124*) the structure of gibberellin A 40 (**409**) (*125*) could be assigned. Table 20 lists chemical shifts of terpenoids (**392**—**408**).

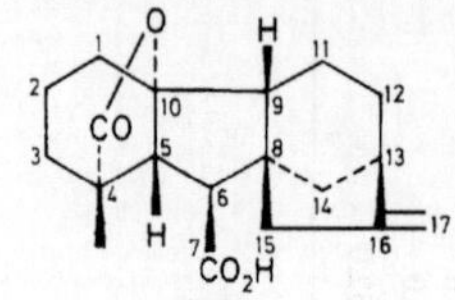

(**392**)

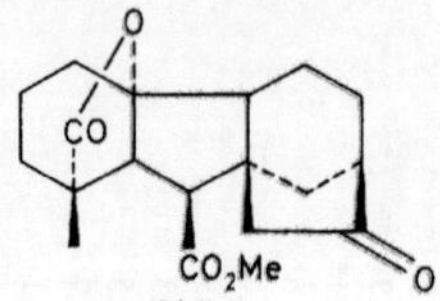

(**393**)

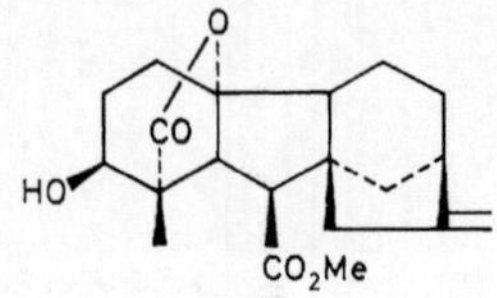

(**394**)

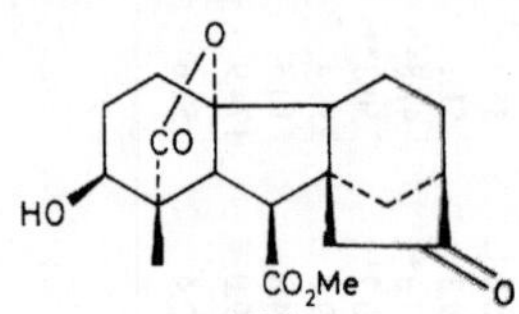

(**395**)

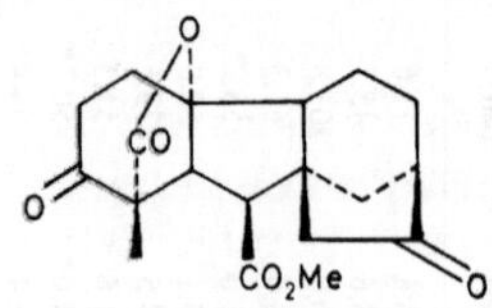

(**396**)

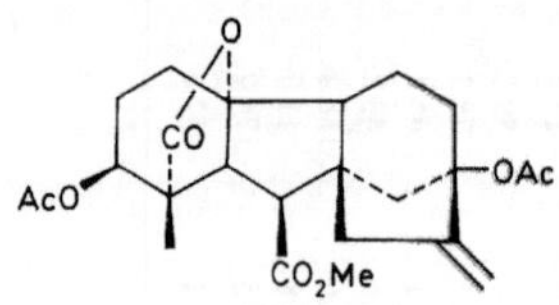

(**397**)

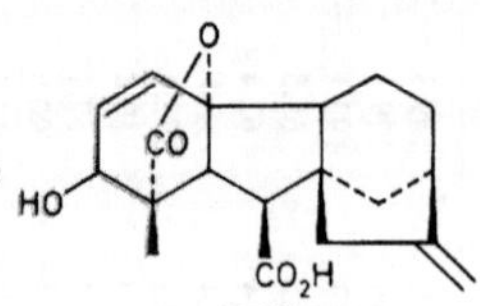

(**398**)

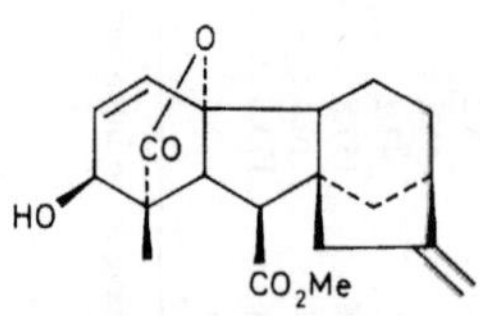

(**399**)

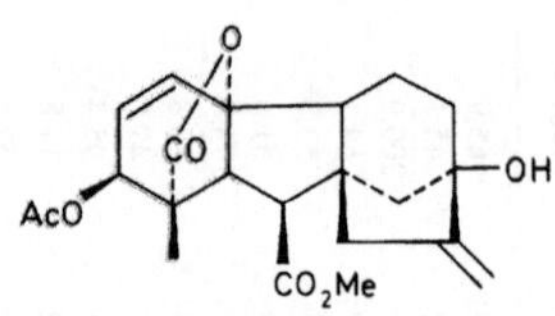

(**400**)

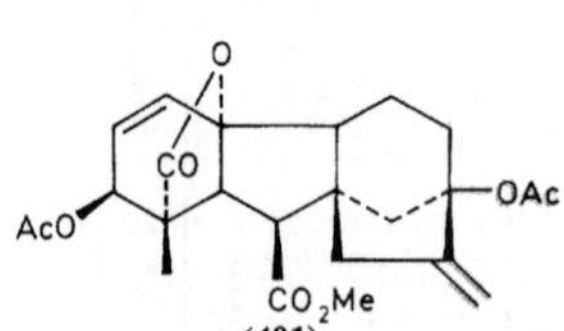

(**401**)

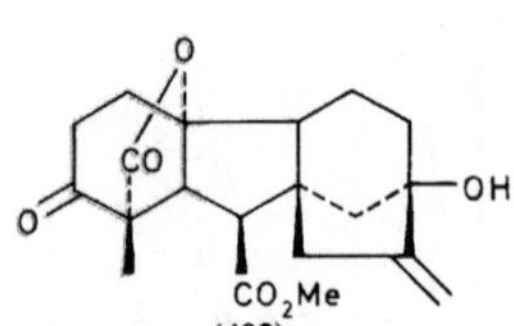

(**402**)

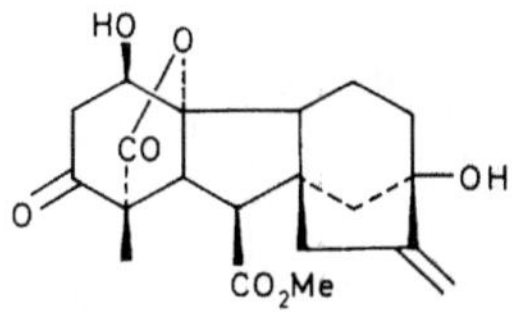

(403)

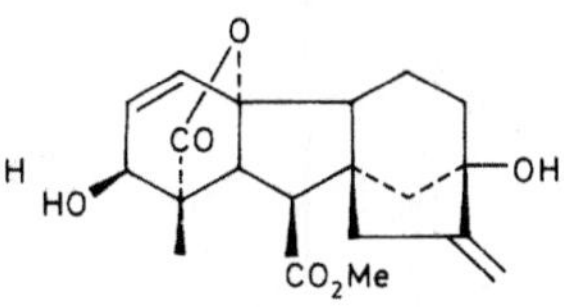

(404)

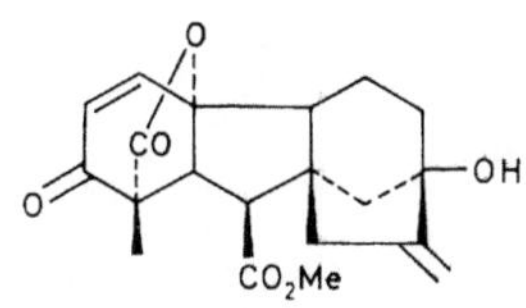

(405)

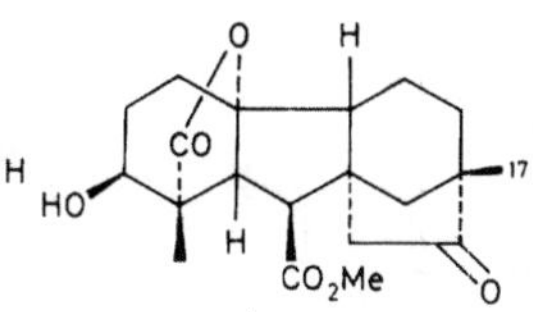

(406)

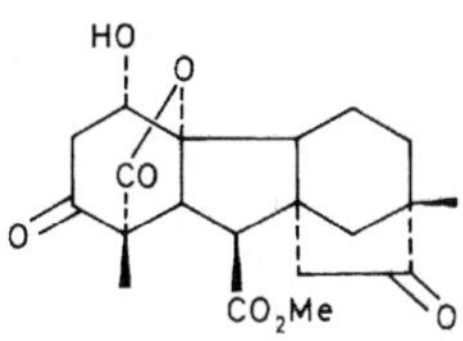

(407)

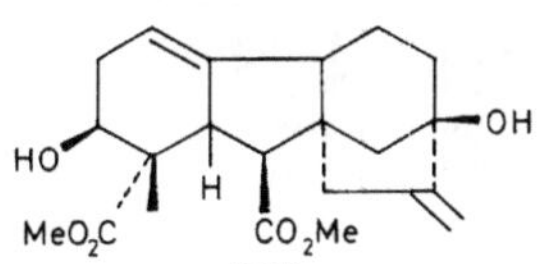

(408)

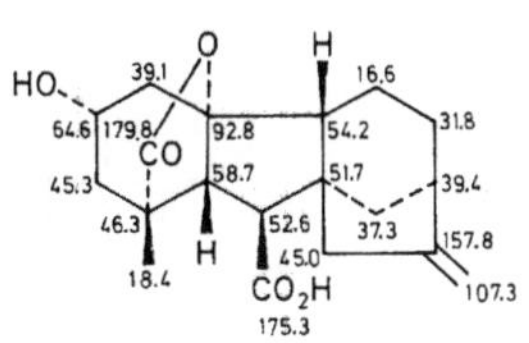

(409) in C_5D_5N

A ^{13}C analysis of various quassinoid bitter principles has been carried out (*126*). Shielding data have been determined for ailanthone **(410)**, chaparrinone **(411)**, glaucarubinone **(412)**, glaucarubin **(413)**, samaderine B **(414)**, quassin **(415)**, brucein B **(416)** and brucein C **(417)**. Their chemical shifts are partly indicated in the formulae and partly tabulated (Table 21).

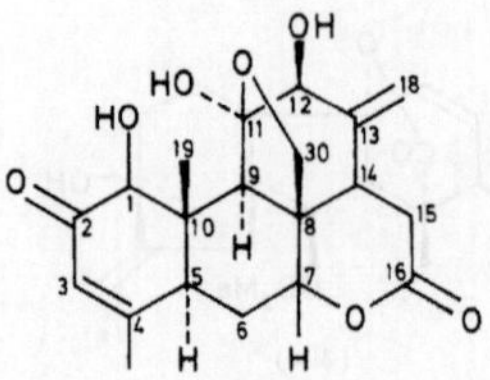

(410)

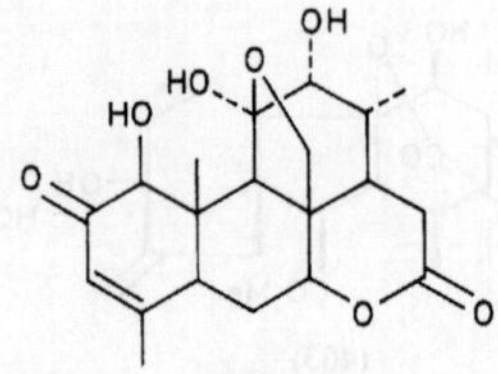

(411)

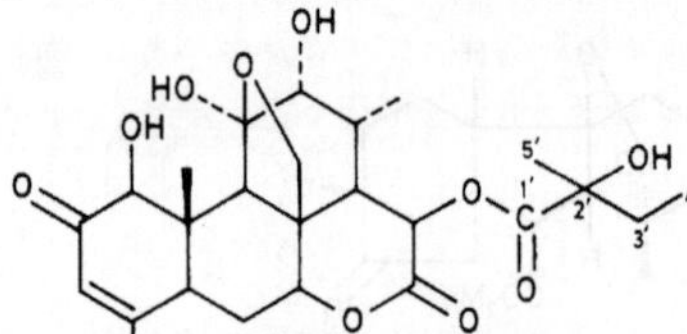

(412)

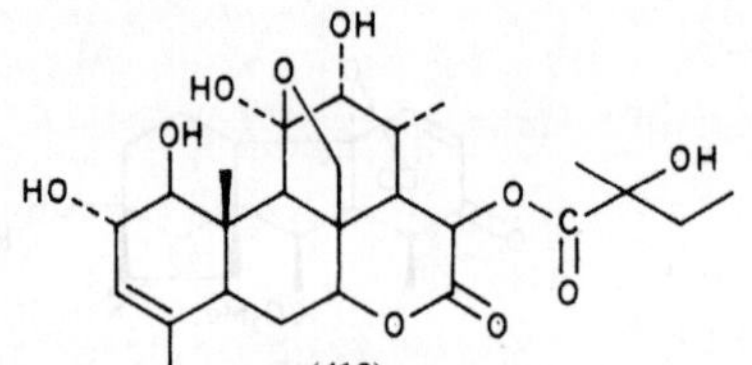

(413)

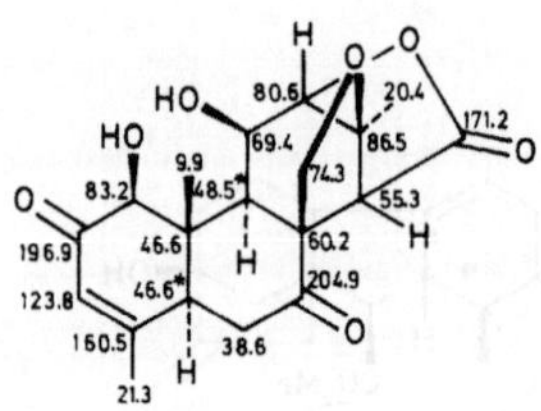

(414)

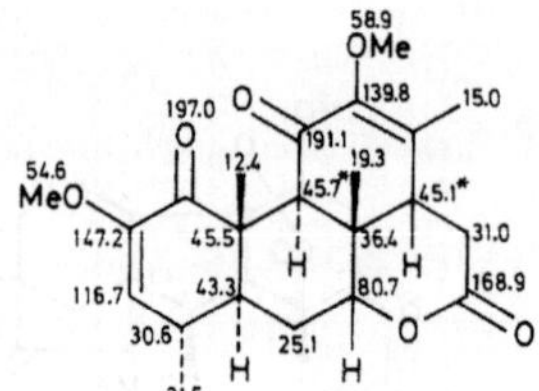

(415)

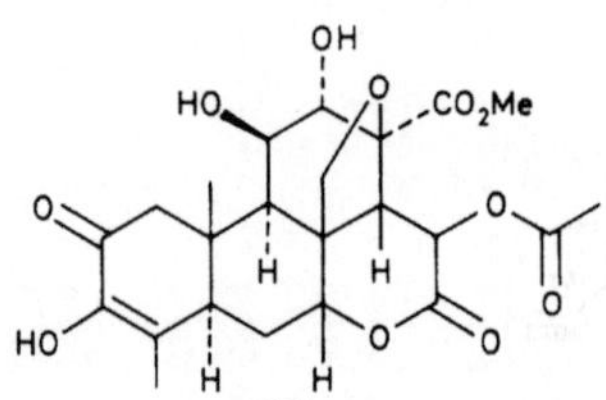

(416)

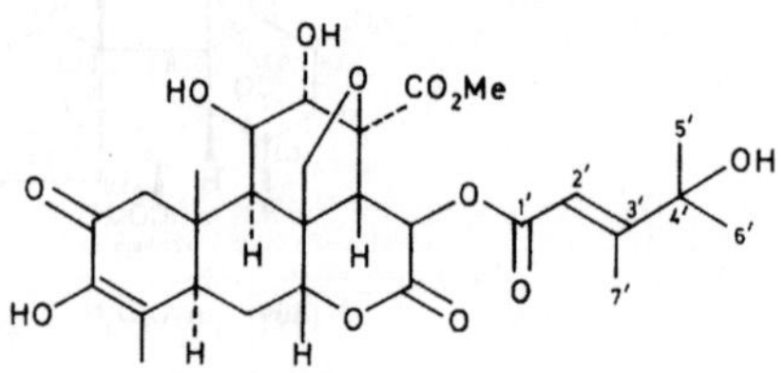

(417)

The 2.3 ppm downfield shift of C-18 in (**412**) and (**413**) relative to that in (**411**) reflects the δ effect arising from the interaction of the peri C-13 and C-15 substituents (cf. Ref. 41). An additional 2 ppm shielding is observed for methyl C-19 in (**410**), (**411**) and (**412**), which possess the oxymethylene bridge spanning C-8 and C-11, when compared to those compounds bearing a β-methyl in position 8. Conformations for molecules **(410)** and **(411)** are represented by formulae **(410a)** and **(411a)**. The 15β-hydrogen in **(411)** experiences a peri interaction with C-18 causing

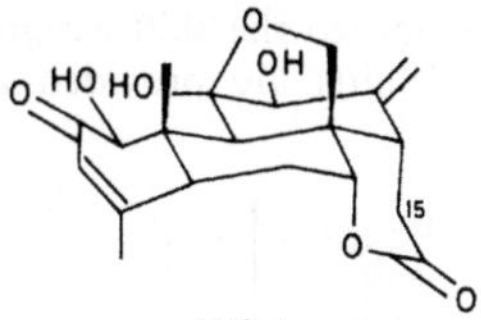

(410a)

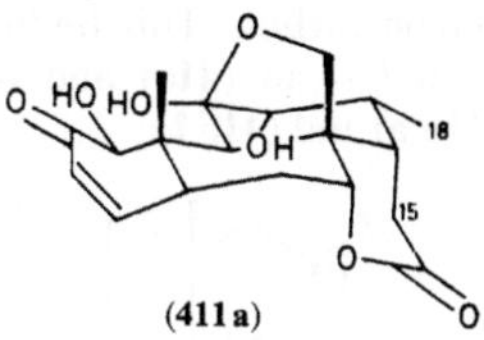

(411a)

Table 21. ^{13}C *Chemical Shifts for Quassinoid Terpenoids* **(410—413)**, **(416)**, **(417)** (*126*)

Carbon	**(410)**	**(411)**	**(412)**	**(413)**	**(416)**	**(417)**
1	85.2	82.5	82.6	82.6	48.7	47.8
2	197.2	197.1	196.8	71.3	192.9	193.0
3	125.0	124.7	124.8	125.7	144.1	144.3
4	162.5	162.3	162.5	133.8	128.3	129.4
5	43.3	43.1	43.9	40.1	39.9*	42.1*
6	25.1	25.0	24.7	24.9	28.7	29.1
7	77.7	77.6**	77.4	77.9	82.8	83.2
8	44.5	44.9*	46.8	76.7	44.7	45.5
9	46.1	41.1	41.1	40.9	40.4*	41.6*
10	44.5	44.3	44.5	44.1*	40.9	41.1
11	108.9	108.9	108.9	109.1	71.4	71.2
12	79.1	77.9**	78.4	78.5	74.7	75.4
13	146.6	30.4	31.4	31.4	81.4	81.6
14	41.1	41.1	44.5	44.6*	48.7	49.7**
15	34.3	29.4	69.8	69.5	67.3	66.7
16	169.1	169.4	166.8	167.1	167.0	168.0
18	119.7	12.5	14.8	14.8	169.9	168.0
19	9.5	9.4	9.9	10.1	15.0	15.2
30	71.2	70.1	70.0	70.1	72.3	73.6
4-Me	22.2	22.1	22.1	20.9	13.3	13.1
1′			174.4	174.4	168.7	165.9
2′			73.9	73.9	20.4	111.5
3′			32.5	32.6		171.3
4′			7.6	7.7		73.6
5′			24.7	24.9		27.9
6′						27.9
7′					52.3	15.2
OMe					52.3	52.7

*, ** Assignments may have to be reversed.

a 4 ppm upfield shift of C-15 relative to **(410)**. The C-9 resonances in **(411)**, **(412)** and **(413)** are shifted upfield as a consequence of a 1,3-diaxial interaction with the C-12 hydroxyl group. Acetylation of a cyclic α-diketone monoenol slightly shields the olefinic oxo carbon and more strongly the carbonyl carbon while exerting strong deshielding on the mono oxygenated olefinic carbon. This finding permitted the shift assignment for C-2, C-3 and C-4 in **(416)** and **(417)**. Shifts are provided in partial formulae **(416 a)** and **(416 b)**.

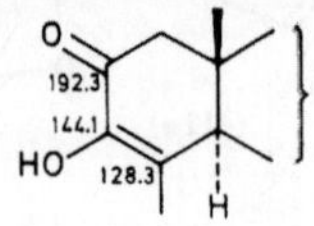

(416a)

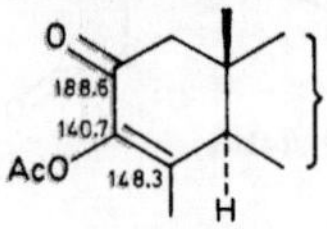

(416b)

In **(416)** C-19 is less shielded by 5.6 ppm than in **(411)**. This was explained in terms of a removal of the γ effect by the 1-hydroxyl group. Similarly the methyl carbon at C-4 resonates at higher field in **(416)** and **(417)**, brought about by the C-3 hydroxyl substituent. The same workers identified the newly isolated quassinoid substance **(420)** as the β-hydroxy derivative of klaineanone **(419)**. Chemical shifts of compounds **(418—420)** are inserted in the formulae (*126*).

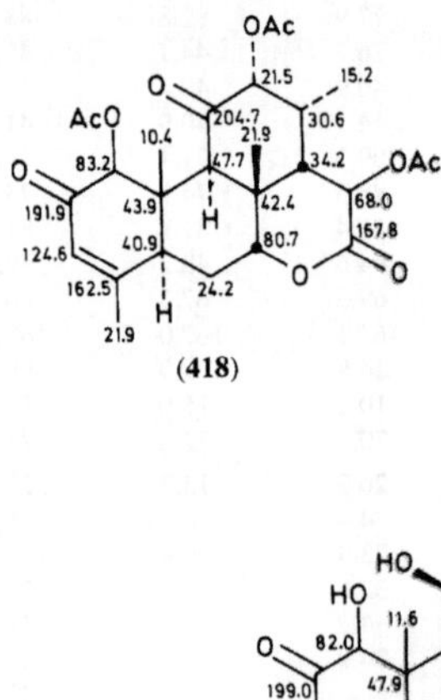

(418)

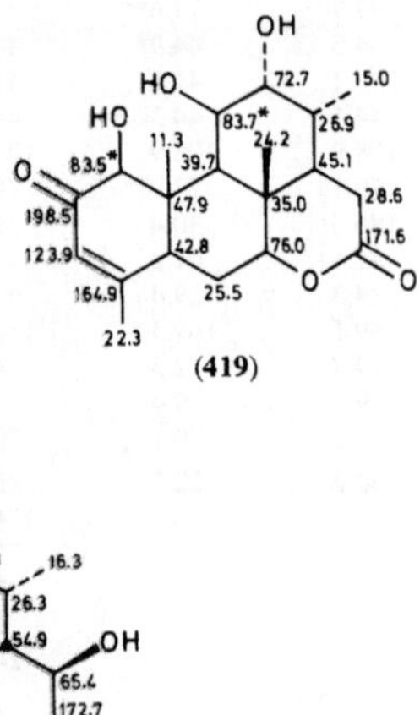

(419)

(420)

References, pp. 195—229

Two novel guaiane-type diterpenes (**421, 422**) together with pachydictiol A (**423**) were found in the digestive gland of the sea hare *Aplysia depilans*. Their structures were found to be in accordance with the ^{13}C spectral data. The two carbinyl carbons could be differentiated

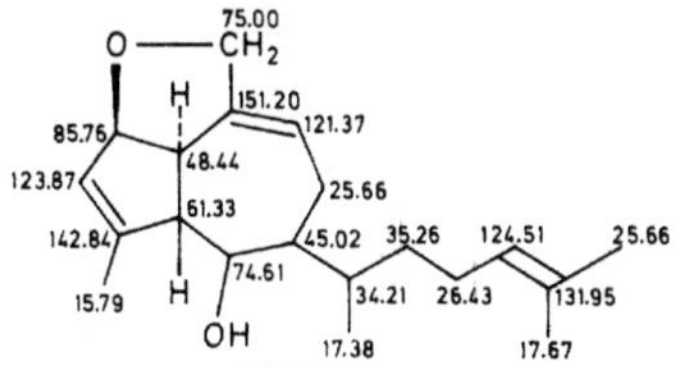

(**421**)

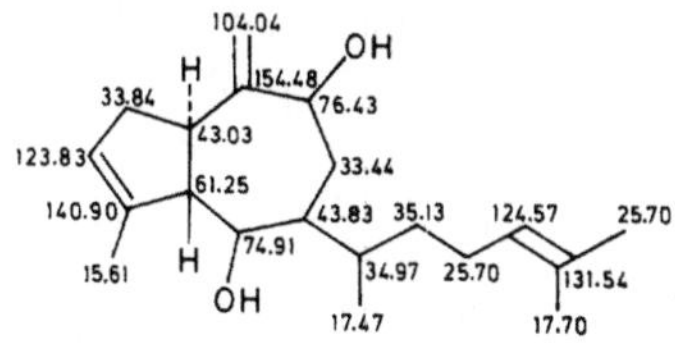

(**422**)

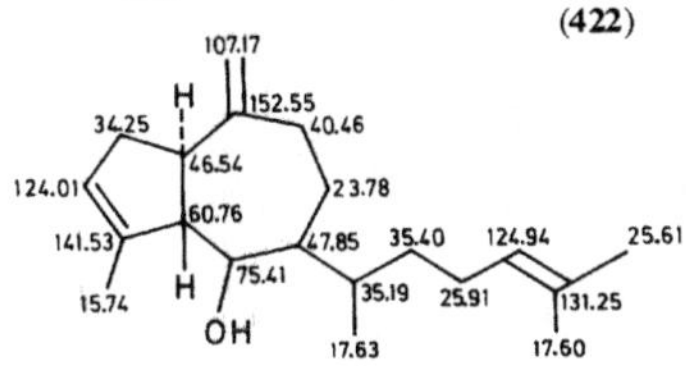

(**423**)

by converting the alcohol into the corresponding trichloroacetyl urethane. The signal which showed no displacement was attributed to the allylic carbinyl where derivatization did not take place due to steric congestion (*127*).

Terpenoids C_n (20 < n < 30)

^{13}C spectra of limonoids (**424—428**) were recorded and assigned in connection with the structure determination of procerin (**429**) (*128*). Chemical shifts for compounds (**424—429**) are provided in Table 22 (**424—427**) and together with the formulae (**428, 429**). A revised structure

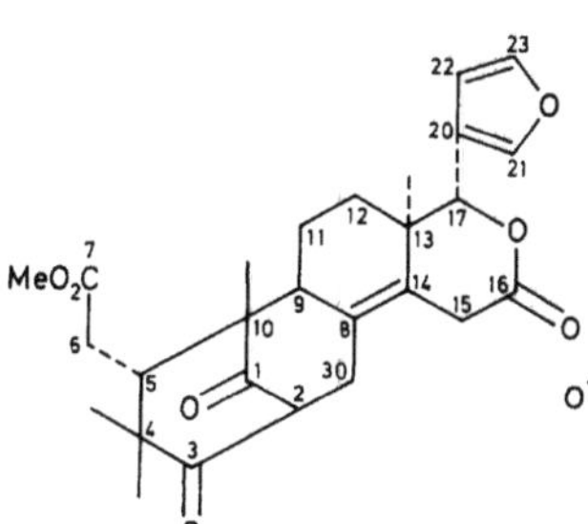

(**424**)

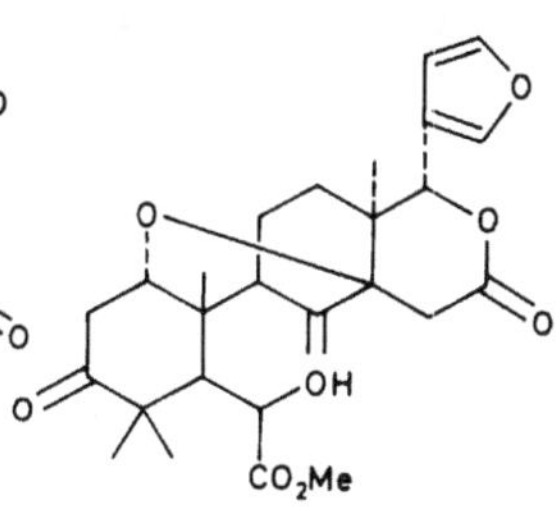

(**425**)

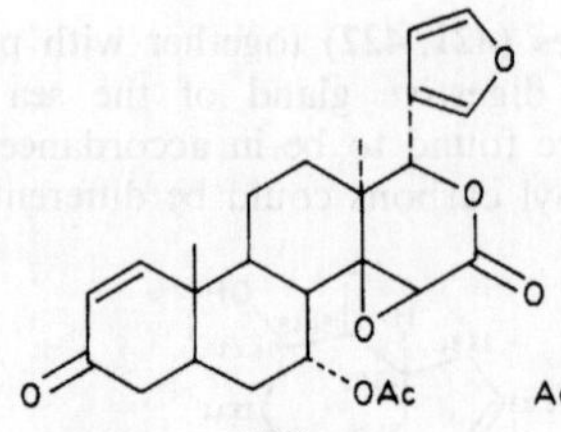

(426)

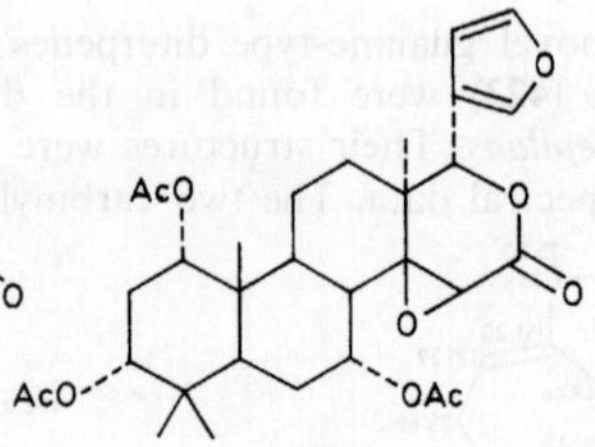

(427)

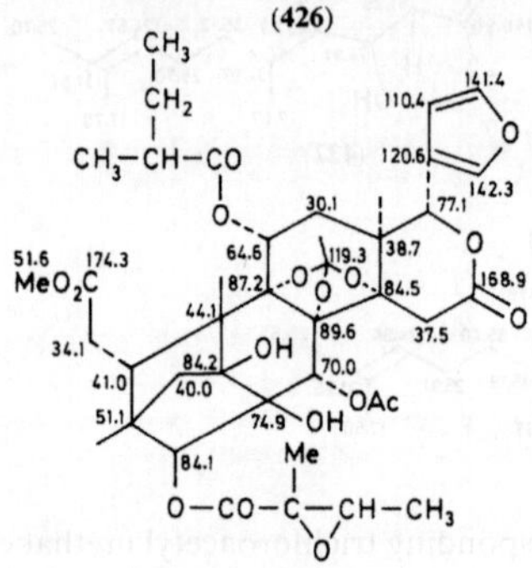

(428)

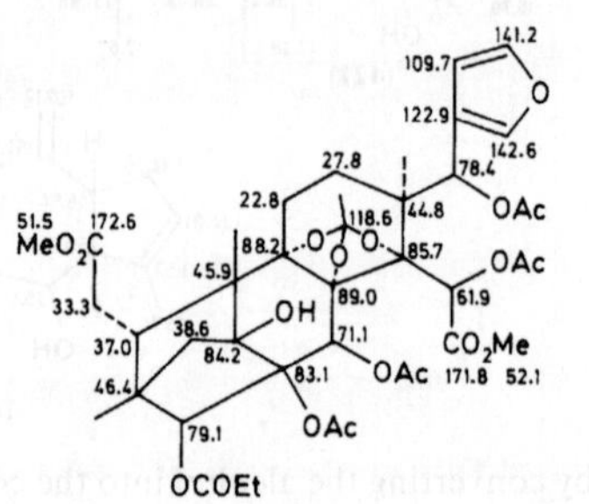

(429)

(430) has recently been proposed for bussein (*129*). The ^{13}C data provided conclusive evidence for the presence of only one ortho ester or hemiortho ester carbon resonating at δ_C 118.9 whereas two such carbons would be required by the old structure **(431)**.

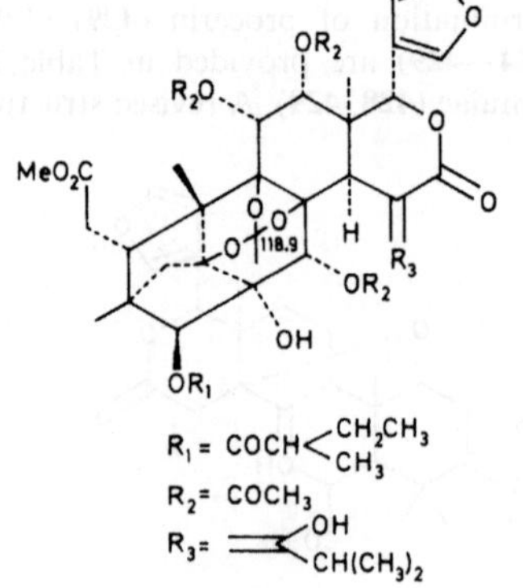

(430)

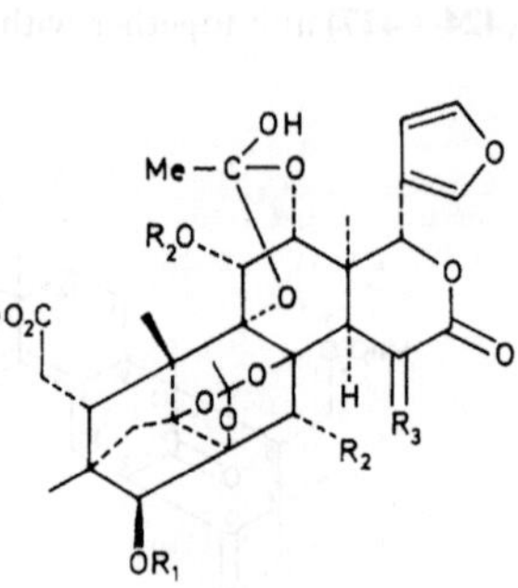

(431)

Table 22. ^{13}C *Chemical Shifts for Limonoids* (**424—427**) (*128*)

Carbon	(**424**)	(**425**)	(**426**)	(**427**)
1	202.7	79.2	157.0	72.3
2	58.0	40.1	125.9	25.6*
3	210.9	212.2	203.8	75.9
4	49.4	49.8	44.0	35.9
5	50.5	51.9	39.5	37.0
6	32.3	73.3	14.9*	22.3*
7	173.5	177.1	73.2	73.8
8	133.8	146.6	42.6	42.2
9	40.2	48.5	46.0	36.3
10	54.3	45.7	40.0	40.7
11	18.6	25.1	17.7*	14.4*
12	28.8	29.7	18.3*	26.1*
13	37.8	42.4	38.7	38.8
14	125.3	81.3	69.8	69.7
15	32.9	34.5	56.9	56.4
16	169.7	170.5	167.4	167.6
17	80.6	80.3	78.2	78.4
20	120.4	121.6	120.5	120.5
21	142.7	143.3	143.0	143.0
22	109.8	110.7	109.8	109.9
23	141.5	141.3	141.4	141.4
30	36.4	112.3		
OMe	52.1	54.2		
CH_3	21.9	25.9	27.2	27.4
	17.8	24.9	26.0	21.6
	17.8	24.5	23.3	21.3
	17.4	15.0	21.7	21.1
			21.0	21.1
			19.7	18.3
				17.3
				16.4
CO			169.8	169.9
				169.7
				169.6

* Ambiguous assignments.

Structure (**432**) belonging to a group of meliacins, isolated from *Khaya anthotheca* heartwood, was derived from a comparison of its ^{13}C spectrum with those of the related systems cedrelone (**433**) and anthothecol (**434**) and their derivatives (*130*). The assignments of C-5 and C-6 in (**433**) and (**434**) are reversed by the present authors on the basis of the unique acetylation effect described for (**416**) (*126*). The ^{13}C spectrum of sendanin was in accordance with the proposed structure (**435**) (*131*).

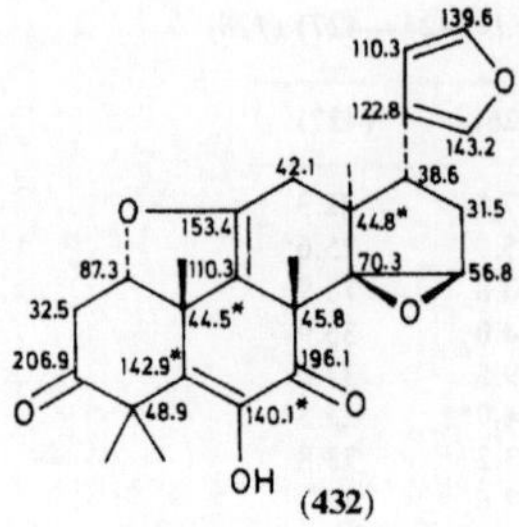

Me : 29.2; 25.5; 21.1; 20.6; 17.6

(432)

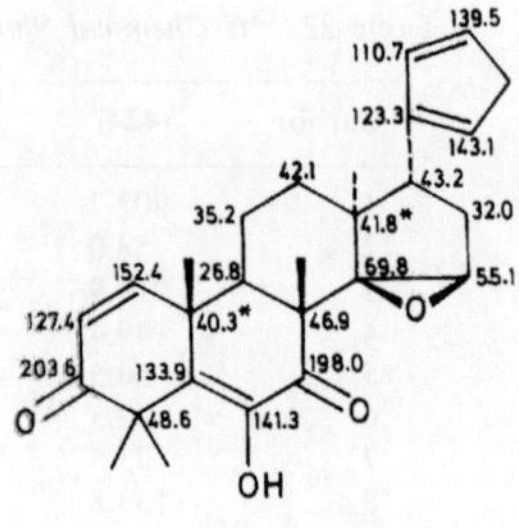

Me : 23.9; 23.1; 21.3; 20.2; 19.5

(433)

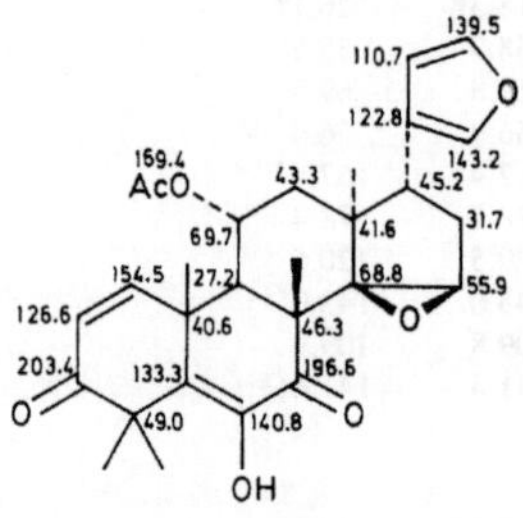

Me : 24.2; 22.8; 21.9; 21.4; 21.1; 19.7

(434)

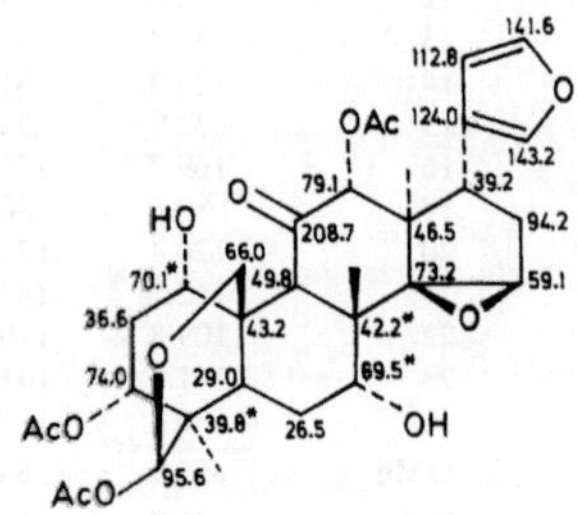

Me : 20.7; 19.2; 15.7

(435)

Structures for cycloepiatalantin (**436**), atalantin (**437**) and dehydro-atalantin (**438**) could be corroborated by correlating their ^{13}C spectra with those of model compounds obacunone (**439**), methylobacunoate (**440**), methyl isoobacunoate (**441**), limonin (**442**), and 7-deacetyl-7-oxogedunin

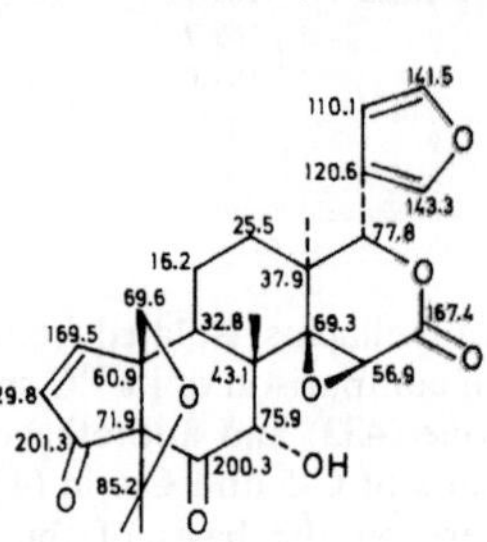

Me : 28.8 25.1; 17.7; 15.5

(436)

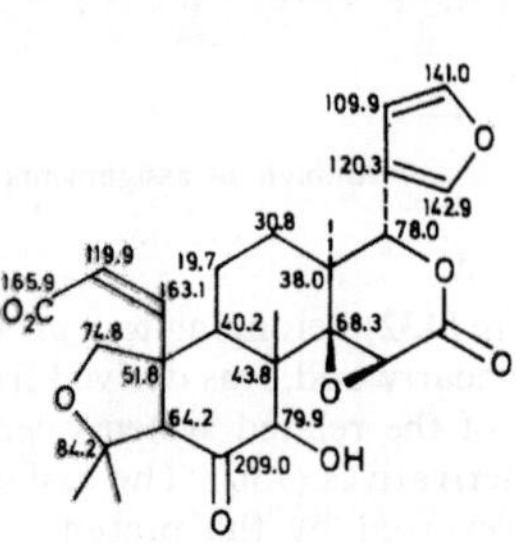

Me : 30.8; 25.0; 20.3; 12.3

(437)

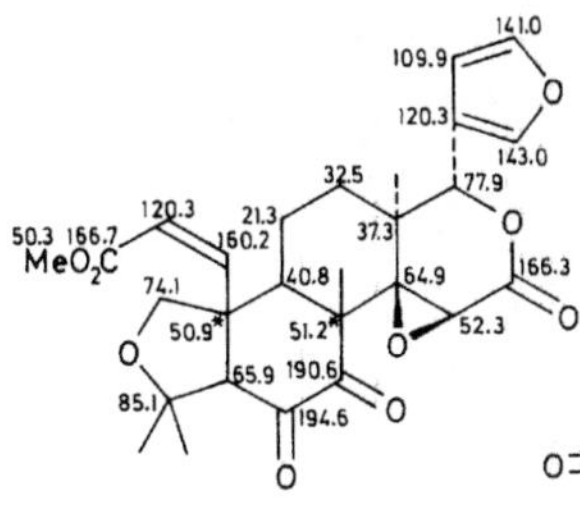

Me : 29.0; 25.3; 19.7; 10.0

(**438**)

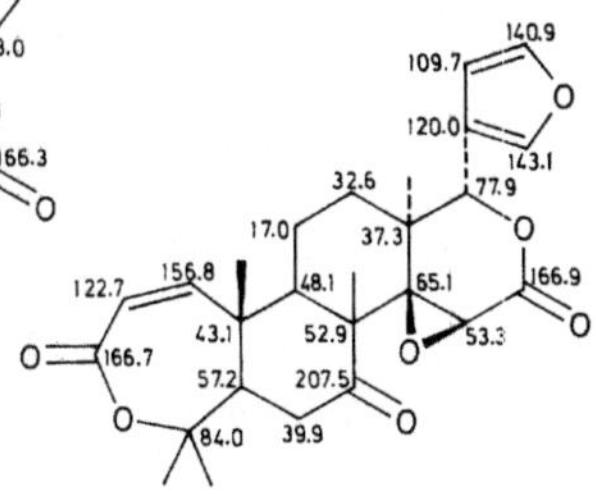

Me: 32.0; 26.7; 21.0; 19.4; 16.4

(**439**)

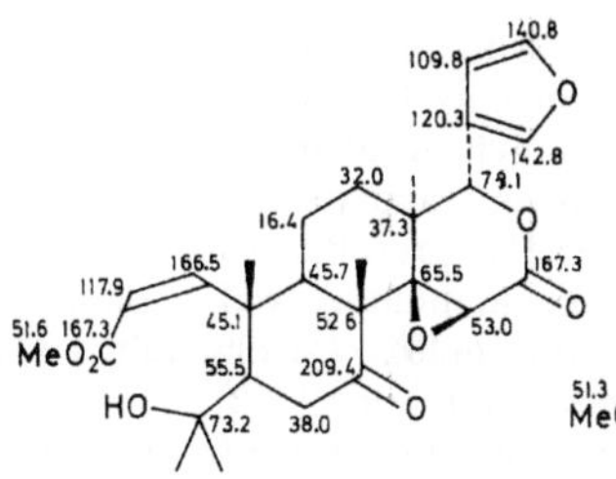

Me : 32.6; 29.4; 20.7; 20.3; 19.4

(**440**)

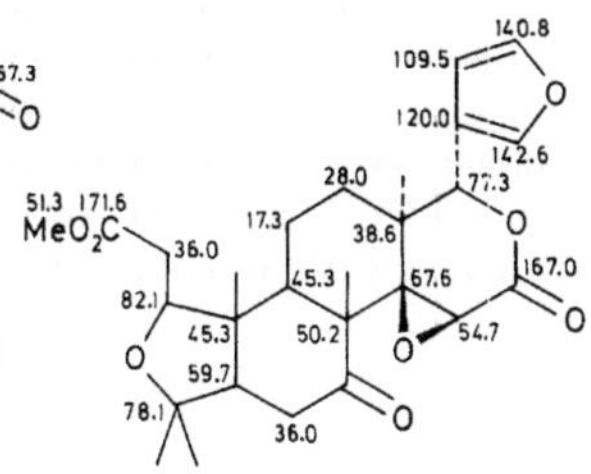

Me : 29.2; 22.9; 19.5; 18.4; 1.8

(**441**)

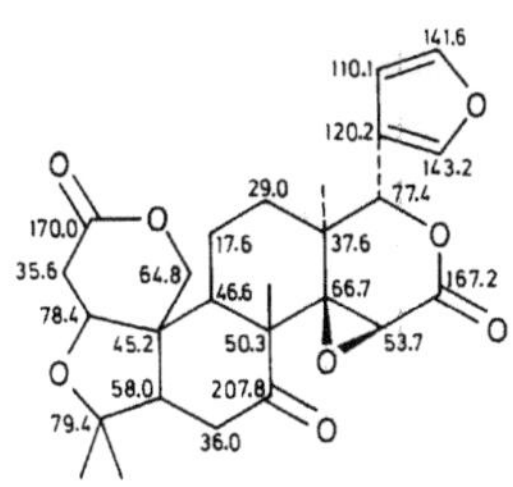

Me : 29.6; 21.3; 19.6; 17.1

(**442**)

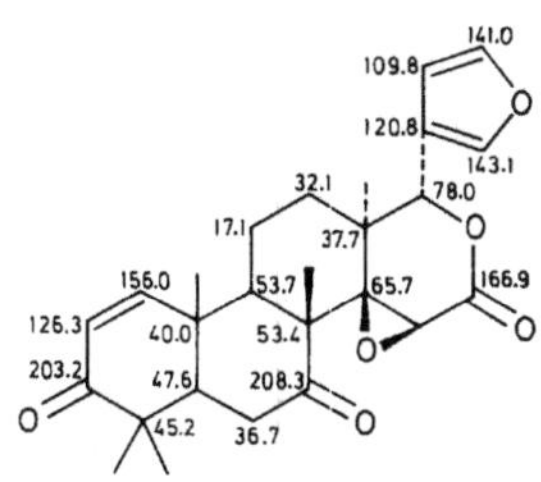

(**443**) Me: 22.0; 20.9; 20.7; 19.7; 17.4

(**443**) (*132*). The inversion-recovery technique in conjunction with SFORD spectra has been employed successfully in a structural study of azadirachtin (**444**) (*133*).

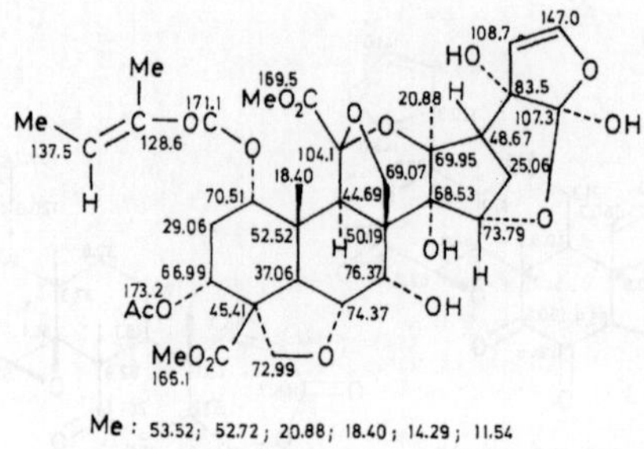

Me : 53.52; 52.72; 20.88; 18.40; 14.29; 11.54

(444)

The technique of incrementally varying the decoupler offset (cf. Sect. 2.1) was employed for checking assignments in a series of hop bitter substances. In this particular case proton chemical shifts were predicted on the basis of the decoupling experiments and these values with the experimental ones, observed in the 300 MHz proton spectra (*134*). Total carbon designations were provided for humulone (**445**), cohumulone (**446**), colupulone (**447**), *trans*-isohumulone (**448**), *cis*-isohumulone (**449**), humulinic acids (**450**) and (**451**), and cohulupone (**452**). Some of the carbonyl and enolic carbons could be assigned on a one-to-one basis.

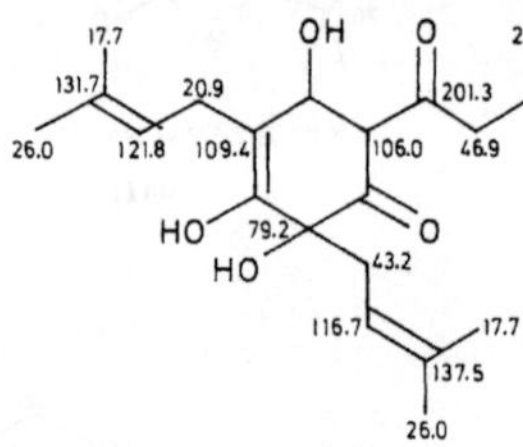

168.4; 191.4; 195.4 : Unassigned

(445)

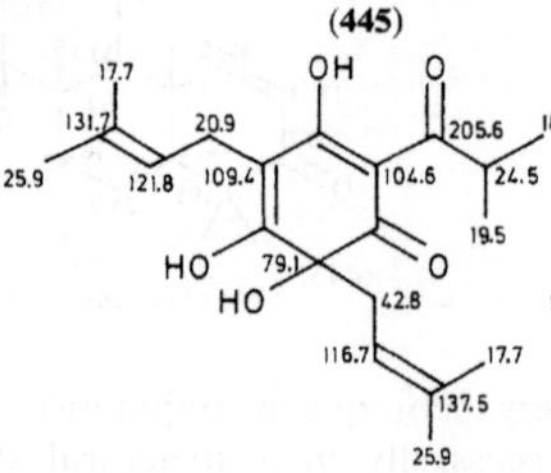

168.1; 191.4; 195.3 : Unassigned

(446)

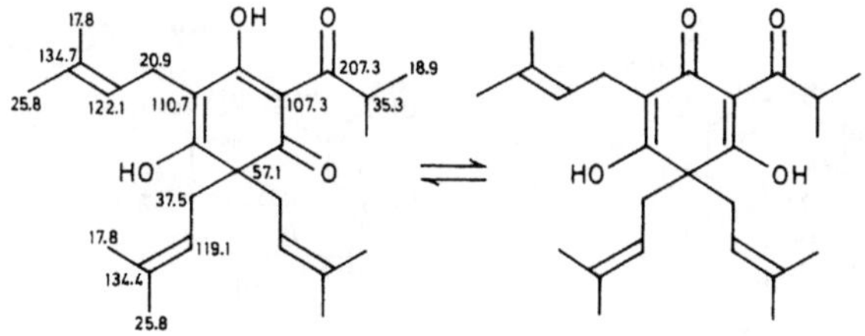

(447)

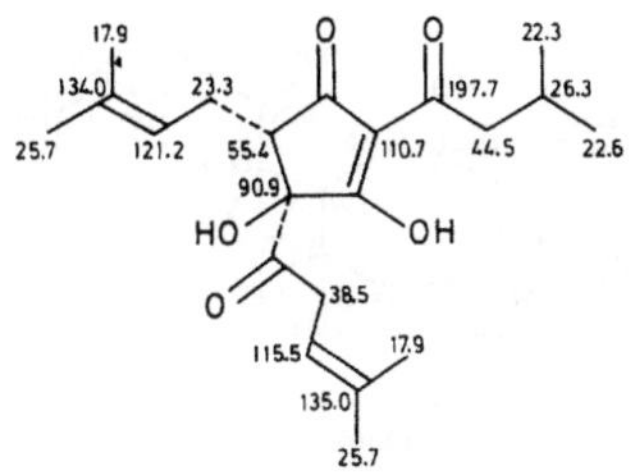

195.6 : 190.3 : 171.8 : Unassigned

(448)

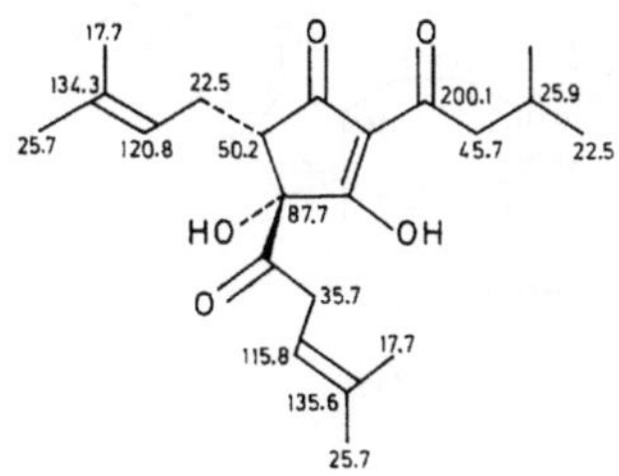

(449)

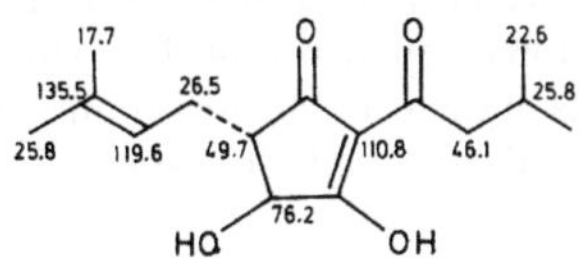

(450)

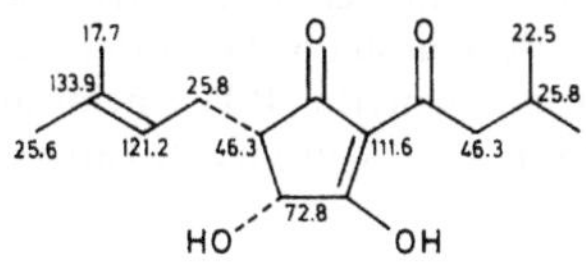

(451)

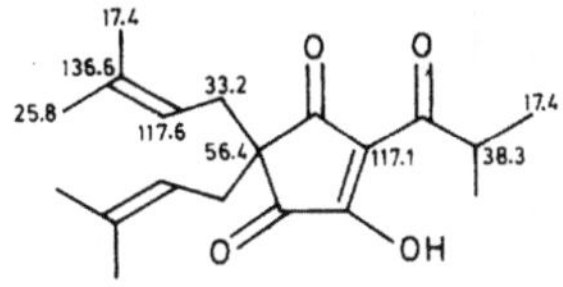

(452)

Formulae (**453**—**456**) provide ^{13}C chemical shifts for the sesterpenes ophiobolin A (**453**), ophiobolin B (**454**), ophiobolin C (**455**) and ophiobolin D (**456**) (*135*).

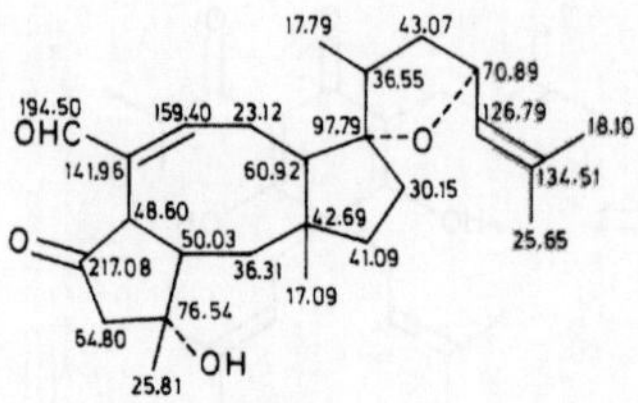

(453)

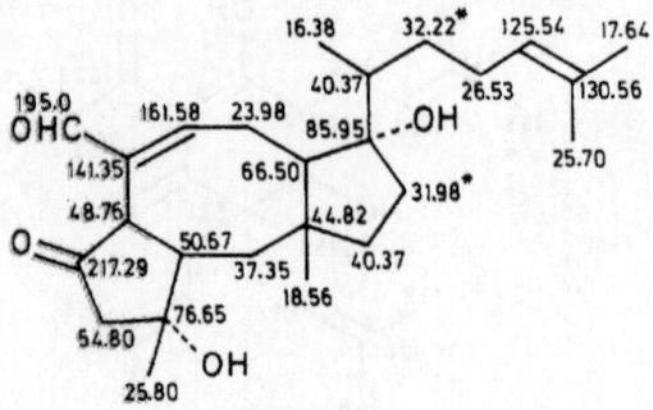

(454)

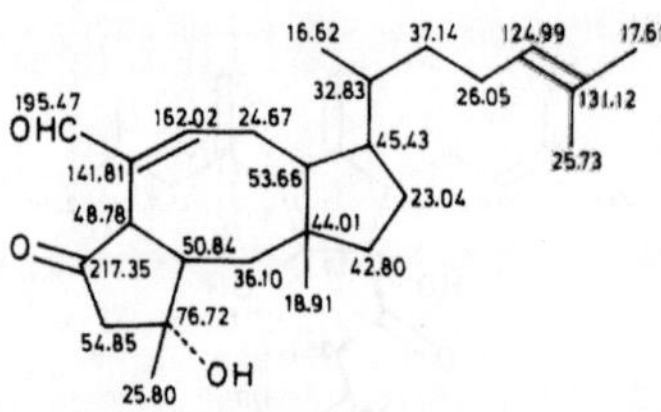

(455)

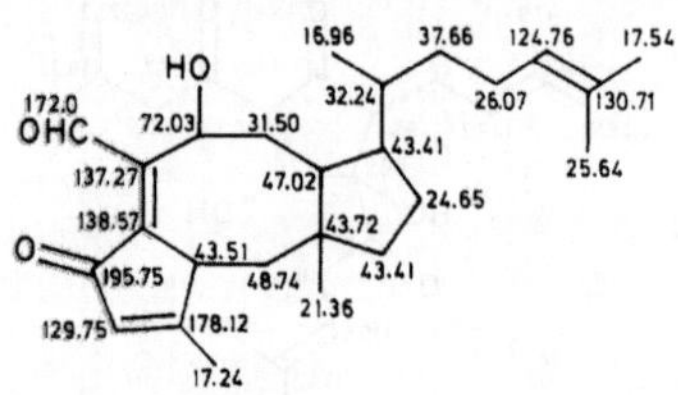

(456)

^{13}C n.m.r. data have also become available for some of the biogenetically important triterpenoids. A thorough analysis of the spectra of tetra- and pentacyclic triterpenes has been reported (*136*). These comprise lanostenol (**457**), lanostanol (**458**), euphenol (**459**), α-amyrin (**460**) and β-amyrin (**461**). Chemical shifts of these and related compounds (vide supra) are collected in Table 23. Signals were identified essentially by

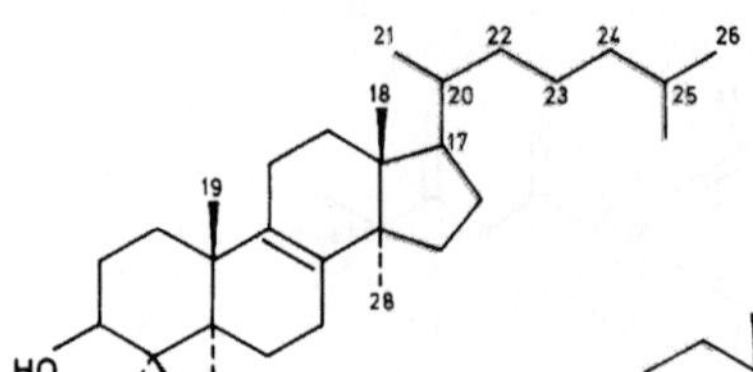

(457)

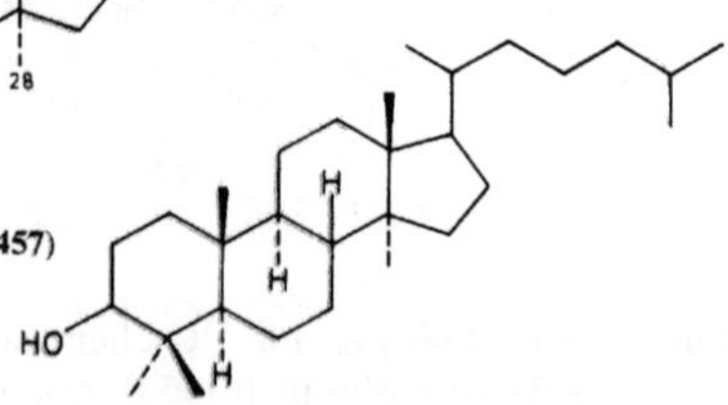

(458)

Table 23. ^{13}C *Chemical Shifts for Triterpenoids* (**457**—**461**) (*136*) *and Related Compounds* (**462**—**464**) (*137*)

Carbon	Lanostenol (**457**)	Lanostenol (**458**)	Euphenol (**459**)	α-Amyrin (**460**)	β-Amyrin (**461**)	Lanostenol*	(**462**)	(**463**)	(**464**)
1	35.8	37.7	35.4	38.95	38.73	37.22	36.76	41.45	36.52
2	27.9	27.7	27.8	27.35	27.29	27.37	23.63	52.00	23.06
3	79.0	79.1	78.9	78.97	78.97	78.35	85.74	91.13	81.78
4	39.0	39.0	37.3	38.78	38.84	36.79	36.65	36.71	34.04
5	50.5	55.1	51.1	55.34	55.29	54.51	54.56	55.32	52.76
6	19.2	18.3	19.1	18.45	18.45	19.90	19.90	20.26	19.77
7	28.3	29.0	28.0	33.07	32.80	28.67	28.48	28.32	28.18
8	134.4	38.7	134.1	40.13	38.78	38.24	38.22	37.95	38.16
9	134.4	48.8	133.6	47.85	47.74	48.17	48.06	48.04	48.01
10	37.2	37.2	39.0	36.95	37.00	38.57	37.79	39.11	36.73
11	21.1	20.1	21.6	23.41	23.57	21.41	21.20	20.52	21.47
12	26.7	21.7	28.1	124.5	121.8	33.28	33.26	33.15	33.26
13	44.6	45.3	44.2	139.6	145.1	44.80	44.80	44.80	44.69
14	49.9	48.0	50.1	42.18	41.80	47.47	47.41	47.41	47.33
15	31.2	33.7	31.0	26.70	26.22	31.83	31.77	31.69	31.58
16	31.0	32.2	29.8	28.16	27.02	27.70	27.73	27.73	27.59
17	50.7	51.1	49.8	33.77	32.47	50.49	50.50	50.46	50.44
18	15.9	16.3	15.7	59.19	47.36	14.21*	14.21**	15.70**	13.97**
19	18.3	13.9	20.2	39.76	46.93	15.48*	16.10**	16.07**	14.16**
20	36.5	36.2	35.7	39.65	31.07	35.82	35.82	35.79	35.79
21	18.8	18.7	19.1	31.29	34.79	18.50	18.48	18.45	18.45
22	36.5	36.7	36.1	41.59	37.22	36.25	36.25	36.19	36.19
23	24.2	24.1	23.9	28.16	28.21	23.87	23.87	23.84	23.84
24	39.6	39.6	39.5	15.65	15.48	39.16	39.13	39.11	39.11
25	28.1	28.0	28.0	15.65	15.59	27.70	27.73	27.73	27.70
26	22.6	22.6	22.8	16.95	16.88	22.63	22.60	22.63	22.60
27	22.8	22.8	22.8	23.25	26.00	22.33	22.33	22.33	22.30
28	24.3	14.4	24.5	28.75	28.43	16.10	16.40	16.37**	16.07**
29	28.1	28.3	28.0	17.48	33.34	27.94	27.83	27.73	54.10
30	15.4	15.6	15.5	21.36	23.73	13.76	13.78	14.28	12.43
CO							159.80	159.80	153.25

* From Ref. (*137*). ** Ambiguous assignments.

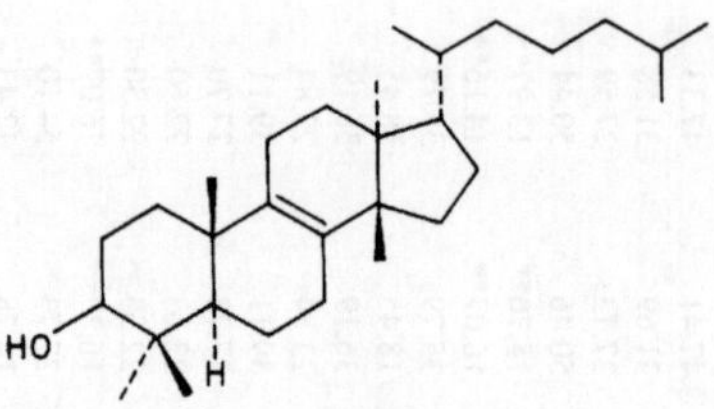

(459)

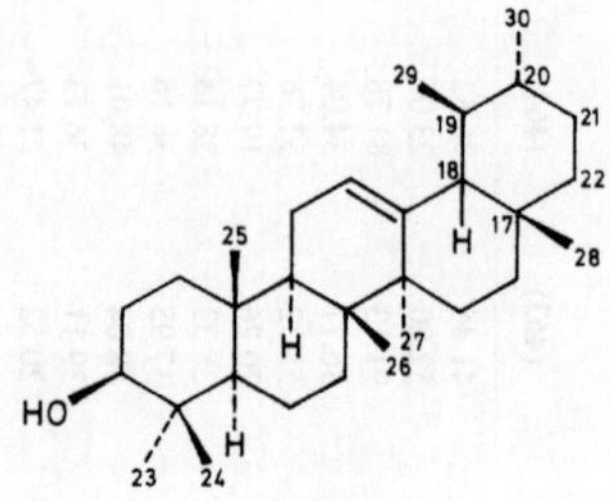

(460)

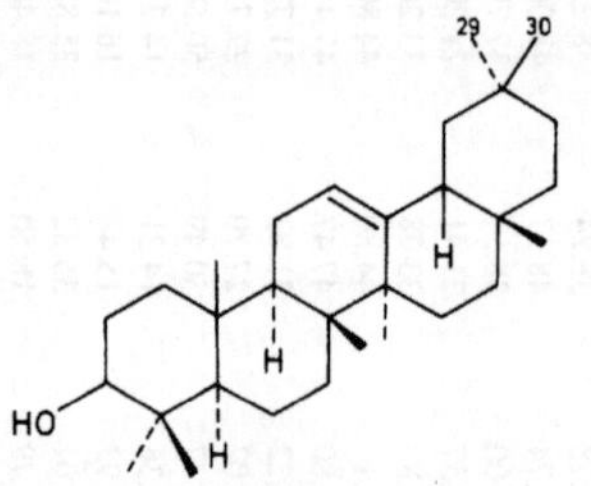

(461)

chemical shifts correlation, aided by T_1 measurements, SFORD as well as noise off-resonance decoupling for quaternary carbons. By shift comparison among derivatives such as 3β-acetate, 3-oxo and 11-oxo, as well as by exploiting the fact that lanostanol (**457**) bears a C-17 sidechain identical to cholesterol, assignment of all carbons except C-15, C-16, C-17 and C-28 could be made. Introduction of a carbonyl function at C-7 induced an upfield shift for C-15 (steric compression shift). Characteristic shifts were also observed upon ketonization and acetylation. The relative shieldings thus produced are portrayed in partial structures (**457 a**) and (**457 b**). The signals due to C-6, C-7, C-8, C-9, C-11 and C-12

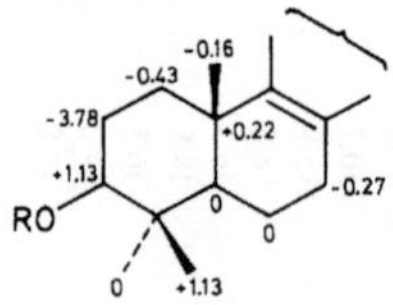

R = H→$COCH_3$

(**457 a**)

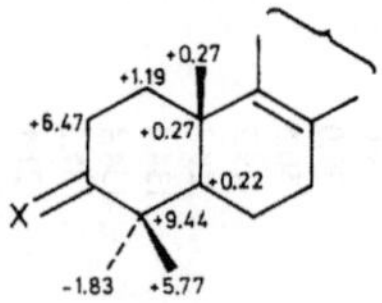

X = H, OH→O

(**457 b**)

in lanostanol (**458**) could be designated by chemical shift comparison with the 11-oxo- and 7,11-dioxo derivatives. The removal of the C-8—C-9 double bond in lanostenol and the concomitant formation of the *trans* B/C ring junction results in an upfield shift of one methyl resonance which identifies it as C-28.

The C-15 and C-16 resonances in euphenol (**459**) are well separated, which reflects the inverse configuration of the carbons at the C/D ring junction.

In α- and β-amyrin the assignment of the ring-A atoms, methyls C-23, C-24 and C-25 was achieved on the basis of the characteristic acetylation and ketonization shifts outlined earlier. Identification of C-6 and C-7 in ring B was possible by comparing the spectra with those of the 3-oxo- and 3-oxo-24-oate derivatives. Replacement of methyl C-24 was accompanied by downfield shifts of ca. 1 ppm at C-5 and C-6 (partial structure (**460 a**)). The 11-oxo compounds served as vehicles for assigning ring-C carbons as depicted by the incremental shifts in partial formulae (**460 b**) and (**461 a**). Introduction of the keto function at C-11 affects methyls

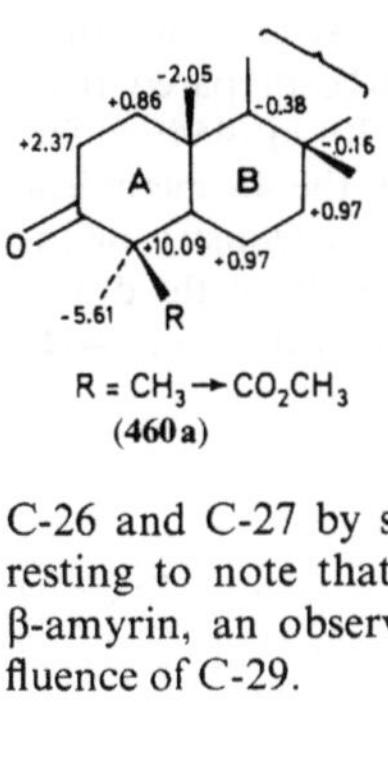

R = $CH_3 \rightarrow CO_2CH_3$

(**460 a**)

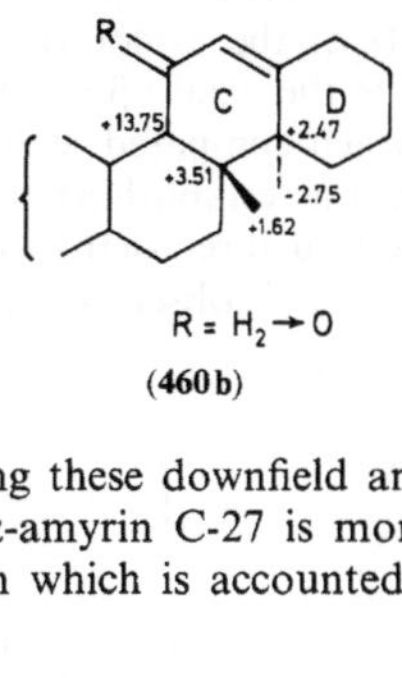

R = $H_2 \rightarrow O$

(**460 b**)

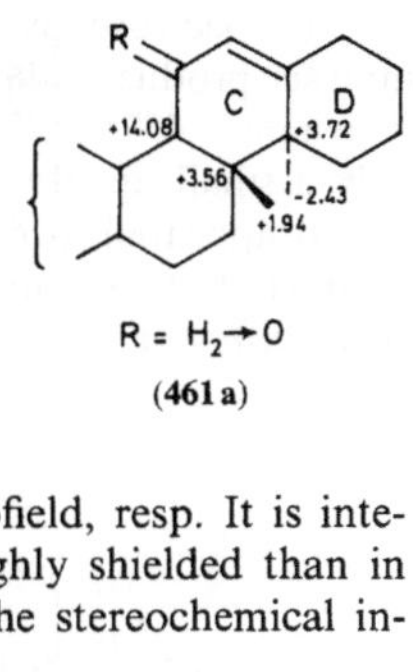

R = $H_2 \rightarrow O$

(**461 a**)

C-26 and C-27 by shifting these downfield and upfield, resp. It is interesting to note that in α-amyrin C-27 is more highly shielded than in β-amyrin, an observation which is accounted to the stereochemical influence of C-29.

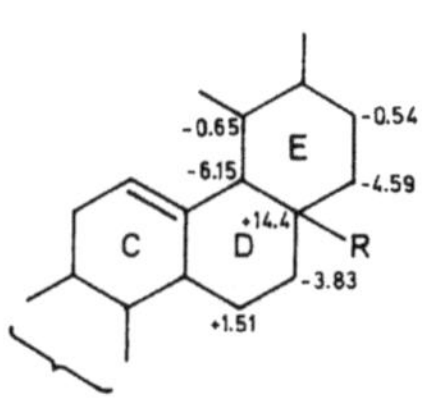

R = $CH_3 \rightarrow CO_2CH_3$

(**460 c**)

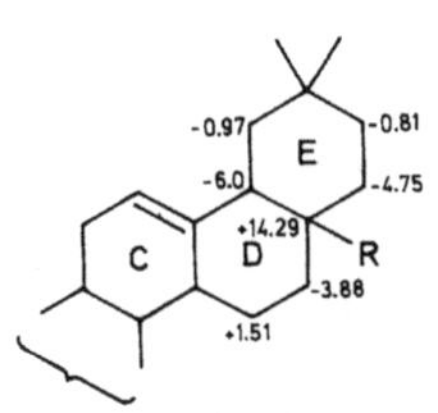

R = $CH_3 \rightarrow CO_2CH_3$

(**461 b**)

The assignment of the ring-D and E carbons was aided by the shifts induced upon carboxylation of C-28. This is indicated in partial structures (**460 c**) and (**461 b**). Comparison of the shifts in glycyrrethic acid methyl ester in which the C-30 methyl of (**461**) is replaced by a carbomethoxy group, with those of the parent compound, corroborated the assignments of the ring-E carbons. Differential shifts are given in partial formula (**461c**).

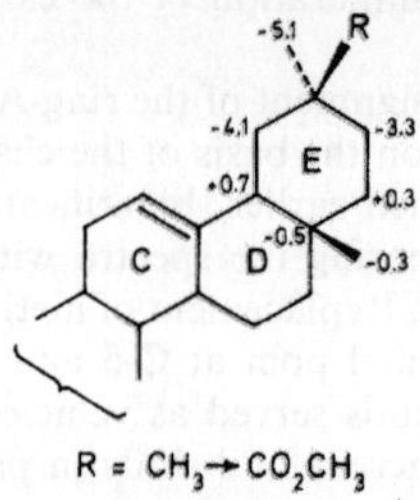

(**461c**)

^{13}C spectroscopy yielded structure and stereochemistry of two thermolysis products (**463**, **464**) of the azidoformate (**462**), prepared from lanostanol (**457**) (*137*). While the axial 4β methyl (C-30) in (**464**) suffers only a small upfield shift when compared with (**462**), the 4α methylene (C-29) appeared at δ_C 54.1, thus establishing the stereochemistry of the product. The oxazolidinone structure (**463**) was derived from the downfield shifts for C-1, C-2 and C-3 observed upon cyclization of (**462**).

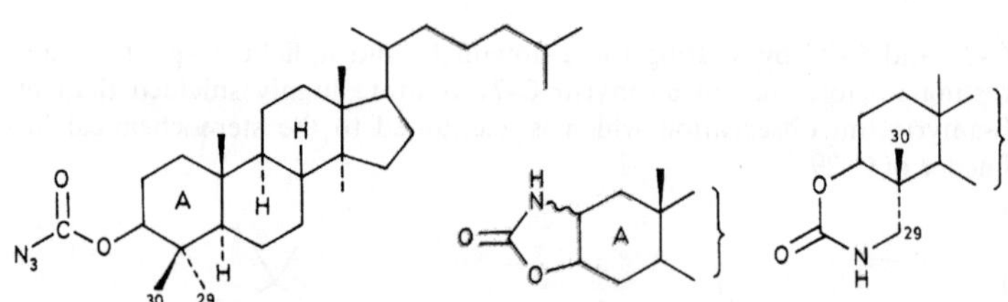

(**462**) (**463**) (**464**)

The assignments given for lanostanol in this work differ considerably from those reported by Knight (**136**) [C-4, C-10, C-12, C-15, C-16, C-18, C-19, C-28 and C-30].

Cycloartanol (**465**) possesses a cyclopropane ring in 9,10 position, as well as a *cis* B/C ring junction and ring-C boat conformation. ^{13}C shift data of artanol and related compounds (**466**—**472**) are listed in

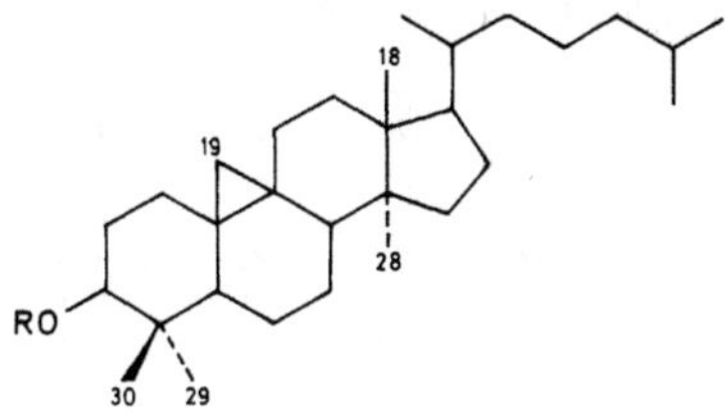

(**465**) R = H
(**466**) R = $COCH_3$

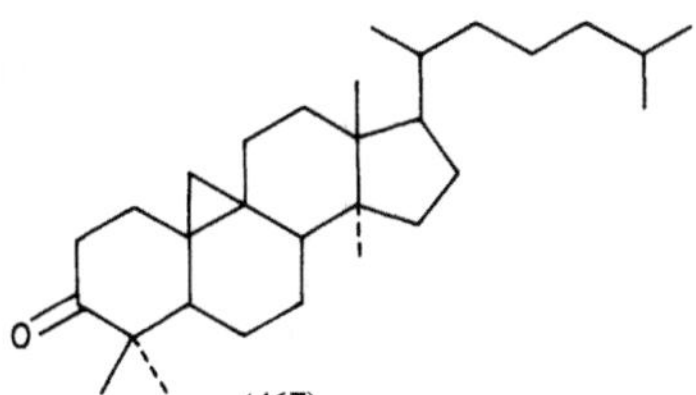

(**467**)

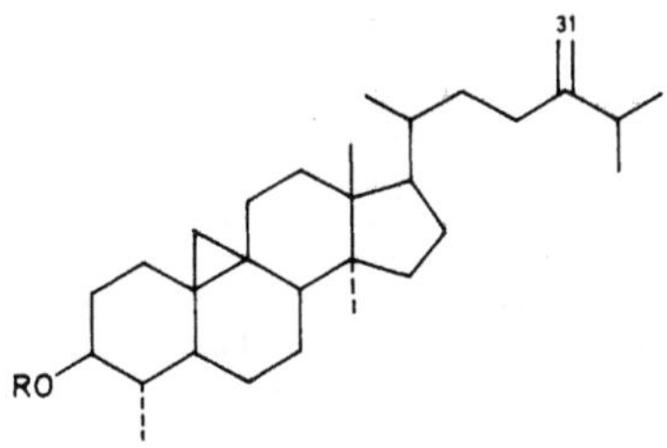

(**468**) R = H
(**469**) R = $COCH_3$

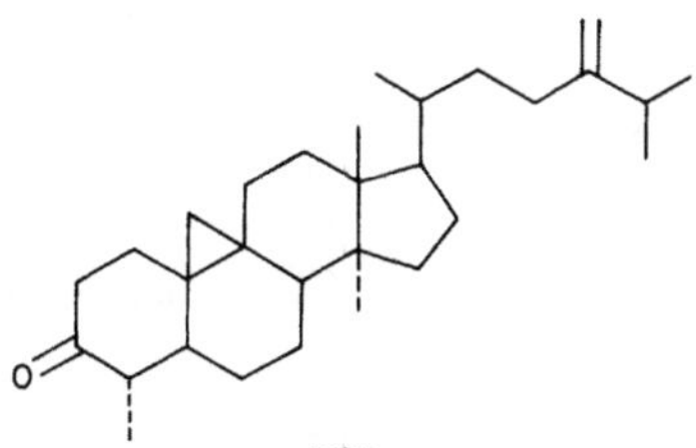

(**470**)

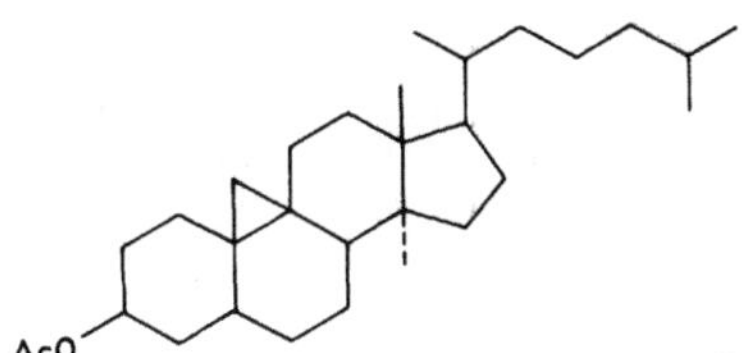

(**471**)

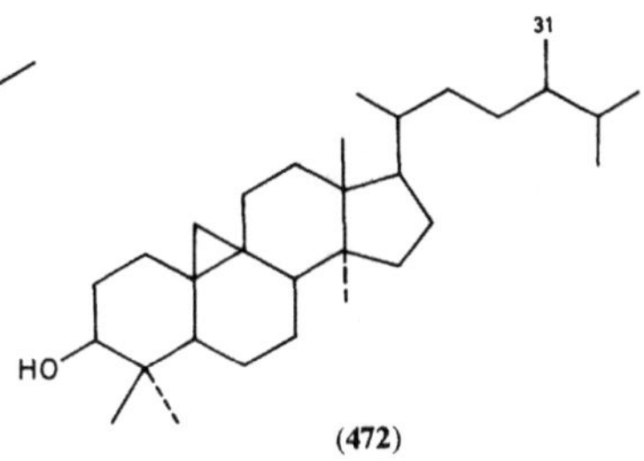

(**472**)

Table 24 (*138*). Cyclopropane carbons C-9 and C-10 were differentiated with the aid of LIS and deuterium isotope shift experiments. The relative shift observed for C-19 when going from (**465**) to cycloeucaneol (**468**) was rationalized in terms of the removal of the δ interaction between the axial C-30 and C-19 (δ_C 29.8→δ_C 26.9).

Table 24. ^{13}C *Chemical Shifts of Cycloartanol* (**465**) *and Related Compounds* (**466—472**) (*138*)

Carbon	(**465**)	(**466**)	(**467**)	(**468**)	(**469**)	(**470**)	(**471**)	(**472**)
1	31.9	31.5	33.3	30.7	30.4*	32.8	31.4	31.9
2	30.3	26.7	37.2	34.8	30.9*	40.8	30.4	30.3
3	78.5	80.3	215.1	76.3	78.5	212.2	73.7	78.5
4	40.3	39.4	50.0	44.5	41.4	49.8	38.5	40.4
5	47.0	47.0	48.3	43.2	43.3	45.9	37.1	47.0
6	21.0	20.8	21.5	24.6	24.6	25.1	28.1*	21.1
7	28.0	28.0	27.9	28.0	28.0	28.0	27.8*	28.0
8	47.8	47.6	47.7	46.7	46.7	46.9	46.1	47.8
9	20.0	20.1	21.0	23.5	23.6	24.9	23.5	20.0
10	26.0	26.0	26.0	29.5	29.3	29.3	29.3	26.0
11	26.0	25.8	26.0	25.1	24.9	25.8	24.6***	26.0
12	35.5	35.4	35.5	35.2	35.3	35.3	35.2	35.5
13	45.1	45.1	45.2	45.2	45.2	45.2	45.3	45.1
14	48.7	48.6	48.5	48.7	48.7	48.7	49.0	48.7
15	32.8	32.8	32.8	32.8	32.8	32.8	32.9	32.9
16	26.5	26.5	26.7	26.9	26.9	26.9	27.1	26.5
17	52.2	52.2	52.2	52.0	52.0	52.1	52.3	52.1
18	17.9*	17.9*	18.1*	17.7*	17.7**	17.9	17.5**	17.9*
19	29.8	29.6	29.5	26.9	26.9	27.1	25.7***	29.6
20	36.0	36.0	36.0	36.0	36.0	36.0	36.2	36.5
21	18.3*	18.3*	18.1*	18.3*	18.3**	18.3	18.5**	18.4*
22	36.4	36.4	36.4	35.0	35.0	35.0	36.5	33.9
23	24.0	24.1	24.1	31.3	31.3	31.3	24.2	31.4
24	39.4	39.4	39.4	156.2	156.1	156.1	39.6	39.0
25	28.0	28.0	27.9	33.7	33.7	33.7	28.1	30.9
26	22.5	22.5	22.5	21.8	21.9	21.8	22.9	20.5
27	22.7	22.7	22.7	21.8	21.9	21.8	22.9	17.6*
28	19.3*	19.2*	19.3*	19.5*	19.1**	19.1	19.0**	19.3
29	25.4	25.3	26.0	14.4	14.4	10.7	—	25.4
30	14.0	15.1	20.6	—	—	—	—	14.0
31				105.6	105.6	105.6	—	15.5*
O<u>C</u>OCH₃		170.0			170.1		170.0	
OCO<u>C</u>H₃		21.2			21.2		21.5	

*, **, *** Assignments may have to be interchanged.

Table 25 presents ^{13}C shift data for a group of triterpenoids related to cycloartanol (*139*). The compounds studied comprise cimigoside (**473**) and actein (**475**), both highly oxygenated polycyclic glucosides, isolated from *Actea racemosa*, as well as their aglycones (**474**) and (**476**) and some further derivatives (**477—479**). Although the conformation of ring C is not settled unambiguously, the marked similarity of ring carbon shifts with those of cycloartanol points to a boat conformation for (**473**) and (**475**) as well. This contrasts with the original authors' suggestion that the

Table 25. ^{13}C *Chemical Shifts for Triterpenes* (**473—479**) (*139*)

Carbon	(473)	(474)	(475)	(476)	(477)	(478)	(479)
1	31.66	31.81	31.91	32.04	31.88	33.67	33.65
2	28.98	30.08	29.18	30.32	26.82	37.37	37.30
3	88.04	77.50	88.12	77.40	79.97	216.30	215.79
4	40.64	40.30	40.70	40.33	39.48	50.26	50.23
5	46.76	46.57	47.16	46.97	47.05	48.40*	48.15
6	20.19	20.38	20.64	20.82	20.76	21.35	21.47
7	25.34	25.55	25.73	25.96	26.06	26.44	25.53
8	45.65	45.71	48.13	48.15	47.47	43.08*	42.26
9	19.77	19.72	19.62	19.56	19.86	20.97	21.19
10	26.39	26.64	26.20	26.43	26.52	26.44	26.36
11	36.69*	36.62*	25.97	25.92	25.83	25.89	25.67
12	76.67	76.77	33.58	33.60	33.36	33.55	31.77
13	47.59	47.60	41.36	41.39	42.22	41.84	40.53
14	48.24	48.25	46.58	46.60	46.37	46.98	53.27
15	43.35	43.37	79.47	79.49	80.51	79.74	211.52
16	72.51	72.55	110.96	110.99	110.78	111.46	103.43
17	55.97	56.00	58.99	58.97	59.45	59.19	56.92
18	12.97	12.97	18.98	18.98	19.37	19.16	20.33
19	29.79	29.91	30.67	30.75	30.84	30.53	29.95
20	25.34	25.33	23.47	23.45	23.39	23.68	23.05
21	20.76	20.73	19.33	19.32	19.86	19.33	19.74
22	36.29*	36.33*	37.75	37.74	37.44	37.67	37.23
23	105.16	105.19	71.23	71.28	71.37	71.74	72.58
24	62.22	62.17	89.46	89.40	88.92	89.06	89.62
25	64.59	62.60	70.42	70.40	70.88	71.31	71.15
26	12.47	12.44	25.04*	25.04*	25.73*	25.93	25.53*
27	97.23	97.29	25.30*	25.33*	25.83*	25.93	25.91*
28	19.49	19.53	11.22	11.23	12.45	11.11	14.24
29	25.34	25.55	25.30	25.59	25.43	22.21	22.24
30	14.91	14.07	14.95	14.20	15.14	20.76	20.69
1′	105.62		105.72				
2′	73.54		73.67				
3′	76.32		76.51				
4′	69.74		69.74				
5′	65.30		65.42				
$\underline{C}H_3CO$	21.49	21.51			21.25 21.34		
$\underline{C}OCH_3$	170.17	170.27			170.84 171.04		

* Assignments may have to be reversed.

12-acetyl group in (**473**) is β-oriented (equatorial) and ring C has chair conformation.

A complete analysis of the ^{13}C spectra of some pentacyclic triterpenes belonging to the class of olean-12-ene compounds has been reported (*140*).

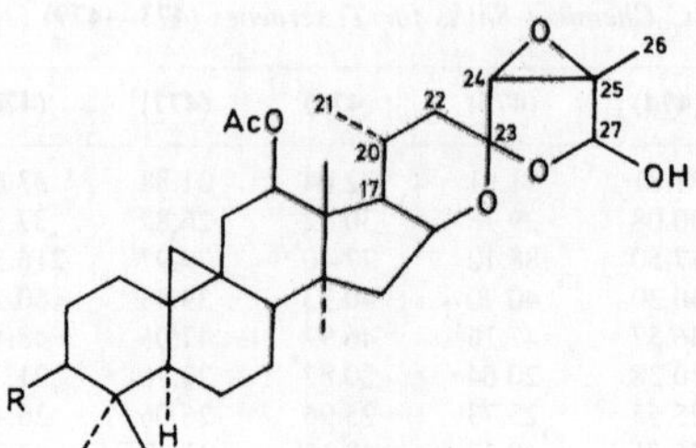

(473) R = 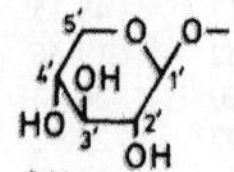

(474) R = OH

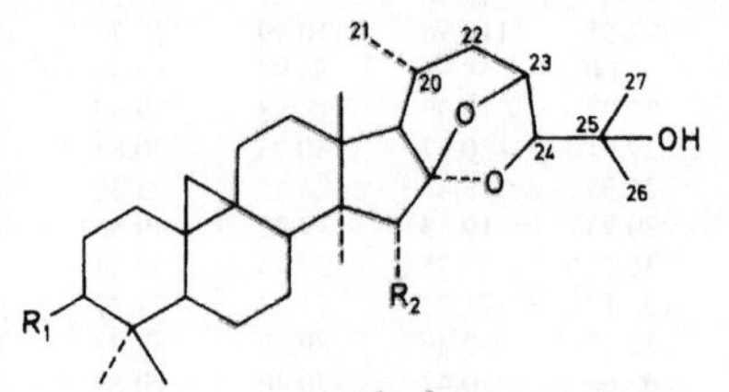

(475) R_1 = 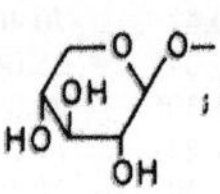 ; R_2 = OH

(476) $R_1 = R_2$ = OH

(477) $R_1 = R_2$ = OAc

(478) R_1 = O ; R_2 = OH

(479) $R_1 = R_2$ = O

The systems studied are oleanolic acid (**480**), hederagenin (**481**), queretaroic acid (**482**), gypsogenin (**483**), quillaic acid (**484**), 22α-hydroxyerythrodiol (**485**), 23-hydroxyerythrodiol (**486**) as well as their acetates. The SFORD spectra were recorded under conditions such that residual splittings arising from long-range $^{13}C—^{1}H$ spin-spin couplings could be observed for some quaternary carbons. One of the split signals could be identified as C-14 which is coupled to C (12)-H by a relatively large trans coupling constant ($^{3}J_{CCH}$). Similarly doublets were found for C-4 in (**483**) and (**484**). In this case the residual splitting originated from the two-bond

Table 26. ^{13}C *Chemical Shifts of Olean-12-enes* (**480**—**486**) *and Their Peracetates (in Parenthesis)* (*140*)

Carbon	(**480**)	(**481**)	(**482**)	(**483**)	(**484**)	(**485**)	(**486**)
1	38.5 (38.1)	38.1 (37.7)	38.7 (38.1)	38.0 (37.7)	38.1 (37.8)	39.0 (38.3)	38.3 (37.9)
2	27.1 (23.6)	26.4 (23.0)	27.0 (23.4)	26.0 (22.5)	26.0 (22.5)	27.0 (23.6)	26.0 (23.0)
3	78.7 (80.7)	76.4 (74.3)	78.9 (80.7)	71.8 (73.2)	71.7 (73.2)	78.8 (80.7)	75.7 (74.6)
4	38.7 (37.5)	41.7 (40.6)	38.7 (37.7)	55.2 (54.2)	55.2 (54.2)	39.0 (37.7)	41.7 (40.6)
5	55.2 (55.2)	49.7 (47.7)	55.3 (55.2)	48.0 (47.7)	48.1 (47.9)	55.6 (55.2)	49.3 (47.9)*
6	18.3 (18.2)	18.5 (18.1)	18.5 (18.2)	20.7 (20.3)	20.7 (20.3)	18.6 (18.2)	18.3 (18.0)
7	32.6 (32.6)	32.4 (32.3)	32.8 (32.6)	32.3 (32.2)	32.3 (32.3)	32.8 (32.5)	32.3 (32.2)
8	39.3 (39.3)	39.3 (39.3)	39.3 (39.3)	39.6 (39.5)	39.7 (39.7)	40.2 (40.0)	39.7 (39.8)
9	47.6 (47.5)	47.5 (47.7)	47.7 (47.5)	47.5 (47.5)	46.7 (46.6)	47.9 (47.5)	47.4 (47.7)*
10	37.0 (36.9)	36.9 (36.8)	37.2 (36.9)	36.0 (35.7)	35.9 (35.8)	37.1 (36.8)	36.8 (36.7)
11	23.1 (23.0)	23.1 (23.0)	23.4 (23.4)	23.0 (22.8)	23.2 (23.2)	23.8 (23.6)	23.6 (23.6)
12	122.1 (122.1)	122.2 (122.0)	122.8 (122.7)	121.9 (121.7)	122.2 (123.0)	123.0 (123.8)	122.1 (122.6)
13	143.4 (143.6)	143.6 (143.6)	143.2 (142.9)	143.6 (143.6)	142.8 (141.9)	143.2 (141.9)	144.2 (143.6)
14	41.6 (41.6)	41.7 (41.6)	41.7 (41.4)	41.7 (41.5)	41.4 (41.0)	41.5 (41.7)	41.7 (41.6)
15	27.7 (27.7)	27.7 (27.7)	27.7 (27.5)	27.6 (27.6)	35.4 (32.3)	25.3 (25.0)	25.6 (25.6)
16	23.4 (23.6)	23.4 (23.4)	23.4 (23.4)	23.3 (23.2)	74.7 (76.0)	14.7 (16.7)	21.9 (22.2)
17	46.6 (46.6)	46.7 (46.6)	46.8 (46.4)	46.6 (46.5)	48.7 (47.5)	43.0 (40.1)	36.8 (35.8)
18	41.3 (41.3)	41.3 (41.3)	40.3 (40.7)	41.3 (41.2)	40.5 (40.3)	42.4 (41.9)	42.5 (42.6)
19	45.8 (45.8)	45.9 (45.8)	40.3 (41.4)	45.8 (45.7)	46.4 (46.1)	46.0 (44.1)	46.6 (46.2)
20	30.6 (30.6)	30.7 (30.6)	35.2 (33.7)	30.7 (30.5)	30.3 (30.3)	31.5 (31.5)	30.9 (30.9)
21	33.8 (33.8)	33.9 (33.8)	32.8* (32.3)*	33.8 (33.7)	35.4 (35.0)	42.4 (38.9)	34.2 (34.0)
22	32.3 (32.3)	32.4 (32.3)	32.1* (31.7)*	32.1 (31.9)	30.3 (30.9)	75.8 (71.7)	30.9 (31.4)
23	28.1 (28.0)	71.3 (65.3)	28.1 (28.0)	207.0 (204.0)	207.0 (204.0)	28.2 (28.0)	70.3 (65.4)
24	15.6* (16.8)	11.6 (13.1)	15.7 (16.7)	8.9 (9.5)	9.0 (9.5)	15.6 (16.7)	11.6 (13.1)
25	15.3* (15.3)	15.7 (15.8)	15.4 (15.4)	15.5 (15.5)	15.7 (15.6)	15.6 (15.5)	15.9 (16.0)
26	16.8 (16.8)	16.9 (16.8)	16.9 (16.7)	16.8 (16.7)	16.9 (16.8)	16.7 (16.7)	16.7 (16.7)
27	26.0 (25.8)	26.0 (25.8)	26.0 (25.9)	26.0 (25.8)	27.0 (26.4)	26.4 (26.2)	26.0 (25.8)
28	177.9 (177.8)	178.2 (177.8)	178.5 (177.4)	177.9 (177.7)	177.2 (175.6)	70.1 (66.0)	69.1 (70.7)
29	33.1 (33.1)	33.1 (33.1)	28.9 (27.8)	33.1 (33.0)	32.7 (33.1)	33.4 (33.0)	33.2 (33.1)
30	23.6 (23.6)	23.6 (23.6)	65.8 (67.7)	23.6 (23.5)	24.6 (24.2)	24.8 (24.3)	23.6 (23.6)
$CO_2\underline{Me}$	51.3 (51.4)	51.6 (51.3)	50.6 (50.2)	51.5 (51.3)	51.9 (52.0)	—	—
$\underline{C}OMe$	(170.5)	(170.1)	(170.6)	(169.8)	(169.6)	(170.0)	(170.4)
		(70.5)	(170.9)		(170.0)	(170.7)	(170.7)
						(170.7)	(171.1)

* Assignments may have to be reversed.

coupling constant involving the formyl proton ($^2J_{CCHO} > 20$ Hz). δ_C values for compounds (**480**—**486**) including their peracetates are tabulated (Table 26).

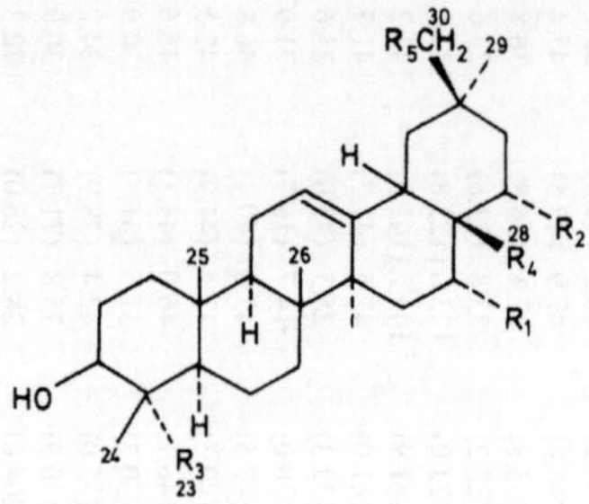

	R_1	R_2	R_3	R_4	R_5
(**480**)	H	H	Me	CO_2Me	H
(**481**)	H	H	CH_2OH	CO_2Me	H
(**482**)	H	H	Me	CO_2Me	OH
(**483**)	H	H	CHO	CO_2Me	H
(**484**)	OH	H	CHO	CO_2Me	H
(**485**)	H	OH	Me	CH_2OH	H
(**486**)	H	H	CH_2OH	CH_2OH	H

Complete ^{13}C chemical shifts have been given for some urs-12-ene triterpenes, methyl ursolate (**487**) and derivatives (**488**—**490**), complemented by α-amyrin and its acetate. The corresponding olean-12-ene systems, methyl oleanolate (**491**), methyl maslinate (**492**) and methyl

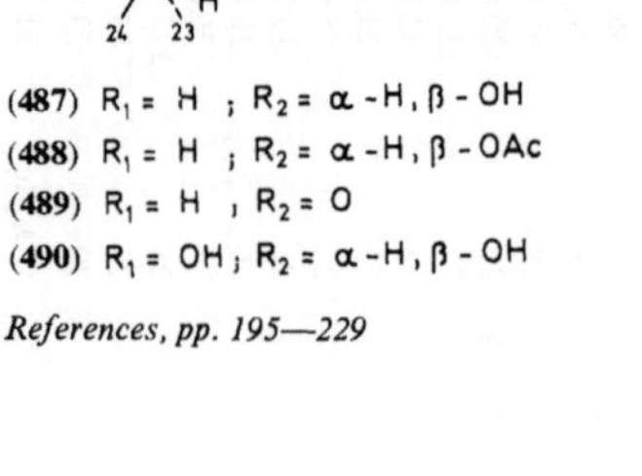

(**487**) R_1 = H ; R_2 = α-H, β-OH
(**488**) R_1 = H ; R_2 = α-H, β-OAc
(**489**) R_1 = H , R_2 = O
(**490**) R_1 = OH ; R_2 = α-H, β-OH

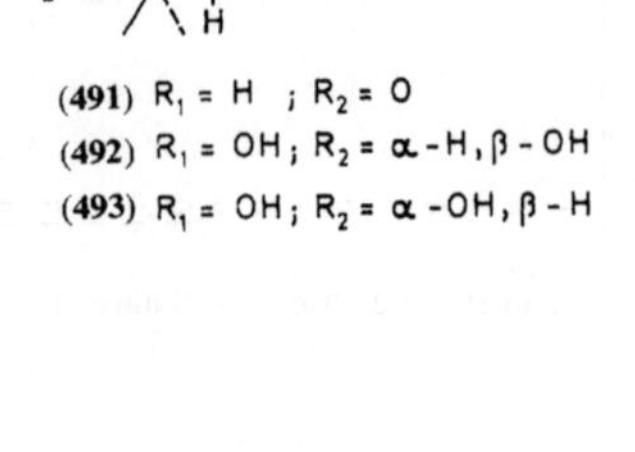

(**491**) R_1 = H ; R_2 = O
(**492**) R_1 = OH ; R_2 = α-H, β-OH
(**493**) R_1 = OH ; R_2 = α-OH, β-H

Table 27. ^{13}C *Chemical Shifts for Urs-12-Enes* (**487—490**) *and Olean-12-Enes* (**491—493**) *and for α- and β-Amyrin and Their Acetates* (*141*)**

Carbon	α-Amyrin	α-Amyrin acetate	(**487**)	(**488**)	(**489**)	(**490**)	β-Amyrin	β-Amyrin acetate	(**491**)	(**492**)	(**493**)
1	38.7	38.4	32.8	38.3	39.3	46.8	38.5	38.2	39.1	46.4	41.7
2	27.2	23.6	27.3	23.6	33.9	68.9	27.0	23.6	34.2	68.8	66.5
3	78.8	80.7	78.8	80.7	216.5	83.8	78.9	80.7	216.8	83.8	78.9
4	38.7	37.6	38.8	37.6	47.1	39.1	38.7	37.6	47.3	39.1	38.5*
5	55.2	55.3	55.4	55.3	55.2	55.4	55.1	55.3	55.2	55.3	48.1
6	18.3	18.3	18.4	18.1	19.5	18.4	18.3	18.3	19.5	18.3	18.1
7	32.9	32.8	33.0	32.8	32.4	32.9	32.6	32.6	32.1	32.6	32.5
8	40.0	40.1	39.6	39.5	39.1	39.6	39.7	39.7	39.1	39.1	39.7*
9	47.7	47.6	47.5	47.4	46.6	47.5	47.6	47.6	46.7	47.5	47.4
10	36.9	36.8	37.0	36.8	36.6	38.3	37.0	36.8	36.6	38.3	38.3*
11	23.3	23.2	23.3	23.2	23.4	23.4	23.4	23.4	23.0	23.1	23.4
12	124.3	124.1	125.5	125.4	125.0	125.3	121.7	121.5	121.9	122.0	122.1
13	139.3	139.4	138.0	138.0	137.9	138.1	145.0	144.9	143.5	143.6	143.8
14	42.0	42.1	42.0	41.9	41.9	42.1	41.7	41.7	41.7	41.7	41.9
15	28.7	28.7	28.2	28.1	27.9	28.0	28.3	28.3	27.6	27.6	27.7
16	26.6	26.7	24.3	24.2	24.1	24.3	26.2	26.2	23.5	23.5	23.2
17	33.7	33.8	48.1	48.0	47.9	48.1	32.5	32.5	46.7	46.6	46.8
18	58.9	59.0	52.8	52.8	52.8	52.8	47.2	47.2	41.2	41.3	41.3
19	39.6	39.7	39.1	38.9	38.8	39.1*	46.8	46.8	45.7	45.8	46.0
20	39.6	39.7	38.8	38.9	38.8	38.9*	31.1	31.1	30.5	30.7	30.7
21	31.2	31.3	30.7	30.7	30.5	30.7	34.8	34.8	33.7	33.8	34.0
22	41.5	41.5	36.7	36.6	36.6	36.7	37.2	37.1	32.1	32.3	32.5
23	28.1	28.1	28.2	28.1	26.5	28.7	28.1	28.1	26.4	28.6	28.5
24	15.6	16.8	15.5*	16.9	21.3	17.0	15.5	16.8	21.3	16.8	21.9
25	15.6	15.7	15.7*	15.5	15.1	17.0	15.5	15.7	14.8	16.8	16.4
26	16.8	16.8	16.9	16.9	16.8	17.0	16.8	16.8	16.7	16.8	17.0
27	23.3	23.2	23.6	23.6	23.4	23.7	26.0	26.0	25.7	26.0	26.2
28	28.1	28.1	177.7	177.6	177.3	177.9	27.3	27.0	177.7	178.0	178.1
29	17.4	17.5	16.9	17.1	16.8	17.0	33.2	33.4	33.0	33.1	33.2
30	21.3	21.4	21.2	21.2	21.1	21.2	23.6	23.6	23.5	23.5	23.6
$CO_2\underline{C}H_3$			51.4	51.3	51.2	51.5			51.3	51.5	51.5
$OCO\underline{C}H_3$		21.2		21.2							
$O\underline{C}OCH_3$		170.4		170.5							

* Assignments may have tc be interchanged. ** Reviewers corrected assignments for C-11, C-16 and C-29 in α-amyrin and related compounds.

3-*epi*-maslinate (**493**) were studied in conjunction with β-amyrin and its acetate as a correlation basis (*141*). Because erroneous assignments were made for C-11, C-16 and C-29 for α-amyrin, these carbons were also incorrectly designated in methyl ursolate (**487**) and derivatives. This illustrates the pitfalls of chemical shift correlation as an assignment technique. The error is propagated through the whole series of related systems. Suitable alterations have therefore been made by the present authors in Table 27 which lists chemical shifts for compounds (**487**—**493**) and α- and β-amyrin and its acetates. SEO et al. (Ref. 141) later reassigned these carbons on the basis of ^{13}C incorporation experiments using doubly labelled acetate as a precursor (*142*). In this revision the assignments for C-15 and C-16 are reversed with respect to those of Ref. 141.

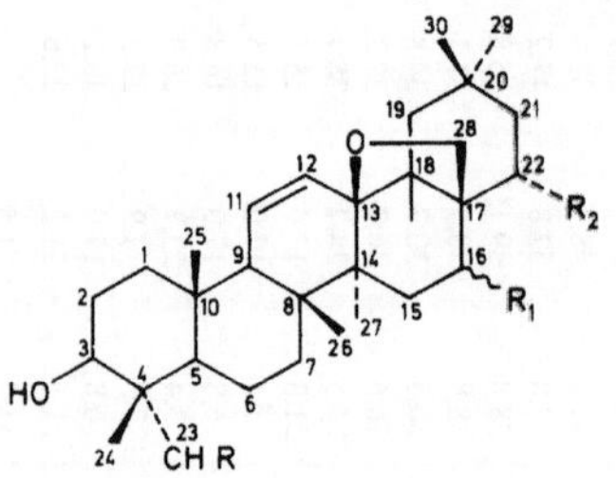

(**494**) R_1 = H ; R_2 = H ; R_3 = H
(**495**) R_1 = α-OH ; R_2 = H ; R_3 = OH
(**496**) R_1 = β-OH ; R_2 = H ; R_3 = H
(**497**) R_1 = β-OH ; R_2 = H ; R_3 = OH
(**498**) R_1 = α-OH ; R_2 = OH ; R_3 = OH

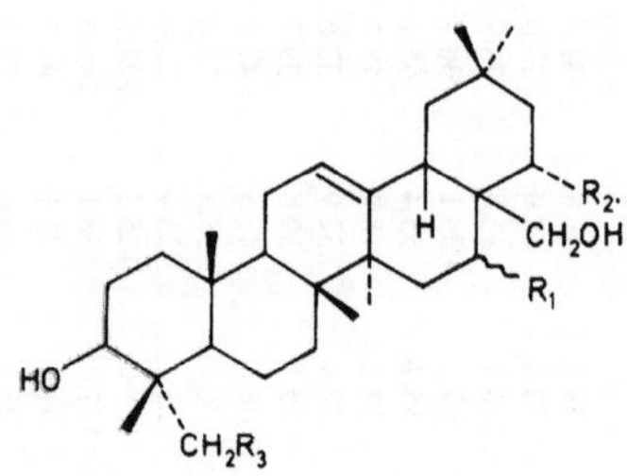

(**499**) R_1 = α-OH ; R_2 = H ; R_3 = OH
(**500**) R_1 = β-OH ; R_2 = H ; R_3 = H
(**501**) R_1 = β-OH ; R_2 = H ; R_3 = OH
(**502**) R_1 = α-OH ; R_2 = OH ; R_3 = OH

The same group of workers reported ^{13}C data for saikogenins: 16-dehydroxysaikogenin (**494**), saikogenin G (**495**), saikogenin E (**496**), saikogenin F (**497**) and 22α-hydroxysaikogenin (**498**) (*143*). Moreover, some related olean-12-ene systems (**499**—**502**) were examined and the effect of C-16 hydroxylation on ^{13}C shieldings discussed in terms of stereochemical interactions (*143*). Table 28 contains chemical shifts for systems (**494**—**502**). While previous X-ray studies pointed to a chair conformation for ring D in 16α- and β-hydroxyolean-12-enes, a slightly deformed chair was inferred for the saikogenins from ^1H n.m.r. spectroscopy. If this were correct, it should manifest itself in a deviation of the substituent effects for 16α-OH and 16β-OH from what would be expected in the case

Table 28. ^{13}C *Chemical Shifts for Saikogenins* (**494**)—(**498**) *and Oleanenes* (**499**—**502**) (*133*)

Carbon	(**494**)	(**495**)	(**496**)	(**497**)	(**498**)	(**499**)	(**500**)	(**501**)	(**502**)
1	38.0	38.0	38.4	38.2	38.2	38.4	38.8	38.5	38.5
2	26.2	26.1	26.9	26.1	26.2	26.4	26.9	26.7	26.3
3	75.9	75.6	78.7	75.4	75.2	76.7	78.9	76.5	76.0
4	41.9	42.0	39.0	42.0	42.2	41.9	38.8	42.0	41.8
5	49.3	49.2	54.9	49.1	49.0	49.9	55.3	49.9	49.3
6	17.7	17.7	17.9	17.8	17.8	18.5	18.4	18.7	18.4
7	31.1	31.1	31.5	31.3	31.2	32.6	32.6	32.6	32.6
8	41.6	41.4	41.8	41.9	41.8	40.0	39.9*	40.0*	39.8
9	53.3	52.5	52.7	52.7	52.8	47.0	46.8	47.0	46.7
10	36.3	36.3	36.4	36.3	36.4	36.9	36.9	37.0	36.0
11	132.5	132.6	132.9	133.0	133.2	23.4	23.6	23.7	23.5
12	130.7	130.5	129.7	129.7	130.3	122.6	122.6	122.7	122.9
13	85.2	85.1	84.1	84.3	85.5	143.1	142.8	143.0	142.7
14	43.9	43.0	45.4	45.5	43.8	41.6	43.7	43.9	41.8
15	25.7*	34.7	35.1*	35.0*	34.3	34.7	35.8	36.2	33.4
16	25.3*	77.3	64.7	64.6	70.5*	74.9	67.7	68.0	67.9
17	41.6	44.9	46.3	46.4	47.5	40.6	40.0*	40.4*	43.9
18	51.1	50.6	51.7	51.9	50.4	42.8	44.7	44.9	42.4
19	37.2	38.0	37.5	37.5	37.4	47.0	46.7	47.0	47.2
20	31.7	31.7	31.5	31.6	33.2	30.4	30.8	30.9	31.4
21	34.9	36.5	34.3*	34.3*	45.7	35.4	33.6	33.7	44.8
22	30.9	30.5	25.1	25.2	74.3*	26.8	26.2	26.2	75.5
23	70.7	70.3	27.8	70.0	69.6	71.9	28.1	71.7	70.8
24	11.1	11.1	15.0	11.2	11.3	11.5	15.6	11.6	11.7
25	18.2	18.3	17.9	18.3	18.4	16.1	15.6	16.0	16.0
26	19.3	19.0	19.5	19.5	19.1	17.2	16.7	16.9	16.2
27	19.3	17.9	20.7	20.7	18.1	27.3	26.9	26.9	27.0
28	77.1	77.3	72.4	72.5	76.9	70.8	70.9	71.1	71.1
29	33.7	33.4	33.6	33.6	33.2	32.8	33.2	33.1	33.2
30	23.6	24.2	23.7	23.8	25.2	25.5	23.9	24.0	24.9

* Assignments may have to be reversed.

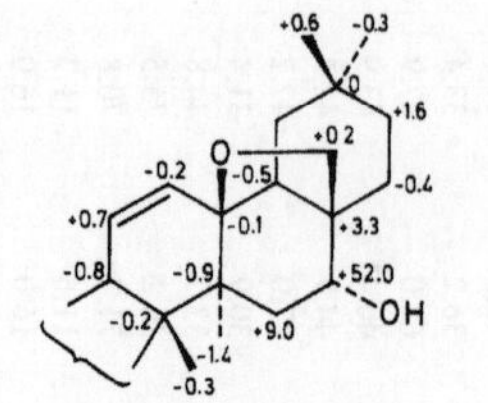

(494 → 494-16α-OH)
(495a)

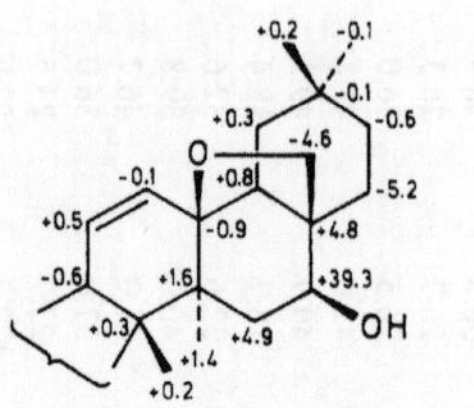

(494 → 494-16β-OH)
(497a)

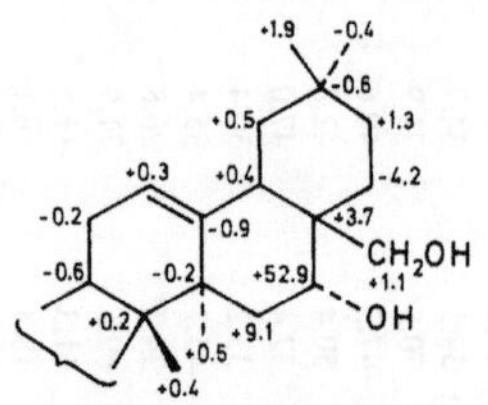

(486 → 486-16α-OH)
(499a)

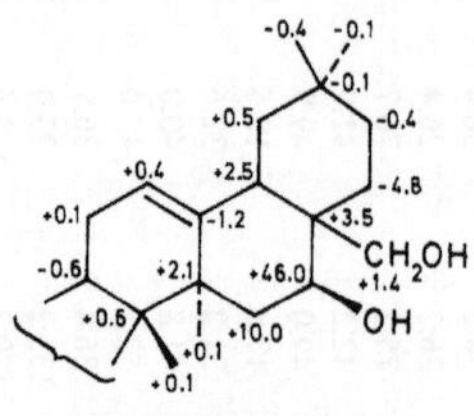

(486 → 486-16β-OH)
(501a)

of a pure chair conformation. The observed C-16 hydroxyl substituent effects are illustrated by the relative shifts indicated in partial formulae (**495a, 497a, 499a** and **501a**). The results for the bridged 13β, 28-epoxyoleanene (saikogenin) indeed point to ring-D distortion, which is reflected by an anomalous "syn-diaxial δ effect", exerted on the C-27 methyl resonance by 16α-OH. As a result C-27 experiences an upfield shift ($\Delta\delta_C$-1.4) instead of the common deshielding in the case of a true diaxial arrangement (**495a**). Introduction of a 16β-hydroxy group in (**494**) produces large upfield shifts at C-29 and C-28 in γ position.

In addition to the saikogenins, ^{13}C data have become available for some glucosides related to the former class of triterpenes, namely saikosaponin D (**503**), saikosaponin A (**504**) and saikosaponin C (**505**). These were used to deduce the structure of saikosaponin F (**506**) (*144*). Chemical shifts for compounds (**503—506**) are found in Table 29.

The structure previously assumed for pristimerin (**507**) was found to be incompatible with its ^{13}C spectrum (*145*). Inversion-recovery ^{13}C spectroscopy permitted classification of the various types of carbons. Partial and group assignments are given along with the formula.

Table 29. [13]*C Chemical Shifts for Saikosaponins* (**503—506**) (*144*)*

Carbon	(**503**)	(**504**)	(**505**)	(**506**)
1	38.8**	38.9	38.8	39.1
2	25.8	25.7	26.4	26.4
3	82.5	82.5	89.2	89.3
4	43.8	43.5	39.7	39.5
5	48.1	48.0	55.7	56.1
6	17.8	17.8	18.1	18.7
7	31.8	31.9	32.2	33.4
8	42.1	42.4	42.4	40.4**
9	53.2	53.2	53.1	47.4
10	36.6	36.6	36.6	37.1
11	131.9	132.0	132.0	24.0
12	131.9	131.0	131.1	122.8
13	85.1	84.0	84.0	144.0
14	43.7	46.0	45.9	44.1
15	35.7	36.2	36.2	36.8
16	77.5	64.4	64.4	67.2
17	45.5	46.9	47.0	41.1**
18	51.5	52.4	52.3	44.9
19	38.7**	38.2	38.1	47.4
20	31.8	31.6	31.6	31.0
21	37.0	35.0	34.9	34.4
22	31.2	25.6	25.7	26.4
23	65.2	65.1	28.1	28.4
24	12.8	12.7	16.3	17.0
25	18.7	18.6	18.1	15.7
26	19.4	19.8	19.8	17.1
27	18.1	20.8	20.8	27.1
28	77.8	72.9	72.9	69.3
29	33.7	33.7	33.7	33.3
30	24.5	23.9	23.9	24.3
1′	105.4	105.3	106.1	106.1
2′	71.1	71.7	75.2	75.2
3′	85.1	85.0	76.8	76.9
4′	71.9	71.8	80.2	80.2
5′	70.9	70.8	75.2	75.5
6′	16.9	16.9	69.2	69.3
1″	105.8	105.7	102.6	102.6
2″	75.5	75.4	72.6	72.6
3″	78.1	78.0	72.2	72.3
4″	71.9	71.8	73.8	73.8
5″	78.1	78.0	70.5	70.5
6″	63.0	62.9	18.1	18.2
1‴			104.8	104.8
2‴			74.7	74.7
3‴			78.2	78.3
4‴			71.8	71.9
5‴			77.9	78.0
6‴			62.9	62.9

* In pyridine-d_5. ** Assignments may have to be reversed.

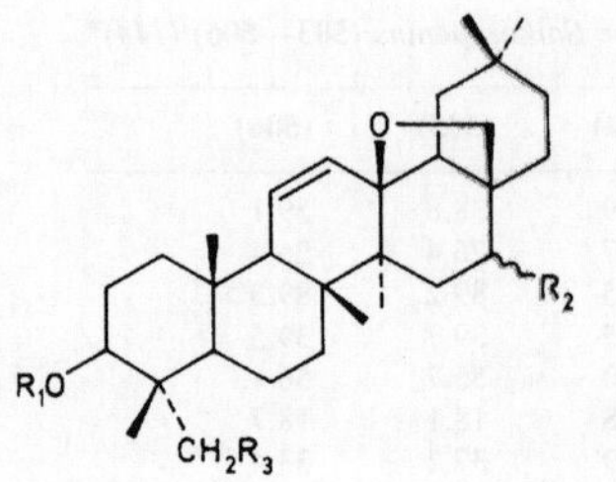

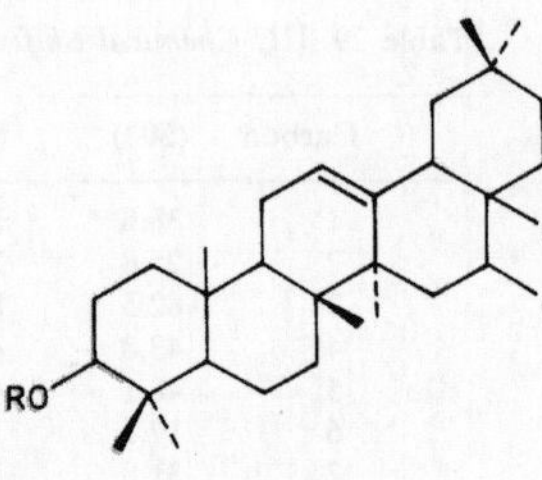

R = Rham - (Glc) - Glc

(506)

(503) R_1 = Glc - Fuc ; R_2 = α-OH ; R_3 = OH

(504) R_1 = Glc - Fuc ; R_2 = β-OH ; R_3 = OH

(505) R_1 = Rham - (Glc) - Glc ; R_2 = β-OH ; R_3 = H

Glc - Fuc =

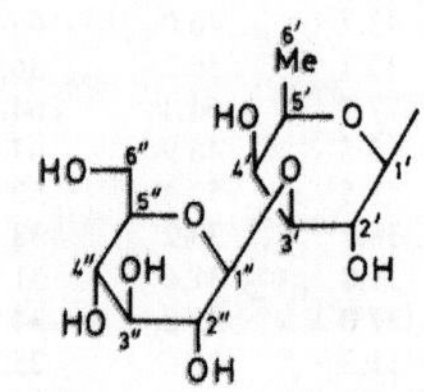

Rham - (Glc) - Glc =

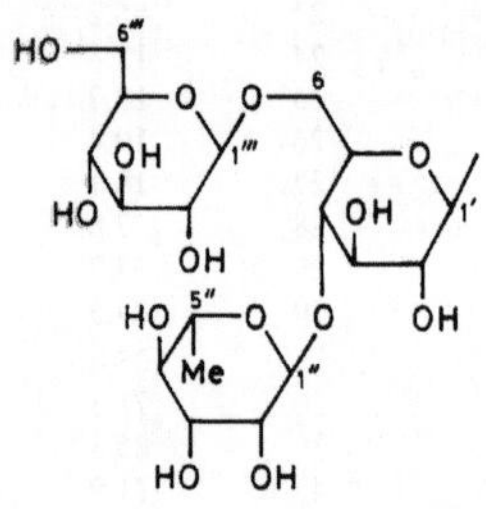

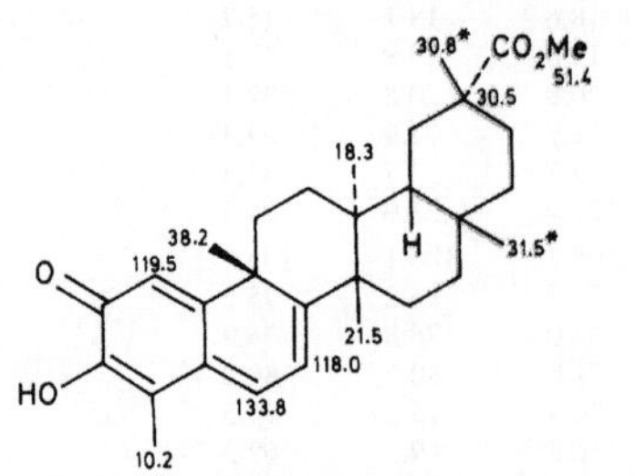

CH_2 : 28.6 ; 29.6 ; 29.8 ; 30.4 ; 33.5 ; 34.8 ; 36.3

C : 38.3 ; 39.3 ; 40.3 ; 44.9

>C= : 117.1 ; 127.2 ; 146.0 ; 164.0 ; 169.8 ; 178.1 ; 178.4

(507)

Steroids

Revised assignments have been reported for cholestane (**508**) and androstane (**509**) (*33*). A large number of keto steroids (*33*) and hydroxy

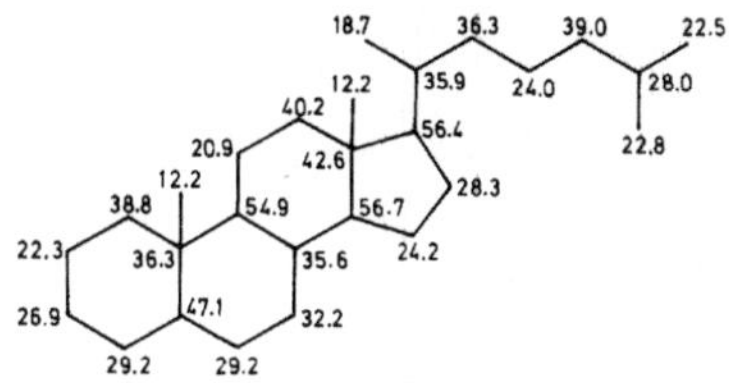

(508)

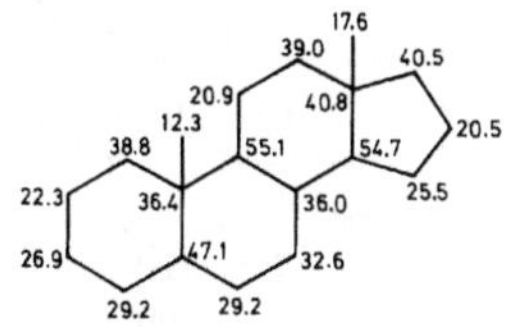

(509)

steroids (*145*) have been studied systematically with the objective of rationalizing the shielding effects exerted by the various substituents. Chemical shift data are provided in the respective formulae: ketoandrostanes (**510**—**521**) (*33*), hydroxyandrostanes and hydroxycholestanes (**522**—**544**) (*146*). Among the assignment procedures employed, specific deuteration and the concomitant isotope shifts turned out to be a particularly potent tool. In summary it was concluded that the shifts induced

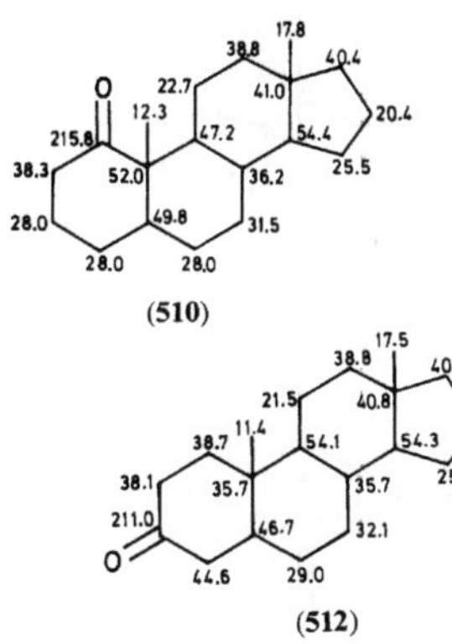

(510)

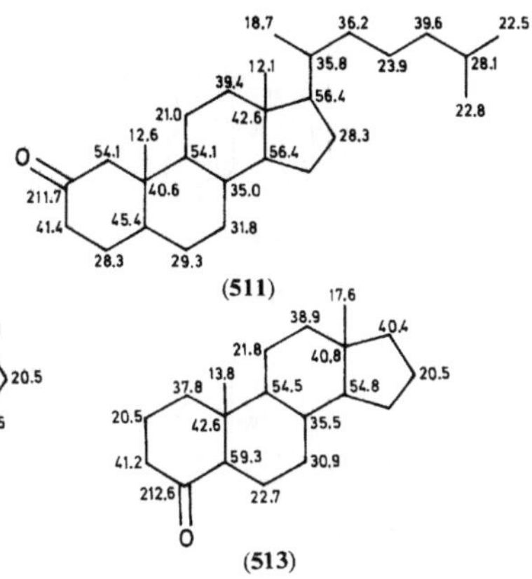

(511)

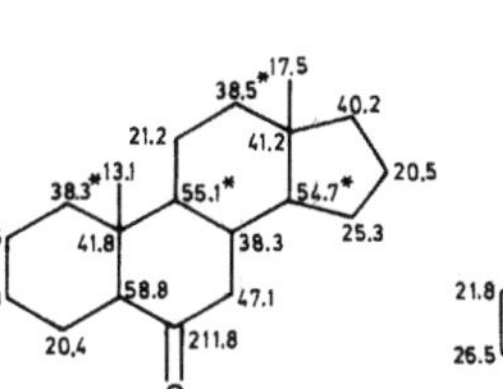

(512)

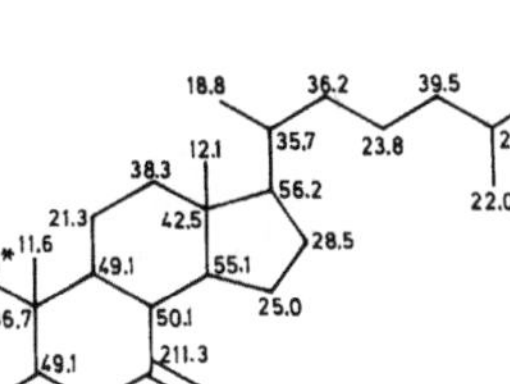

(513)

(514)

(515)

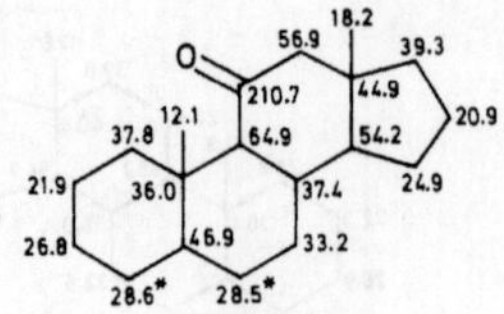

(516)

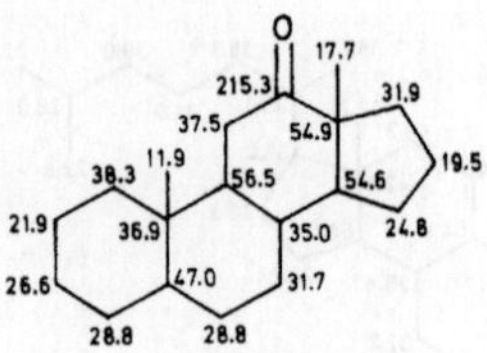

(517)

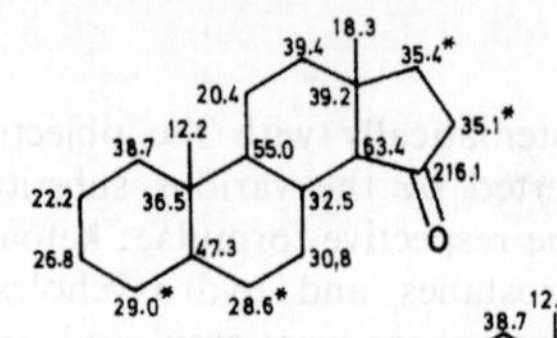

(518)

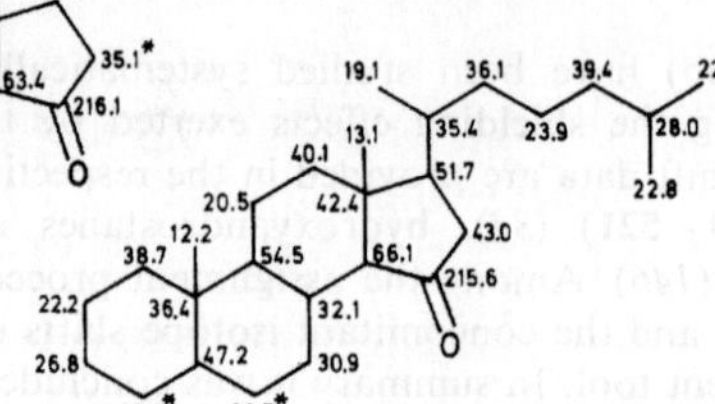

(519)

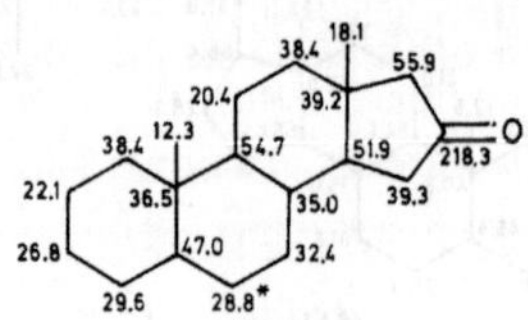

(520)

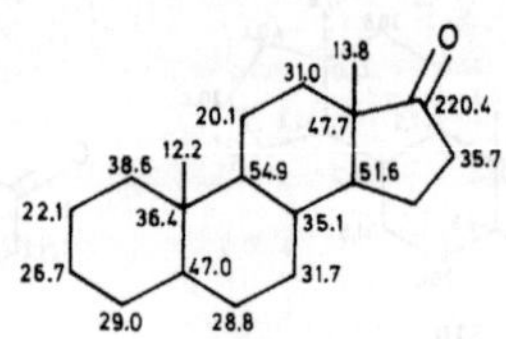

(521)

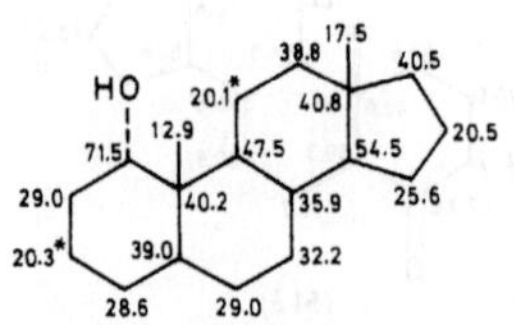

(522)

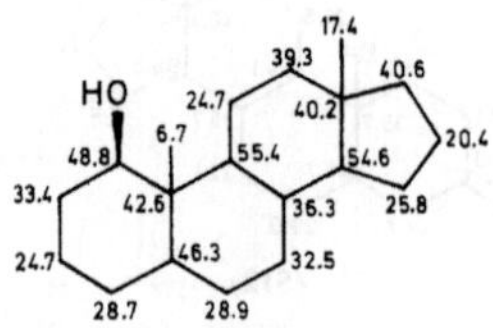

(523)

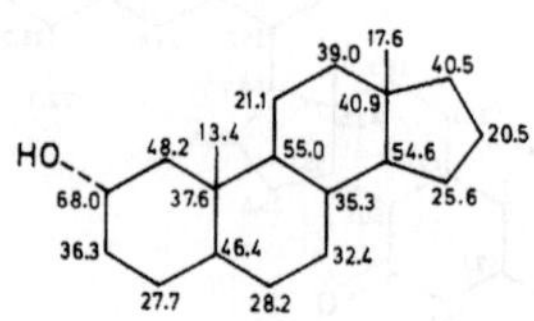

(524)

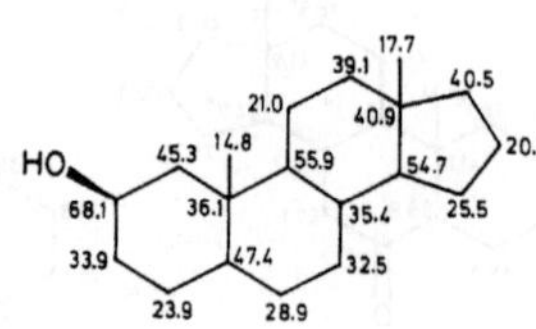

(525)

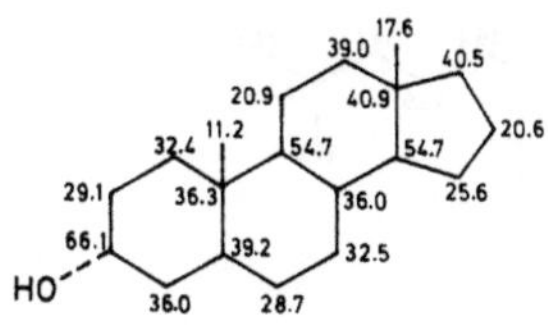

(526)

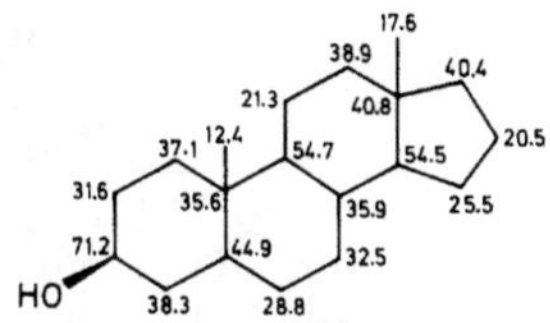

(527)

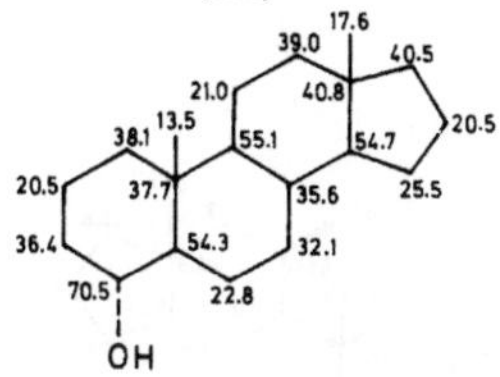

(528)

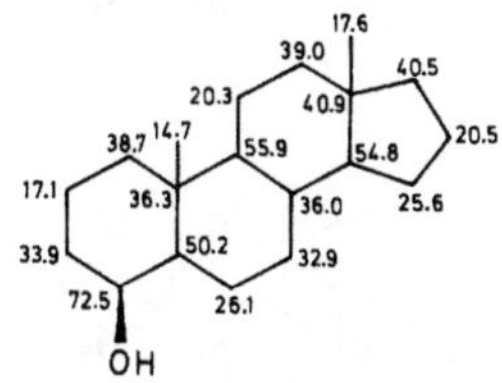

(529)

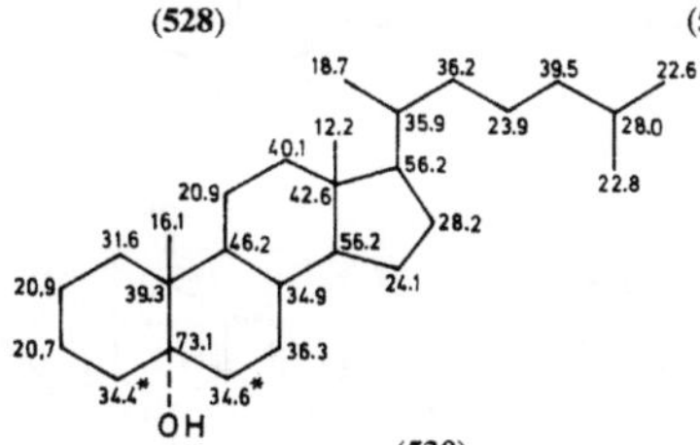

(530)

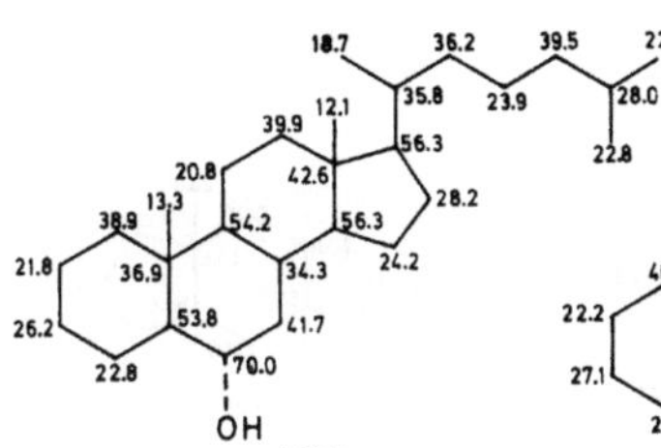

(531)

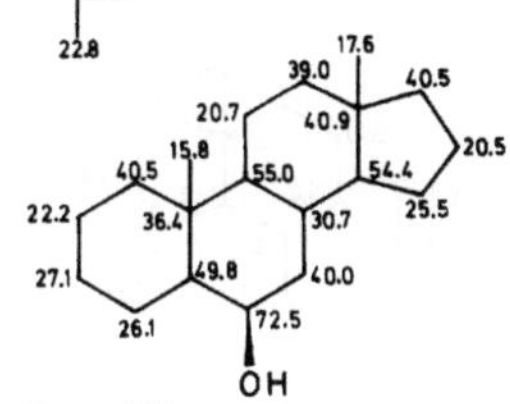

(532)

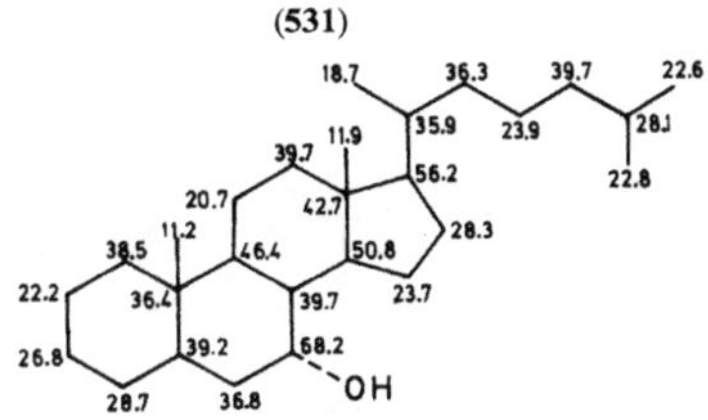

(533)

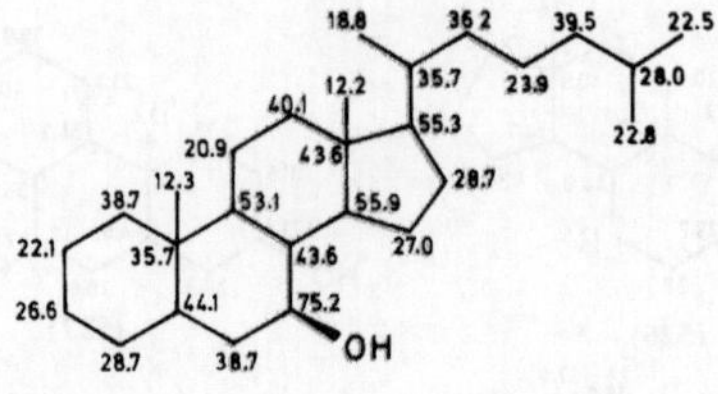

(534)

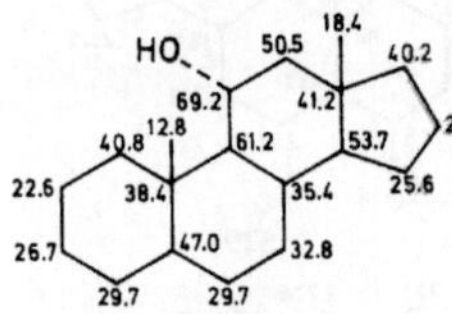

(535)

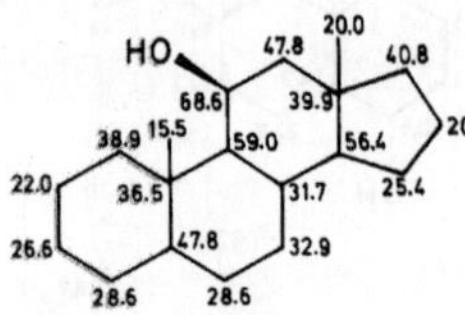

(536)

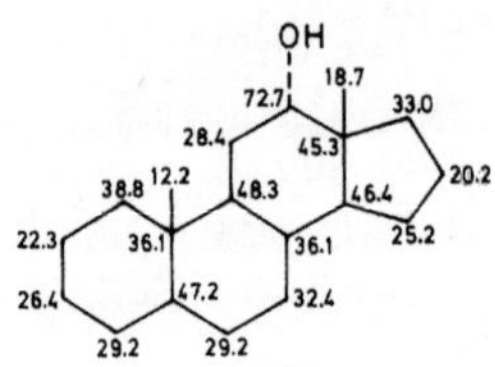

(537)

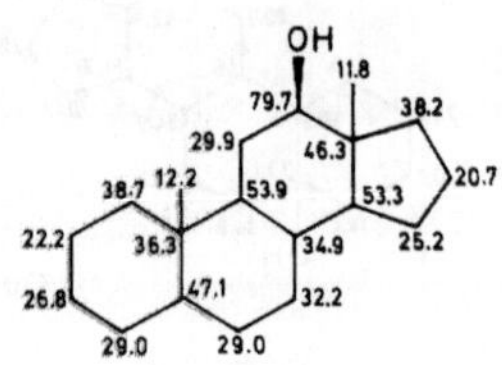

(538)

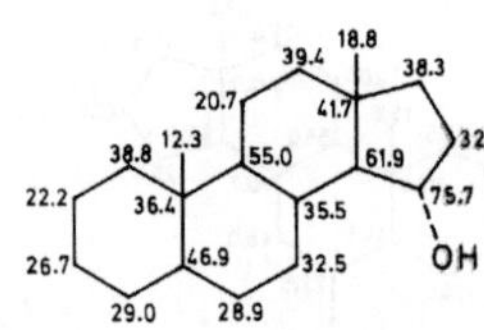

(539)

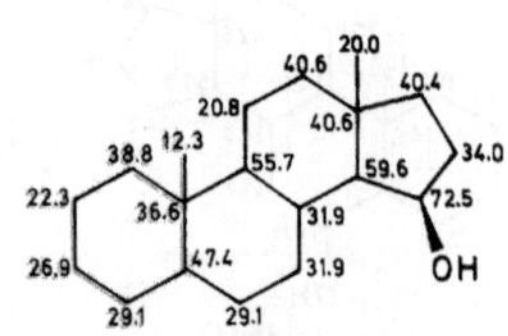

(540)

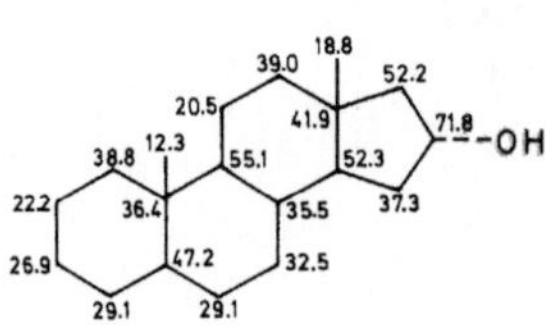

(541)

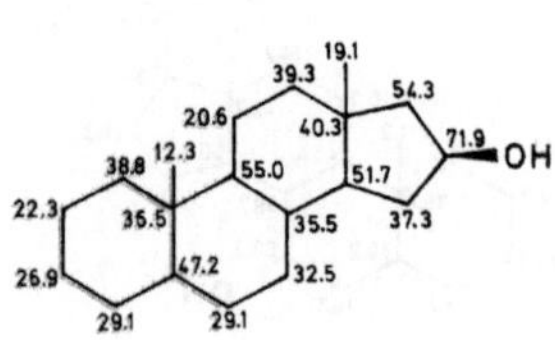

(542)

References, pp. 195—229

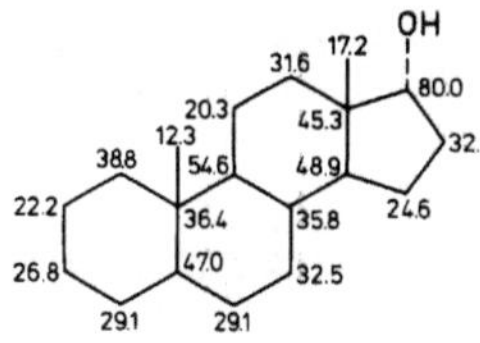

(543)

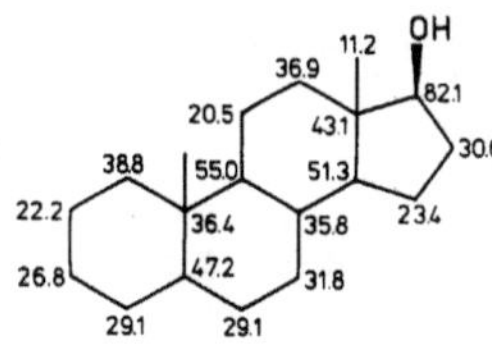

(544)

at the neighboring carbons upon introduction of a carbonyl or hydroxyl substituent, were a function of the relative orientation of these carbons with respect to the substitient. It was suggested that substituent effects on chemical shifts may serve as an aid in stereochemical studies of steroids and similar rigid cyclic systems as well as in identifying the site of substitution and in establishing the configuration of the substituted carbon.

Further members of the cholestane series examined by ^{13}C n.m.r. comprise some 6β-substituted 5α-cholestan-3β-diols (*147*) (**545**—**554**) and the androstane derivative (**555**). γ and δ effects produced by the 6β substituent were shown to depend on both the nature of the substituent and on its geometric orientation.

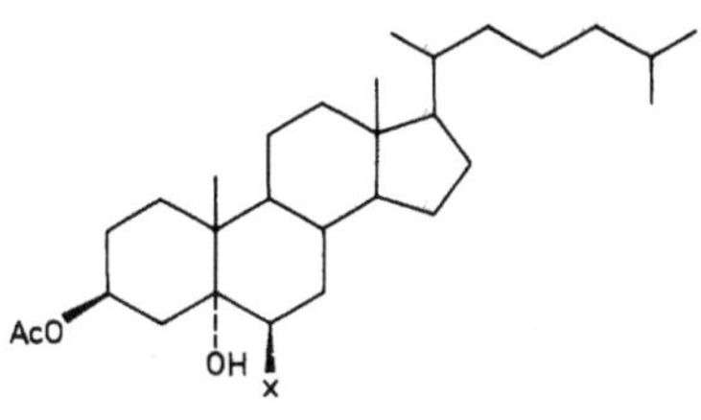

(**545**) X = H (**551**) X = Me
(**546**) X = D (**552**) X = OH
(**547**) X = F (**553**) X = OAc
(**548**) X = Cl (**554**) X = OMe
(**549**) X = Br
(**550**) X = I

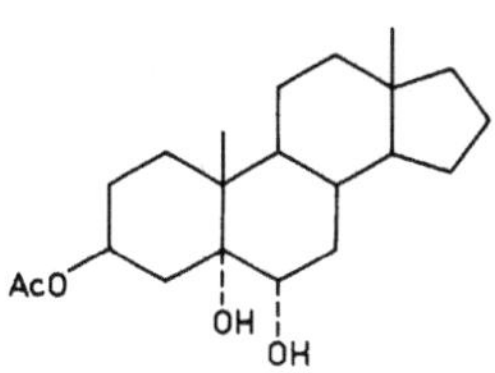

(**555**)

Remarkable is the fact that the δ-carbon-induced shift for C-19 (deshielding) increases with the bulkiness of the C-6 substituent as may be seen from the shift data in Table 30.

Table 30. ^{13}*C Chemical Shifts for 6β Substituted 5α-Cholestan-3β-Diols* (**545—554**) *and the Androstane Derivative* (**555**) (*147*)

Carbon	(545)	(546)	(547)	(548)	(549)	(550)	(551)	(552)	(553)	(554)	(555)
1	30.5	30.6	31.5	32.5	32.9	32.9	33.4	32.2	31.9	31.8	31.0
2	26.8	26.9	26.6	26.6	26.6	26.5	26.9	26.8	26.8	26.8	26.7
3	70.9	70.9	70.8	71.0	71.5	71.5	71.7	71.5	71.0	71.4	71.7
4	40.0	40.1	36.6	38.1	39.1	39.9	38.5	37.0	36.7	37.2	34.4
5	74.9	74.9	74.1 A	76.6	76.1	75.4	77.2	75.7	74.8	75.8	76.5
6	34.5	—	95.7 B	64.0	56.4	34.5	41.8	76.2	76.4	85.7	70.7
7	25.9	25.9	31.9 C	35.5	36.4	38.6	33.4	34.6	31.5	29.1	35.0
8	34.8	34.8	30.5	30.2	30.6	31.6	30.6	30.3	30.8	30.6	33.7
9	45.6	45.6	44.8	45.5	45.1	45.5	46.0	45.4	44.9	45.4	44.3
10	38.8	38.8	38.1 D	39.1	39.4	41.5	39.1	38.4	38.5	38.5	39.3
11	21.3	21.4	21.1	21.1	21.2	21.2	21.2	21.2	21.1	21.1	21.2
12	40.0	40.1	39.9	39.9	40.0	40.0	40.0	40.0	40.0	40.0	39.9
13	42.7	42.7	42.8	42.8	42.8	42.8	42.8	42.8	42.8	42.8	42.7
14	56.1	56.2	55.8	55.4	55.3	55.1	55.8	55.9	55.9	55.9	55.9
15	24.1	24.1	24.1	24.1	24.1	24.2	24.3	24.2	24.1	24.2	24.2
16	28.3	28.3	28.2	28.2	28.3	28.2	28.2	28.3	28.3	28.2	28.3
17	56.3	56.3	56.4	56.3	56.5	56.4	56.4	56.4	56.4	56.4	56.3
18	12.1	12.1	12.1	12.2	12.2	12.2	12.2	12.2	12.2	12.1	12.1
19	16.1	16.1	15.8 E	18.3	19.0	20.7	18.3	16.7	16.4	16.1	15.5
20	35.9	35.9	35.9	35.8	35.9	35.8	35.9	35.9	35.9	35.8	35.8
21	18.7	18.7	18.7	18.7	18.7	18.7	18.7	18.7	18.7	18.7	18.7
22	36.2	36.2	36.2	36.2	36.3	36.2	36.2	36.3	36.3	36.2	36.2
23	23.9	24.0	24.0	24.0	24.1	24.0	24.0	24.0	24.1	24.2	24.2
24	39.5	39.6	39.5	39.5	39.5	39.5	39.5	39.6	39.6	39.5	39.6
25	28.0	28.0	28.0	28.0	28.0	28.0	28.0	28.1	28.0	28.0	28.0
26	22.6	22.6	22.6	22.6	22.6	22.6	22.6	22.6	22.6	22.6	22.6
27	22.8	22.8	22.8	22.8	22.8	22.8	22.8	22.8	22.8	22.8	22.8
$OCOCH_3$	21.3	21.3	21.4	21.4	21.4	21.4	21.4	21.4	21.5	21.5	21.5
Other							17.4		21.5	57.7	

A, B, C, D: Doublets due to ^{13}C-^{11}F spin-spin coupling; A: $^{2}J_{CCF} = 20$ Hz, B: $^{1}J_{CF} = 180$ Hz, C: $^{2}J_{CCF} = 20$ Hz, D: $^{3}J_{CCCF} = 2$ Hz, E: $^{4}J_{CCCCF} = 10$ Hz.

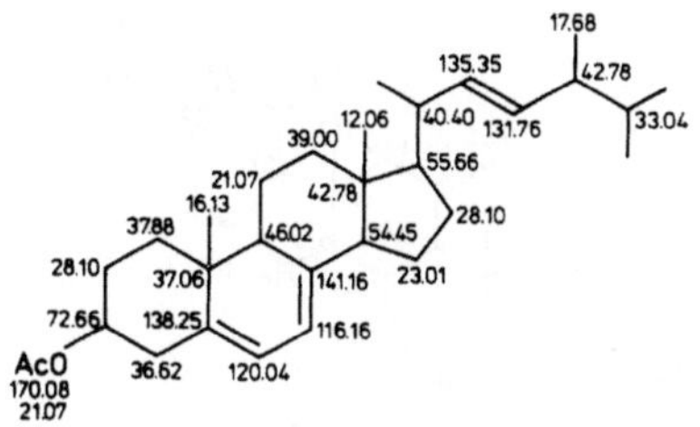

C-21,-26,-27: 19.67; 19.96; 21.07

(**556**)

A further revision of early assignments resulted from the observed labelling pattern in ergosterol (**556**), isolated from *Saccharomyces cerevisiae* inoculated with [1-^{13}C] acetate (*148*). This experiment has led to re-assignment of carbons 5, 8, 12, 16, 22 and 23. Full chemical shift analyses have been presented for some ergostadienes and -trienes (**557**—**561**) (*149*). Their chemical shifts are delineated in the formulae.

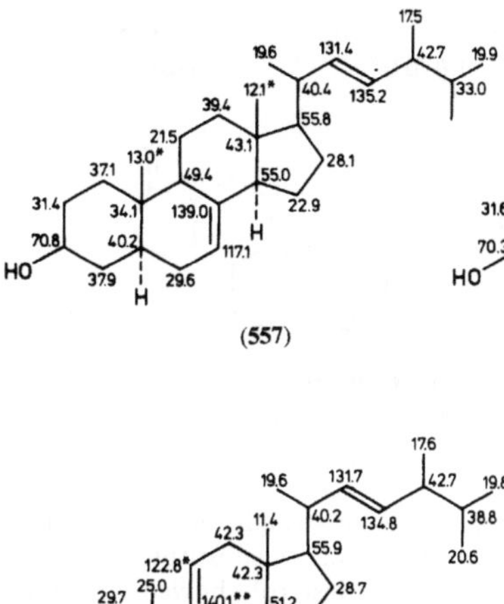

(**557**)

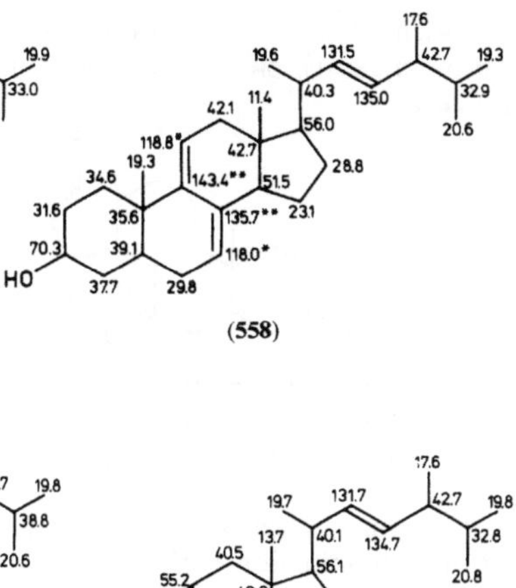

(**558**)

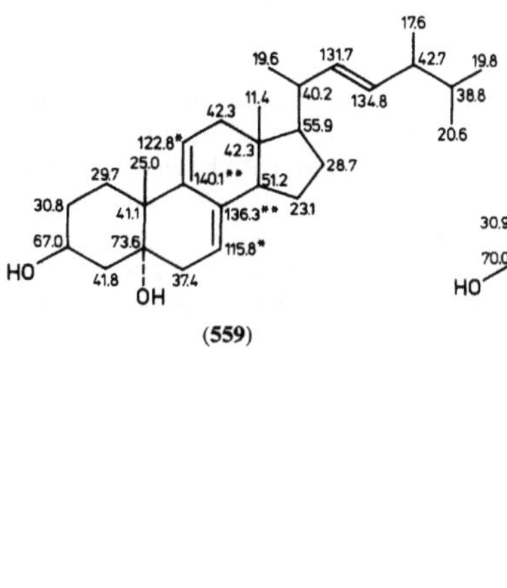

(**559**)

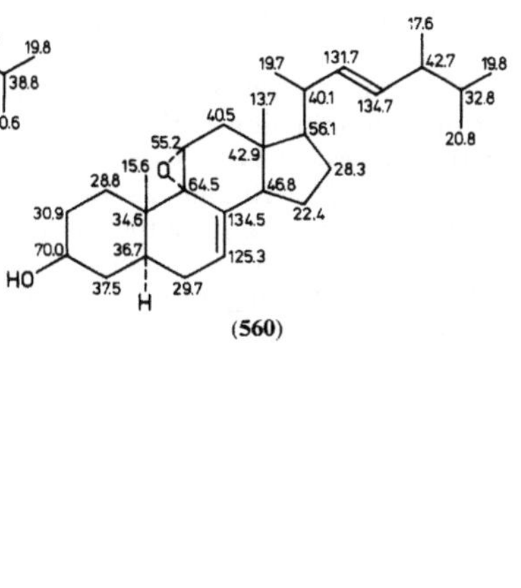

(**560**)

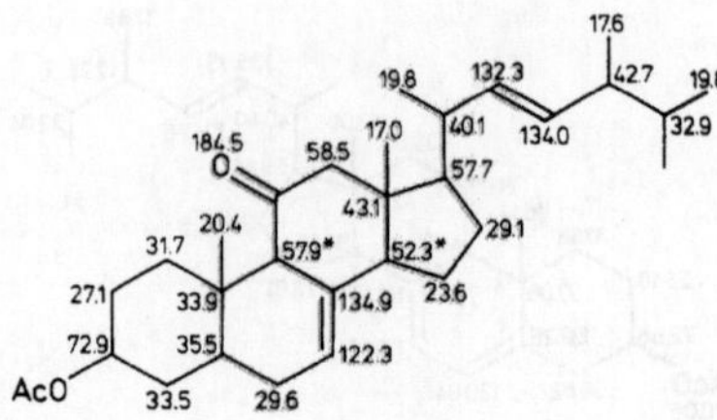

(561)

Unsaturated 3- and 7-oxoandrostanes (562—574) have been the subject of a further ^{13}C n. m. r. study (*150*). Chemical shifts are portrayed in Table 31.

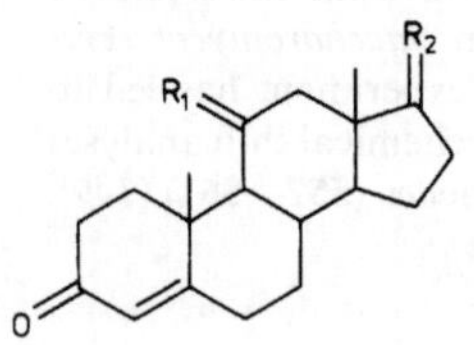

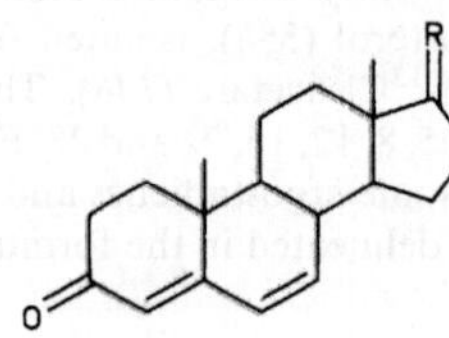

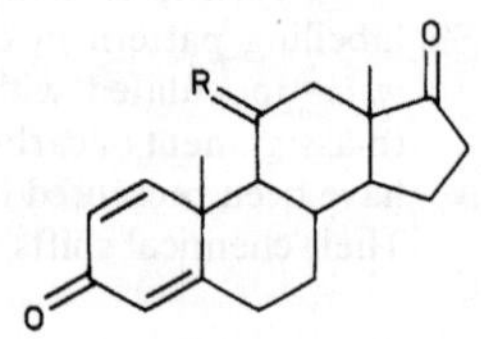

(562) $R_1=H_2$; $R_2=O$
(563) $R_1=H_2$; $R_2=\alpha$-H,β-OH
(567) $R_1=R_2=O$

(564) R = O
(565) R = α-H,β-OAc

(566) $R=H_2$
(568) R = O

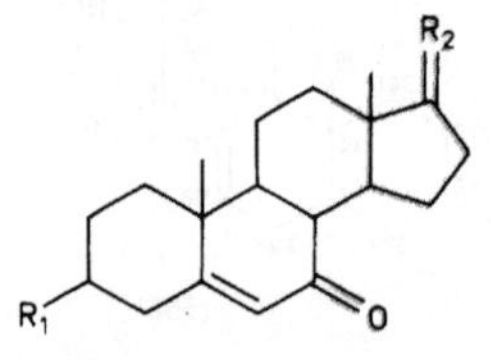

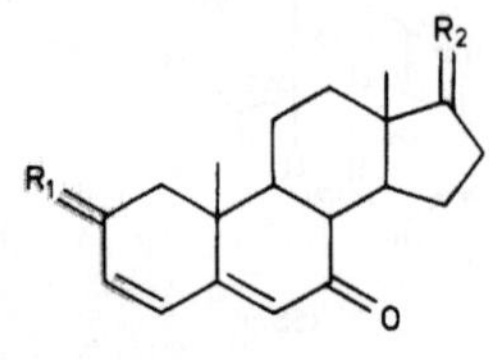

(569) R_1 = H ; R_2 = O
(570) R_1 = H ; $R_2=\alpha$-H,β-OAc
(571) R_1 = OAc; $R_2=\alpha$-H,β-OAc

(572) $R_1=R_2=H_2$
(573) R_1 = H ; $R_2=\alpha$-H,β-OAc
(574) R_1 = O ; $R_2=\alpha$-H,β-OAc

^{13}C n.m.r. spectroscopy revealed itself as a straightforward method for the determination of the configuration at C-22 in substituted steroids of the cholestane type. Comparison with cholesterol showed shielding effects on C-22, -20 and -23 and deshielding of C-17, -21 and -24 as a result of introducing substituents such as OH, NH_2 and N_3 at position 22. The observed β effects are characteristic and unequivocal and directly reflect the C-22 configuration: + 4.5 ppm for the 22-S series and + 6.8 ppm for the 22-R series. The effects on C-23 are even more pronounced:

Tabelle 31. ^{13}C *Chemical Shifts for Unsaturated Androstane Derivatives* (**562—574**) (*150*)

Compound	Carbon																		
	1	2	3	4	5	6	7	8	9	10	11	12	13	14	15	16	17	18	19
(**562**)	35.5	33.7	198.9	123.9	170.1	32.3	31.1	34.9	53.6	38.4	20.1	30.5	47.3	50.6	21.5	35.5	220.0	13.5	17.2
(**563**)	35.6	33.8	199.4	123.6	171.0	32.7	31.5	35.0	53.9	38.6	20.6	36.4	42.7	50.4	23.2	30.1	81.0	11.0	17.3
(**564**)	33.9	33.9	199.2	124.1	162.9	128.7	138.3	37.0	48.7	36.1	20.0	31.3	48.3	50.7	21.4	35.6	219.3	13.7	16.3
(**565**)	33.9	33.9	199.1	123.7	163.2	128.1	139.9	37.3	48.0	36.5	20.1	36.0	43.4	50.6	23.1	27.4	82.0	12.0	16.3
(**566**)	155.2	127.6	186.0	124.0	168.2	32.3	31.2	35.0	52.3	43.4	22.0	32.5	47.6	50.4	21.8	35.5	219.6	13.8	18.7
(**567**)	34.6	33.6	199.2	124.6	167.8	31.9	30.8	36.2	63.2	38.2	207.4	50.3	50.3	49.6	21.5	35.9	216.7	14.6	17.2
(**568**)	154.8	127.6	185.9	124.7	165.9	32.1	31.9	35.9	60.6	42.3	207.4	50.3	50.0	49.1	21.6	35.9	216.4	14.7	18.9
(**569**)	39.0	24.0	26.8	32.7	169.2	124.3	200.8	44.3	45.7	39.2	20.1	30.7	47.8	50.4	21.8	35.5	220.0	13.7	17.4
(**570**)	39.3	23.4	26.8	32.8	168.9	124.4	201.4	45.0	45.0	39.3	20.4	35.9	43.1	50.2	26.0	27.6	82.0	12.1	17.4
(**571**)	37.7	27.2*	71.9	35.7	164.2	126.3	200.8	44.9	44.9	38.3	20.6	35.9	43.0	49.7	25.8	27.4	81.7	12.0	17.2
(**572**)	32.9	23.4	136.6	127.7	161.1	124.1	202.0	46.4	48.8	36.3	20.7	39.3	41.7	49.9	27.7	21.2	38.0	17.4	16.6
(**573**)	32.8	23.4	136.9	127.6	161.3	123.9	201.5	45.5	45.5	36.3	20.7	36.0	43.4	49.6	26.0	27.6	82.0	12.1	16.6
(**574**)	49.2	197.7	143.6	131.3	157.2	129.5	200.7	45.0*	45.3*	39.7	20.6	35.6	43.4	49.2	25.7	27.5	81.7	12.1	19.5

* Assignments may have to be reversed.

+9.8 ppm for the S series and +3.6 ppm for the R series. Substituent shielding effects are illustrated by the data for cholesterol derivatives (**575**) and (**576**) (*151*) (Table 32).

Table 32. *C-18 and Side Chain* ^{13}C *Chemical Shifts for 22-Hydroxy Cholesterol Epimers* (**575, 576**) (*151*)

	Carbon							
	17	18	20	21	22	23	24	25
Chol-esterol	56.3	11.9	35.8	18.8	36.3	23.9	39.6	28.0
(**575**)	52.6	11.8	40.3	11.6	73.8	33.3	35.7	27.8
(**576**)	53.2	11.9	42.6	12.5	74.0	27.5	36.1	27.9

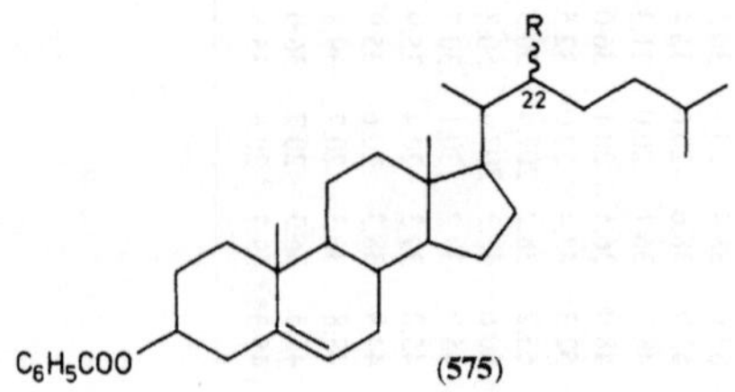

(**575**)
(**576**)

	17	18	20	21	22	23	24	25
R = H	56.3	11.9	35.8	18.8	36.3	23.9	39.6	28.0
(**575**) R = 22S-OH	52.6	11.8	40.3	11.6	73.8	33.3	35.7	27.8
(**576**) R = 22R-OH	53.2	11.9	42.6	12.5	74.0	27.5	36.1	27.9

The biological conversion of cholesterol to pregnenolone has been assumed to proceed via (**577**) (20R, 22R-dihydroxycholesterol); (**578**), (**579**), and (**580**) occurring as intermediates. Epoxides (**580 a—580 d**), which were synthesized and their ^{13}C spectra recorded, exhibit characteristic chemical shifts, permitting unambiguous assignment of sidechain stereochemistry (*152, 153*). ^{13}C chemical shifts for steroids (**577—579**) and four epimers of 20,22-epoxycholesterol (**580a—580c**) are listed in Table 33.

Full chemical shifts designations have been given for a host of cardenolides (**581—590**) (*154*), some 3β,5α,6-trihydroxyandrostanes (**591a—c** and **592a—c**) (*155*), as well as for several steroidal sapogenins

References, pp. 195—229

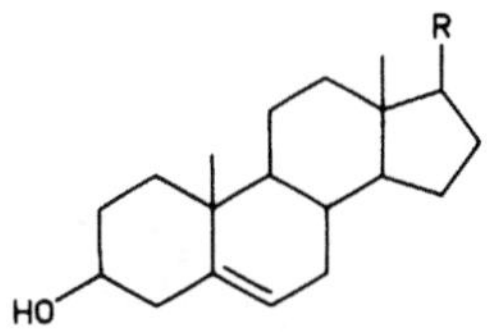

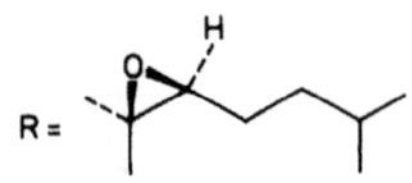

(**580a**) (20R, 22S)

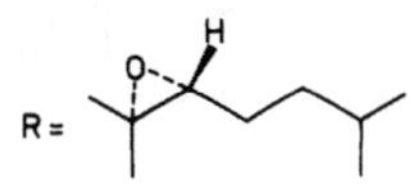

(**580b**) (20S, 22R)

(**577**) R =

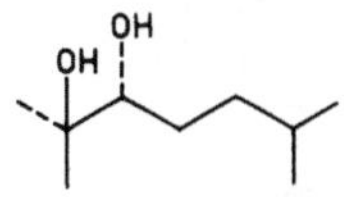

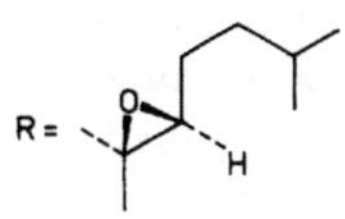

(**580c**) (20R, 22R)

(**578**) R =

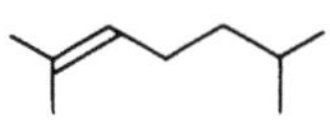

(**579**) R =

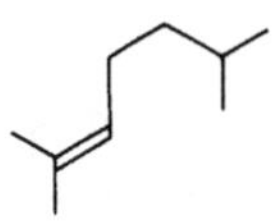

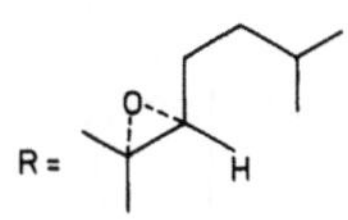

(**580d**) (20S, 22S)

Table 33. *Partial* ^{13}C *Shieldings for Cholesterol Derivatives* (**578**—**580**) (*152*)

Carbon	Chol-esterol	(**578**)	(**579**)	(**580a**)	(**580b**)	(**580c**)	(**580d**)
13	42.3	45.9	43.6	42.5	42.4	43.2	42.5
14	56.8	55.9	56.5	56.4	56.2	56.7	56.8
15	24.3	24.7	24.5	24.3	24.7	23.9	24.1
16	27.9	26.3	26.0	26.9	26.3	26.2	26.6
17	56.3	51.2	59.1	53.3	50.5	56.7	56.8
18	11.8	14.0	13.0	14.7	14.4	13.2	13.5
20	35.8	134.2	134.1	61.1	61.0	60.0	60.4
21	18.7	22.8	17.8	22.6	22.7	17.3	20.3
22	36.5	129.6	126.1	66.5	65.1	60.4	59.8
23	23.8	24.7	24.9	23.4	24.7	21.9	22.4
24	39.9	38.5	38.9	36.4	35.8	35.6	35.7

(**593**—**600**) (*156*). ^{13}C shieldings for these systems are collected in Tables 34—36.

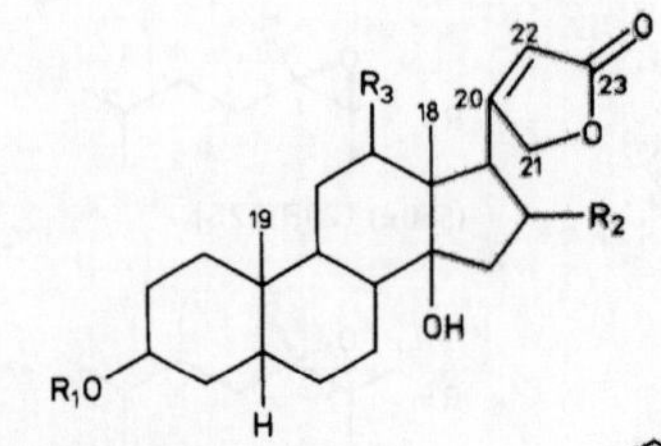

(**581**) $R_1 = R_2 = R_3 = H$

(**582**) $R_1 = R_3 = H$; $R_2 = OH$

(**583**) $R_1 = R_3 = H$; $R_2 = OAc$

(**584**) $R_1 = Ac$; $R_2 = OAc$; $R_3 = H$

(**585**) $R_1 = Ac$; $R_2 = R_3 = H$

(**589**) $R_1 = R_2 = H$; $R_3 = OH$

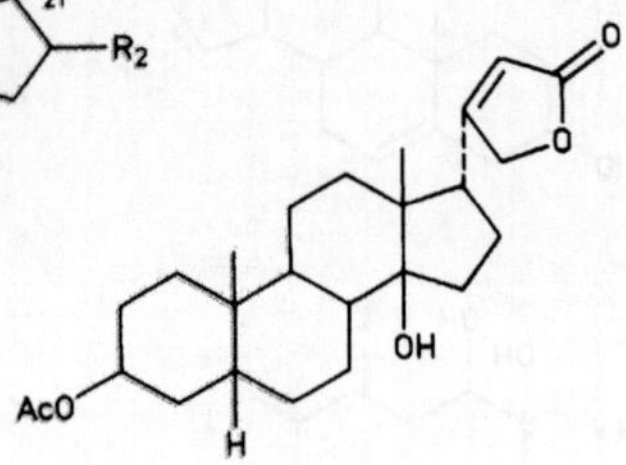

(**586**)

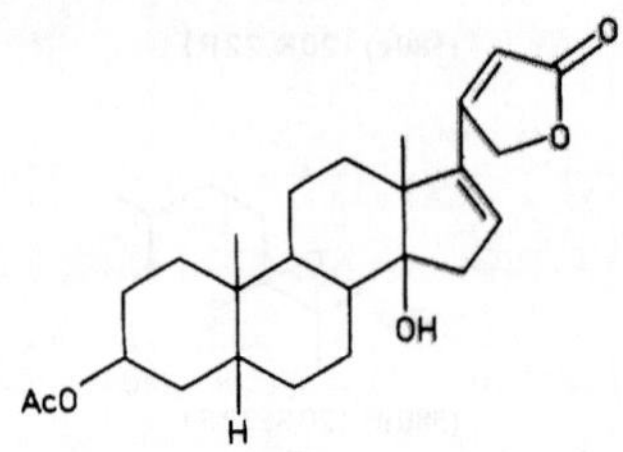

(**587**)

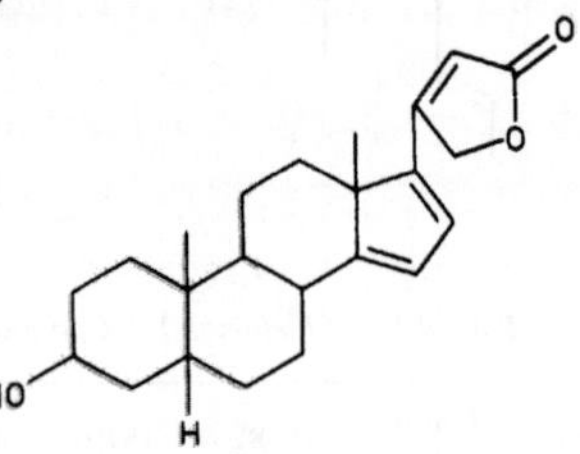

(**588**)

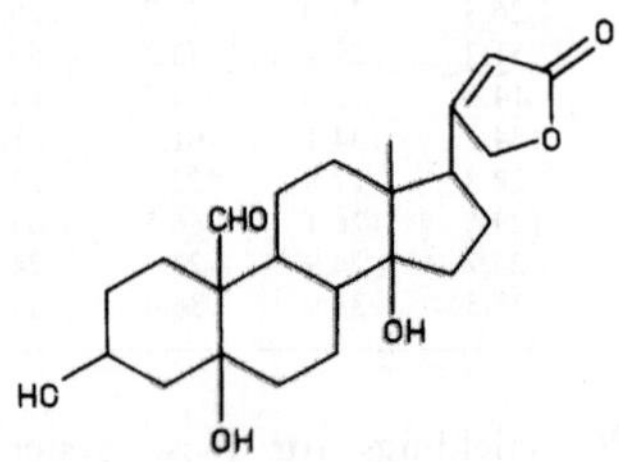

(**590**)

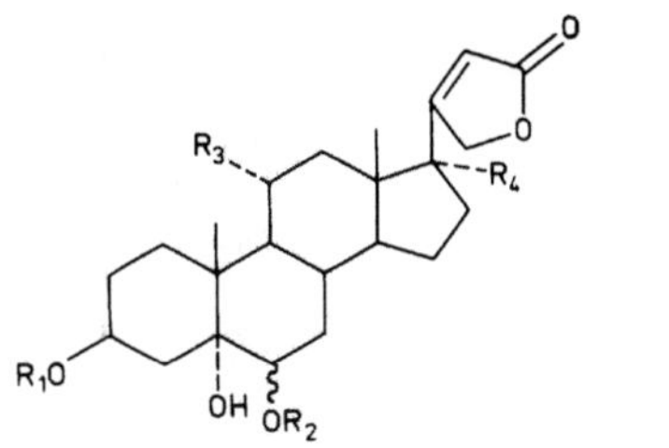

		R_1	R_2	R_3	R_4
(591 a—c) 6β - OR_2	a	H	Ac	H	H
(592 a—c) 6α - OR_2	b	H	Ac	OH	H
	c	H	Ac	H	OH

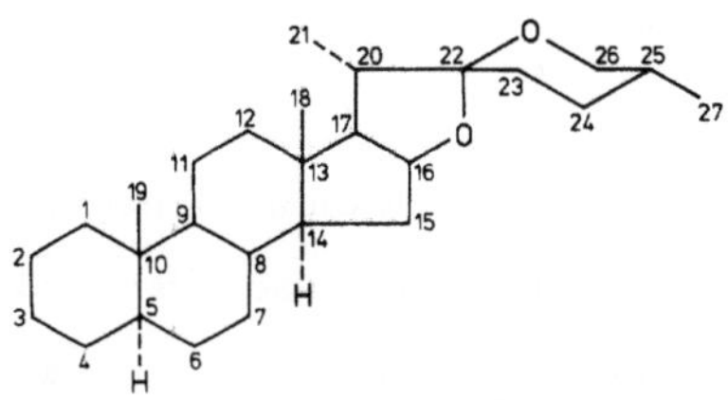

(593) (25 R) - 5α - spirostane (desoxytigogenin)
(594) (25 R) - 5α - spirostan - 3 - ol (tigogenin)
(595) (25 S) - 5α - spirostan - 25 - ol
(596) Δ^2 - (25 R) - 5α - spirosten - 15β - ol
(597) Δ^5 - (25 R) - 5α - spirosten - 3β - ol (diosgenin)
(598) 3β - hydroxy - (25R) - 5α - spirostan - 12 - one (hecogenin)
(599) (25 R) - 5β - spirostan - 3β - ol (smilagenin)
(600) (25 S) - 5β - spirostan - 3β - ol (sarsasapogenin)

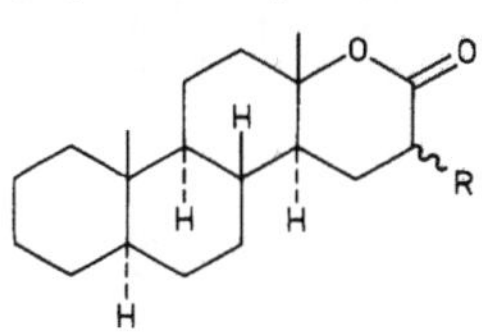

(601 a) R = H
(601 b) R = β - OAc
(601 c) R = α - OAc
(601 d) R = β - OH
(601 e) R = α - OH

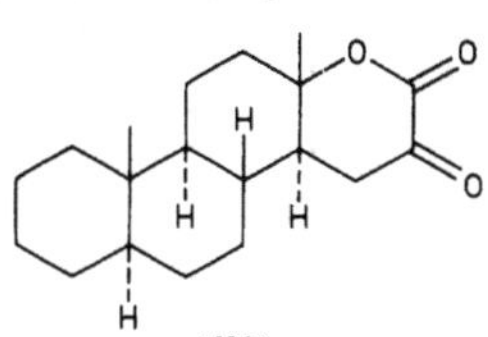

(602)

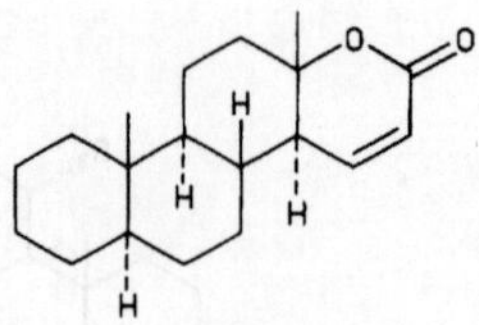

(603)

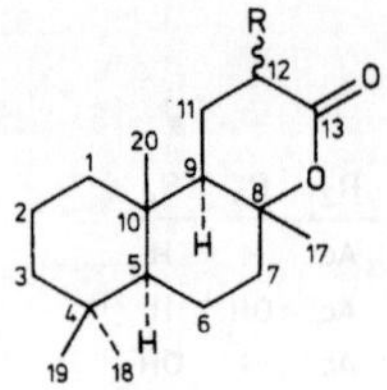

(604a) R = H

(604b) R = β - OAc

(604c) R = α - OAc

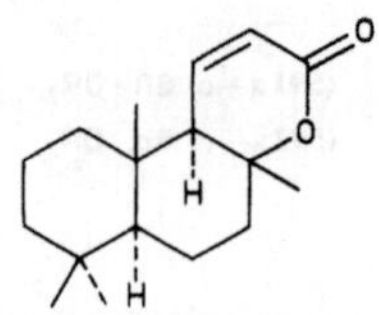

(605)

Table 34. ^{13}*C Chemical Shifts for Cardenolides* (**581—590**) (*154*)

Carbon	(**581**)	(**582**)	(**583**)	(**584**)	(**585**)	(**586**)	(**587**)	(**588**)	(**589**)	(**590**)
1	30.0	30.0	30.0	30.7	30.8	30.8	30.8	30.7	30.0	24.8
2	28.0	28.0	27.9	25.2	25.4	25.3	25.4	27.9	27.9	27.4[a]
3	66.8	66.8	66.8	71.1	71.4	71.3	71.3	66.7	66.6	67.2
4	33.5	33.5	33.4	30.7	30.8	30.8	30.8	33.5	33.3	38.1
5	35.9[a]	36.4	36.4	37.2	37.4	37.4	37.3	36.8[a]	36.4	75.3
6	27.1	27.0	26.9	26.6	26.8	26.8	26.6	26.6	26.9	37.0
7	21.6[b]	21.4[a]	21.2[a]	20.9[a]	21.6	20.6[a]	20.2[a]	24.0	21.9	18.1[b]
8	41.9	41.8	41.8	41.6	41.8	41.5	41.2	36.7[a]	41.3	42.2[c]
9	35.8[a]	35.8	35.9	35.8	36.1	36.2	36.8	45.1	32.6	40.2[c]
10	35.8	35.8	35.6	35.4	35.8	35.5	35.4	36.2	35.5	55.8
11	21.7[b]	21.9[a]	21.3[a]	21.4[a]	21.6	21.2[a]	21.3[a]	21.4	30.0	22.8[b]
12	40.4	41.2	41.0	40.9	40.3	31.3	40.6	37.7	74.8	40.2
13	50.3	50.4	50.7	50.5	50.3	49.5	52.6	54.2	56.4	50.1
14	85.6	85.2	84.1	83.3	85.6	86.1	85.7	146.3[c]	85.8	85.3
15	33.0	42.6	39.5	39.3	33.0	31.3	38.8	108.3[d]	33.0	32.2
16	27.3	72.8	75.0	74.7	27.3	24.8	133.8	135.8[d]	27.9	27.5[a]
17	51.5	58.8	56.8	56.6	51.5	48.9	161.2	158.0[c]	46.1	51.4
18	16.1	16.9	16.1	16.1	16.0	18.5	16.6	20.1	9.4	16.2
19	23.9	23.9	23.9	23.8	23.9	24.0	24.1	24.0	23.8	195.7
20	177.1[c]	171.8[b]	171.5[b]	171.5[b]	177.1[a]	173.6[b]	172.8[b]	173.5[b]	177.1[a]	177.2[d]
21	74.5	76.7	76.8	76.5	74.7	74.8	72.6	72.1	74.6	74.8
22	117.4	119.6	121.3	121.1	117.4	116.6	111.7	119.5	117.0	117.8
23	176.3[c]	175.3[b]	175.8[b]	175.4[b]	176.3[a]	175.8[b]	176.3[b]	176.8[b]	176.3[a]	176.6[d]

$OCO\underline{C}H_3$, 21.3 ± 0.3; $O\underline{C}OCH_3$, 171.4 ± 0.4 ppm in (**583—587**).

[a, b, c, d] Assignments may have to be interchanged.

Table 35. ^{13}C *Chemical Shifts for Androstane Derivatives* (**591 a**—**c**) *and* (**592 a**—**c**) (*155*)*

Carbon	(**591 a**)	(**591 b**)	(**591 c**)	(**592 a**)	(**592 b**)	(**592 c**)
1	32.37	34.05	32.39	31.51	32.94	31.87
2	30.50	30.82	30.50**	30.42	30.80	31.87
3	67.01	66.75	66.99	66.82	66.64	66.59
4	40.14	40.40	40.11	38.21	38.35	39.51
5	74.67	75.43	74.78	75.96	76.65	75.53
6	76.81	76.84	76.86	74.72	74.65	74.90
7	31.54	31.52	31.72	31.33	31.38	32.19**
8	31.54	30.48	31.72	34.37	33.17	34.84
9	45.18	51.80	44.96	44.75	51.32	44.95
10	38.78	40.40	38.77	40.14	41.93	40.50
11	21.22	68.01	21.03	21.29	68.03	21.31
12	38.48	49.43	30.58**	38.35	49.48	30.73
13	45.08	45.01	48.96	45.07	44.96	48.88
14	56.06	55.00	49.33	56.04	55.14	50.37
15	24.58	24.58	23.85	24.44	24.47	23.84
16	26.26	26.21	37.30	26.20	26.24	37.40
17	51.19	50.84	82.70	51.11	50.87	82.32
18	13.66	14.41	16.25	13.52	14.36	15.99
19	16.57	16.79	16.62	15.90	16.27	15.89
20	173.03	172.43	175.45	172.86	172.50	175.22
21	74.35	74.22	73.46	74.30	74.28	73.03
22	115.95	116.08	115.67	116.01	116.13	115.92
23	175.66	175.50	175.45	175.59	175.60	174.18**
3′	171.40	171.26	171.36	171.59	171.70	170.53
3″	21.51	21.48	21.48	21.06	21.06	20.82

* In pyridine-d_5.
** Assignments may have to be reversed.

The ^{13}C spectra of the steroidal and terpenoid δ-lactones, deoxy-andrololactone (**601 a**) and its C-16 epimeric acetoxy and hydroxy derivatives (**601 b**—**e**) furnished some interesting stereochemical information. Similar conclusions could also be drawn from ambreinolide (**604 a**) and its C-12 acetoxy epimers (**604 b**) and (**604 c**) (*157*). When compared to similar steroids C-7 in (**601 a**) is found to resonate at higher field (cf. androstan-17-one **521**). This is ascribed to D-ring extension which results in a more severe steric interaction with the axial hydrogen C(15)-H. The C-16 shieldings in compounds (**601 b**—**e**) are again found to be sensitive to the configuration of this carbon (Table 37). The shift difference for C-16 in the quasi equatorial and quasi axial arrangement amounts to ca 3 ppm. The deshielding of C-15 which occurs upon introduction of the β-acetoxy group in (**601 b**) and (**604 b**) is not in

Table 36. ^{13}C *Chemical Shifts for Steroidal Sapogenins* (**593—600**) (*156*)

Carbon	(**593**)	(**594**)	(**595**)	(**596**)	(**597**)	(**598**)	(**599**)	(**600**)
1	38.7	37.0	38.6	39.7	37.2	36.5	29.9	29.9
2	22.2	31.5	22.2	125.7	31.6	31.2	27.8	27.8
3	26.8	71.2	26.8	125.5	71.5	70.7	67.0	67.0
4	29.0	38.2	29.0	28.7	42.2	37.8	33.6	33.6
5	47.1	44.9	47.0	41.6	140.8	44.6	36.6	36.5
6	29.0	28.6	29.0	30.3*	121.3	28.3	26.5	26.6
7	32.4	32.3	32.4	31.4***	32.0	31.4*	26.5	26.6
8	35.2	35.2	35.2	31.2***	31.4	34.4	35.3	35.3
9	54.8	54.4	54.8	54.6	50.1	55.5	40.3	40.3
10	36.3	35.6	36.4	34.9	36.6	36.0	35.3	35.3
11	20.7	21.1	20.6	21.0	20.9	37.8	20.9	20.9
12	40.2	40.1	40.1	42.6	39.8	213.0	39.9	39.9
13	40.6	40.6	40.6	40.7	40.2	55.0	40.7	40.6
14	56.5	56.3	56.5	61.3**	56.5	55.8	56.5	56.4
15	31.8	31.8	31.7	69.6	31.8	31.5*	31.8	31.7
16	80.8	80.7	81.3	82.1	80.7	79.1	80.9	80.9
17	62.3	62.2	62.0	60.7**	62.1	53.5	62.4	62.1
18	16.5	16.5	16.5	19.1	16.3	16.0	16.4	16.5
19	12.3	12.4	12.3	11.7	19.4	12.0	23.8	23.9
20	41.6	41.6	41.5	42.6	41.6	42.2	41.6	42.1
21	14.5	14.5	14.4	14.2	14.5	13.2	14.4	14.3
22	109.0	109.0	108.8	109.9	109.1	109.0	109.1	109.5
23	31.4	31.4	24.7	31.4	31.4	31.2	31.4	27.1
24	28.9	28.8	32.7	28.6	28.8	28.8	28.8	25.8*
25	30.3	30.3	66.6	30.2*	30.3	30.2	30.3	26.0*
26	66.7	66.7	68.9	67.1	66.7	66.8	66.8	65.0
27	17.1	17.1	27.0	17.1	17.1	17.1	17.1	16.1

*, **, *** Assignments may have to be interchanged.

accord with the commonly observed β effect in chair-like conformations. While the β effect normally is larger for an equatorial than for an axial substituent, the reverse behaviour observed here points to a quasi axial orientation of C(15)-OH in the β epimer (**601 b**). This interpretation is further supported by the absence of a γ effect at C-14 in either epimer. The suggested conformations are illustrated by stereoformulae (**A**) and (**B**).

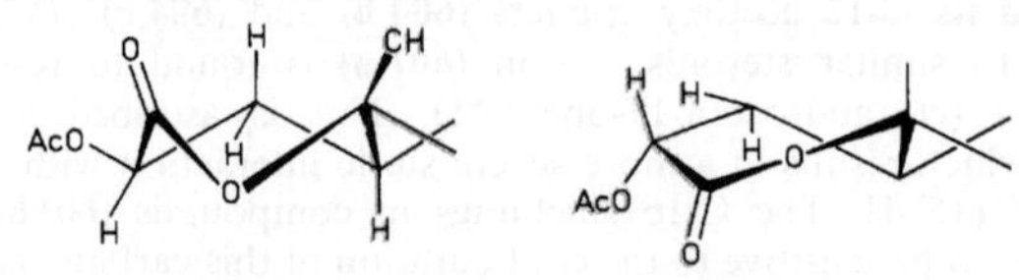

(**601 b**)
(**604 b**)
(**A**)

(**601 c**)
(**604 c**)
(**B**)

Table 37. ^{13}C *Chemical Shifts for Steroidal δ-Lactones* (**601—603**) (*157*)

Carbon	(**601a**)	(**601b**)	(**601c**)	(**601d**)	(**601e**)	(**602**)	(**603**)
1	38.4	38.4	38.5	38.4	38.3	38.5*	38.4
2	22.0	22.0	21.9	21.9	21.9	21.9	22.0
3	26.6	26.7***	26.6	26.6	26.5	26.5	26.6
4	28.7	28.7**	28.5	28.5*	28.6	28.4**	28.7*
5	46.4	46.3*	46.3	46.3	46.3	46.3	46.7
6	28.7	28.5**	28.5	28.5*	28.6	28.6**	28.5*
7	30.6	30.6	30.5	30.7	30.7	30.2	30.4
8	37.9	37.2	38.5	37.1	38.6	38.5	35.2
9	53.5	53.4	53.2	53.5	53.0	52.8	54.2
10	36.2	36.2	36.2	36.2	36.1	36.2	36.4
11	21.6	21.6	21.6	21.6	21.6	21.6	21.5
12	39.4	38.9	39.5	38.8	39.4	38.8*	38.4
13	83.1	85.4	84.4	86.3	85.1	84.7	83.3
14	46.4	46.7*	46.3	46.3	46.3	45.8	48.6
15	19.8	26.8***	29.1	28.2*	30.5	38.1	145.4
16	28.7	68.6	65.6	68.3	64.5	190.3	121.6
17	170.9	168.5	168.3	174.8	174.8	157.4	163.7
18	20.1	20.7	19.5	21.1	19.3	20.6	18.5
19	12.0	12.0	12.0	12.1	12.0	12.0	12.0

*, **, *** Assignments may have to be interchanged.

Table 38. ^{13}C *Chemical Shifts for Terpenoid δ-Lactones* (**604**) *and* (**605**) (*157*)

Carbon	(**604a**)	(**604b**)	(**604c**)	(**605**)
1	39.2	39.3	38.8	38.6
2	18.4	18.4	18.2	18.1
3	41.8	41.8	41.7*	41.8
4	33.2	33.3	33.1	33.1
5	56.0	56.0	55.8	56.1
6	19.7	19.8	19.6	19.7
7	41.3	40.8	41.4*	39.8
8	83.5	85.8	84.9	84.0
9	53.7	53.5	53.1	55.4
10	37.2	37.4	37.3	36.5
11	15.8	24.0	24.6	145.4
12	29.0	69.1	65.5	122.2
13	171.2	168.5	168.4	163.9
17	22.9	23.5	22.7	21.0
18	33.3	33.3	33.3	33.1
19	21.5	21.3	21.5	21.4
20	15.1	15.7	14.4	17.0

* Assignments may have to be reversed.

^{13}C shieldings for the corresponding terpenoid δ-lactones (**604 a—c**) and (**605**) are found in Table 38. Conformers (**A**) and (**B**) may explain the lack of a γ effect for C-14 in either epimeric acetate. Furthermore, the twisted boat form (**B**) allows for a substantial decrease in the C-18 methyl eclipsing effect at C-15, thus accounting for its greater deshielding in epimer (**601 c**). The larger shielding of C-16 in (**601 c**) may be attributed to the eclipsing effect of C-18. Similar trends are observed in (**604**) where introduction of the acetoxy group does not produce a γ effect-induced shielding of C-9, thus pointing to conformations (**A**) and (**B**) for the lactone moiety in the two epimers (*157*).

Carotenoids and Related Terpenoids

Several reports have dealt with ^{13}C n. m. r. spectroscopy of retinals (*53, 54, 158, 159*). Systems studied include all-*trans*-retinal (**606**) (*53, 54, 158, 159*), 9-*cis*-retinal (**607**) (*53, 54*), 11-*cis*-retinal (**608**) (**53**, *54*) and 13-*cis*-retinal (**609**) (*53, 54*). Double bond positions were ascertained

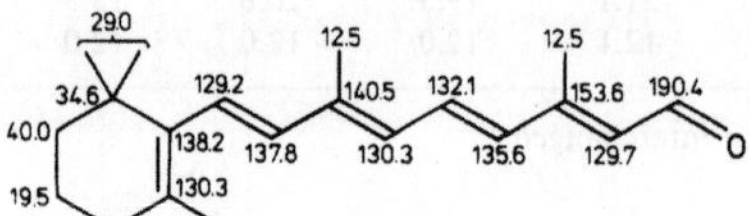

(**606**)

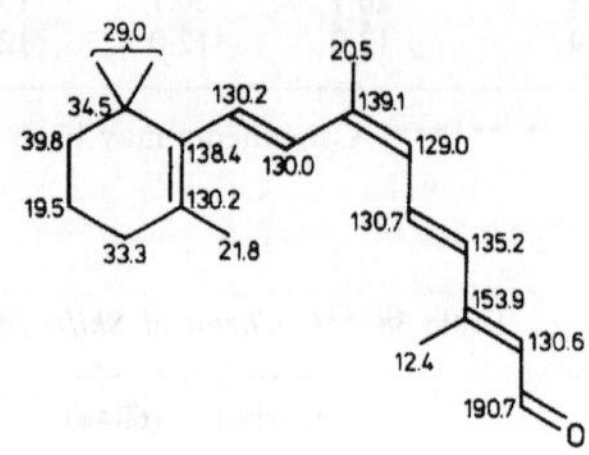

(**607**)

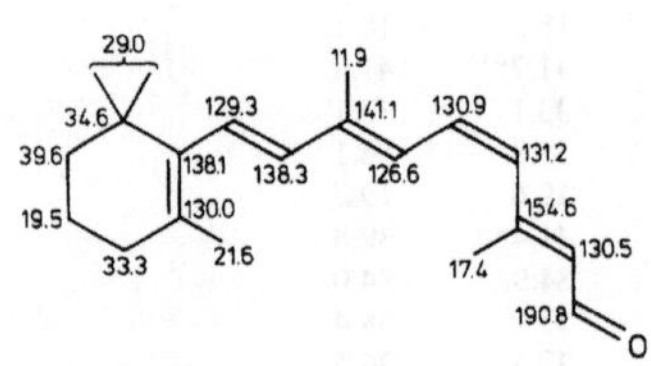

(**608**)

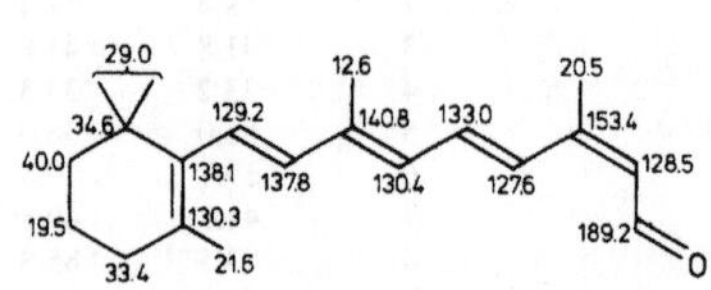

(**609**)

on the basis of *cis-trans* isomerization shifts (*53, 54*). The data indicate that (**606**), (**607**) and (**609**) in solution exist essentially with a planar polyene chain. In contrast, 11-*cis* retinal (**608**), the chromophore of the visual pigment rhodopsin, is not planar along the complete unsaturated hydrocarbon chain, but exhibits some distortion. Conformational information was mainly obtained from evaluation of the spin-lattice relaxation data as previously discussed (cf. Sect. 2). Spin-lattice relaxation times (in sec) are given along with formulae **(606 a)** and **(608 a)**. The data

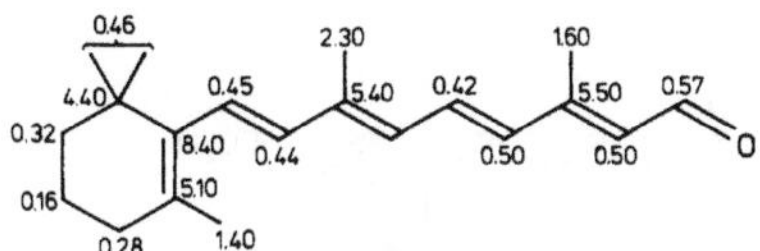

(606a)

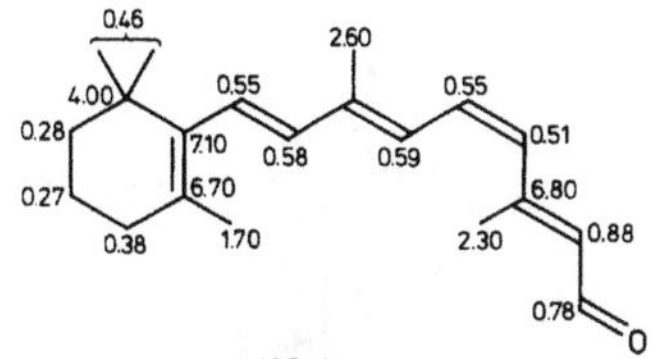

(608a)

T_1 Relaxation Times in Seconds

indicate a high degree of motional anisotropy for both systems **(606)** and **(608)** where one of the principle axes for rotational diffusion is collinear with the polyene axis (*53*, *54*). C-14 in the protonated propylamine Schiff base of **(606)** was found to resonate in a spectral region which is normally free from intense protein peaks. ^{13}C labelling studies of the rhodopsin chromophore were interpreted on this basis (*158*).

A comprehensive ^{13}C study of vitamin A derivatives, carotenoids and related compounds included chemical shift data for 35 all-*trans* vitamin A compounds, 17 mono-*cis* vitamin A compounds, 11-*cis* and 13-*cis* retinol as well as 23 different β-*apo* carotenoids and was complemented by 30 further reference compounds (*160*). In this context correct ^{13}C chemical shift designations were given for α-ionone **(610)***. ^{13}C

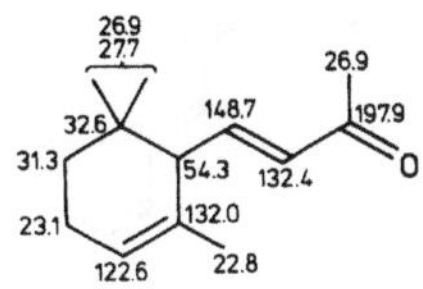

(610)

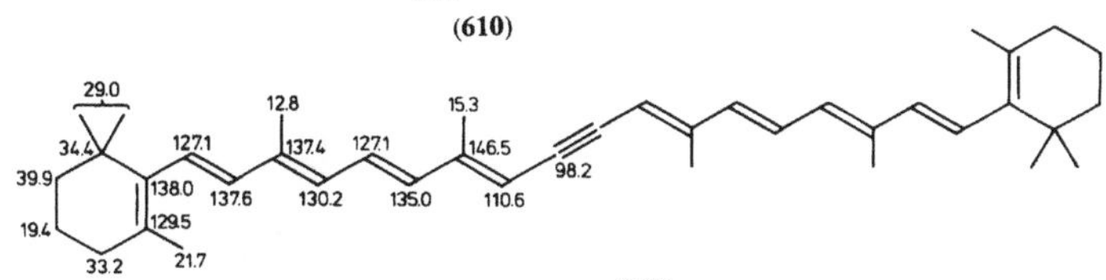

(611)

* Englert (*160*) failed to point out some erroneous assignments first reported by Roberts *et al.*

n.m.r. spectroscopy supported the structure of 15,15′-dehydro-β-carotene (**611**), which is an intermediate in the synthesis of β-carotene (*161*).

The site of methylation in azafrin methyl ester monomethyl ether could be established unequivocally on the basis of the ^{13}C spectra, which

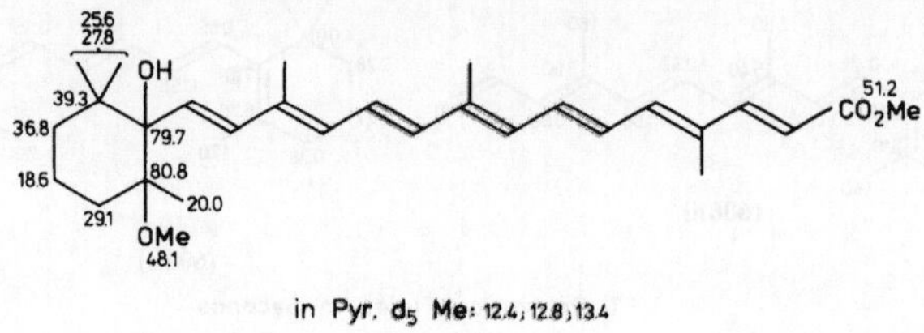

(**612**)

permitted structure (**612**) to be attributed to this compound (*162*). For this purpose unambiguous distinction had to be made between quaternary carbons C-5 and C-6. This was accomplished by selectively irradiating the high-field protons ($\delta_H \sim 1.2$—2.2). While one of the signals collapsed into a singlet, the other one appeared as a quartet resulting from long-range coupling to protons C (7)-H ($^2J_{CCH}$) and C (8)-H ($^3J_{CCCH}$) thus identifying this latter as C-6. Since methylation of azafrin methyl ester resulted in a 5 ppm deshielding of C-5 but only insignificantly affected C-6, definitive proof was given that methylation had occurred at the C-5 hydroxyl.

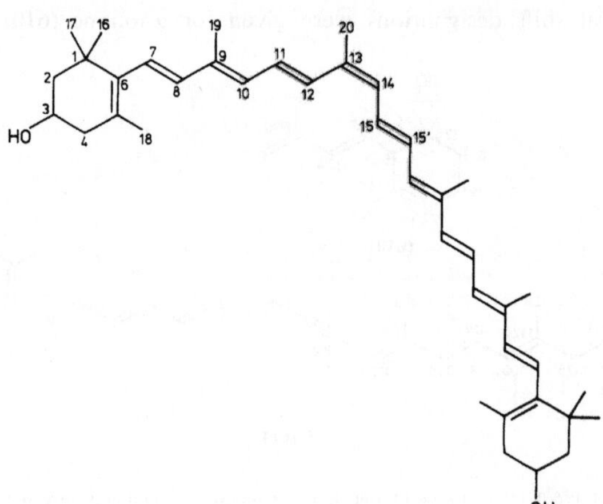

(**613**)

The potential of ^{13}C n.m.r. spectroscopy as a stereochemical probe has been illustrated in a study of some mono *cis* C_{40} carotenoids (*163*). The data indicated that *neo*-zeaxanthin A (**613**) and *neo*-capsorubin A (**614**) have 13-*cis* configuration whilst *neo*-zeaxanthin B (**615**) and *neo*-capsorubin B (**616**) are *cis* linked at position 9. The ^{13}C data of the corresponding all-*trans* carotenoids are also available (**617**, **618**). Double bond positions were located on the basis of known *cis*/*trans* isomerization shifts (*53*, *54*). ^{13}C chemical shifts for systems (**613**—**618**) are collected in Table 39.

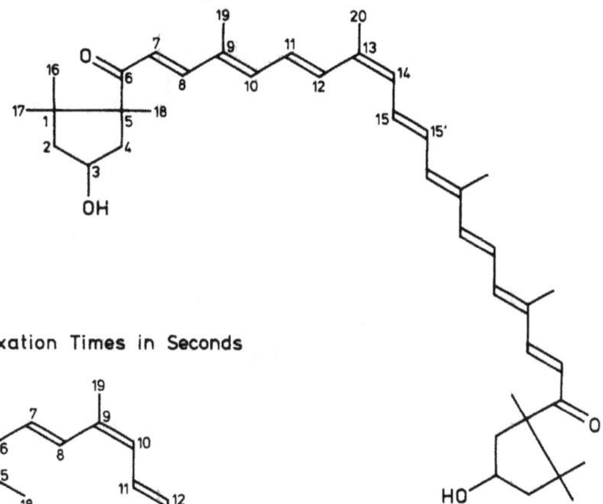

(614)

T1 Relaxation Times in Seconds

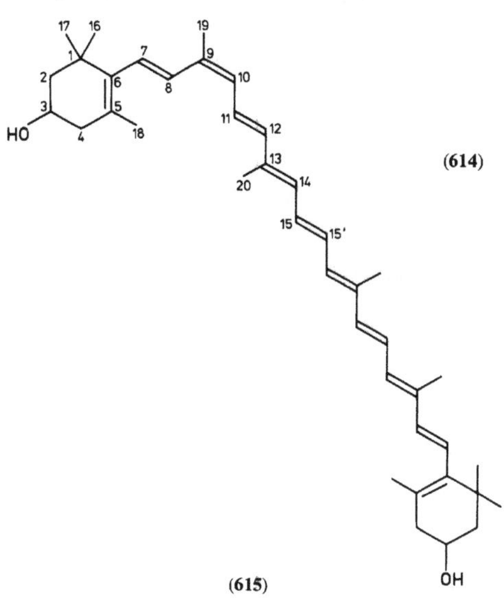

(615)

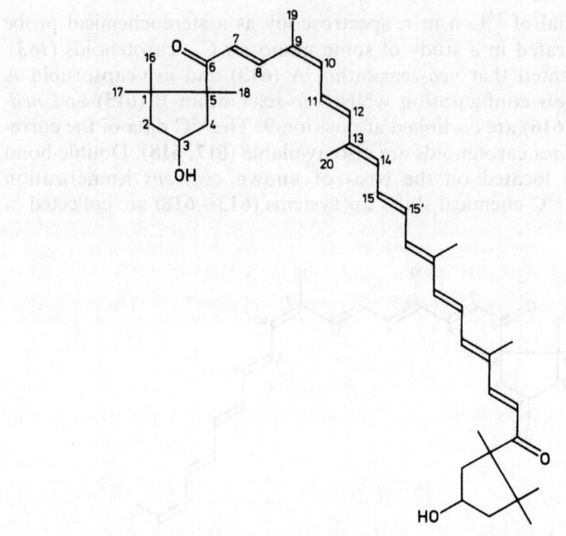

(616)

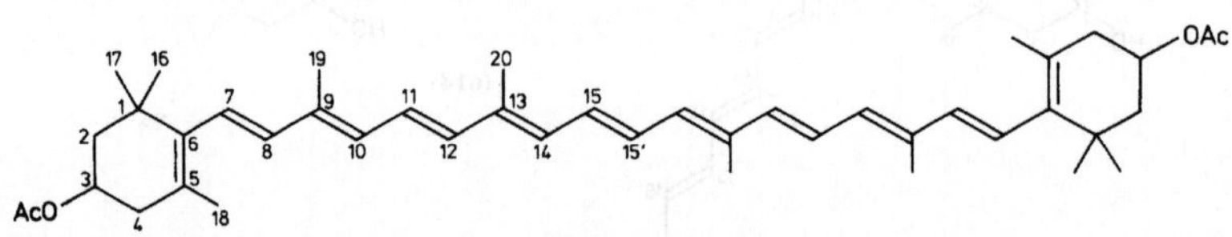

(617)

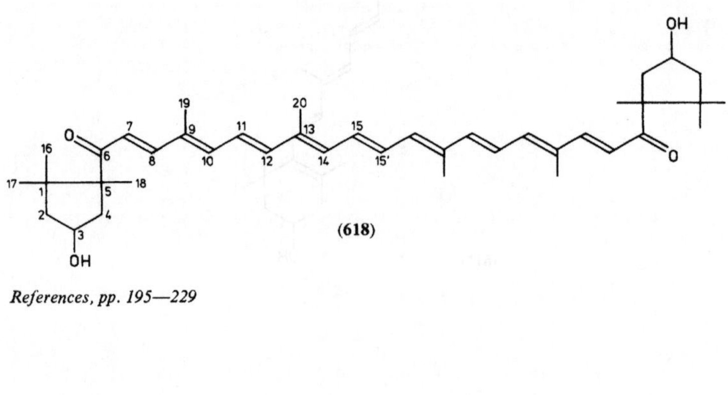

(618)

Table 39. ^{13}C *Chemical Shifts for Tetraterpenes* (**613—618**) (*163*)

Carbon	(**613**) *trans* end	(**613**) *cis* end	(**614**) *trans* end	(**614**) *cis* end	(**615**) *trans* end	(**615**) *cis* end	(**616**) *trans* end	(**616**) *cis* end	(**617**)	(**618**)
1	37.18	37.18	44.0	44.0	37.16	37.16	44.03	44.03	36.73	44.42
2	48.66	48.66	50.93	50.93	48.64	48.64	50.95	50.95	44.24	51.03
3	65.15	65.15	70.37	70.37	65.16	65.16	70.40	70.40	68.47	70.35
4	42.72	42.72	45.36	45.36	42.71	42.71	45.41	45.41	38.57	45.46
5	126.37	126.26	59.01	59.01	126.22	126.47	59.00	59.01	125.66	59.01
6	137.92	137.92	203.0	203.0	137.90	138.15	202.98	203.55	138.03	203.0
7	125.71	125.71	121.19	121.55	125.68	127.52	121.17	122.58	125.36	121.27
8	138.58	138.58	146.85	146.75	138.56	130.86	146.87	138.10	138.73	146.87
9	135.63	135.71	133.97	134.07	135.67	134.28	133.98	132.15	135.55	134.10
10	131.38	131.38	140.38	140.58	131.38	129.91	140.59	139.09	131.52	140.50
11	124.96	126.05	124.52	125.88	124.98	123.79	124.54	123.70	124.96	124.67
12	137.70	129.43	141.84	133.38	137.65	136.38	141.87	140.59	137.74	141.75
13	136.43	134.92	136.73	135.24	136.38	136.90	136.70	137.26	136.46	137.00
14	132.57	131.02	134.82	133.28	132.63	132.49	135.07	134.35	132.68	134.99
15	130.19	128.93	131.05	129.90	130.05	130.14	130.86	131.42	130.18	103.25
16	28.82	28.82	25.12	25.12	28.89	28.89	25.13	25.13	28.58	25.14
17	30.34	30.34	25.93	25.93	30.34	30.34	25.90	25.90	30.07	25.94
18	21.59	21.59	21.34	21.34	21.59	21.69	21.35	21.27	21.40	21.37
19	12.78	12.78	12.87	12.87	12.83	20.70	12.89	20.44	12.75	12.83
20	12.78	20.68	12.87	20.61	12.83	12.83	12.84	13.00	12.75	12.83
OCO<u>Me</u>									21.39	
O<u>C</u>O									170.63	

^{13}C chemical shifts have also been reported for a number of isomeric C_{20} pentaenes (**619—624**) (*164*). Table 40 provides chemical shift assignments for these compounds.

Table 40. ^{13}C *Chemical Shifts for Isomeric Pentaenes* (**619—624**) (*164*)

Carbon	(**619**)	(**620**)	(**621**)	(**622**)	(**623**)	(**624**)
1	25.67	25.85	25.73	25.73	25.67	25.79
2	131.52	132.57	131.80	131.87	131.75	132.28
3	124.03	123.15	124.10	123.97	124.03	123.39
4	26.67	26.84	26.96	26.90	26.90	27.02
5	40.12	40.58	32.19	32.34	32.63	32.34
6	137.95	139.65	138.30	139.53	137.89	139.53
7	125.44	120.00	126.25	120.87	126.20	120.82
8	127.18	123.27	126.84	122.80	127.01	122.86
9	127.18	123.27	126.84	122.80	127.01	122.86
10	125.44	120.00	126.25	120.87	125.43	120.02
11	137.95	139.65	138.30	139.53	137.66	139.76
12	40.12	40.58	32.69	32.34	40.12	40.53
13	26.67	26.84	26.96	26.90	26.67	27.02
14	124.03	123.15	124.10	123.97	124.03	123.15
15	131.52	132.57	131.80	131.87	131.52	131.51
16	25.67	25.85	25.73	25.73	25.67	25.79
17	17.66	17.89	17.72	17.66	17.66	17.84
18	16.72	16.55	23.98	24.33	16.72	16.49
19	16.72	16.55	23.98	24.33	23.98	24.33
20	17.66	17.89	17.72	17.66	17.66	17.84

(**619**) all-trans
(**620**) 6-trans, 8-cis, 10-trans
(**621**) 6-cis, 8-trans, 10-cis
(**622**) all-cis
(**623**) 6-cis, 8-trans, 10-cis
(**624**) 6-cis, 8-cis, 10-trans

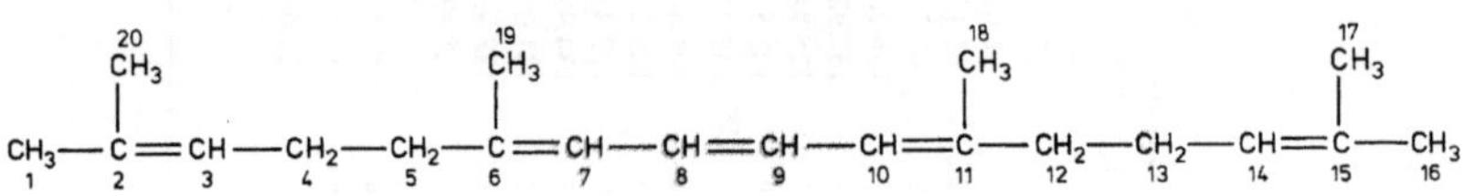

3.2. Alkaloids

Signal identification in the ^{13}C n.m.r. spectra of alkaloids is accomplished essentially by the methods previously outlined (cf. Sect. 2), i. e. mainly by chemical shift comparison with model compounds.

Complete spectral assignments have been reported for the *Apidosperma* alkaloids tabersonine (**625**), vincadifformine (**626**), vandrikidine (**627**), vandrikine (**628**), hazuntinine (**629**), and vindoline (**630**) (*165*). ^{13}C chemical shifts are portrayed in the formulae. Inspection of the chemical

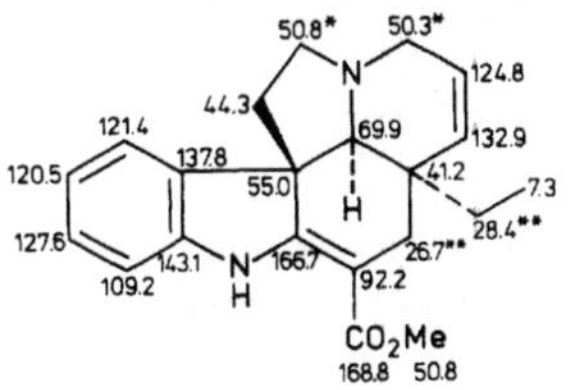

(**625**)

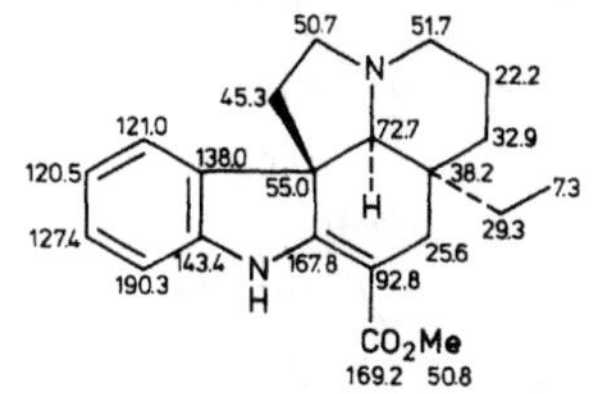

(**626**)

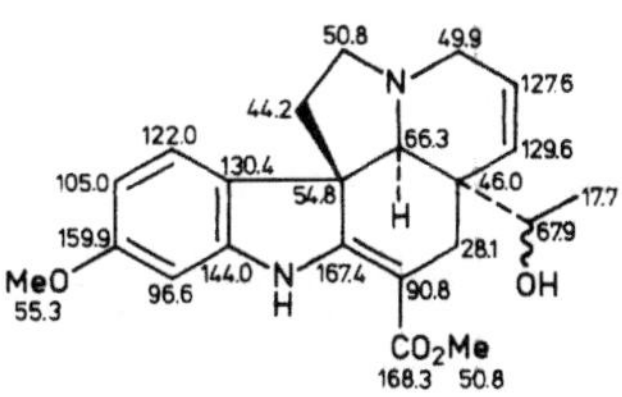

(**627**)

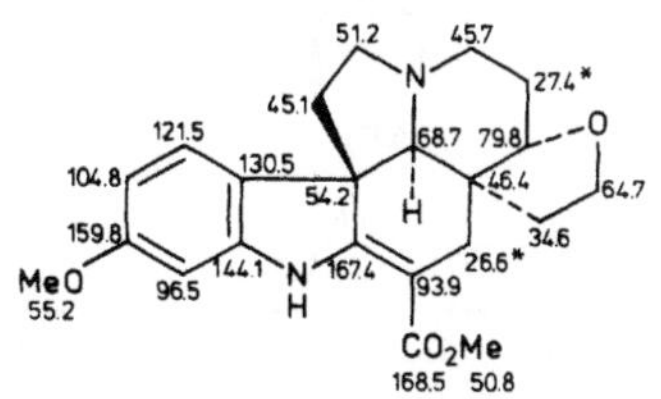

(**628**)

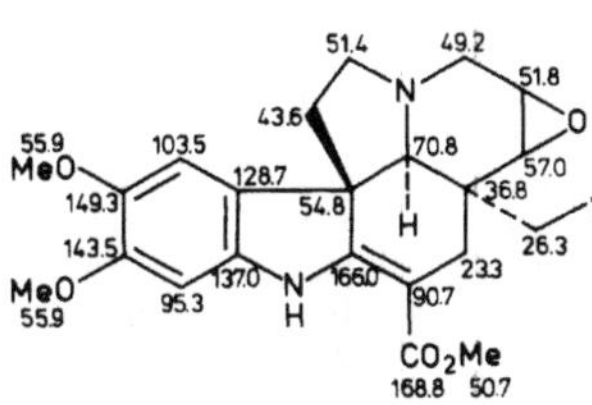

(**629**)

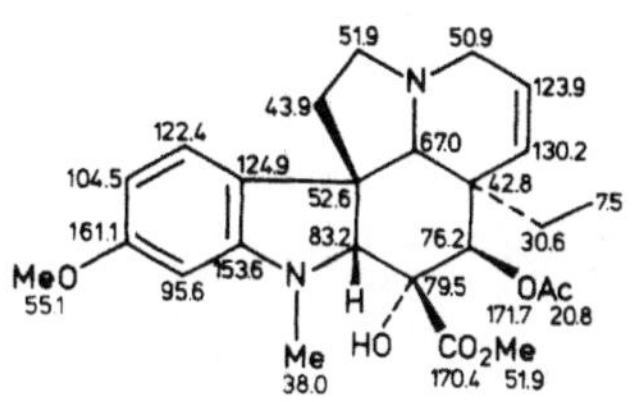

(**630**)

shift data in alkaloids and model compounds indicates that introduction of a double bond in a piperidine nucleus shields the homoallylic carbons. This *endocyclic homoallylic effect* is exhibited by C-21 (δ_C 69.9) in (**625**) when compared with its dihydro derivative (**626**) (δ_C 72.7). The observed effect has been suggested to serve as a criterion for locating the double bond.

Use was made of the *Aspidosperma* alkaloid data in a reinvestigation of the structure of vindolinine (*166*). This alkaloid, which is produced by various *Catharantus* species had earlier been allocated structure **(631)**, on the basis of its mass spectrum. However, instead of the aminomethine (C-2)

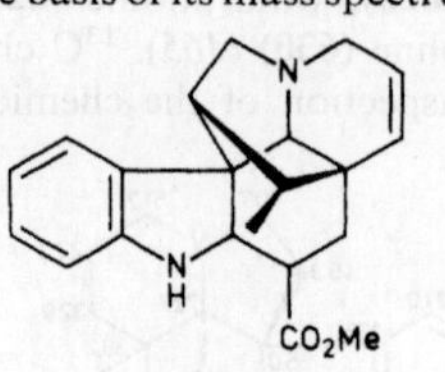

(631)

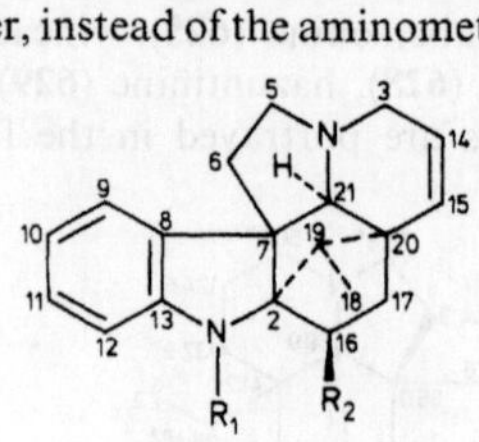

(632) R_1=H ; R_2=CO_2Me
(633) R_1=Me ; R_2=CO_2Me
(634) R_1=H ; R_2=CH_2OH
(635) 16-epi-632
(636) 14,15-dihydro-632

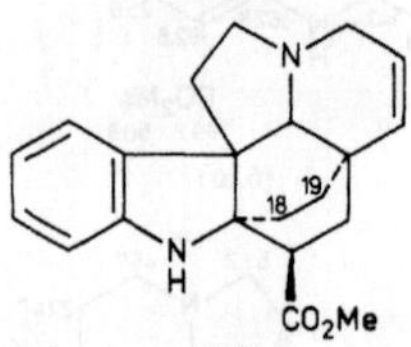

(637)

Table 41. ^{13}*C Chemical Shifts for Vindolinine* **(632)** *and Some Related Alkaloids* **(633—637)** (*166*)

Carbon	(632)	(633)	(634)	(635)	(636)	(637)
2	81.4	84.4	82.2	80.5	80.6	66.5
3	58.0	58.0	58.2	57.4	55.0	49.0
5	50.3	50.0	50.0	50.1	48.1	50.0
6	36.3	36.0	36.8	35.0	37.3	36.4
7	59.8	58.8	59.0	60.7	60.3	56.1
8	139.8	135.8	138.6	135.7	140.1	139.5
9	123.6	123.0	123.8	123.1	123.6	121.1
10	121.0	117.8	120.8	118.9	121.1	119.0
11	127.2	127.7	127.1*	126.9	127.2	126.8
12	112.0	105.6	110.8	109.0	112.7	110.9
13	149.4	150.2	148.6	148.7	149.5	149.0
14	128.5	127.7	127.6*	128.2	20.7	126.5
15	130.7	130.8	132.0	130.6	31.2	132.5
16	39.2	37.0	36.8	39.4	40.2	43.4
17	29.1	28.0	30.0	31.9	29.0	29.6
18	7.4	9.0	7.4	7.8	7.5	31.6
19	48.4	47.0	46.6	44.8	51.0	34.0
20	46.2	45.6	48.0	47.8	44.5	35.0
21	78.0	77.0	78.0	76.4	78.8	66.8
CO	174.2	174.0	63.6**	174.5	175.0	173.7
OMe	51.8	52.0		51.7	52.0	51.6
NMe		30.0				

* Assignments may have to be reversed.
** $\underline{C}H_2OH$.

signal, a nonprotonated resonance at 81.4 ppm was detected. This finding pointed to C-20—C-2 rather than C-20—C-6 bridging and led to the revised structure (**632**) (*166*). Since the structures of related alkaloids had been derived from the erroneous structure of vindolinine, incorrect structures were at first also reported for tuboxenine, 19-*epi*-tuboxenine, pseudo-kopsinine, melobaline, 19-*epi*-melobaline and the dimeric alkaloids pycnanthine and pleiomutinine. Ref. (*166*) therefore demanded a reinvestigation of these compounds. Chemical shifts for vindolinine (**632**) and some related alkaloids (**633**—**637**) are compiled in Table 41. A structure revision has also been given for melobaline (**638**) (*167*) and the dimeric 14′,15′-dihydropycnanthine (**639**) (*168*). Formulae (**638**) and (**639**) represent corrected structures for these two substances.

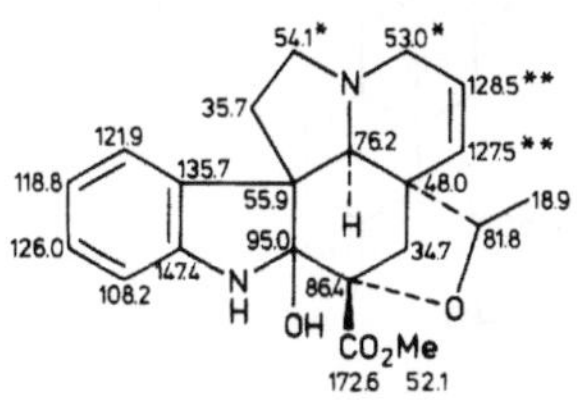

(**638**)

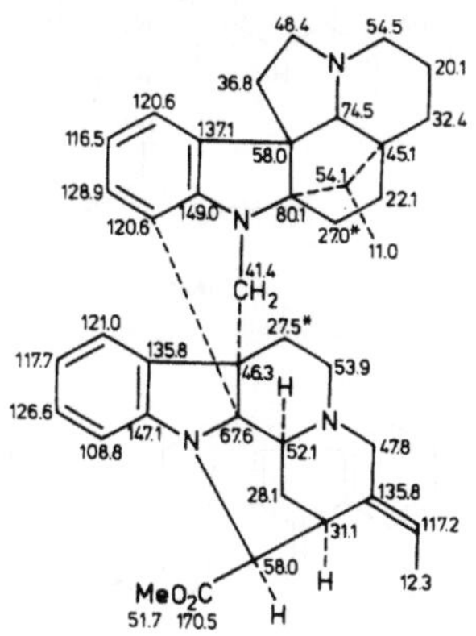

(**639**)

A host of yohimboid and ajmalicinoid alkaloids has recently been examined by ^{13}C n. m. r. (*169*). These indole alkaloids are categorized on the basis of their ring-D stereochemistry. The following configurations are distinguished: *normal* (**A**), *pseudo* (**B**), *allo* (**C**) and *epiallo* (**D**).

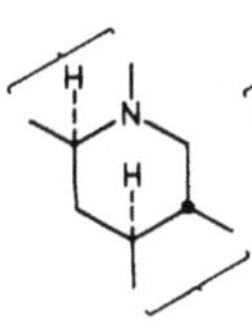

(**A**)

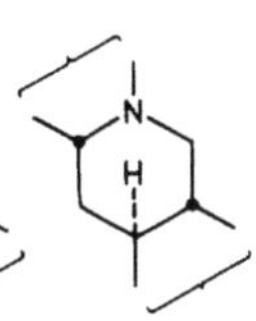

(**B**)

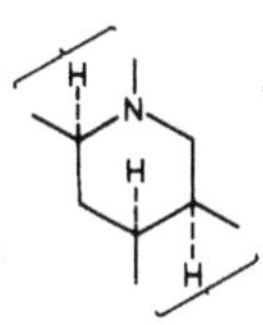

(**C**)

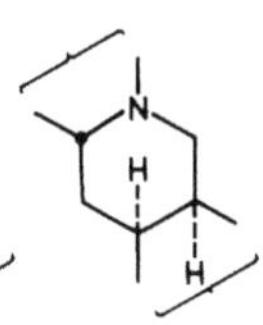

(**D**)

Yohimbin (**640**), β-yohimbin (**641**) and corynanthine (**642**) all belong to the *normal* type. Pseudoyohimbine (**643**) has *pseudo* configuration while α-yohimbine (**644**) and *allo*yohimbine (**645**) are of the *allo* type. 3-*epi*-α-yohimbine (**646**), reserpine (**647**), raunescine (**648**) and isoraunescine (**649**) possess *epiallo* configuration. Ajmalicine (**650**) again exhibits *normal* ring-D stereochemistry while 3-*iso*-19-epiajmalicine (**651**) is of the *pseudo*, tetrahydroalstonine (**652**) and rauniticine (**653**) are of *allo* type. Akuammigine (**654**) finally, has *epiallo* configuration.

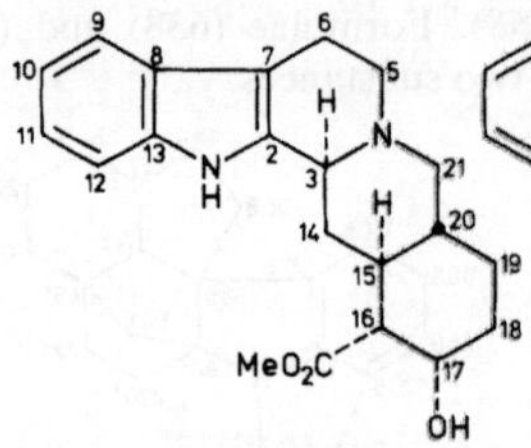

(**640**)

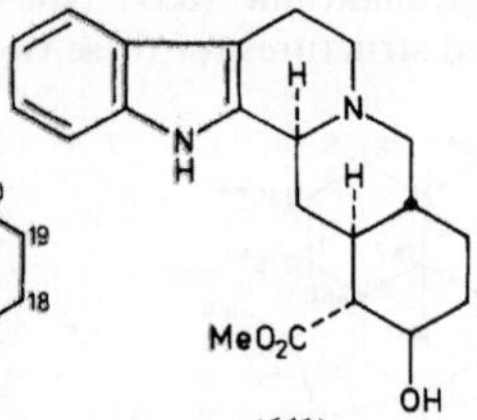

(**641**)

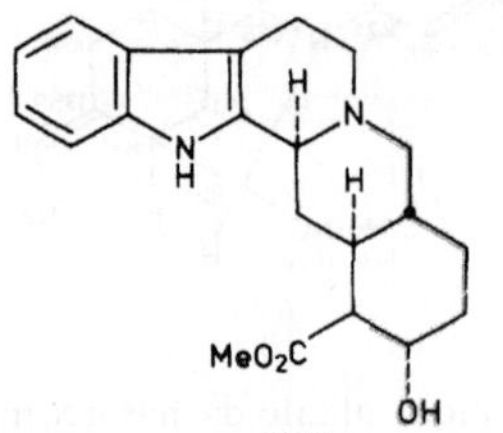

(**642**)

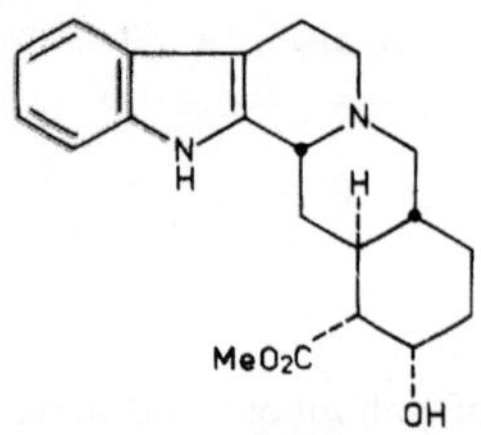

(**643**)

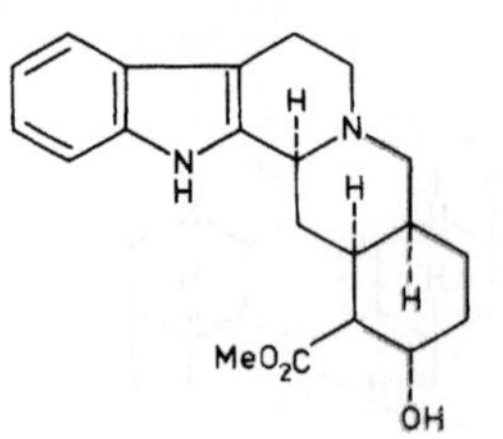

(**644**)

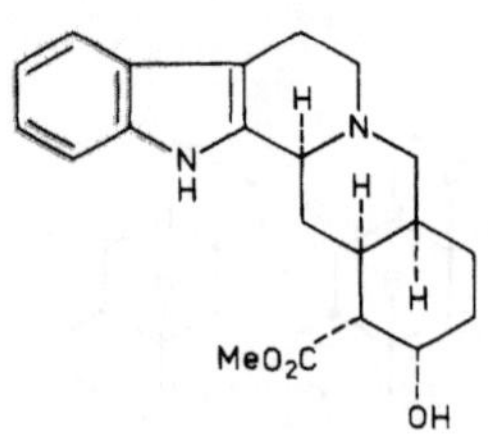

(**645**)

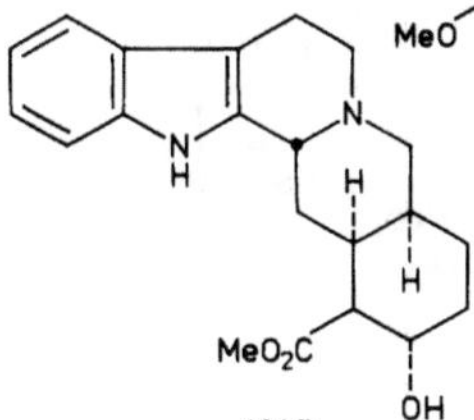

(646)

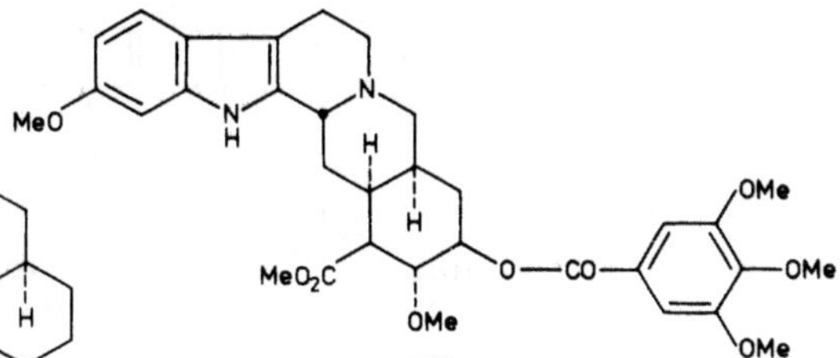

(647)

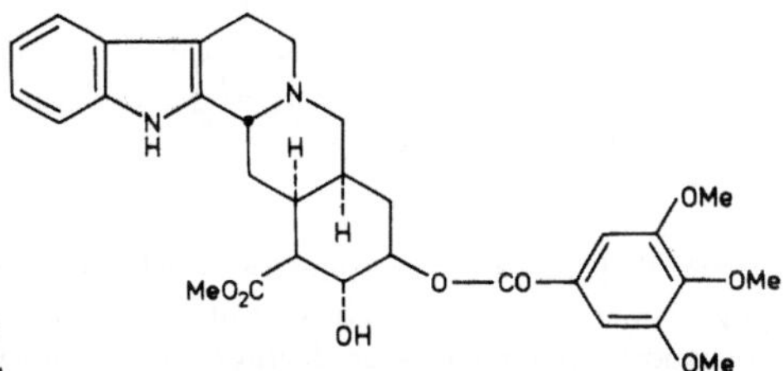

(648)

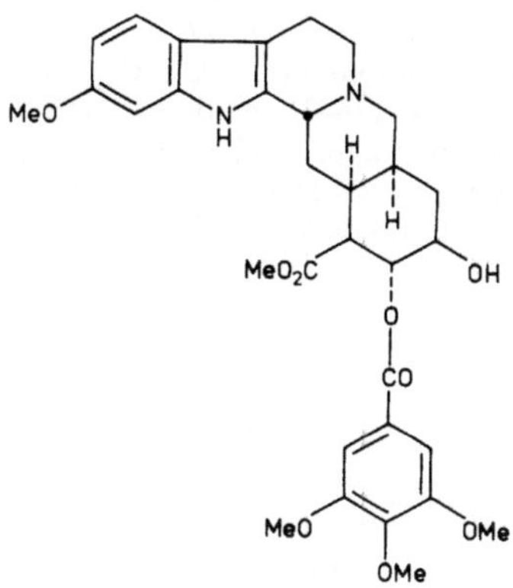

(649)

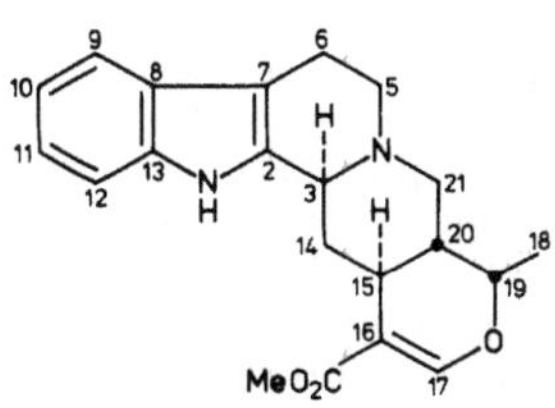

(650)

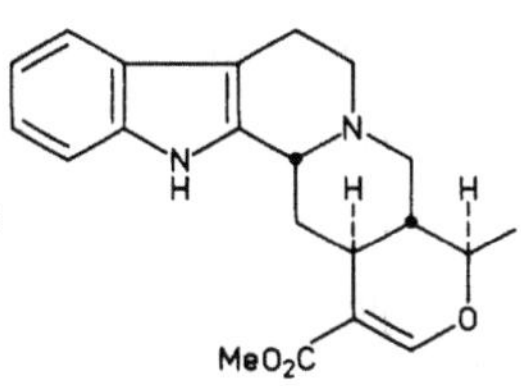

(651)

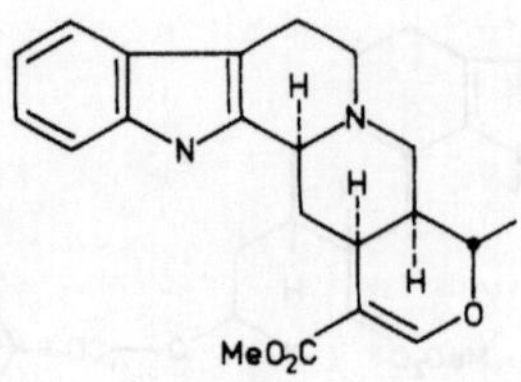

(652)

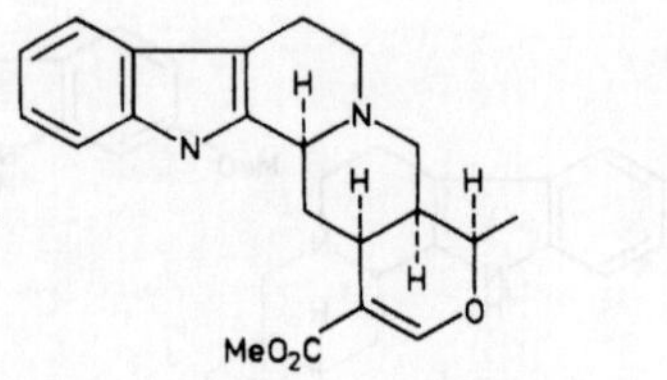

(653)

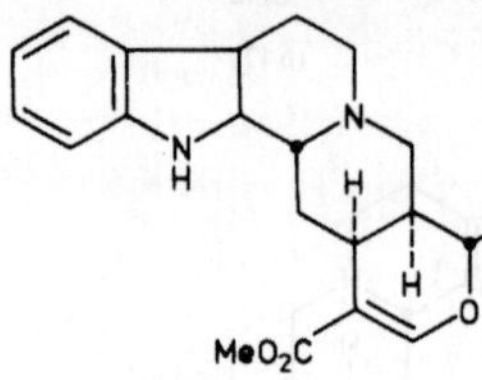

(654)

We will now next consider the shielding implications of the various types of stereochemistry. The transition from *normal* to *pseudo* configuration is accompanied by large chemical shift variations for C-3, C-6, C-15 and C-21. The *cis*-quinolizideine form as it occurs in (**643**), imposes a γ effect on C-15 and C-21, both axially disposed to ring-C, and a reciprocal γ effect on C-6 and C-21 (cf. conformational partial formula representing the pseudo configuration below). No such 1,3-diaxial interactions are

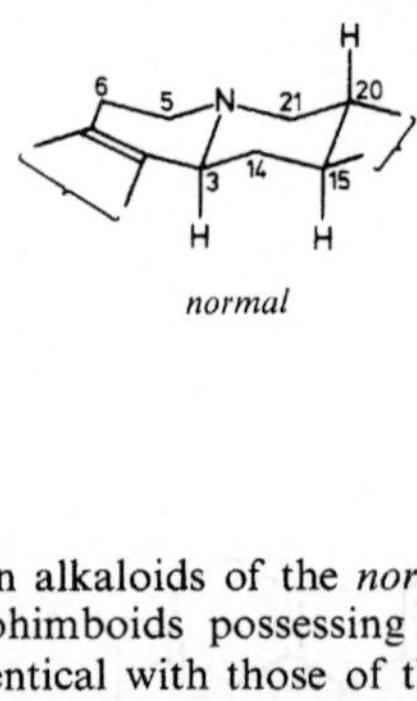

normal

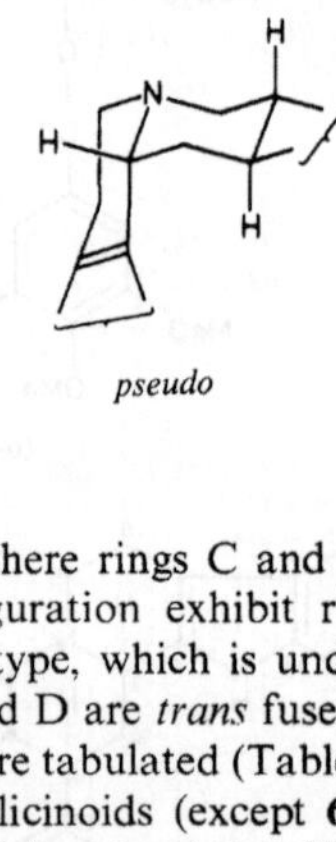

pseudo

possible in alkaloids of the *normal* type where rings C and D are *trans* fused. Yohimboids possessing *allo* configuration exhibit ring-C shifts nearly identical with those of the *normal* type, which is understandable because for both configurations rings C and D are *trans* fused. Chemical shifts for yohimboid alkaloids (**640**—**649**) are tabulated (Table 42).

The ring-C chemical shifts of all ajmalicinoids (except **654**) fit into the patterns of either *trans*- or *cis*-quinolizideine structures. The spectrum of akuammigine (**654**) deviates from the remaining ajmalicinoids in that

Table 42. ^{13}C *Chemical Shift for Yohimboid Alkaloids* (**640—649**) (*169*)

Carbon	(**640**)	(**641**)	(**642**)[a]	(**643**)[a]	(**644**)	(**645**)	(**646**)[a]	(**647**)[b]	(**648**)[c]	(**649**)[c]
2	134.3	134.0	135.8	134.0	134.3	134.4	131.7	130.2	131.3	131.3
3	59.8	59.0	60.5	53.7	60.1	60.1	53.7	53.6	53.5	53.4
5	52.1	52.3	52.6	50.7	53.2	52.8	50.8	51.1	50.8	50.6
6	21.5	21.3	21.6	16.4	21.7	21.3	16.5	16.7	16.4	16.3
7	107.5	107.4	106.3	105.9	108.1	107.1	107.3	107.7	107.5	106.8
8	127.0	126.9	127.0	127.2	127.1	126.8	127.2	121.9	127.3	127.0
9	117.7	117.7	117.5	117.2	117.9	117.5	117.6	118.2	117.6	117.3
10	118.8	118.8	118.4	119.1	119.1	118.6	118.9	108.7	119.1	118.7
11	120.8	120.9	120.4	120.1	121.1	120.5	121.0	155.8	121.2	120.8
12	110.6	110.7	111.1	111.1	110.6	110.6	110.8	95.0	110.7	110.8
13	135.8	135.8	136.1	135.5	135.7	135.8	135.6	136.1	135.5	135.6
14	33.8	33.8[d]	33.6	32.2	27.6	31.0	23.6[d]	24.1	23.7	23.5
15	35.4	41.6	34.7	32.4	37.9	37.4	32.5	32.2	31.9	32.5
16	52.6	57.1	51.1	52.4	54.6	50.6	54.1	51.6	52.1	49.9
17	66.9	71.6	65.9	66.6	66.0	66.7	65.7	77.8[d]	68.3	73.8[d]
18	31.4	33.5	28.2	30.9	33.2	30.2	33.5	77.7[d]	76.9	72.9[d]
19	23.1	27.5	23.5	23.0	24.5	24.8	23.9[d]	29.6	29.1	32.5
20	40.2	39.1	36.5	39.5	36.4	32.0	35.6	33.8	33.7	34.1
21	61.0	60.5	62.0	51.5	60.4	59.6	49.4	48.8	48.7	48.6
C=O	175.1	175.0	172.7	172.9	174.4	174.0	174.7	172.5	172.8	171.8
OMe	51.7	51.6	51.1	51.2	51.8	51.5	51.7	51.6	51.7	51.7

[a] In DMSO-d_6.
[b] δ_C (11-OMe) 55.6, δ_C (17-OMe) 60.5.
[c] δ_C for 3,4,5-trimethoxybenzoyl group: 124.9 (C-1), 106.7 (C-2), 152.5 (C-3), 141.9 (C-4), 56.0 (3-OMe), 60.6 (4-OMe).
[d] Assignments may have to be reversed.

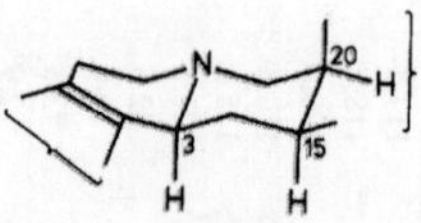

allo

the C-2 and C-6 resonances showed extensive broadening, while no signal at all could be observed for C-21. Moreover, the chemical shift of C-6 was found to be incompatible with either of the two possible confor-

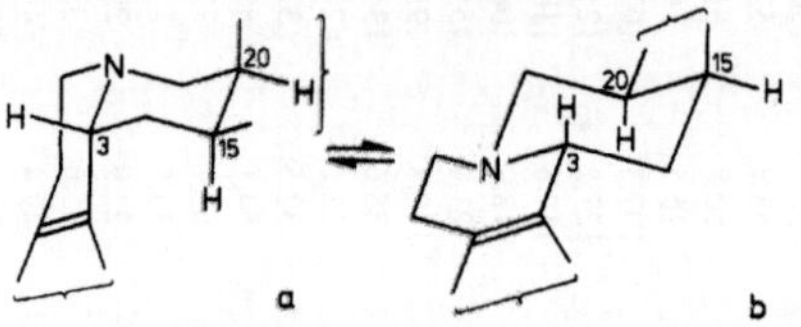

epiallo

Table 43. ^{13}C *Chemical Shifts for Ajmalicinoid Compounds* (**650—654**) (*169*)

Carbon	(**650**)	(**651**)	(**652**)	(**653**)	(**654**)
2	134.0	132.4	134.4	134.3	132.8[a]
3	59.8	53.8	52.6	58.0	54.5
5	52.7	50.9	53.3	52.8	52.2
6	21.3	16.8	21.7	21.1	19.2[a]
7	106.1*	107.4*	107.6	107.1*	106.8*
8	126.6	127.3	126.9	127.0	127.2
9	117.3	117.6	117.8	117.7	117.7
10	118.4	119.1	119.0	119.1	119.1
11	120.5	121.3	120.9	121.0	121.2
12	110.6	111.1	110.6	110.6	110.8
13	135.9	135.7	135.8	135.8	135.7
14	32.1	31.2	34.2	32.5	30.6
15	30.1	30.8	31.2	29.5	25.7
16	106.5*	107.7*	109.3	107.7*	107.6*
17	154.5	155.9	155.5	154.3	154.8
18	14.5	18.0	18.4	19.1	18.4
19	73.3	75.3	72.3	76.4	73.2
20	40.2	43.8	38.3	34.2	37.2
21	56.2	46.8	56.0	53.7	50.3[b]
CO	167.3	167.2	167.8	168.0	167.5
OMe	50.6	50.9	51.0	51.0	50.9

* Assignment may have to be reversed.
[a] Broadened signal.
[b] No signal observed at ambient temperature. Chemical shift indicated was obtained at high temperature.

References, pp. 195—229

mations of the *epiallo* system. This behaviour was explained in terms of a dynamic equilibrium involving rapid interconversion between conformer

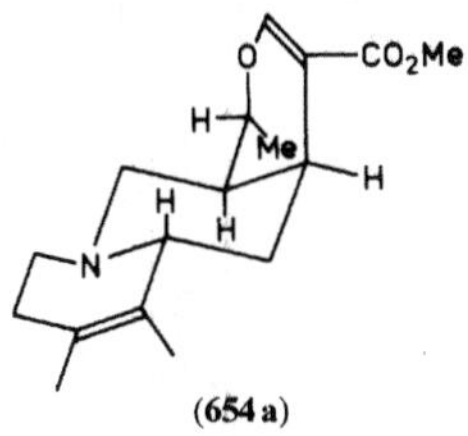

(654 a)

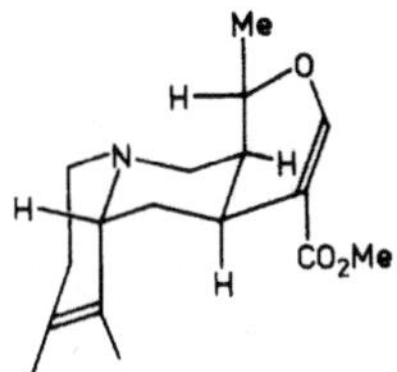

(654 b)

(654 a) and **(654 b)**. By increasing the sample temperature both the signals pertaining to C-2 and C-6 sharpen and C-21 becomes observable. Conversely, slow-exchange limit spectra corresponding to two sets of signals were obtained below room temperature (*169*). From this study it was inferred that ^{13}C shieldings for C-3 and C-6 are stereospecific and thus lend themselves to determining stereochemistry and conformation of yohimboids and ajmalicinoids (*169*). The chemical shift range for these two carbons in the various possible stereochemical environments is given in Table 44.

Table 44. *Characteristic ^{13}C Shielding Ranges for Carbons 3 and 6 in Yohimboids and Ajmalicinoids* (*169*)

Carbon	normal/*allo*	*pseudo*/*epiallo* C/D *cis*	*epiallo*	
			C/D *trans*	C/D *cis* and *trans*
3	60 ±1	53.5±0.5	54.5±0.5	54± 1
6	21.5±0.5	16.5±0.5	21.5±0.5	16–22

Two alkaloids, ochrolifuanine A and B were shown to possess structures **(655)** and **(656)**. These two natural alkaloids were studied by ^{13}C n. m. r. together with the synthetic epimers **(657)** and **(658)**. The assignments were corroborated by Yb $(dpm)_3$-induced shifts where it was shown that the rare earth reagent coordinated exclusively at the secondary nitrogen (*170*). The consistency of C-3 chemical shifts reflects the *trans* quinolizidine conformation for all four isomers. The dissimilarity of the C-17 shieldings, on the other hand, was interpreted as stemming from differen-

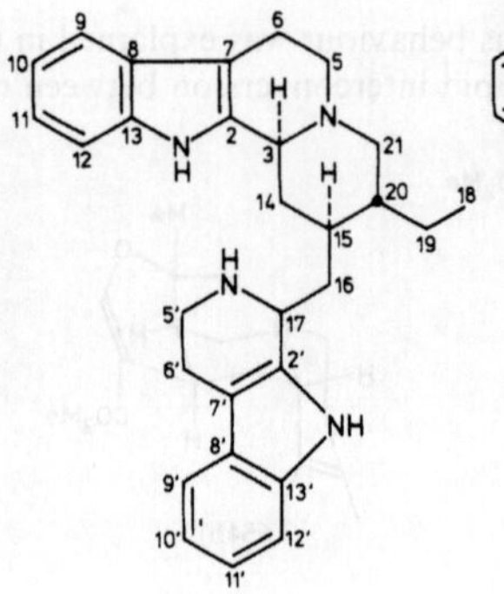

(655) 17β-H
(656) 17α-H

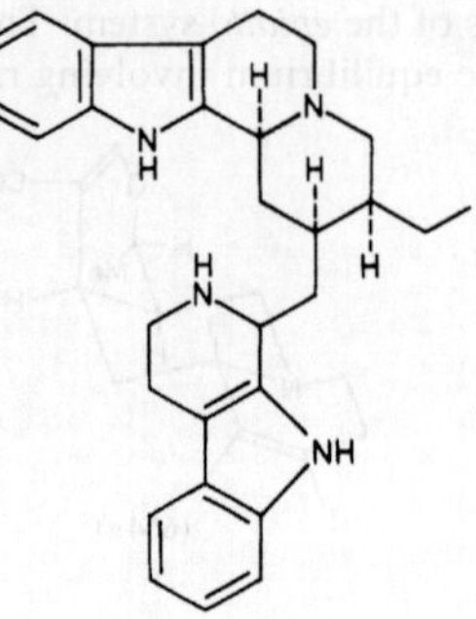

(657) 17β-H
(658) 17α-H

cies in the rotamer populations for the equatorial C-15 substituent. Preferred conformations were discussed in terms of the observed ^{13}C chemical shifts (*170*). Chemical shifts for alkaloids (**655—658**) are provided in Table 45.

Table 45. ^{13}C *Chemical Shifts for Alkaloids* (**655—658**) (*170*)

Carbon	(655)	(656)	(657)	(658)
2, 2′	124.7, 135.5	134.6, 135.5	135.2, 135.4	135.1, 135.4
3	59.3	59.5	59.4	60.3
5	52.6	52.9	53.1	53.1
6	21.5	21.6	21.5	21.6
7, 7′	107.3, 108.1	107.3, 108.6	107.9, 108.7	107.3, 108.4
8, 8′	127.0, 127.3	127.0, 127.2	127.3, 127.4	127.1, 127.2
9, 9′	117.7, 117.9	117.7, 117.9	117.9, 118.0	117.7, 117.7
10, 10′	121.0, 121.3	120.9, 121.6	120.9, 121.4	120.6, 121.2
11, 11′	118.9, 119.0	118.9, 119.3	119.1, 119.2	118.8, 119.0
12, 12′	110.6, 110.9	110.8, 110.8	110.6, 110.6	110.6, 110.6
13, 13′	135.9, 136.1	135.8, 135.9	135.7, 136.0	135.8, 135.8
14	34.3	36.4	31.1	32.4
15	35.8	37.8	35.1	36.1
16	38.1	38.4	38.4	37.8
17	48.8	51.9	49.8	30.0
18	11.0	11.2	12.5	12.4
19	23.2	23.8	18.6	17.5
20	42.2	42.5	41.3	38.3
21	59.9	60.1	57.3	57.5
5′	42.2	42.0	42.3	42.2
6′	22.4	22.4	22.5	22.3

A ^{13}C chemical shift analysis was also carried out for a group of isomeric indole alkaloids isolated from a Malaysian species of *Uncaria* (*171*). The compounds studied were roxburghine D (**659**), E (**660**), C (**661**)

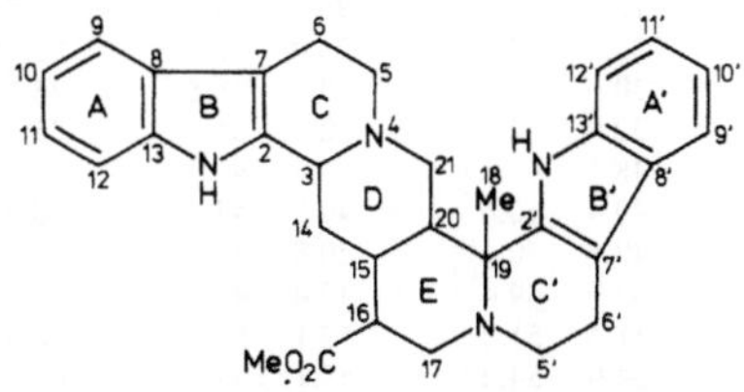

(**659**) H(3β), H(15α), H(20β), C(18α)
(**660**) H(3β), H(15α), H(20β), C(18β)
(**661**) H(3α), H(15α), H(20β), C(18α)
(**662**) H(3α), H(15α), H(20β), C(18β)

and B (**662**). The ^{13}C data confirmed earlier configurational assignments for compounds (**659**—**661**) and identified roxburghine B (**662**) as being of the *epiallo* type. Chemical shift assignment has mainly been accomplished by comparison with model compounds and other ajmalicinoid substances. Chemical shifts for roxburghines (**659**—**661**) are given in Table 46.

The ^{13}C spectrum of geissoschizin (**663**), an alkaloid which for a long time had been considered as a biogenetic intermediate leading to three types of classical indole alkaloids, pointed to a 3α-quinolizidine structure. The unusual shielding found for C-6 (20.4 ppm instead of 16.5 ppm expected on the basis of the data in Ref. *169*) was accounted to a boat conformation for ring C (*172*).

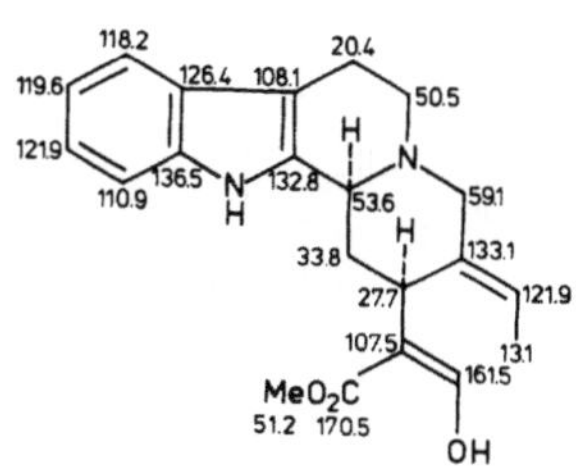

(**663**)

Table 46. ^{13}C *Chemical Shifts for Roxburghines* (**659**—**661**) (*171*)

Carbon	(**659**)[a]		(**660**)[a]		(**661**)
6	17.6	17.1	17.5	17.0	23.3*
Me(18)	18.1	18.7	26.9	26.6	18.7
6′	23.0	22.6	23.0	22.3	22.7*
15	30.6	30.4	29.5	28.8	36.1
14	32.9	32.4	32.8	31.9	35.3
21	48.1	47.8	48.6	46.7	57.9
20	49.4	50.6	49.3	48.4	50.0
OMe	49.4	50.4	50.8	50.7	50.1
5′	49.9	49.4	47.4	47.7	50.7
5	51.4	51.2	52.1	51.3	54.0
3	54.0	54.2	55.1	54.2	60.6
19	57.7	57.6	58.1	57.3	58.5
7′	106.0	107.4	107.6	107.0	107.7
7	108.1	110.2	109.7	109.2	109.9
12, 12′	110.2	111.4	112.2	111.6	111.7
	110.4	111.8	112.4	111.8	111.9
9, 9′	116.4	118.1	118.7	118.0	118.1
	117.0	118.7	118.8	118.0	118.8
10, 10′	117.5	119.5	119.8	119.7	119.2
	118.0	120.1	120.0	119.7	119.7
11, 11′	119.5	121.5	122.1	121.9	121.1
	120.4	122.5	122.4	122.2	122.2
8, 8′	125.5	126.7	127.7	126.9	127.3
	126.8	128.1	128.7	127.6	128.1
13, 13′	135.3	136.2	137.7*	136.5*	137.4*
	135.3	136.2	137.4*	136.1*	137.3*
2′	135.3	136.2	137.2*	134.0*	137.2*
2	132.9	133.6	132.3	132.2	136.5*
C=O	165.4	168.2	167.8	167.8	167.7
16	95.1	95.9	105.7	104.7	96.2
17	144.8	146.9	149.1	148.3	146.9

[a] In acetone-d_6.
* Assignments may have to be interchanged.

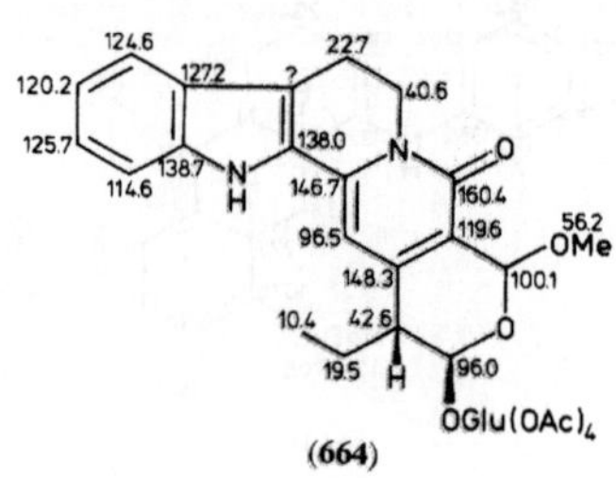

(**664**)

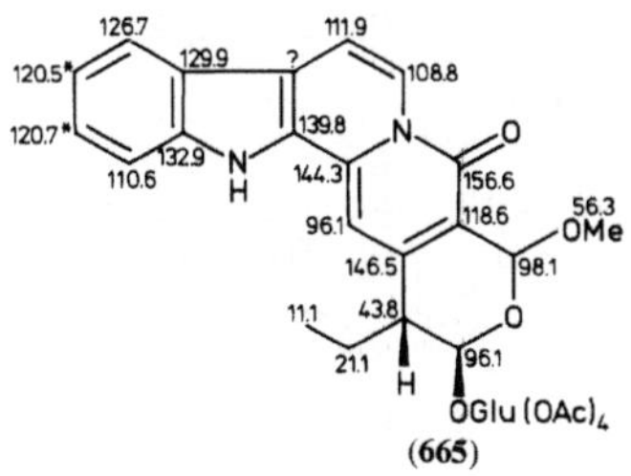

(665)

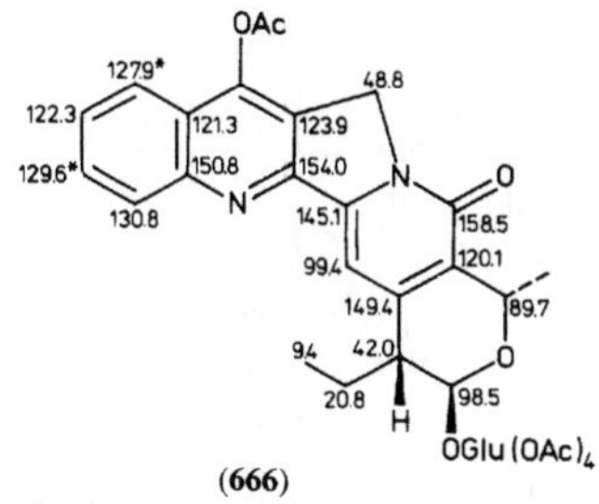

(666)

The structures of alkaloid glucosides (**664**—**666**) were found to be in accord with their ^{13}C spectral data (*173*).

Two communications dealt with the conversion of tabersonine into indolo [2,3-*b*] quinolizine (**667**) whose structure could be determined by means of various spectroscopic techniques and X-ray analysis (*174*, *175*).

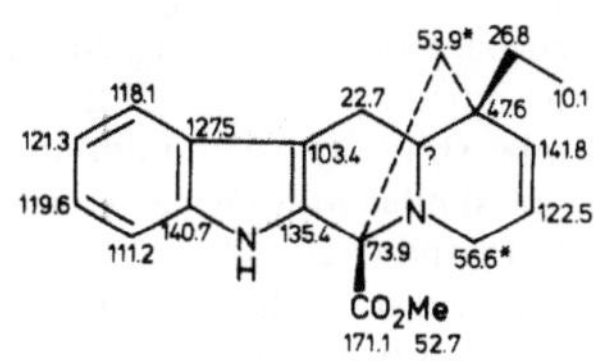

(667)

Chemical and spectroscopic evidence led to structures (**668**) and (**669**) for two novel indole alkaloids designated ibophyllidine and iboxyphylline, resp. The latter structure accomodates a 7-membered heterocycle. ^{13}C chemical shifts are indicated in the formulae (*176*).

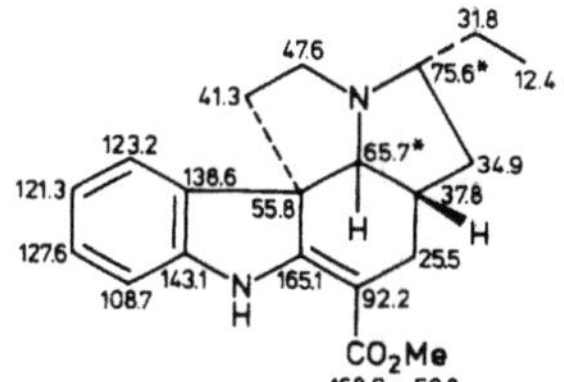

(668)

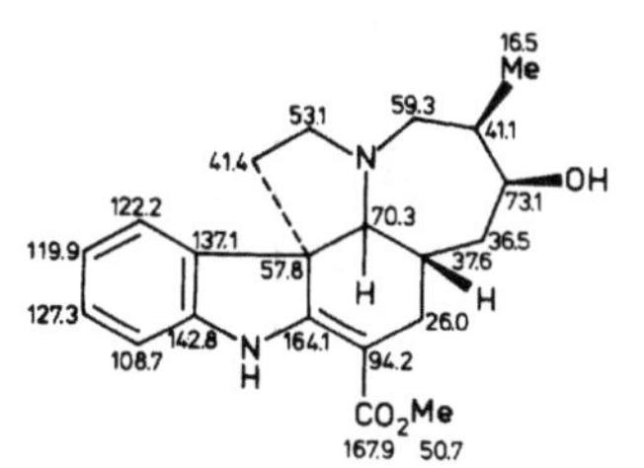

(669)

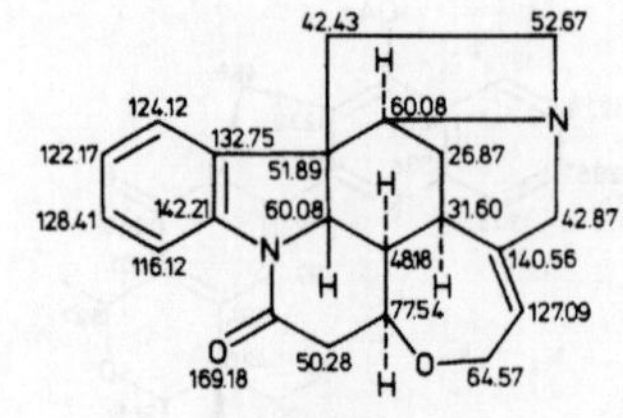

(670)

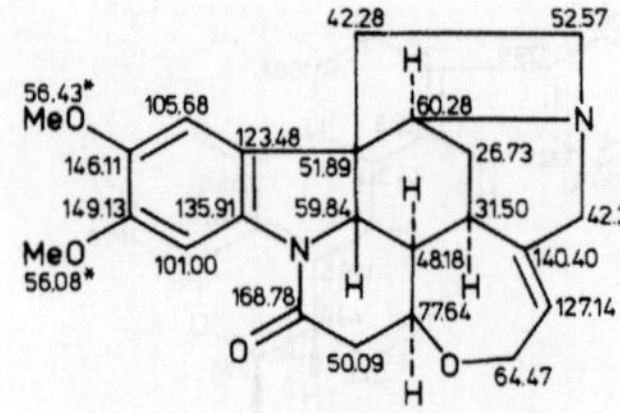

(671)

^{13}C spectral data have recently been reported for strychnine (**670**) and brucine (**671**) as well as for their ions, obtained upon addition of trifluoroacetic acid (*177*). Assignments were made with the aid of residual splitting arguments and protonation shifts. In model amines and amides it was found that amino and amido protonation induces upfield and downfield shifts, resp., on the adjacent carbons. Among the methylene carbons, which could not be assigned in a straightforward manner (C-11, -15, -17, -18 and -20) the resonance at δ_C 50.28 in strychnine moved downfield after addition of TFA ($\Delta\delta_C$ 2.6), thus identifying it as C-11. C-15 was recognized because of its expected large shielding. C-17, -18 and -20, finally, were assigned on the basis of the predicted relative order of the SFORD multiplet splittings. Although an earlier ^{13}C analysis of brucine (*58*) made use of similar arguments, a discrepancy remains in the assignments of C-11, C-18 and C-20.

In the course of a structural study of vincarodine (**672**), its ^{13}C spectrum was analyzed (*178*). Further members of the same family of alkaloids for which ^{13}C n.m.r. data are available, include cuanzine (**673**)

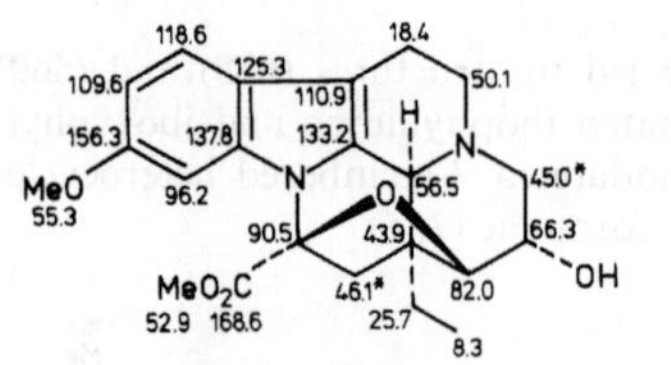

(672)

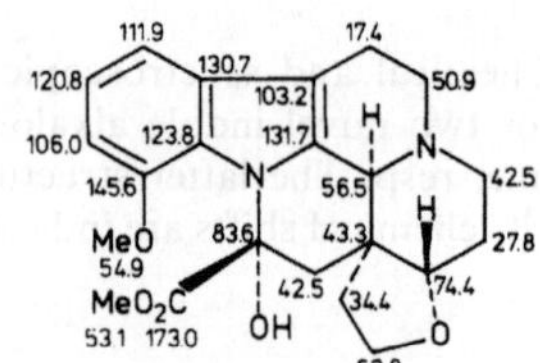

(673)

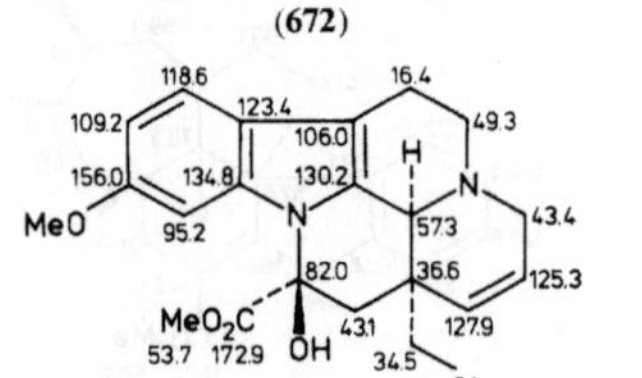

(674)

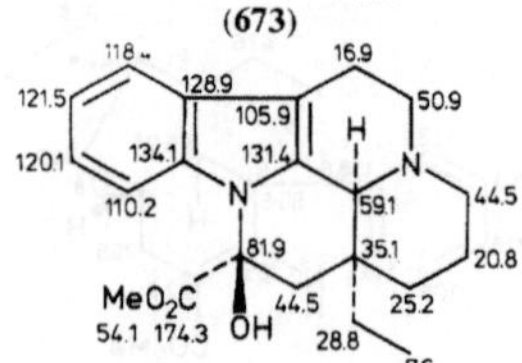

(675)

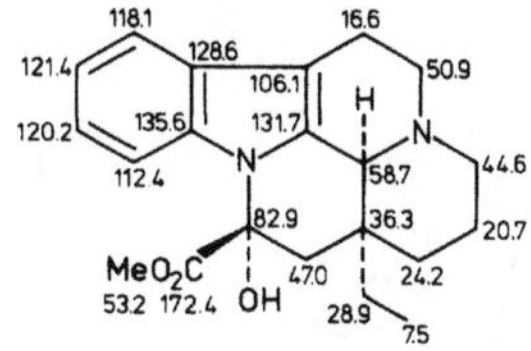

(676)

(*179*) and the related compounds (**674**—**676**) (*178*, *179*). ^{13}C spectroscopy was employed in structure investigations of pandine (**677**) and its derivative (**678**) (*180*), andraginine (**679**) (*181*) as well as the rearrangement product (**680**) from catharanthine N-oxide (*182*).

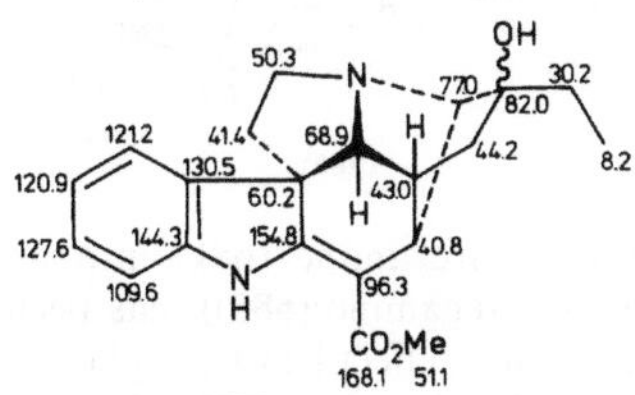

(677)

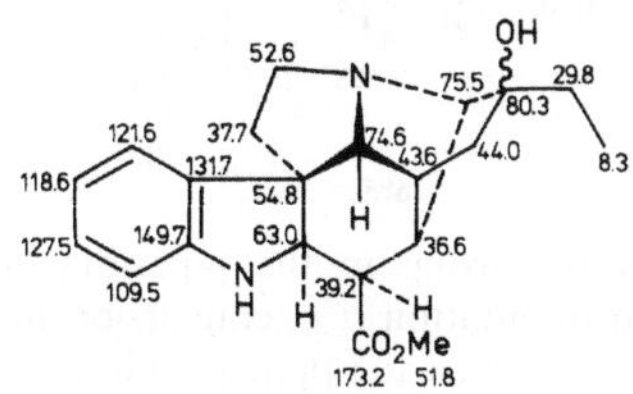

(678)

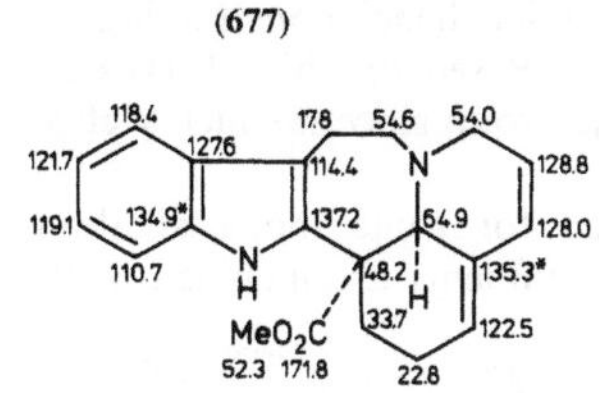

(679)

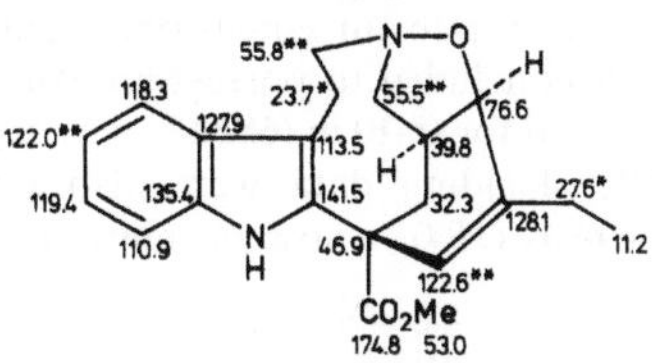

(680)

A thorough ^{13}C n.m.r. analysis has been performed on a series of vobasine-like indole alkaloids (*183*). Compounds studied in this context comprise systems (**681**—**686**). Conversion of the equatorial carbo-

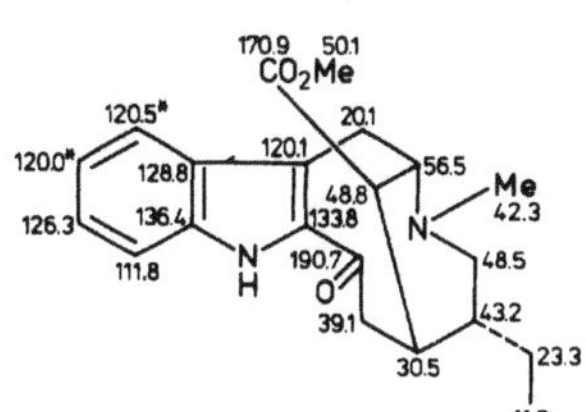

(681)

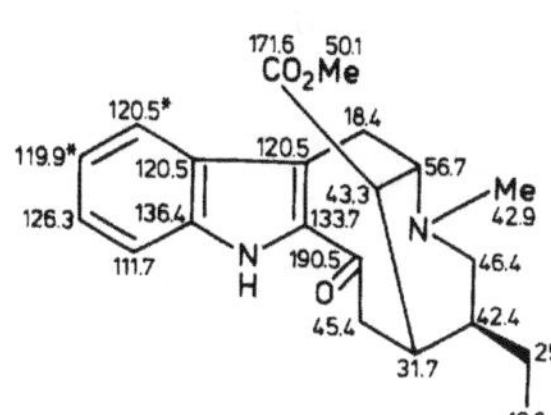

(682)

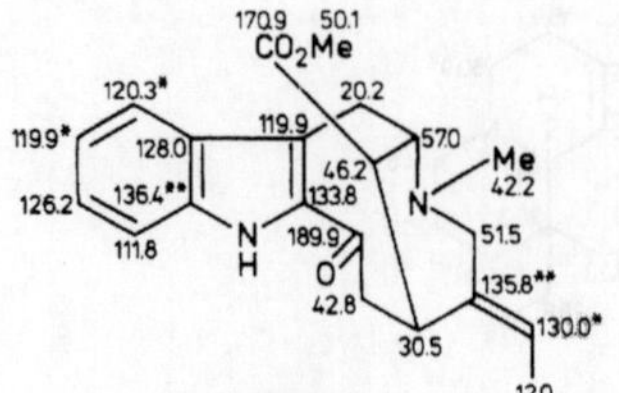

(683)

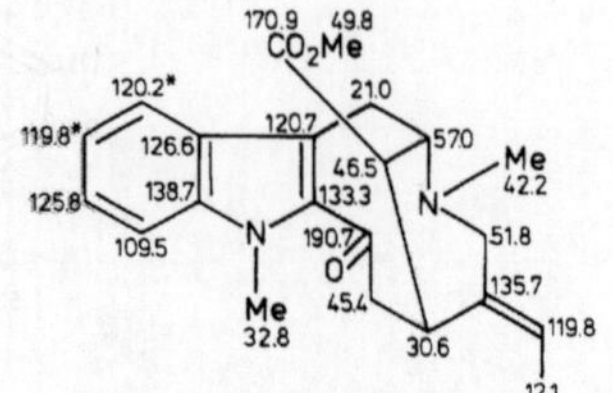

(684)

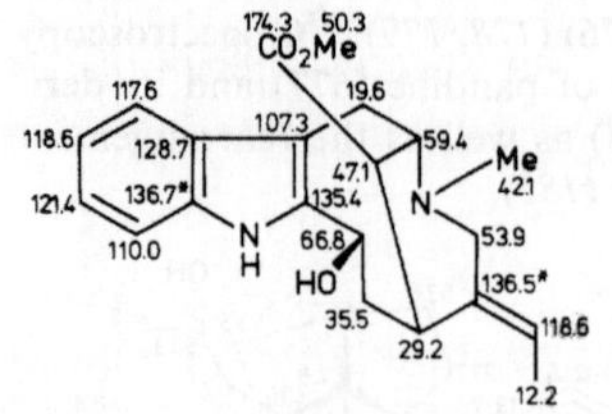

(685)

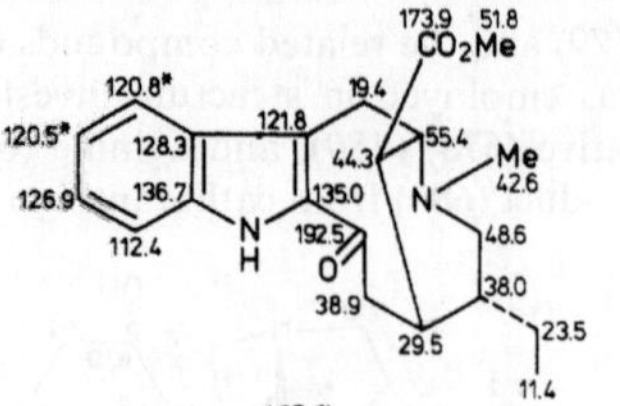

(686)

methoxy group in the piperidine moiety of dregamine (**681**) into an axial orientation (i.e. conversion into 16-*epi*-dregamine (**686**)), has been found to cause a 5 ppm shielding on C-20. This γ effect can be explained only by assuming proton C(20)-H to be axial, hence constraining the ethyl group into an equatorial orientation. Based on this observation (**682**) is concluded to possess an axial ethyl group since no such γ effect is observed for C-20 in (**682**).

^{13}C shielding data were also reported for some iboga alkaloids (**687**—**694**) (*184*), whose chemical shift assignments are included in the

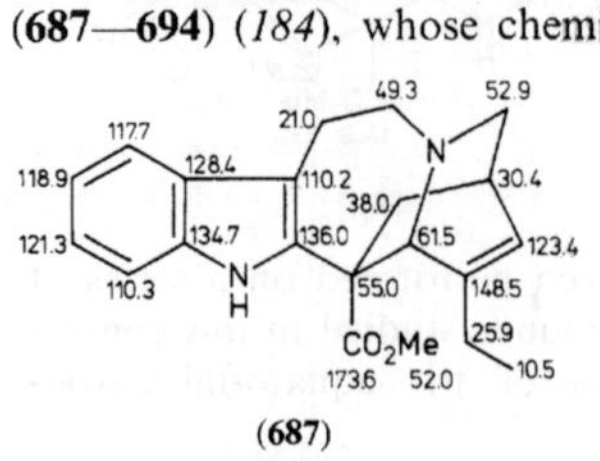

(687)

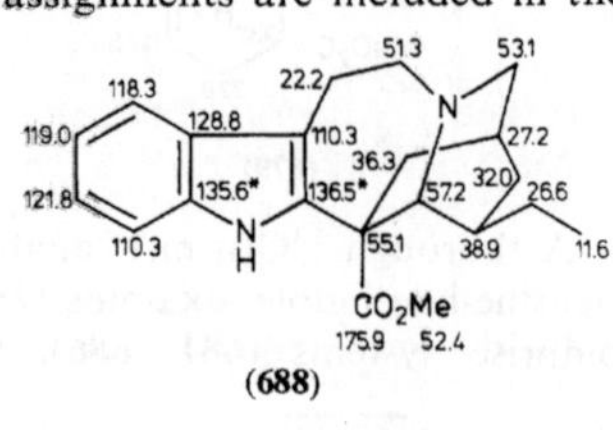

(688)

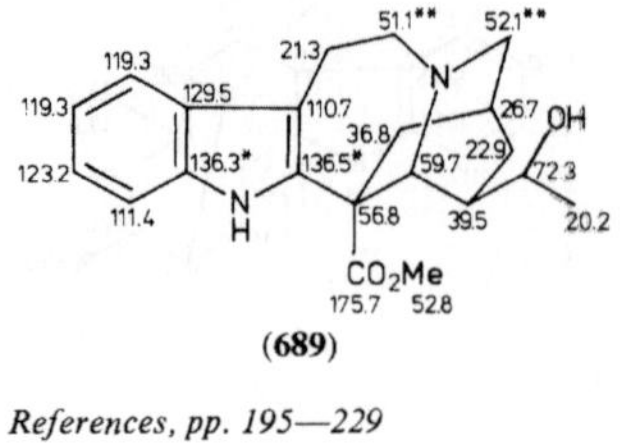

(689)

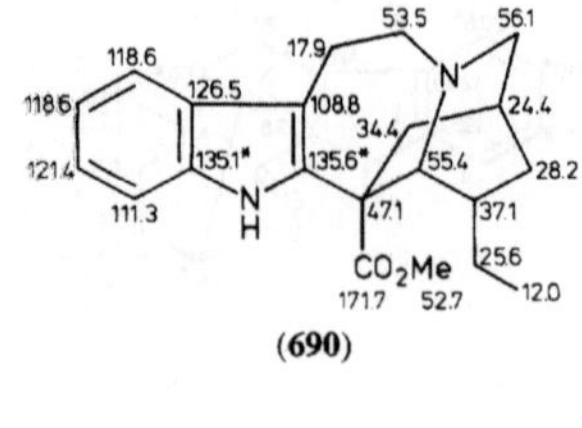

(690)

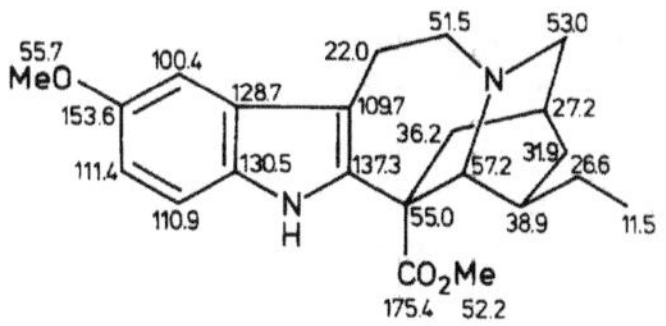

(691)

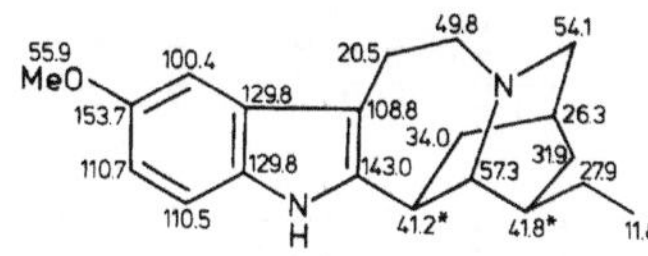

(692)

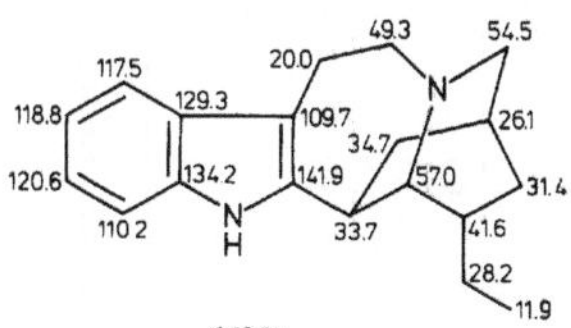

(693)

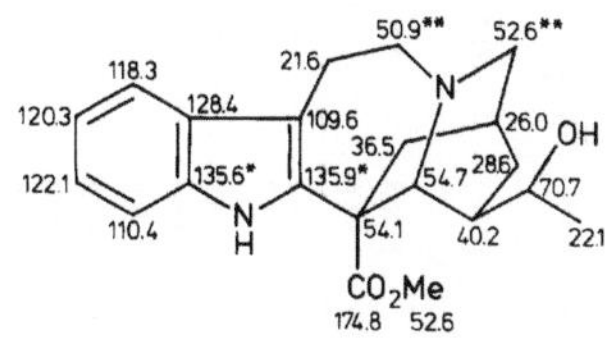

(694)

formulae. The general shift pattern for dihydrocatharanthine **(690)** was found to deviate completely from that of the remaining iboga alkaloids. It was therefore suggested that **(690)** possesses an unusual conformation. Evaluation of stereochemical shifts (γ effects) permitted deduction of conformation **(A)** for compounds **(687—689)** and **(691)**, while for **(690)** the data pointed to a conformation of the type depicted by stereostructure **(B)**.

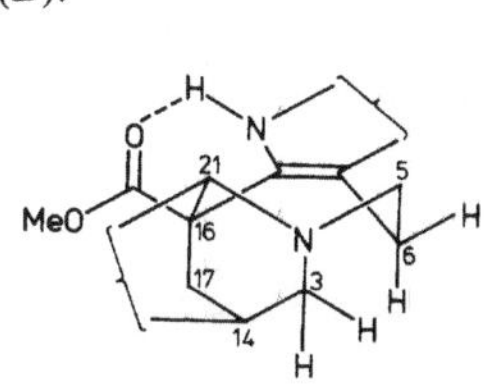

(A)

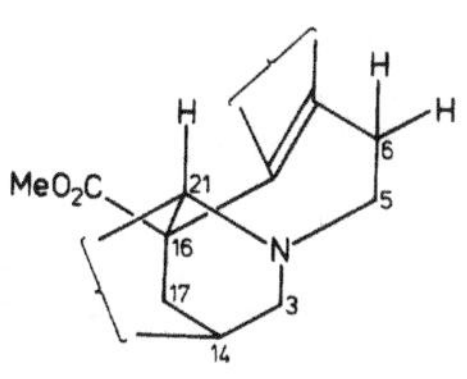

(B)

Heyneanine **(689)** and *epi*-heyneanine **(694)** exhibit nearly identical ^{13}C shieldings, with the exception of C-15 and C-21. For the two compounds, which are epimers with respect to the carbinyl configuration, sidechain conformations **(689 A)** and **(694 A)** have been suggested. In

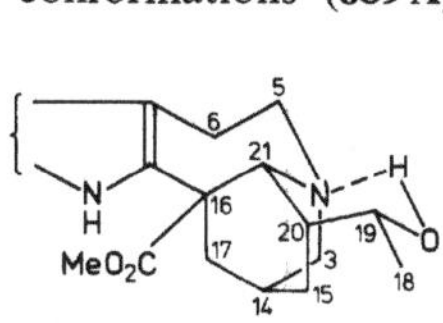

(689 A)

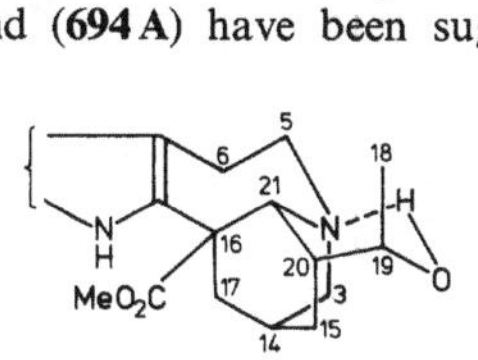

(694 A)

Table 47. ^{13}C *Chemical Shifts for* 16β-*Substituted and Related Unsubstituted Quebrachamines and Cleavamines* (*185*)

Carbon	(695)	(697)	(698)	(700)	(703)	(705)	(706)	(711)	(714)	(715)	(713)	(709)[a]
2	139.7	135.2*	141.2	135.1*	138.1	138.4	**	139.4	138.5	**	135.0	139.0
3	54.9	55.0	55.1	54.4	47.6	47.0	47.3	51.7	50.6	50.9	50.6	48.4
5	53.2	52.7	53.0	51.5	51.7	51.2	51.3	53.2	52.3	52.0	52.5	52.8
6	21.7	21.8	22.0	21.3	22.2	21.7	21.7	24.1*	22.7	22.4	22.1	21.5
7	108.3	109.4	109.1	109.1	109.1	109.5	109.0	108.7	108.0	109.2	109.5	108.2
8	128.6	127.7	127.8	127.6	128.1	127.9	128.4	128.8	127.4	127.7	127.8	127.6
9	117.1	117.4	117.2	117.3	117.3	117.5	117.5	117.3	116.8	117.5	117.4	116.7
10	118.4	118.5	118.4	118.5	118.5	118.6	118.5	118.5	118.4	119.0	118.6	117.5
11	119.9	120.6	120.1	120.6	120.4	120.9	120.8	120.1	120.4	121.4	120.7	119.3
12	109.9	110.5	110.3	110.4	110.3	110.6	110.4	109.8	110.8	111.1	110.5	110.2
13	134.5	134.9*	134.8	134.7*	134.8	135.0	**	134.7	135.2	134.0	135.0	135.2
14	22.6	22.3	22.5	124.6	34.0	34.1	33.4	33.8	30.1	30.5	32.8*	35.3*
15	33.4	33.9	34.1	135.4	124.7	124.0	124.4	37.6	40.4	39.5	36.7*	36.0
16	22.4	37.8	33.7	38.1	36.6*	39.3	75.8	23.3*	22.7	39.3	39.0	38.5
17	34.7	38.6	35.8	44.1	36.2*	39.1	41.8	31.9	31.5	36.1	36.3*	34.1
18	7.8	7.4	7.6	8.3	12.3	12.3	12.3	11.4	6.9	6.9	11.3	12.2
19	32.0	30.6	31.0	29.3	27.3	27.3	27.3	27.5	32.3	32.6	27.3	28.0
20	36.9	37.9	37.6	40.9	**	138.4	138.2	32.9	71.6	71.2	33.1*	32.6*
21	56.6	56.7	56.8	51.5	57.1	57.5	57.5	61.2	65.8	66.1	61.3	56.3
C=O		176.2		175.6		175.8				175.2	176.1	
OMe		52.0		52.0		52.2	57.4			52.2	52.1	
OCH_2			67.4		66.6							65.2

[a] In DMSO-d_6. * Assignments may have to be interchanged. ** Quaternary carbon signal, not detected.

(694 A) the 1,3-diaxial interaction between C-18 and C (21)-H should be accompanied by an upfield shift for C-21 relative to **(689 A)**, where no such interaction is present. This is indeed observed (δ_C 54.7 vs. 59.7). Conversely C-18 is expected to sterically interfere with C-15 in **(689 A)**, but not in **(694 A)** (δ_C 22.9 vs. 28.6).

Conformational information was also obtained from a ^{13}C n.m.r. study of some 16α- and 16β-substituted derivatives of quebrachamine **(695—698)**, 14,15-dehydroquebrachamine **(699, 700)**, clevamine **(701** to **707)** and 15,12β-dihydroclevamine **(708—715)** (*185*). Chemical shifts for

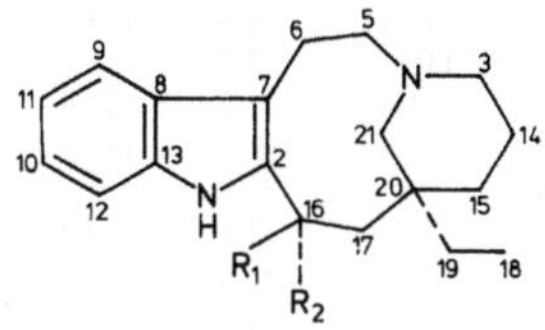

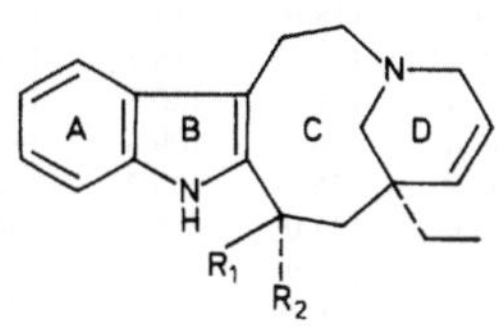

(695) $R_1 = R_2 = H$

(696) $R_1 = H, R_2 = CO_2Me$

(697) $R_1 = CO_2Me, R_2 = H$

(698) $R_1 = CH_2OH, R_2 = H$

(699) $R_1 = H, R_2 = CO_2Me$

(700) $R_1 = CO_2Me, R_2 = H$

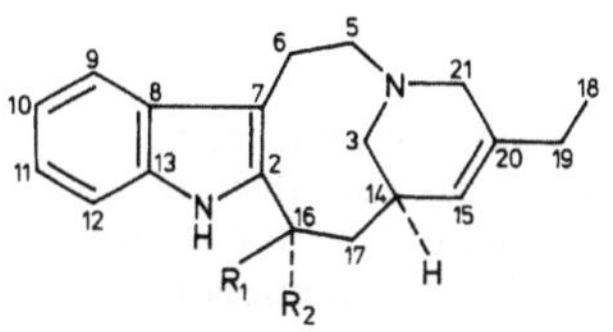

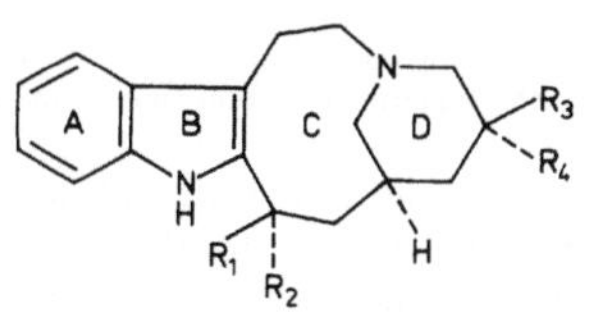

(701) $R_1 = R_2 = H$

(702) $R_1 = H, R_2 = CO_2Me$

(703) $R_1 = CH_2OH, R_2 = H$

(704) $R_1 = H, R_2 = OMe$

(705) $R_1 = CO_2Me, R_2 = H$

(706) $R_1 = OMe, R_2 = H$

(707) $R_1 = H, R_2 = CH_2OH$

(708) $R_1 = R_2 = R_4 = H, R_3 = Et$

(709) $R_1 = CH_2OH, R_2 = R_4 = H, R_3 = Et$

(710) $R_1 = R_4 = H, R_2 = CO_2Me, R_3 = Et$

(711) $R_1 = R_2 = R_3 = H, R_4 = Et$

(712) $R_1 = R_3 = H, R_2 = CO_2Me, R_4 = Et$

(713) $R_1 = CO_2Me, R_2 = R_3 = H, R_4 = Et$

(714) $R_1 = R_2 = H, R_3 = OH, R_4 = Et$

(715) $R_1 = CO_2Me, R_2 = H, R_3 = OH, R_4 = Et$

16β-substituted systems are listed in Table 47; those related to 16α-substituted quebrachamines and cleavamines can be found in Table 48.

Table 48. ^{13}C *Chemical Shifts for 16α-Substituted and Related Unsubstituted Quebrachamines and Clea Amines (185)*

Carbon	(696)	(699)	(701)	(702)	(704)	(712)	(710)	(708)
2	133.7	134.4	139.2	134.2	**	133.7	133.8	138.4
3	53.8*	52.0	53.5*	52.5*	53.1*	55.8	51.2*	51.4*
5	54.0*	53.7	53.8*	53.2*	53.9*	54.1	51.8*	52.2*
6	26.2	26.0	26.1	26.0	26.1	26.5	26.4	26.0
7	111.5	111.5	109.5	110.9	112.2	111.4	111.8	109.6
8	127.6	127.8	128.5	127.5	**	127.6	127.6	128.3
9	117.9	118.0	117.6	117.7	118.1	118.0	118.1	117.6
10	118.7	118.7	118.5	118.5	118.6	118.7	118.8	118.6
11	121.4	121.3	120.3	121.0	121.4	121.2	121.4	120.5
12	110.6	110.5	109.8	110.3	110.4	110.5	110.5	109.8
13	135.7	135.7	135.2	135.5	**	135.6	135.7	d
14	23.6	127.0	35.3	34.3	34.4	31.1	34.8*	35.0*
15	37.3	132.9	122.3	121.5	122.0	39.0*	31.0	31.2
16	40.9	39.1	22.4	38.3	72.8	42.0	37.5	21.3
17	42.8	43.4	34.1	37.5	41.5	40.3*	38.5	33.7
18	7.3	7.7	12.6	12.3	12.6	11.4	11.7	11.7
19	35.6	33.1	27.6	27.4	27.5	27.7	28.6	28.7
20	35.6	39.5	140.4	140.8	141.9	36.1	32.1*	32.8*
21	60.8	58.6	55.1	54.9	54.9	60.6	58.9	58.7
C=O	175.6	175.6		175.3		175.4	175.3	
OMe	51.9	51.8		51.8	55.7	52.0	52.0	

* Assignments may have to be interchanged.
** Quaternary carbon signal, not detected.

In principle, four different conformational modes of attachment of the 9-membered ring to the piperidine moiety are conceivable; however, diequatorial and diaxial disposition of substituents C-5 and C-17 had to be rejected for energetic reasons. The most likely conformers therefore are those designated **(A)** and **(B)**. In the case of conformation A, the

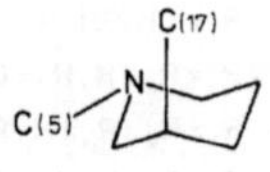

(A)

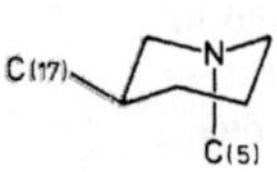

(B)

methylene protons α to the piperidine nitrogen should be highly anisochronous because of the *trans*-diaxial disposition of the axial methylene proton and the nitrogen lone pair. In the SFORD spectra double doublets rather than triplets should therefore be obtained for these methylene resonances (cf. Sect. 2, p. 4). Such a SFORD pattern was indeed observed for C-21 in all but one **(709)** of these compounds, thus assigning conformation (**A**) to these systems. On the basis of the chemical shift assignments listed in Tables 47 and 48, the structure and stereochemistry of a synthetic coupling product of vindoline and cleavamine could be determined. The structure of this substance was identified as 16′β(10-vindolyl) cleavamine (**716**) (*185*).

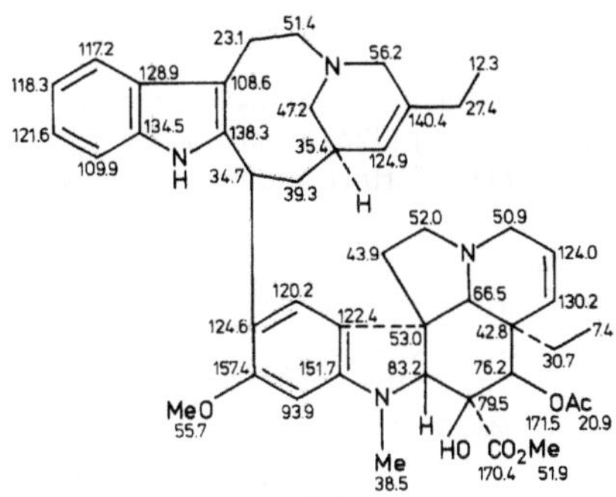

(**716**)

The ^{13}C spectra of pandoline (**717**) and *epi*-pandoline (**718**) showed markedly different shieldings for C-13, C-16 and C-19, but the data did not permit the C-20 configuration to be established. However, chemical shift comparison with the corresponding *seco* derivatives (**719—721**) furnished the desired stereochemical information (*186*).

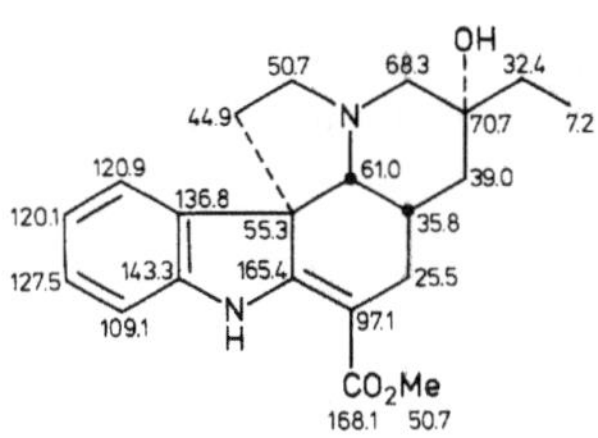

(**717**)

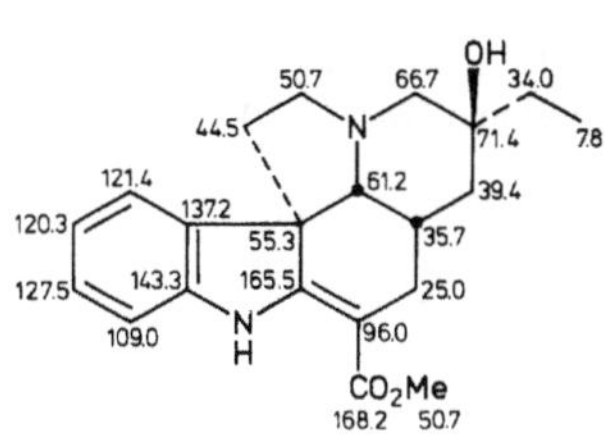

(**718**)

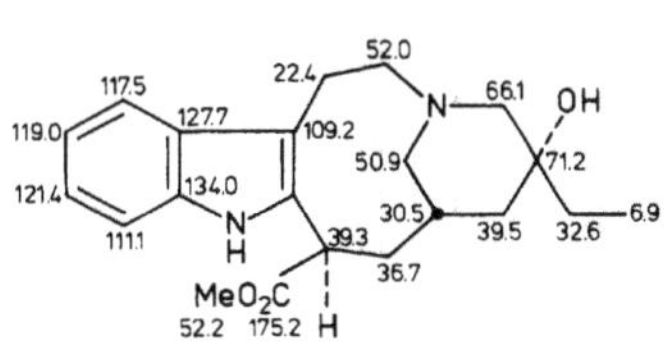

(**719**)

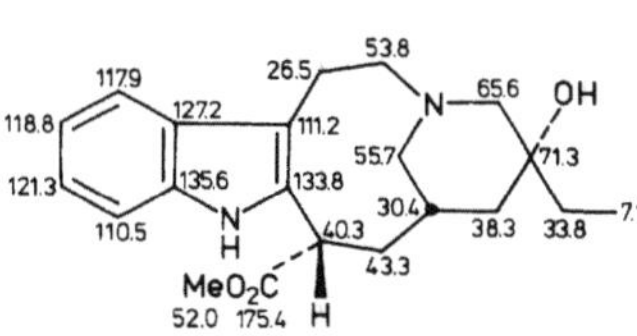

(**720**)

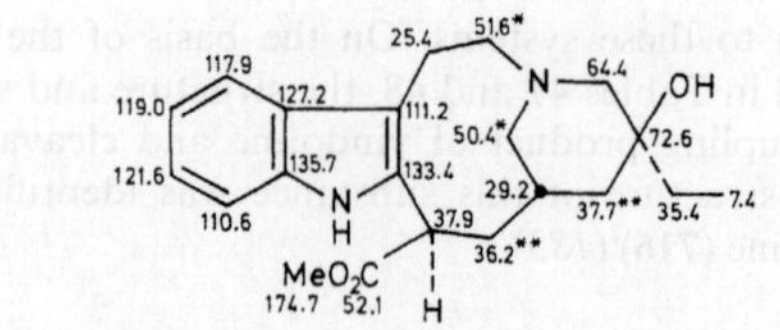

(721)

Complete ^{13}C spectral assignments have been given for vincaleucoblastine (= vinblastine, VLB) (**722**) as well as for some related dimeric indole alkaloids (*187*, *188*). Further structural studies of dimeric indole

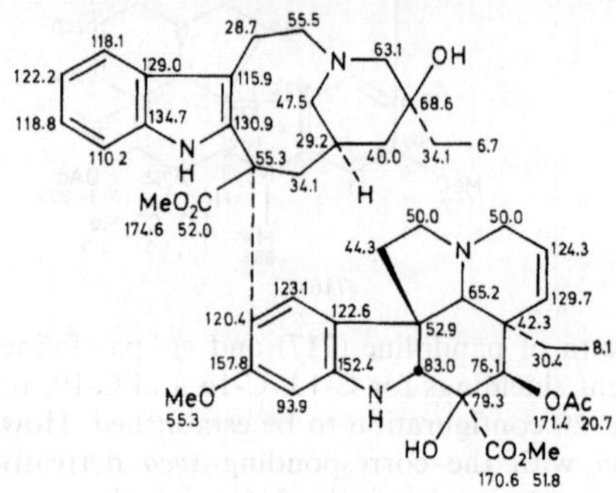

(722)

alkaloids include the following substances: vincathicin (**723**) (*189*), criophylline (**724**) (*190*), pleiocorine (**725**) (*191*), bonafousine (**726**) (*192*),

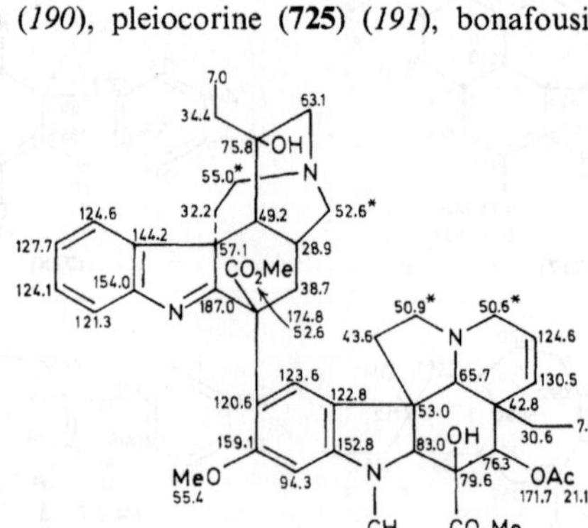

(723)

12-bis[11-hydroxycoronaridinyl] (**727**) (*193*), tabernaelegantines A—D and tabernaelegantinine A (**728**) and B (**729**). Chemical shifts for these systems are displayed in the respective formulae.

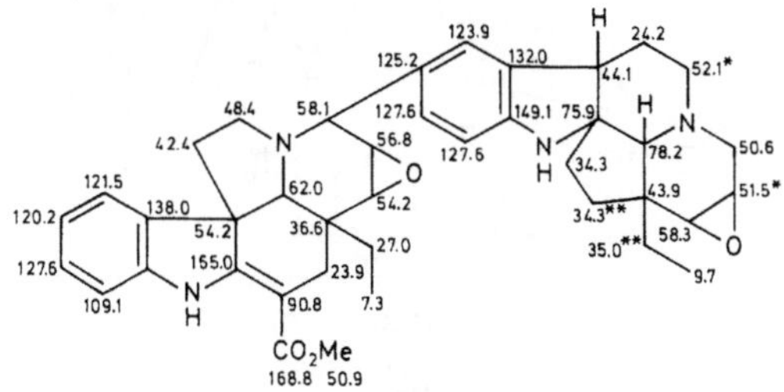

(**724**)

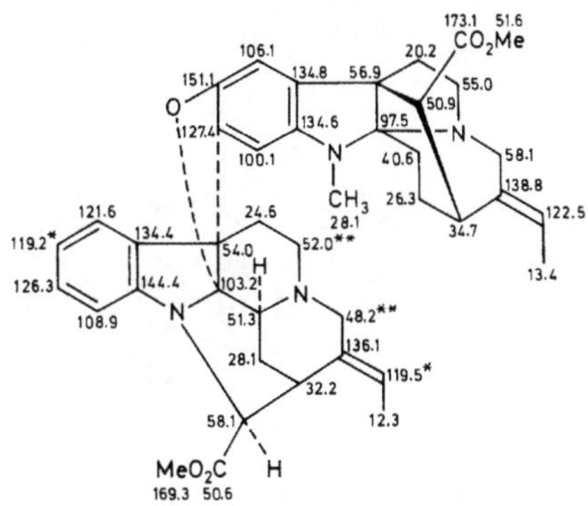

(**725**)

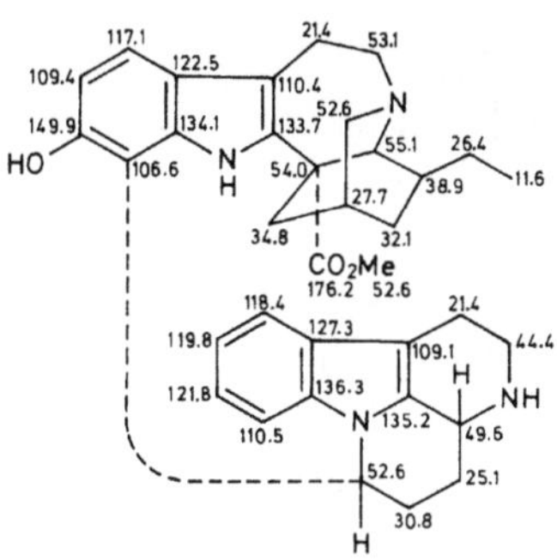

(**726**)

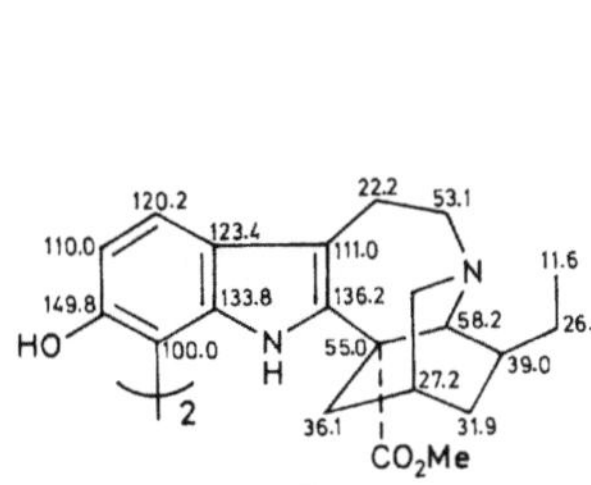

(**727**)

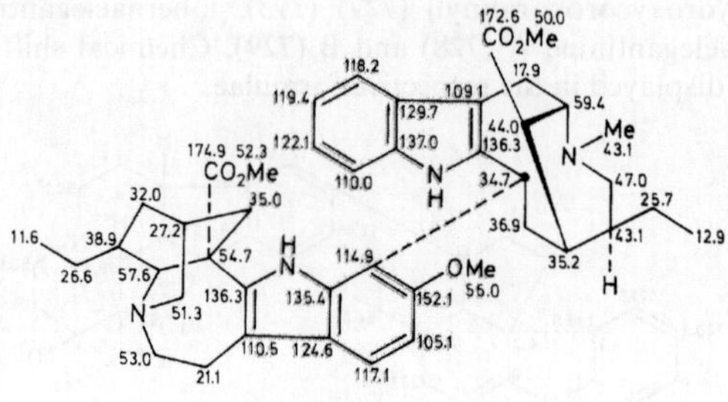

(728)

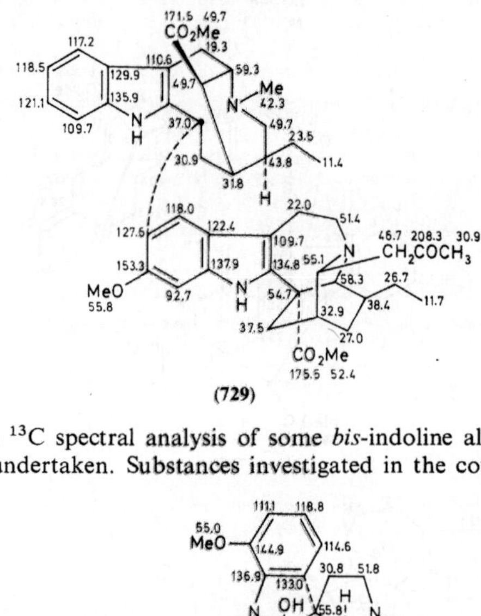

(729)

A complete ^{13}C spectral analysis of some *bis*-indoline alkaloids has recently been undertaken. Substances investigated in the course of this

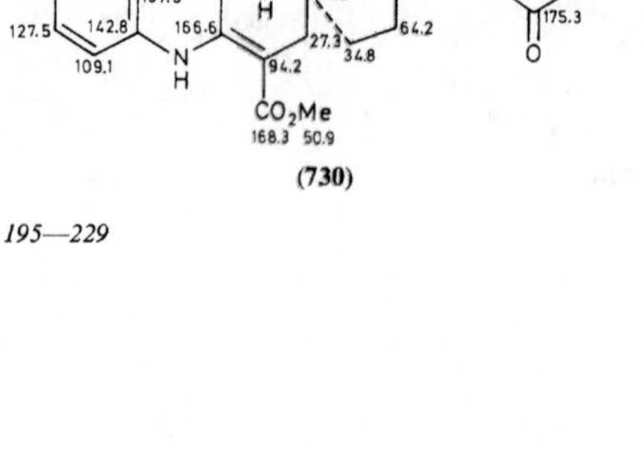

(730)

work include vobtusine, 2′-deoxyvobtusine and vobtusine lactone (**730**). For this latter ^{13}C chemical shifts are given along with the formula. On the basis of this data set, the structures of related alkaloids vobtusine 3-lactam, vobtusine 3-lactam N_b'-oxide, voafolidine, voafoline, isovoafoline, folicanginine, subsessiline and subsessiline lactone could be determined (*195*).

Oxaline, a novel alkaloid from *Penicillium oxalicum*, to which structure (**731**) was assigned, has been thoroughly studied by ^{13}C n.m.r. spectroscopy (*30*). Unequivocal distinction between quaternary carbons was made on the basis of proton-coupled spectra and selective double resonance experiments (cf. Sect. 2, p. 11).

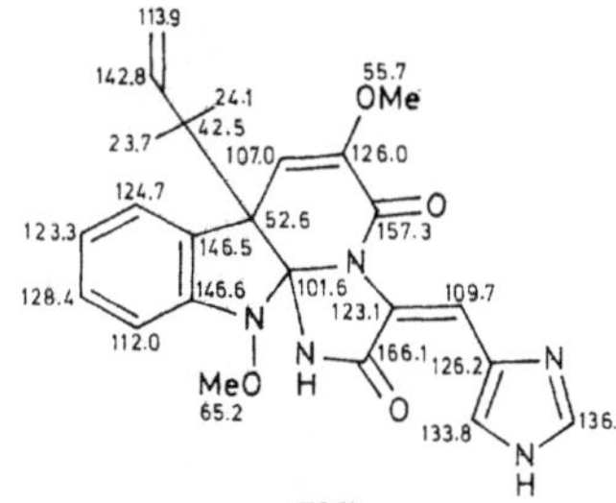

(**731**)

A further class of indole alkaloids whose ^{13}C n.m.r. data have become available includes the ergot alkaloids. Among the compounds studied are methyl lysergate, its dihydro derivative, ergonovine, ergonovinine, ergotamine, ergotaminine, ergokryptidine, agroclavine, elymoclavine acetate, festuclavine and fumigaclavine B (*196*) as well as the methyl-10-methoxy-dihydrolysergates (*197*). The stereochemistry of these systems has been discussed in terms of the observed shieldings. Because of the vast amount of available data, reproduction of chemical shifts is confined to one member (methyl lysergate **732**) of the family.

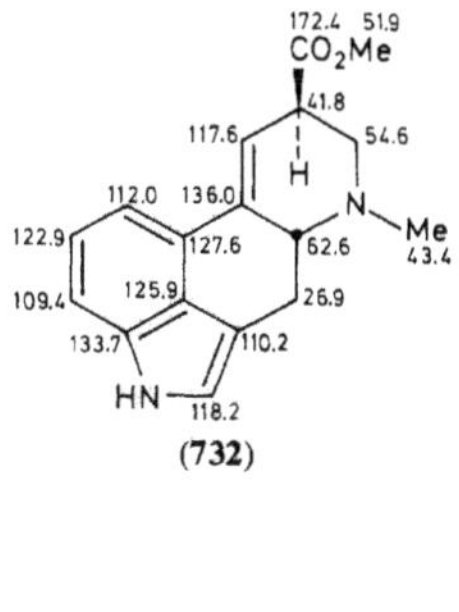

(**732**)

Epimers among the cinchona alkaloids were found to exhibit characteristically different shift patterns (*198*). Quinine (**733**) and quinidine (**734**), which differ in their C-2 configuration, have markedly different C-2 and C-6 shieldings, which are straightforwardly interpreted in terms of *γ-gauche* effects. Chemical shifts for molecules (**733**—**735**) are incorporated in the pertinent formulae. One of the human biotransfor-

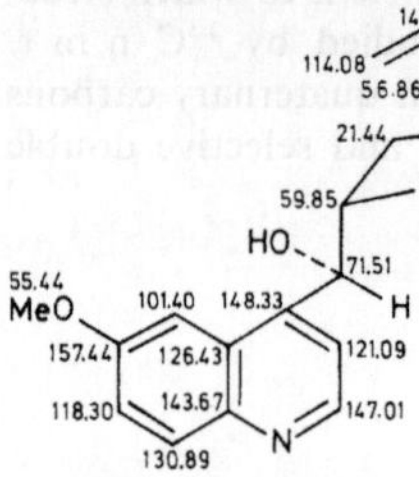

(**733**)

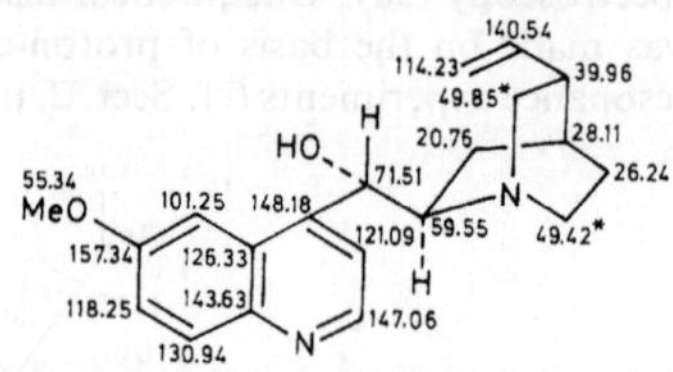

(**734**)

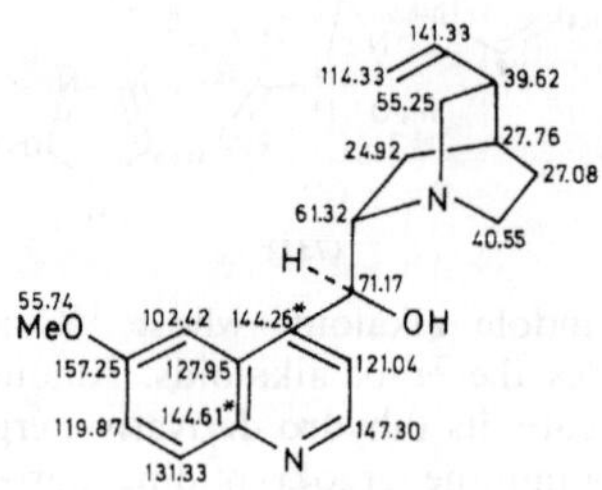

(**735**)

mation products of quinidine has recently been identified as 3-hydroxyquinidine. In order to classify the stereochemistry at C-3, epimers (**736**) and (**737**) were synthesized and their ^{13}C n.m.r. spectra analyzed in conjunction with related compounds. The data permitted the absolute configuration at C-3 to be determined and the biotransformation product could be assigned 3 S configuration (*199*).

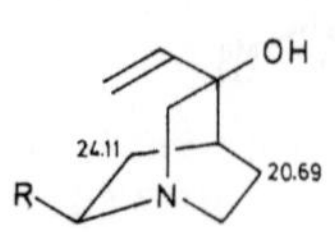

(**736**)

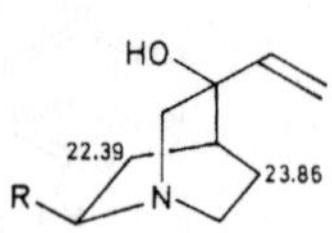

(**737**)

Full ^{13}C chemical shift assignments have also been given for quinoline alkaloids meloscine (**738**), *epi*-meloscine (**739**), scandine and meloscandine (*200*). The results were employed for studying some structurally related systems such as scandomelonine (**740**), *epi*-scandomelonine, scandomeline and *epi*-scandomeline (*201*). Structures for compounds (**738**—**740**) are reproduced along with chemical shift assignments.

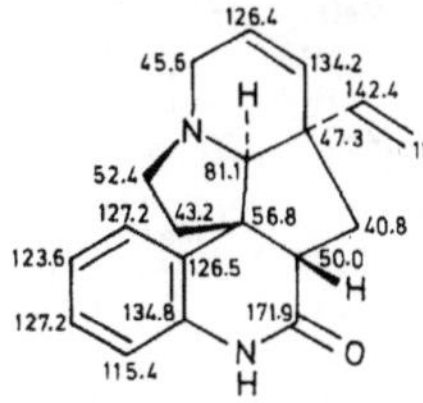

(**738**)

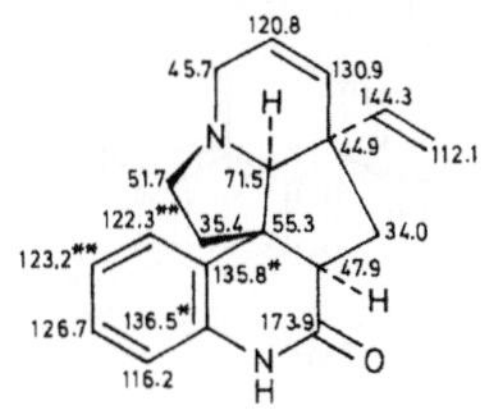

(**739**)

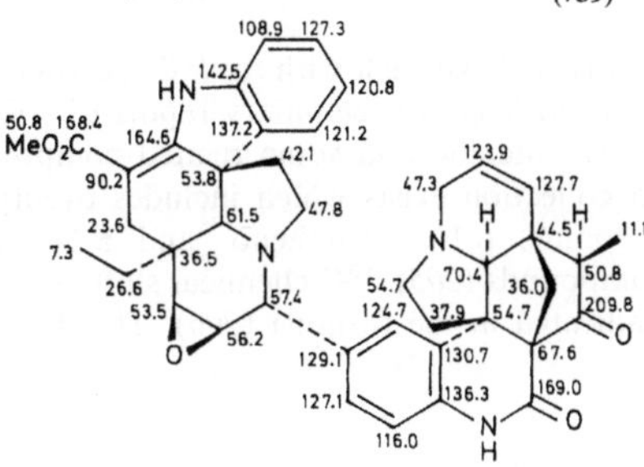

(**740**)

Several groups of investigators have dealt with ^{13}C n.m.r. of isoquinoline-related alkaloids (*170*, *202*—*204*). The ability of ^{13}C n.m.r. to distinguish between diastereomeric pairs and to permit unequivocal

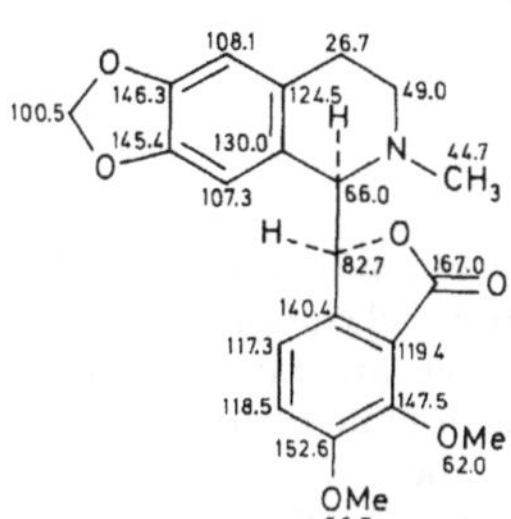

(**741**)

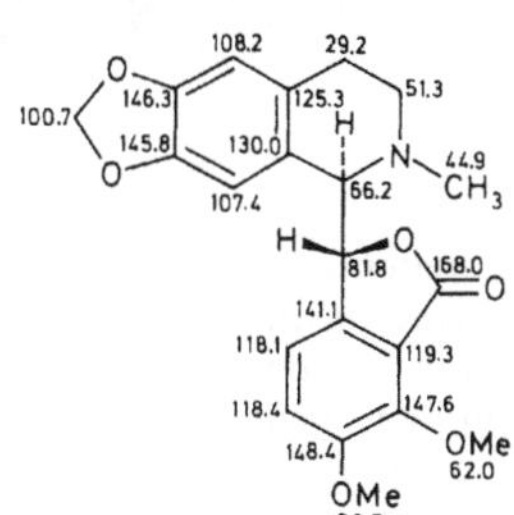

(**742**)

configurational assignment has been exploited for differentiating epimers (**741**) and (**742**) as well as for protoberberines (**743**) and (**744**) (*202, 204*). A further isoquinoline alkaloid studied by ^{13}C n. m. r. is emetin (*170*).

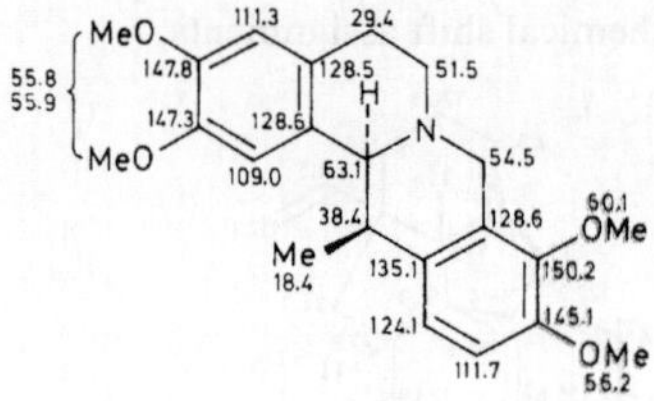

(**743**)

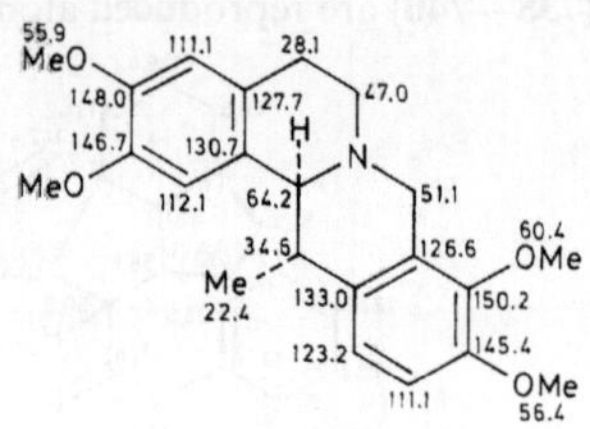

(**744**)

Several groups of workers have dealt with the ^{13}C spectra of morphine alkaloids (*205—207*). Aside from a preliminary report of ^{13}C shift data for codeine, thebaine, sinomenine and some related compounds (*206*), a comprehensive data collection exists which includes twenty-five morphine, 14-hydroxymorphine, 6,14-*endo*-etheno- and 6,14-*endo*-ethanotetrahydrothebaine compounds (*205*), ^{13}C chemical shift assignments for morphine (**745**) are indicated in the formula (*206*). The designations of

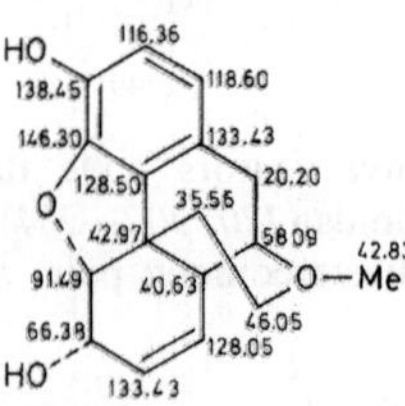

(**745**)

the aromatic quaternary carbons are in accordance with those derived earlier for codeine on the basis of spin-lattice relaxation measurements (*58*). Among the protonated sp^2 carbons, the aromatic carbons were identified upon hydrogenation of the olefinic double bond. A one-to-one assignment of carbons C-1 and C-2, and C-7 and C-8, could be made as a consequence of the characteristic upfield shifts (γ effect) induced upon conversion of morphine into its 3,6-diacetate (heroin).

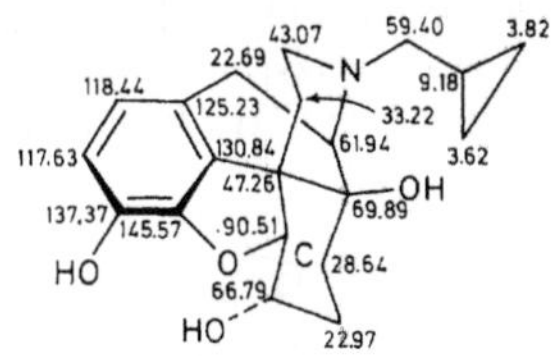

(746)

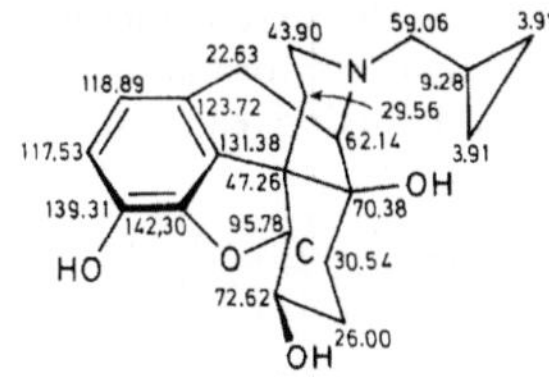

(747)

α- and β-Naltrexol (**746, 747**) are biotransformation products of naltrexone, a potent narcotic antagonist, which differs from (**746**) and (**747**) by bearing a carbonyl function at C-6. The ^{13}C data confirmed earlier conclusions based on ^{1}H n.m.r. that ring C in the naltrexols existed in a chair conformation. Evidence for this corroborative finding was the upfield shift observed for C-6 in going from (**747**) to (**746**) (*207*). ^{13}C chemical shifts have also become available for some amaryllidaceae alkaloids (*208*) and for physostigmine (**748**) and related compounds (*209*).

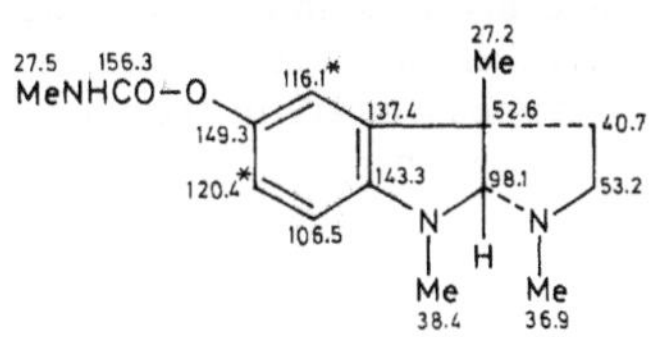

(**748**)

A re-investigation of the ^{13}C chemical shift assignments for tropine (**749**) and pseudotropine (**750**) has recently been undertaken in aqueous and chloroform solution (*210*). SFORD pattern analysis (cf. Sect. 2) led to reversal of some of the assignments reported earlier for data obtained in aqueous solution (*211*). The technique was also exploited for

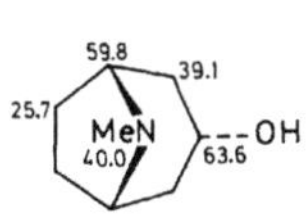

in $CDCl_3$

(**749a**)

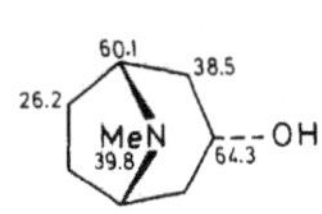

in H_2O

(**749b**)

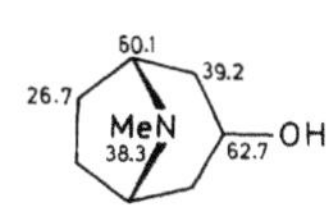

in $CDCl_3$

(**750**)

determining the stereostructures of five alkaloids of the 2-benzyltropine type. A chair conformation with equatorial benzyl and axial hydroxyl group has been derived from the data obtained for (**751**). From a chemical

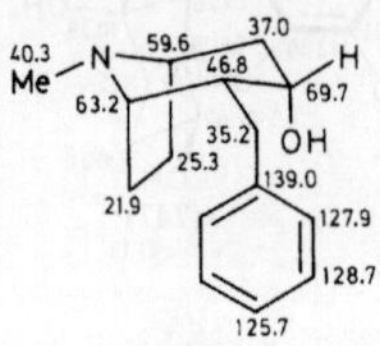

(**751**)

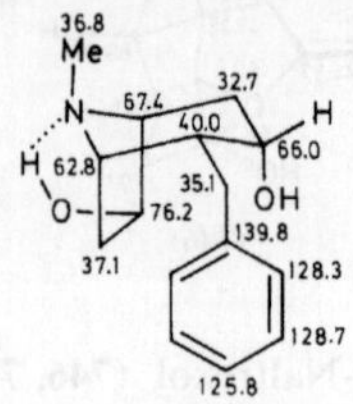

(**752**)

shift comparison among carbons C-2 and C-4 and N-CH_3 in (**751**) and (**752**) emerges the noteworthy and unusual fact that the N-Me group in (**752**) is axially disposed. It appears that hydrogen bonding involving C(6)-OH constrains the N-Me group into an axial conformation (*210*).

The ^{13}C n.m.r. spectra of *Lycopodium* alkaloids lycopodine (**753**), dihydrolycopodine (**754**), *epi*dihydrolycopodine (**755**), flabelliformine, clavolonine, lycodoline (**756**), alkaloid L-23 (**757**) and α-lofoline have been analyzed and the results discussed in terms of stereochemistry (*212*). In this context the C-12 stereochemistry in sauroxine (**758**) and α-obscurine (**759**) was confirmed. The stereochemistry of the N-methyl group in N-methyllycodine could als be ascertained.

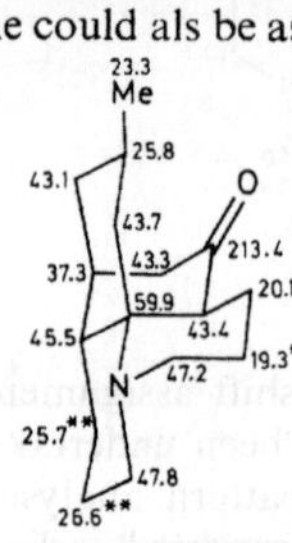

(**753**)

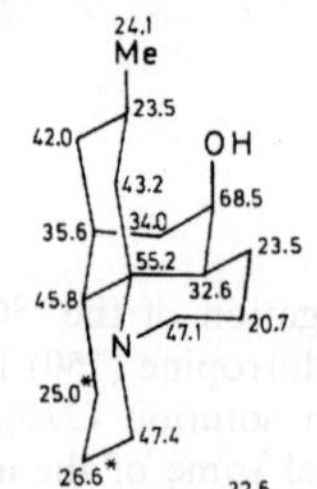

(**754**)

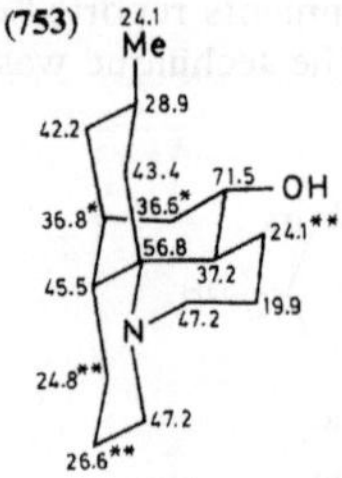

(**755**)

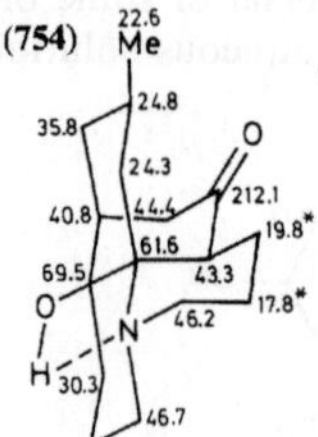

(**756**)

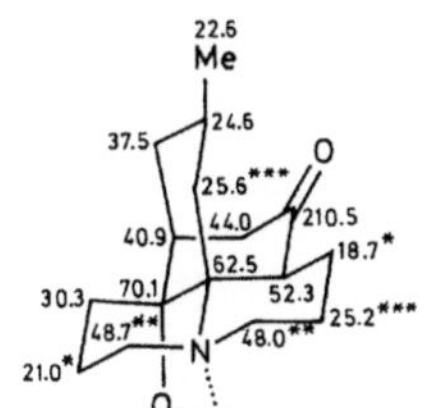

(757)

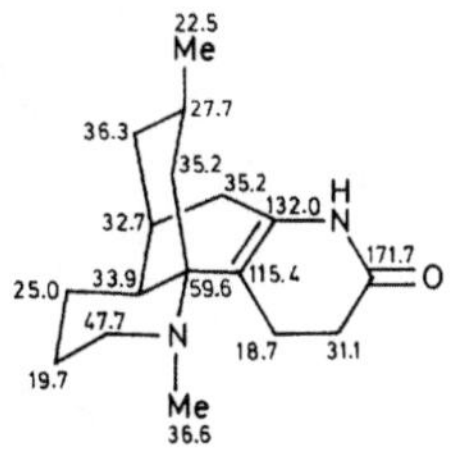

(758)

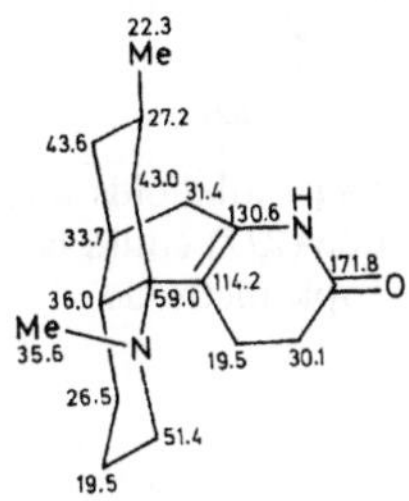

(759)

Other alkaloids for which ^{13}C chemical shifts have recently been reported comprise a new quinolizidine alkaloid from *Cadia purpurea* (**760**) (*213*) as well as the sesquiterpene alkaloids catheludin-2 (**761**) (*214*), isolated from *Catna edulis,* and polyathenol (**762**) (*215*). The structures of the

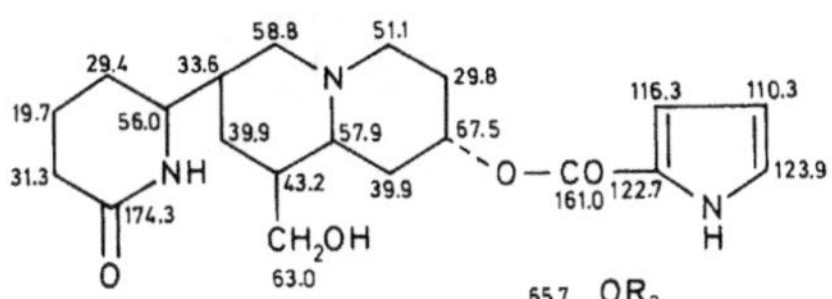

(760)

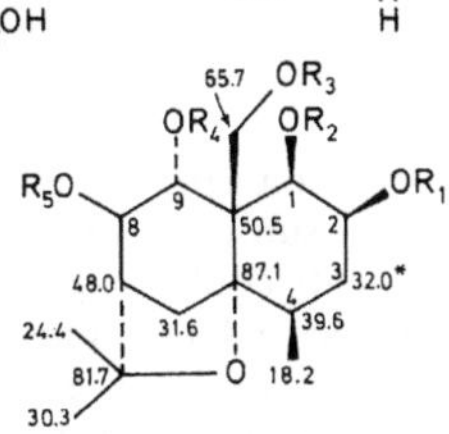

C_1, C_2, C_8, C_9 : 70.6, 71.2, 72.3, 76.5

$R_1 = R_2 = Ac$, $R_3 = R_5$ = nicotinoyl, R_4 = PhCO

(761)

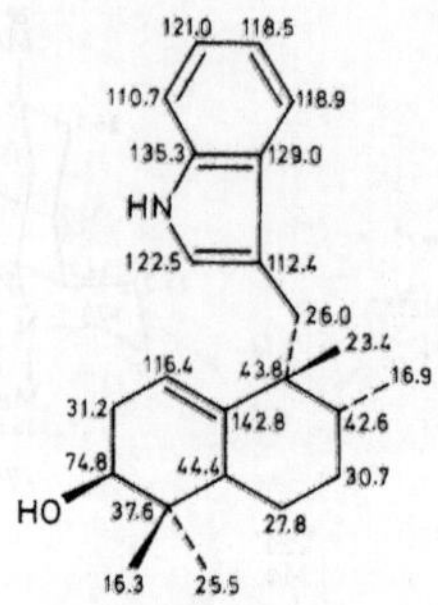

(762)

new ent-kaurene-type diterpene alkaloids anopterine (**763**) (*216*), anopterimine (**764**) and its N-oxide (*217*) could be determined with the aid of joint chemical and spectroscopic methods.

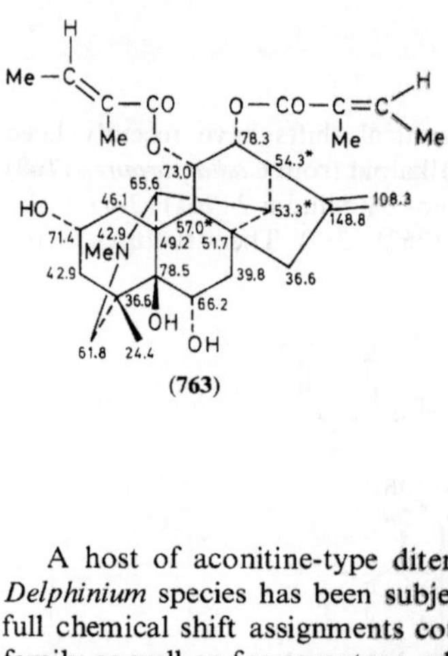

(763)

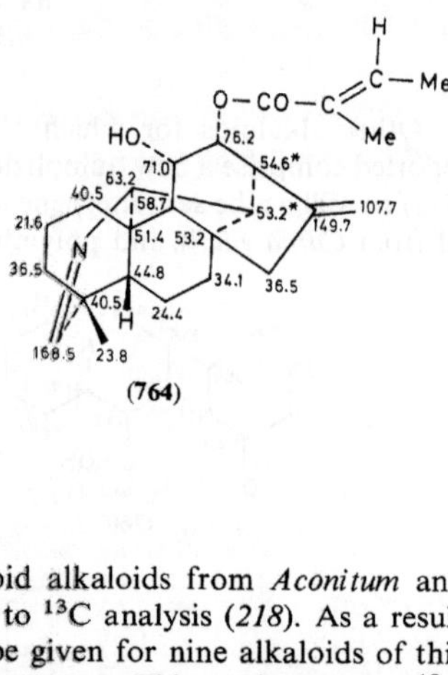

(764)

A host of aconitine-type diterpenoid alkaloids from *Aconitum* and *Delphinium* species has been subjected to ^{13}C analysis (*218*). As a result full chemical shift assignments could be given for nine alkaloids of this family as well as for seventeen related systems. This study revealed ^{13}C n.m.r. as the most powerful tool for establishing compound identy whereas identical i.r. and x-ray powder patterns were earlier found to be inadequate for this purpose. ^{13}C chemical shift data are reproduced for two representatives: aconitine (**765**) and neoline (**766**). Anomalous substitution shifts were observed upon acetylation or methylation of the

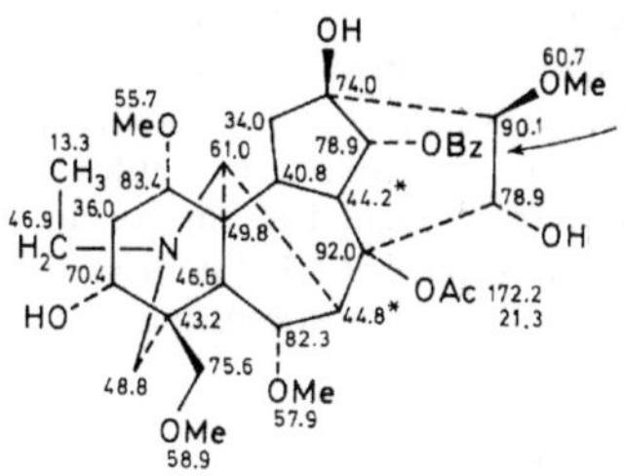

(765)

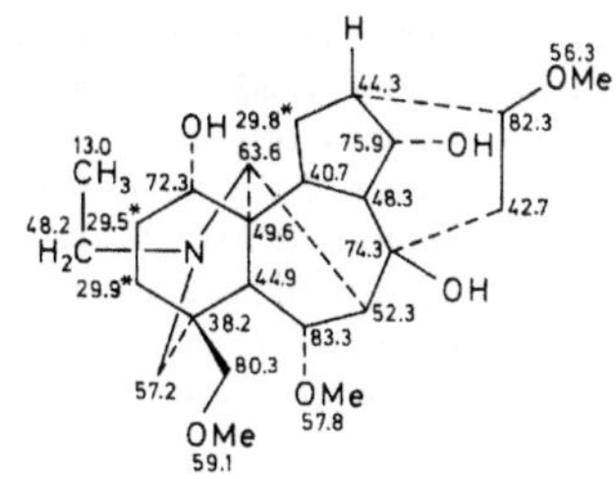

(766)

C-1 α-OH. Acetylation, for example, was found to cause a 5.5 ppm downfield shift at C-1, whereas in 3β-cholestanol this shift amounts to only 2.9 ppm. Likewise, methylation involves a 13.8 ppm deshielding as compared to an expected 10 ppm downfield shift. In contrast, the anticipated substituent shifts were observed for compounds bearing a β-hydroxy substituent at C-1. The anomalous behaviour exhibited by the C-1 α-OH series was interpreted in terms of the boat conformation as depicted by

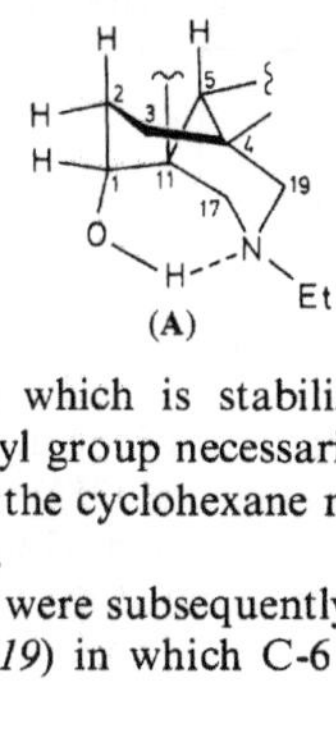

(A)

partial stereostructure (**A**) which is stabilized by a hydrogen bond. Esterification of the hydroxyl group necessarily entails disruption of this hydrogen bond, converting the cyclohexane moiety into the energetically more favorable chair form.

The results of this study were subsequently applied to revise the structure of alkaloid A (**767**) (*219*) in which C-6 previously was believed to

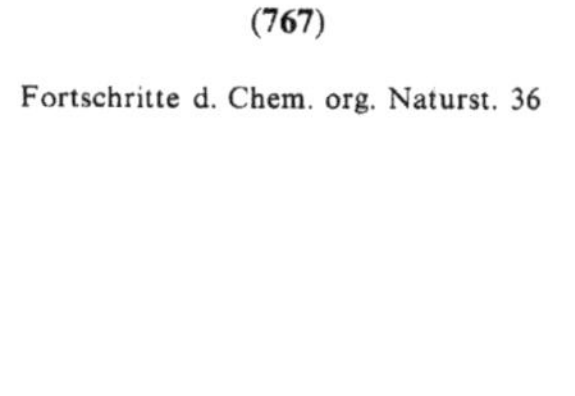

(767)

(768)

carry a methoxyl and C-8 an acetoxy group. Moreover, chemical shifts for alkaloid B (**768**) were reassigned. Using combined ^{13}C and proton n.m.r. spectroscopy the same group of workers succeeded in establishing the structures of the novel *bis*-diterpene alkaloids staphidine, staphinine and staphimine (*220*). Partial assignments are indicated in the formula for staphigine (**769**) (*221*), a dimeric spirocyclic diterpene alkaloid. Further representatives for this class of compounds include staphirine (*221*), staphisagnine and staphisagrine (*222*).

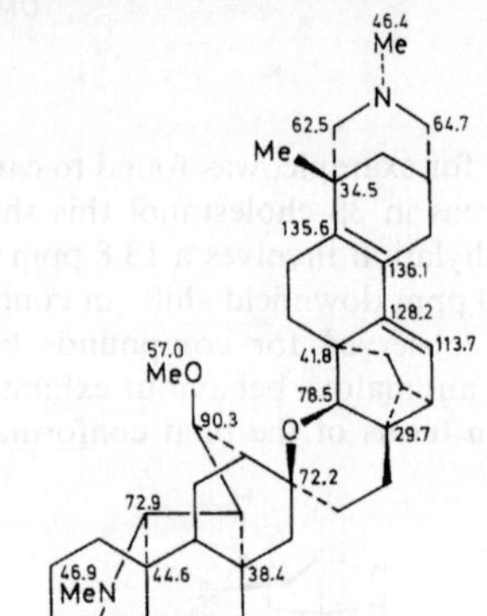

(**769**)

The ^{13}C chemical shift patterns of the new steroidal alkaloids 25-isosolafloridine (**770**) and solacallinidine (**771**) were found to follow closely those of 5α-cholestan-3β-ol and 3-amino-5α-cholestane (*223*). The structure of (**770**) was further corroborated by X-ray crystallography.

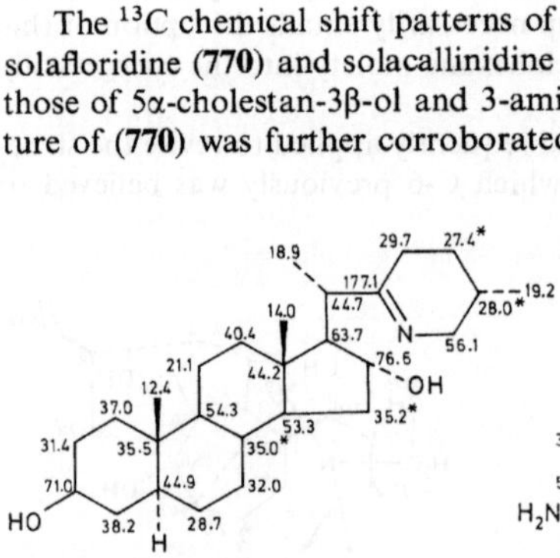

(**770**)

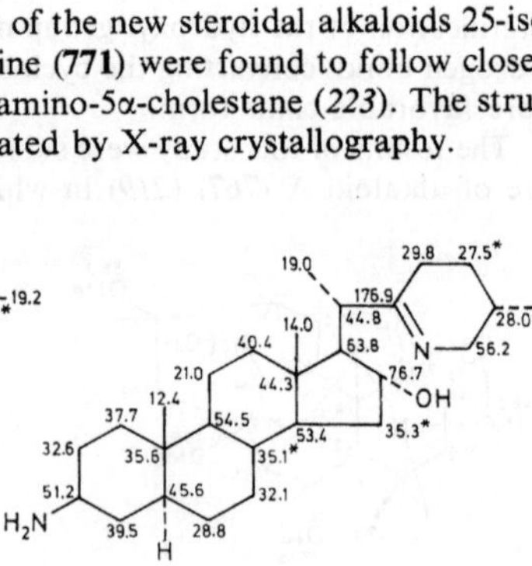

(**771**)

3.3. Purines, Pteridines, Flavonoids, and Related Substances

The ^{13}C n. m. r. spectra of purines and related compounds have been examined by several groups of workers (*224—226*). An extensive analysis of the spectra of purine (**772**) and twenty-six substituted purines, in terms of chemical shifts and spin-spin coupling constants, has been the subject

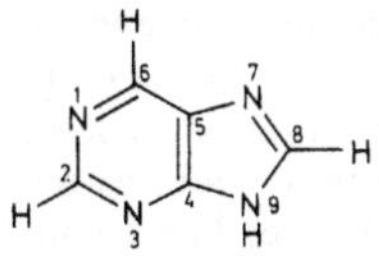

(**772**)

Table 49. ^{13}*C Chemical Shifts for Purine* (**772**) *and 2,6-Substituted Purines* (*224*)[a]

Substituent 2	Substituent 6	Carbon 2	Carbon 4	Carbon 5	Carbon 6	Carbon 8	CH_3
–H	–H	152.10	154.77	130.46	145.50	146.09	
$-NH_2$	–H	160.59	155.11	125.52	147.69	141.62	
–F	–H	158.31	158.22	128.79	147.25	147.98	
–Cl	–H	152.69	157.68	129.06	146.91	147.79	
–H	–Me	151.76	153.87	129.59	155.67	144.51	19.48
–H	–OMe	151.30	155.06	118.14	159.26	142.59	53.66
–H	–SMe	151.57	150.21	129.40	158.70	143.13	11.30
–H	$-NH_2$	152.37	151.30	117.61	155.30	139.29	
–H	–NHMe	152.42	150.0	118.2	154.72	138.76	27.22
–H	$-NMe_2$	151.76	151.23	118.97	154.29	137.74	37.84
–H	$-NEt_2$	151.91	151.08	118.46	153.07	137.86	13.49
–H	–Cl	151.50	154.16	129.23	147.78	146.18	
–H	–Br	151.52	152.98	131.97	140.12	145.94	
–H	–I	151.72*	150.02*	120.18*	136.53*	145.02	
–H	–CN	152.20	154.97	133.47*	127.80*	149.29	114.43
–H	$-NMe_3^+$	150.33*	151.62*	137.74	156.59*	147.32*	54.27
–SMe	–SMe	163.75	151.83	127.86	159.78	141.99	
$-NH_2$	$-NH_2$	160.20	152.77	112.50	155.78	135.91	
–Cl	–Cl	150.98	156.23	128.51	148.10	147.44	
–Me	$-NH_2$	160.72	151.8	115.8	154.94	138.62	25.33
$-NH_2$	–Me	160.11	154.34	124.40	157.25	140.02	19.02
–SMe	$-NH_2$	163.89	152.20	115.42	154.92	138.47	16.63
$-NH_2$	–SMe	159.65	151.65	123.97	159.20	138.74	10.80
–Cl	–OMe	151.13	156.95	116.83	159.57	143.83	54.70
–F	$-NH_2$	158.77	153.36	115.53	156.81	140.06	
–Cl	$-NH_2$	152.78	152.78	116.20	155.94	140.17	
–Et	–Cl	165.01	155.11	127.70	147.59	145.99	12.61

[a] In DMSO-d_6.
* Ambiguous assignments.

of Ref. (*224*). Chemical shifts for these systems are reproduced in Table 49. The assignments of the proton-bearing carbons in purine, reported earlier on the basis of selectively deuterated molecules (cf. Ref. 5 in *224*) could be corroborated by comparing the one-bond ^{13}C-^{1}H coupling constants with those observed from ^{13}C satellites in the proton spectra. Among the one-bond coupling constants $^{1}J_{C(8)\text{-}H}$ is largest (~209—214 Hz) and its magnitude appears to be quite intensitive to substitution. This is followed by $^{1}J_{C(2)\text{-}H}$ (~198—212 Hz) and $^{1}J_{C(6)\text{-}H}$ (~179—189 Hz). Assignments were further aided by the observation of long-range couplings. C-2, for example, in 2-substituted purines, could readily be identified on the basis of its characteristic transoid three-bond coupling constant $^{3}J_{C(2)N\,C(6)H}$ (~11—14 Hz).

The ^{13}C n. m. r. spectra of a large number of pyrrolo [2,3-*d*] pyrimidines (**773**—**783**) and purines (**784**—**800**) have been studied with the objective of establishing a data basis for the investigation of tautomerism in purines (*226*). Chemical shifts for these systems are given in Table 50 and 51.

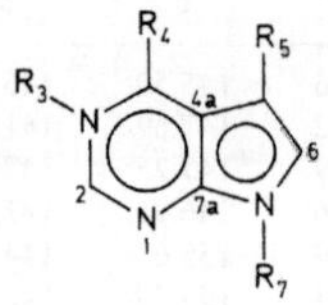

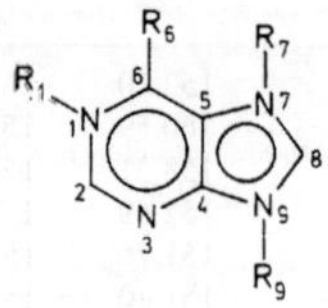

	R_3	R_4	R_5	R_7
(**773**)	-	H	H	H
(**774**)	-	H	H	Ribose
(**775**)	-	NH_2	H	H
(**776**)	-	NH_2	H	CH_3
(**777**)	-	NH_2	H	Ribose
(**778**)	-	NH_2	CN	Ribose
(**779**)	-	NH_2	$CONH_2$	Ribose
(**780**)	H	O	H	H
(**781**)	H	O	H	Ribose
(**782**)	H	S	H	H
(**783**)	H	S	H	Ribose

	R_1	R_6	R_7	R_9
(**784**)	-	H	CH_3	-
(**785**)	-	H	-	CH_3
(**786**)	-	H	-	Ribose
(**787**)	-	NH_2	CH_3	-
(**788**)	-	NH_2	Ribose	-
(**789**)	-	NH_2	-	CH_3
(**790**)	-	NH_2	-	Ribose
(**791**)	H	O	CH_3	-
(**792**)	H	O	Ribose	-
(**793**)	H	O	-	Ribose
(**794**)	CH_3	O	-	Ribose
(**795**)	-	OCH_3	-	Ribose
(**796**)	H	S	CH_3	-
(**797**)	H	S	Ribose	-
(**798**)	H	S	-	Ribose
(**799**)	CH_3	S	-	Ribose
(**800**)	-	SCH_3	-	Ribose

Table 50. ^{13}C *Chemical Shifts in Pyrrolo*[2,3-*d*]*Pyrimidine Derivatives* (**773**—**783**) (*226*)

Compound	Aglycon						Ribose				
	2	4	4[a]	5	6	7[a]	1′	2′	3′	4′	5′
(**773**)	150.8	148.8	118.2	99.4	127.2	151.2					
(**774**)	151.0	149.6	119.3	100.5	127.8	150.7	86.9	74.2	70.8	85.3	61.8
(**775**)	151.1	157.2	102.3	99.2	121.5	150.0					
(**776**)	151.6	157.4	102.3	98.3	124.9	149.9					
(**777**)	151.3	157.4	103.2	99.8	122.8	149.5	87.8	73.7	70.7	85.1	61.8
(**778**)	153.7	157.2	101.5	83.2	132.6	150.3	88.1	74.5	70.4	85.7	61.4
(**779**)	153.0	158.3	101.4	111.2	126.0	151.1	87.7	74.2	70.9	85.6	62.1
(**780**)	143.3	158.7	107.8	102.2	120.5	148.2					
(**781**)	144.0	158.5	108.6	102.7	121.4	148.0	87.3	74.5	70.8	85.3	61.8
(**782**)	142.8	176.3	119.9	104.4	123.8	143.6					
(**783**)	143.5	176.6	120.7	104.9	124.2	143.4	87.1	74.6	70.7	85.4	61.1

[a] In DMSO-d_6.

Table 51. ^{13}C *Chemical Shifts in Purine Derivatives* (**784—800**) (*226*)[a]

Compound	Aglycon						Ribose				
	CH_3	2	4	5	6	8	1′	2′	3′	4′	5′
(**784**)	31.6	152.0	159.8	125.7	140.7	149.7					
(**785**)	29.3	151.8	151.3	133.4	147.4	147.4					
(**786**)		152.2	151.0	134.2	148.3	145.5	87.7	73.9	70.4	85.8	61.4
(**787**)	33.7	152.3	159.8	111.7	151.9	145.9					
(**788**)		152.8	160.7	110.2	151.7	144.6	89.4	75.0	69.0	86.4	60.5
(**789**)	29.3	152.5	149.9	118.7	155.9	141.4					
(**790**)		152.6	149.2	119.5	156.3	140.2	88.2	73.7	70.9	86.1	61.9
(**791**)	33.3	144.3	157.0	115.4	154.6	144.3					
(**792**)		144.8	157.7	114.7	154.1	142.4	89.4	75.1	69.7	85.4	61.0
(**793**)		146.1	148.4	124.6	156.8	139.1	87.8	74.4	70.5	85.9	61.5
(**794**)	33.5	148.7	147.6	123.6	156.4	139.2	87.5	74.2	70.4	85.7	61.4
(**795**)	54.0	151.6	151.8	121.2	160.4	142.3	87.8	73.8	70.5	85.8	61.4
(**796**)	34.6	144.7	152.6	125.8	170.4	148.3					
(**797**)		144.9	153.3	125.3	169.8	144.9	89.1	75.7	68.9	84.6	60.3
(**798**)		145.4	144.1	135.6	176.1	141.4	87.9	74.5	70.4	85.9	61.3
(**799**)	40.4	148.4	142.0	135.7	177.4	141.6	87.6	74.3	70.2	85.7	61.2
(**800**)	11.2	151.5	148.0	131.3	160.4	143.0	88.0	73.9	70.3	85.8	61.3

[a] In DMSO-d_6.

The general purpose of this study was to derive the chemical shifts of various tautomeric species. This was essentially achieved by blocking intramolecular tautomeric exchange through derivatization. For this purpose the labile hydrogen was replaced by either a β-D-furanosyl moiety or a methyl group. By use of suitable corrections accounting for the effect of substitution, the bridgehead carbon chemical shifts for each tautomer in the unsubstituted purines could be estimated with high accuracy. Since at ambient temperature the tautomeric exchange in these systems is fast on the n.m.r. time scale, only one set of chemical shifts was usually observed as weighted averages of the contributing structures.

This procedure allowed a quantitative determination of the tautomeric populations for purine (**772**), adenine (**801**), hypoxanthine (**802**), 6-mercaptopurine (**803**) as well as for some related purines (*227*). Percentage tautomeric populations for compounds (**801**—**803**) are indicated in the formulae. Whereas purine was found to consist of 60% N (9) H and 40% N (7) H tautomers, the introduction of the C-6 amino group in adenine (**801**) entails a significant decrease in the N (7) H tautomer. The lactim equilibrium in hypoxanthin (**802 a⇌b** and **802 c⇌d**) was found to lie almost completely on the lactam side. In 6-mercaptopurine (**803**) the dominant tautomeric forms are N (1) H-N (7) H (79%) and N (1) H-N (9) H (21%) and virtually no thiol.

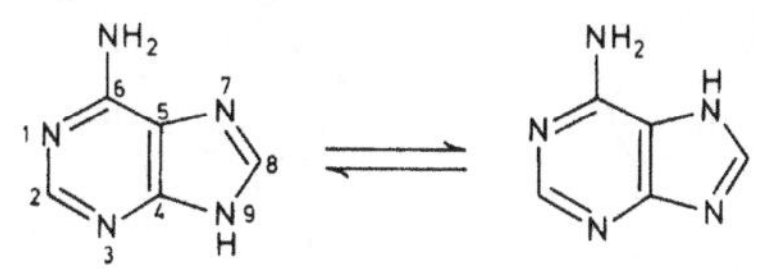

85% (**801a**) 15% (**801b**)

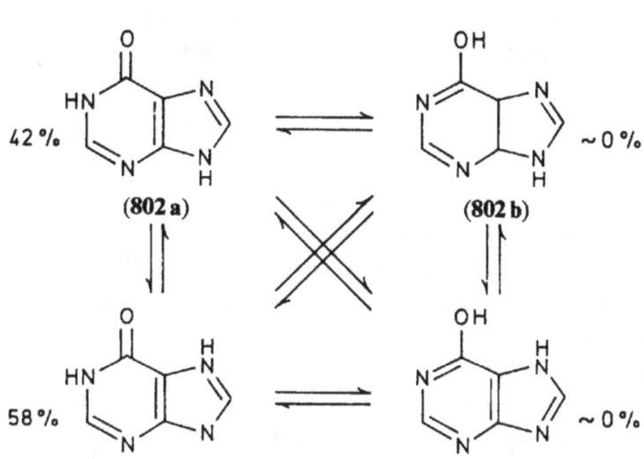

42% (**802a**) ~0% (**802b**)

58% (**802c**) ~0% (**802d**)

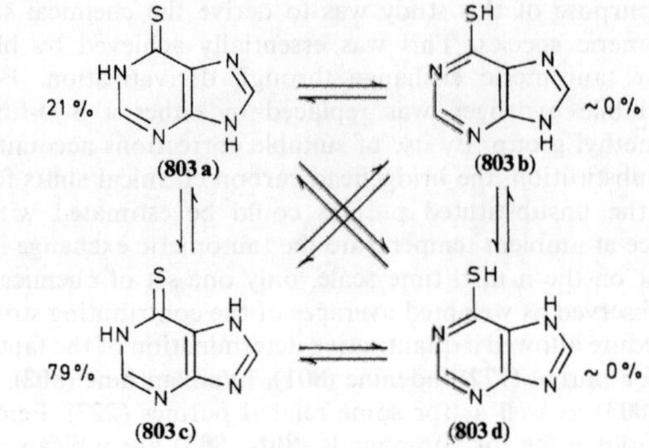

(803 a) (803 b) (803 c) (803 d)

In contrast to adenosine and inosine, which are N-substituted, antibiotics formycin (**804**) and formycin B (**805**) possess a C-glycosyl linkage, which enable these systems to undergo prototropic tautomerism (N (1) H ⇌ N (2) H) (*228*). In addition to earlier reports

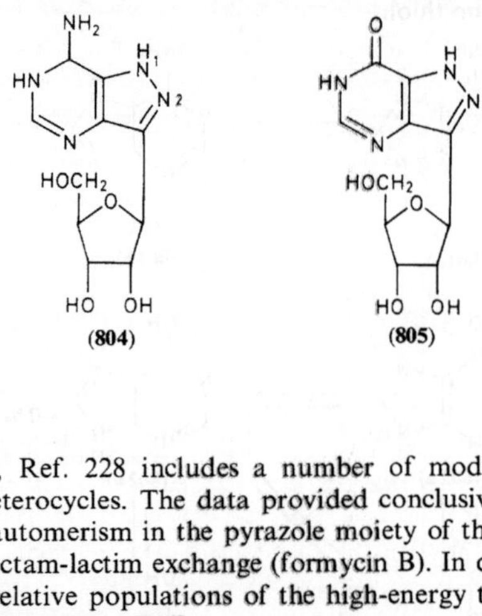

(**804**) (**805**)

on this topic, Ref. 228 includes a number of model pyrazolo [4,3-*d*] pyrimidine heterocycles. The data provided conclusive evidence for the presence of tautomerism in the pyrazole moiety of the molecule and no appreciable lactam-lactim exchange (formycin B). In dimethylsulphoxide solution the relative populations of the high-energy tautomer (N (2) H) were 15% (**804**) and 23% (**805**).

^{13}C chemical shifts were reported for xanthine (**806**), theophylline (**807**), theobromine (**808**) and caffeine (**809**) (*229*). From the data protonation and methylation shifts could be derived.

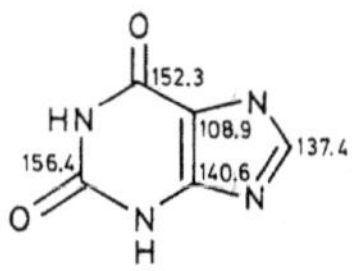

in D_2O

(806)

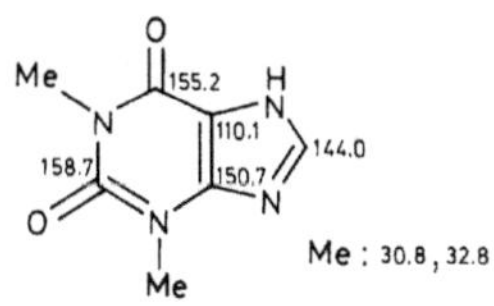

in D_2O

(807)

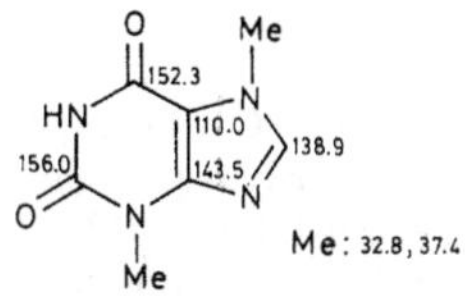

in D_2O

(808)

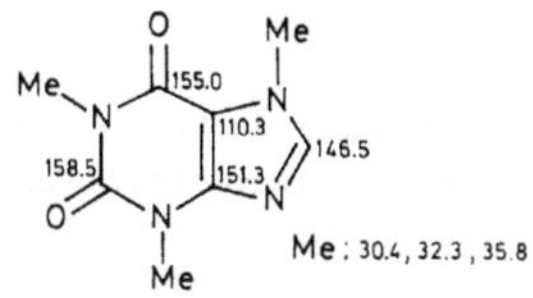

in D_2O

(809)

The structure of saxitoxin, the paralytic shellfish poison, has recently been determined by X-ray crystallography. Its ^{13}C spectrum furthermore showed that the compound exists in solution as the carbamate **(810)** rather than as the earlier suggested cyclol structure **(810 a)** (*230*). This

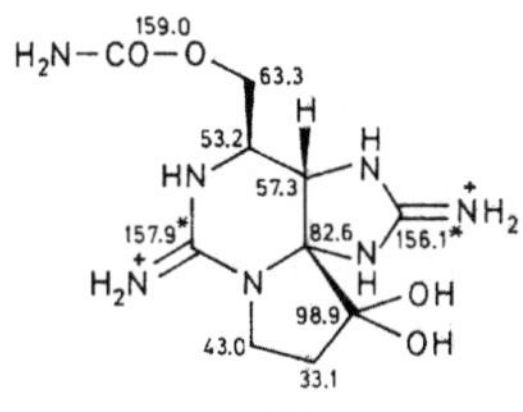

(810)

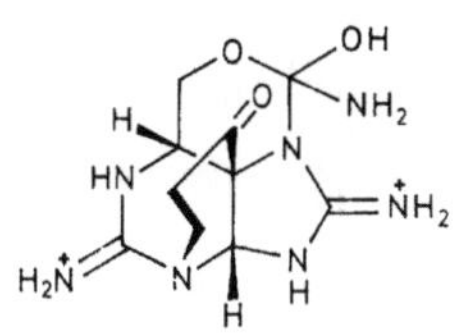

(810 a)

has been concluded from the absence of a ketone signal and from the presence of a line at δ_C 159.0, typical of a carbamate function. By comparison with model compounds, a chemical shift of 104 ppm was estimated for the corresponding carbon in the hypothetical structure **(810 a)**. The ^{13}C n.m.r. spectrum of gonyautoxin II showed close resemblance to the one of saxitoxin, permitting structure **(811)** to be assigned to the former (*231*).

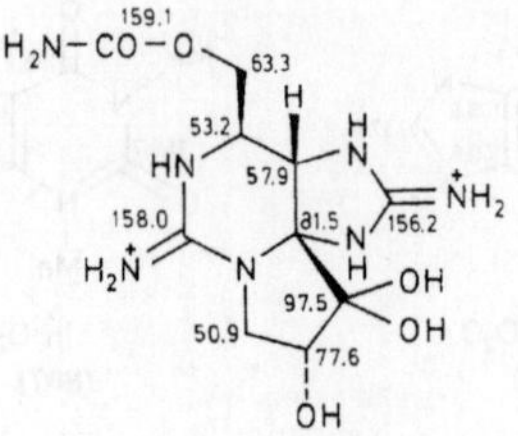

in D_2O

(811)

Several groups of workers have dealt with ^{13}C n. m. r. spectra of natural products related to pterin (**812**) (*232—234*). Because of the low solubility of the neutral species, pterin (**812**), 6-methylpterin (**813**), 7-methylpterin (**814**), lumazine (**815**), xanthopterine (**816**), isoxanthopterine (**817**), and leucopterin (**818**), were examined in alkaline or acidic solution as their anions and cations, resp. (*232*). Chemical shifts for these systems are

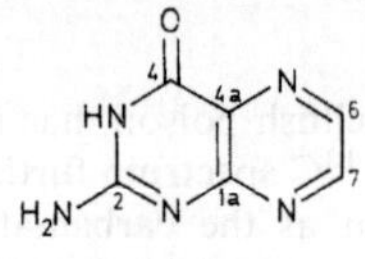

(812)

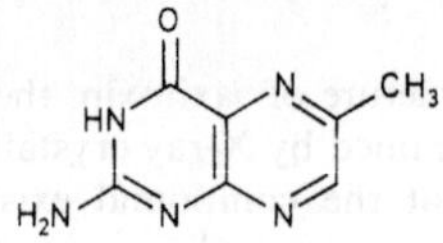

(813)

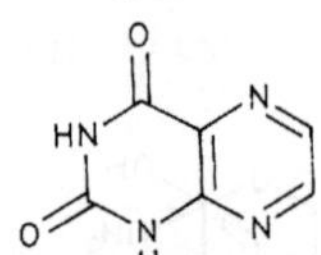

(814)

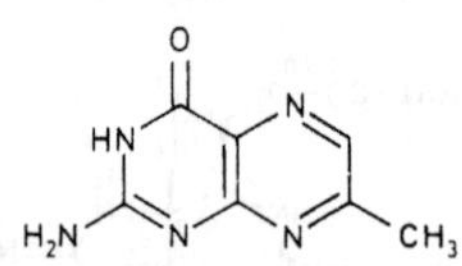

(815)

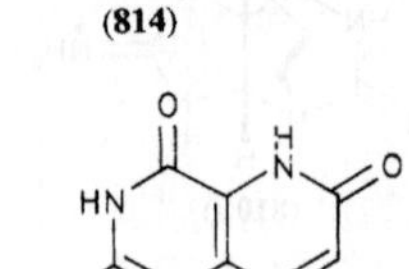

(816)

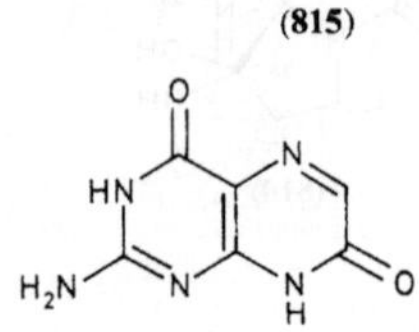

(817)

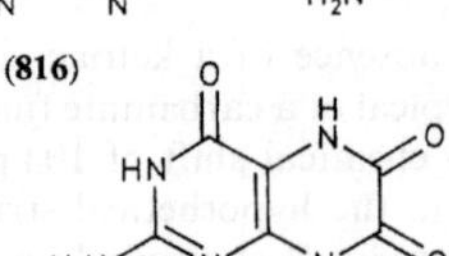

(818)

Table 52. ^{13}C *Chemical Shifts for Pterin* (**812**) *and Related Compounds* (**813—818**) *in Alkaline and Acidic Aqueous Solution* (*232*)

Compound	Solvent	Carbon 2	Carbon 4	Carbon 4a	Carbon 6	Carbon 7	Carbon 1a	Ionization State
(812)	1m NaOD	165.4	174.1	130.5	139.7	149.5	157.7	(−)
	CF_3COOH	158.7[a]	152.1	126.5	143.7	151.6	147.6	(+)
	H_2SO_4	153.0	152.2	116.8	136.7	158.8	150.2	(++)
	FSO_2H	151.7	151.2	155.9	136.5	159.0	149.7	(++)
(813)	CF_3COOH	[a]	151.8	125.0	155.7 CH_3: 47.8	152.1	145.5	(+)
(814)	CF_3COOH	[a]	152.0	123.4	144.2	[a] CH_3: 46.5	147.1	(+)
(815)	CF_3COOH	161.7	151.3	126.5	141.7	150.2	148.7	(+)
	FSO_3H	155.4	151.9	118.2	137.5	157.9	148.8	(++)
(816)	1m NaOD	165.0	174.5	125.9	161.3	147.0	149.0	(−−)
	CF_3COOH	155.8	150.6	112.7	156.9	156.9	137.3	(+)
	H_2SO_4	154.3	151.2	111.7	155.2	152.0	142.1	(++)
	FSO_2H	153.4	151.0	111.1	154.6	151.5	141.9	(++)
(817)	1m NaOD	164.9	174.5	118.0	138.5	170.8	159.6	(−−)
	H_2SO_4	153.4	152.0	109.4	133.3	164.9	149.4	(+)
	FSO_3H	152.4	151.7	109.6	131.9	165.9	149.4	(+)
(818)	1m NaOD	161.8	172.5	133.8	160.6	168.9	154.0	(−−−)
	H_2SO_4		156.7; 154.2; 152.3; 150.4[b]	99.4			137.1	?

[a] Resonances obscured by solvent peaks.
[b] These four lines were not assigned.

compiled in Table 52. Complete ^{13}C chemical shift assignments were given for folic acid (**819**) (*234*) as well as for its 7,8-dihydro-(*233*, *234*) and 5,6,7,8-tetrahydro derivatives (*233*), 10-methylfolate and amethopterin (*234*).

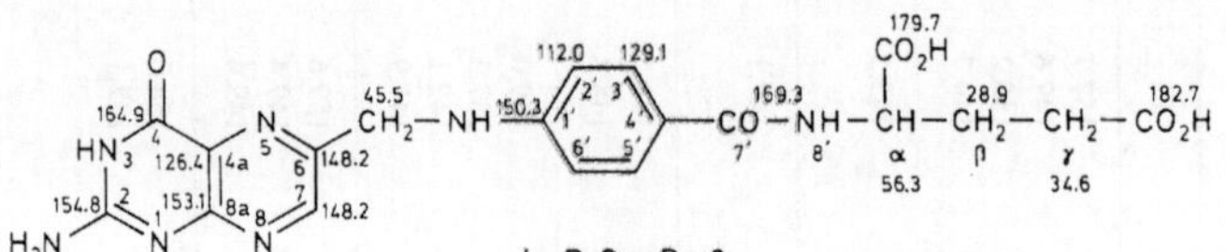

(**819**)

A host of substances related to coumarin (**820**)[1] has been examined by ^{13}C n.m.r. spectrometry. In many cases spectral assignments were assisted by spin-spin coupling constants extracted from proton-coupled spectra (*235*—*242*). The structure of dalrubone, an extractive from *Dalea* species (*241*), was found to be compatible with (**821**)[2].

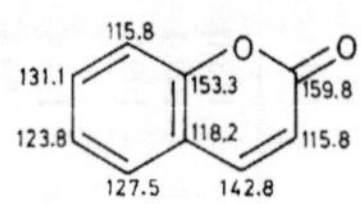

(**820**)

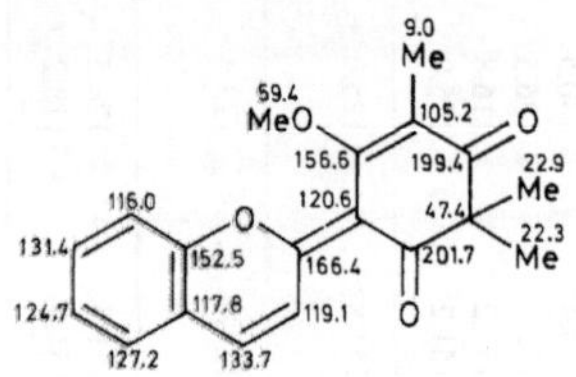

(**821**)

A large body of ^{13}C data has also become available for the group of natural products related to flavone (*25*, *26*, *37*, *242*—*245*). A comprehensive collection of chemical shifts involving twelve flavonones, six flavones and ten isoflavonones has recently been presented (*242*). Chemical shifts for the parent compounds flavone (**822**) (*37*) and flavanone (**823**) (*242*) are indicated in the respective formulae. Ref. (*245*) finally is concerned with some flavone glycosides.

[1] The early assignment of C-6 in (**820**) (JOHNSON, L. F., and W. C. JANKOWSKI: "Carbon-13 NMR Spectra Catalog", Wiley, New York, N. Y. 1972) was revised by the present authors.

[2] The chemical shift assignments for dalrubone were revised by the present authors. Original incorrect assignments were based on those of the reference cited in footnote 1) (spectrum 333).

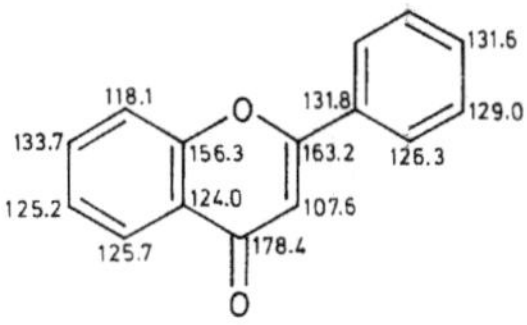

(822)

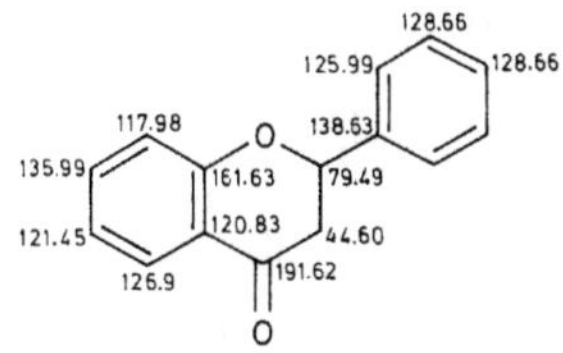

(823)

Several natural products related to 2-pyrone (*246*, *247*) and 4-pyrone (*248*) have been examined. The latter study includes ^{13}C-^{1}H spin-spin coupling data. Chemical shifts for simple systems **(824)** and **(825)** are

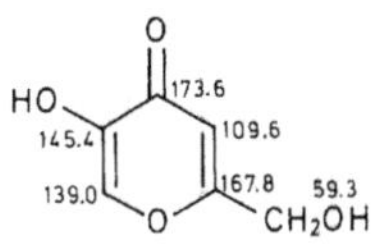

(824)

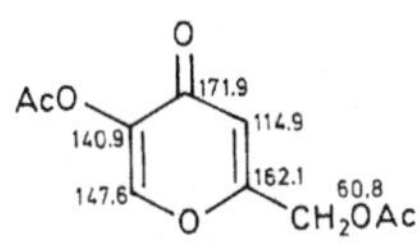

(825)

given for reasons of completeness. The ^{13}C n.m.r. spectra of rotenone **(826)** and some of its relatives were examined and full chemical shifts reported. However, most of the closely spaced aromatic resonances were solely identified on the basis of the assumed additivity of substituent effects in benzenoid systems and are therefore likely to have a relatively low confidence level. Chemical shifts for rotenone **(826)**, amorphigenin **(827)**, rotenonic acid **(828)**, elliptone **(829)**, α-toxicarol **(830)** and malaccol **(831)** are collected in Table 53 (*249*). In addition the paper reports chemical shifts for seventeen derivatives of these compounds.

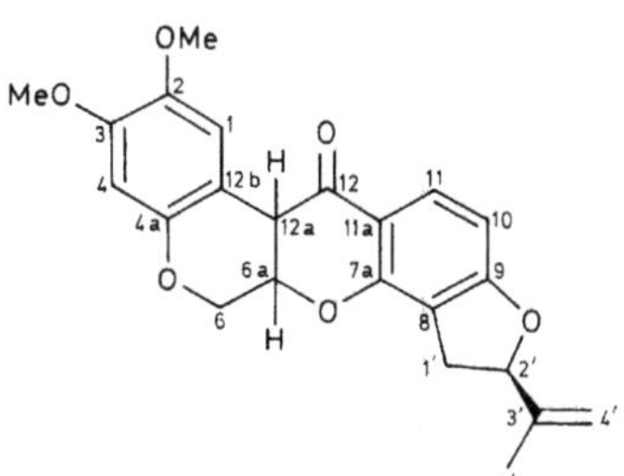

(826)

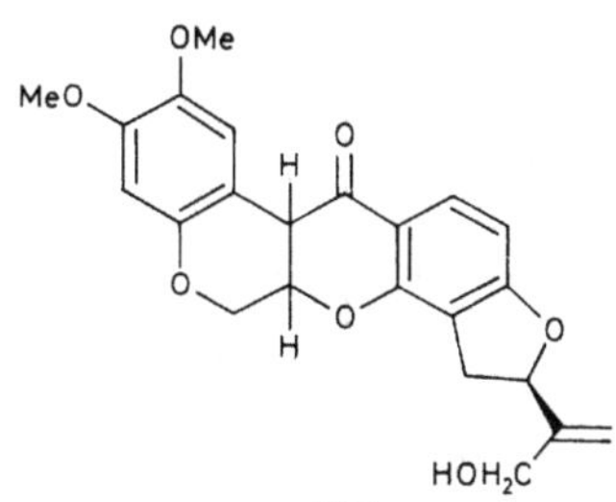

(827)

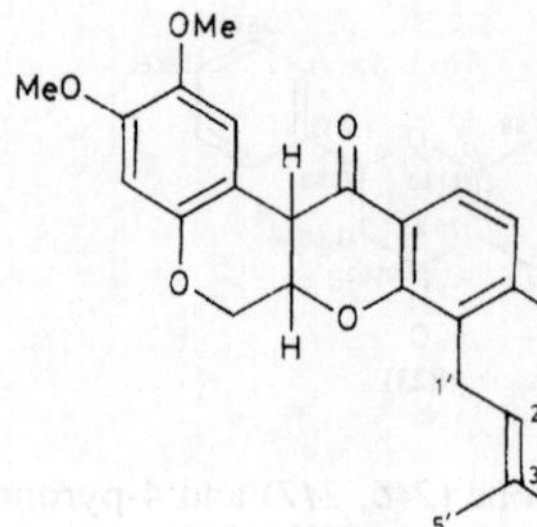

(828)

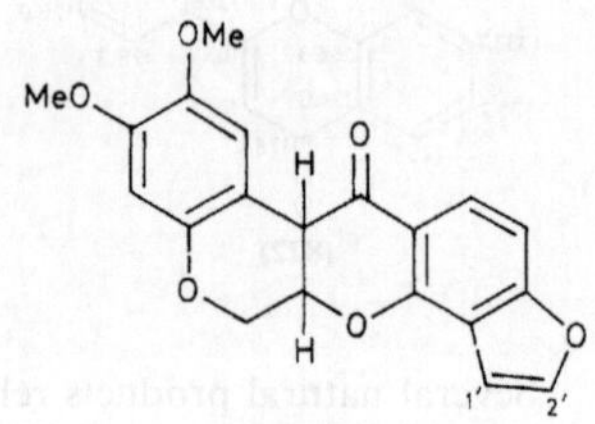

(829)

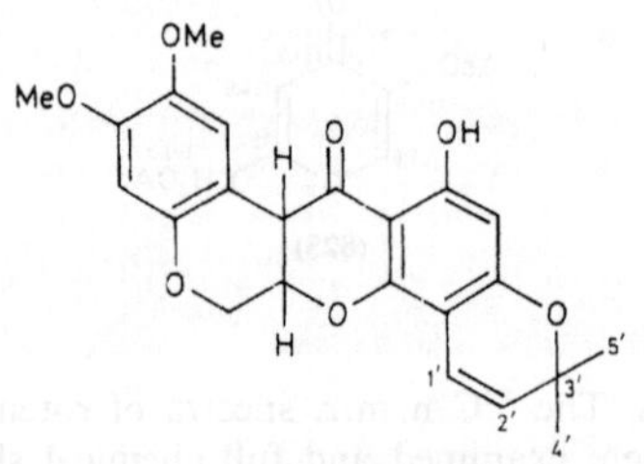

(830)

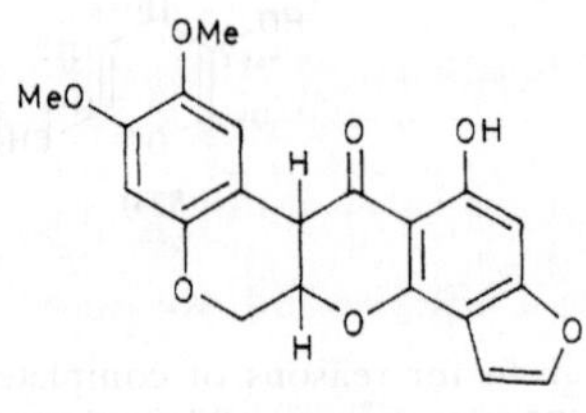

(831)

3.4. Carbohydrates

^{13}C n. m. r. spectroscopy of carbohydrates has been reviewed annually since 1973 (*250*). A fairly comprehensive coverage therefore exists for this class of natural products. For this reason the authors of the present review felt relieved from reproducing early data.

Since carbohydrate carbons all resonate in the two relatively narrow bands of 60—80 ppm and 95—105 ppm (anomeric carbons), a number of the earlier, rather speculative assignments were recently re-examined with the aid of the more elaborate assignment techniques discussed in Section 2. Unambiguous recognition of the carbon resonances in most common monosaccharides has been possible through the use of ^{2}H substitution and ^{13}C labelling studies (*35* and *32*, resp.). The former type of isotopic labelling results in the well known geminal and vicinal isotope effects on the ^{13}C shieldings. ^{13}C enrichment, on the other hand, permits ^{13}C-^{13}C spin-spin couplings to be extracted, from which stereochemical

Table 53. ^{13}C *Chemical Shifts for Rotenoids* (**826**—**831**) (*249*)

Carbon	(**826**)	(**827**)	(**828**)	(**829**)	(**830**)	(**831**)
1	110.4	111.7	110.4	109.1	110.1	110.1
2	143.9	145.2	143.6	141.7	143.7	143.9
3	149.5	150.4	149.3	147.4	149.4	149.8
4	100.9	101.6	100.8	99.6	101.0	101.1
4a	147.4	148.3	147.6	145.2	147.1	147.4
6	66.3	66.8	66.3	65.1	65.8	65.8
6a	72.2	73.1	72.1	71.8	71.7	72.5
7a	156.1	158.6	162.2	157.6	162.6	161.9
8	113.0	113.5*	112.6	111.7	101.0	102.6
9	167.4	167.6	160.1	159.0	164.3	160.8
10	104.7	105.2	110.8	104.9	97.5	93.0
11	130.0	130.5	127.0	121.9	155.7	160.8
11a	114.7	114.6	114.7	115.0	101.5	101.1
12	188.9	189.3	188.9	186.6	194.1	193.9
12a	44.6	45.3	44.2	44.0	43.4	44.1
12b	104.7	105.2	104.7	103.0	104.3	104.3
1′	31.3	32.5	22.1	103.0	**	104.3
2′	87.3	86.2	121.1	142.5	**	143.9
3′	143.0	147.8	143.2		78.1	
4′	112.6	112.7*	25.8		28.2	
5′	17.2	63.5	17.8		28.2	
OMe	55.8	56.1	55.8	54.9	56.1	55.8
OMe	56.3	56.9	56.2	55.5	56.1	56.3

* Ambiguous assignments.
** Chemical shifts not reported.

information becomes available. The approach chosen in Ref. (*32*) has been outlined briefly in Section 2.3.

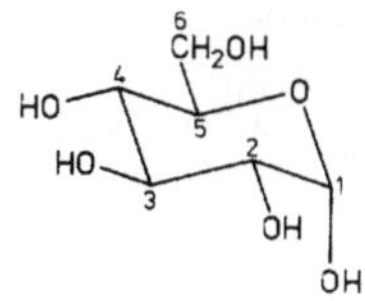

(**832**)

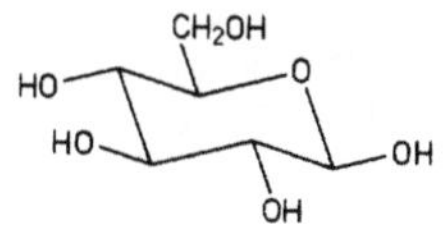

(**833**)

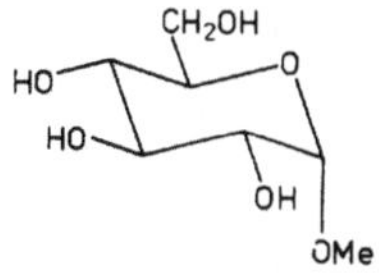

(**834**)

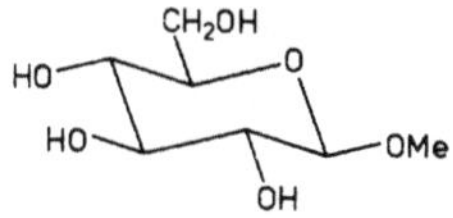

(**835**)

Table 54. ^{13}C *Chemical Shifts for Carbohydrates* (**832—855**) (*32, 35*)*

Compound	Carbon 1	2	3	4	5	6	OMe	Ref
(**832**)	93.6	73.2	74.5	71.4	73.0	62.3		32
(**833**)	97.4	75.9	77.5	71.3	77.4	62.2		32
(**834**)	100.6	72.7	74.7	71.2	73.0	62.2	56.5	32
(**835**)	104.6	74.6	77.4	71.2	77.3	62.4	58.5	32
(**836**)	95.5	72.3	71.9	68.5	73.9	62.6		32
(**837**)	95.2	72.8	74.8	68.3	77.6	62.6		32
(**838**)	102.2	71.4	72.1	68.3	73.9	62.5	56.1	32
(**839**)	102.3	71.7	74.5	68.4	77.6	62.6		32
(**840**)	93.8	70.0	70.8	70.9	72.0	62.8		32
(**841**)	98.0	73.6	74.4	70.4	76.6	62.6		32
(**842**)	93.8	69.8	71.0	73.5	67.8	17.2		32
(**843**)	97.8	73.4	74.6	73.1	72.3	17.2		32
(**844**)	93.3	72.5	73.9	70.4	62.1			35
(**845**)	97.6	75.1	76.9	70.3	66.3			35
(**846**)	100.6	72.3	74.3	70.4	62.0		56.0	35
(**847**)	105.1	74.0	76.9	70.4	66.3			35
(**848**)	93.1	72.9	73.6	76.4	68.6	18.0		35
(**849**)	96.8	75.6	76.6	76.1	73.0	18.0		35
(**850**)	100.3	72.6	73.9	76.2	68.7	17.6	56.2	35
(**851**)	104.3	74.5	76.7	76.2	73.0	17.8		35
(**852**)	95.0	71.9	71.1	73.3	69.4	18.0		35
(**853**)	94.6	72.4	73.8	72.9	73.1	18.0		35
(**854**)	97.8	73.0	73.5	69.6	67.5			35
(**855**)	93.7	69.6	69.8	69.8	63.6			35

* In H_2O or D_2O, externally referenced to TMS.

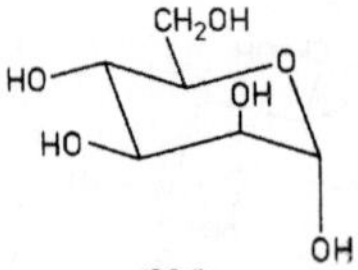

(**836**)

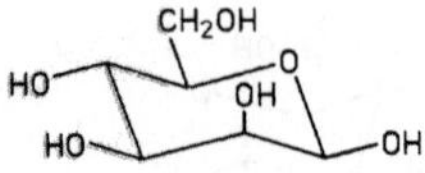

(**837**)

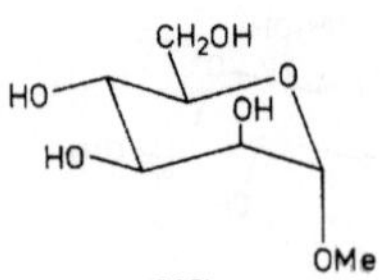

(**838**)

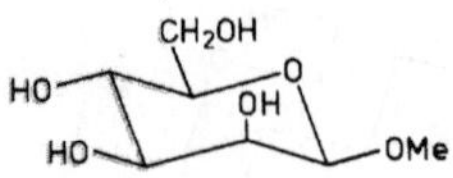

(**839**)

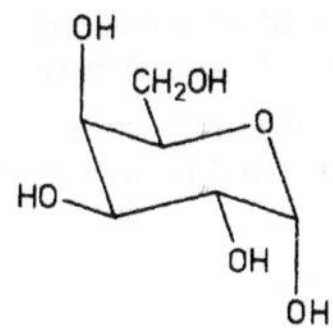

(840)

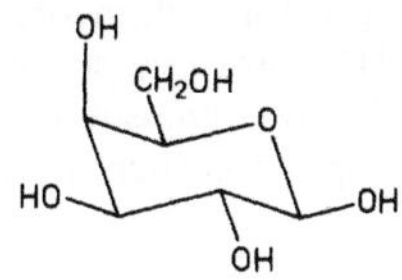

(841)

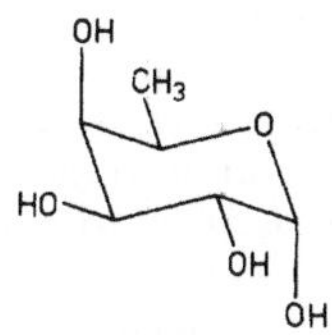

(842)

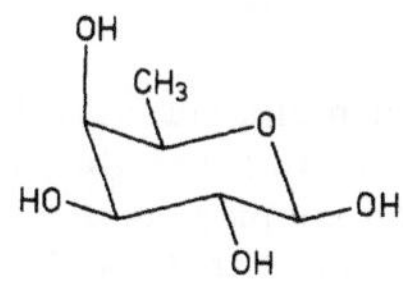

(843)

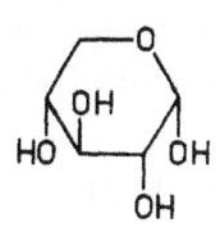

(844)

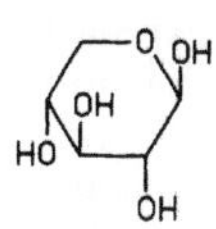

(845)

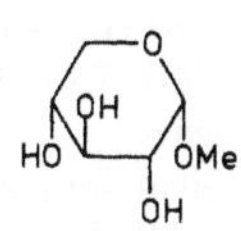

(846)

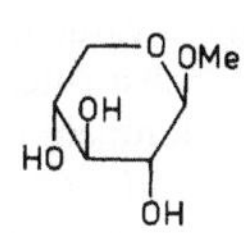

(847)

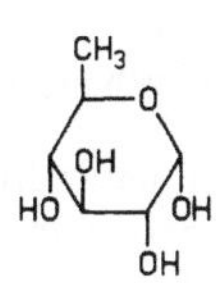

(848)

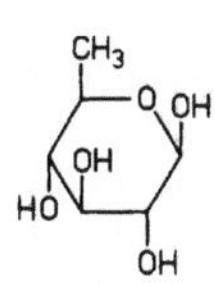

(849)

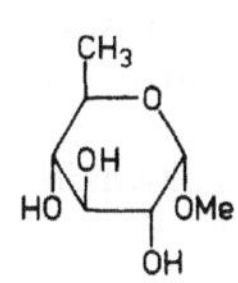

(850)

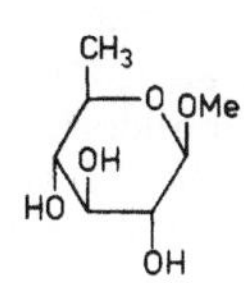

(851)

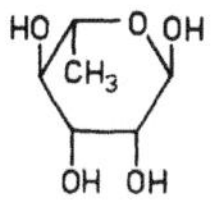

(852)

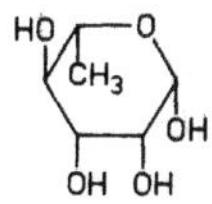

(853)

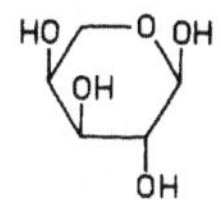

(854)

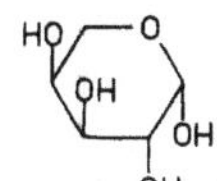

(855)

Chemical shifts for carbohydrates (**832**—**855**) are compiled in Table 54 (*32, 35*). ^{13}C-^{1}H spin-spin coupling as well as off-resonance patterns served to identify the signals in the spectra of some 1,6-anhydropyranoses and related compounds such as 1,6-anhydro-β-D-idopyranose (**856**) and

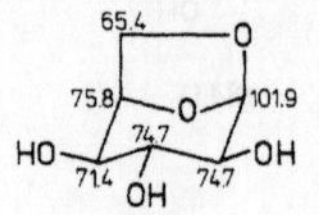

(856)

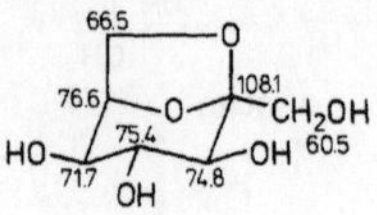

(857)

2,7-anhydroheptulopyranose (**857**) (*251*). Methylenes C-1 and C-7 in (**857**) could be differentiated on the basis of the second-order patterns observed in the proton coupled spectra. These were ascribed to strong coupling between protons C(6) H and C(7) H.

Reports of characteristically different one-bond ^{13}C-^{1}H coupling constants for the anomeric carbon in the α and β anomer of some hexapyranoses corroborate earlier findings by MATWIYOFF *et al.* (*32*). These authors state that $^{1}J_{CH}$ is in the order of 160 Hz in the β-anomeric series while a value near 170 Hz can be expected for the α anomers (*252*). $^{1}J_{CH}$ thus constitutes a very valuable diagnostic criterion for configurational assignment of sugars.

Proton and ^{13}C chemical shift assignments resulted from the spectra of specifically labelled trimethyl ethers of methyl-β-D-arabinopyranoside, methyl-α-D-lyxopyranoside, methyl-β-D-pyranoside and methyl-α-D-xylopyranoside (*253*). Substitution of the hydroxyl hydrogens by CH_3CO-, $ClCH_2CO$-, Cl_3CCO-, CH_3-, CH_3SO_2 was found to induce characteristic shifts on the various carbohydrate carbons permitting their unambiguous recognition (*254*). ^{13}C n.m.r. was utilized in structural studies of chlorodeoxy sugars (*255*) and some 2- and 3- substituted 4,6-O-benzylidene-D-glucopyranosides (*256*).

A number of branched-chain sugars, which occur naturally as the glycoside constituents of antibiotics, as cell wall components of some aquatic plants and in nucleotides from *Azobacter vinilandi*, have been examined by ^{13}C n.m.r. spectrometry with the objective of determining the configuration at the branching carbon (*257*). On the basis of the presence or absence of a steric compression shift for the branching atom (usually a methyl carbon at C-4), the configuration at the site of branching could be established unequivocally.

The effects of glucosylation on the shieldings of glycoside and aglycone carbons have been thoroughly investigated (*258, 259*). The latter study disclosed increased shielding of C-1 in alcoholic β-D-glucopyranosides when going from the primary to the tertiary pyranosides ($\delta_{C\text{-}1}$ 104.4,

102.0—102.5 and 98.0—100.5). The carbinyl carbon of the aglycone moiety was usually found to be displaced to lower field by ca. 7 ppm (*259*). These data were subsequently applied to determine the site of glycosylation in some arabinosyl flavones (*260*).

Complete spectral assignments have been reported for the ^{13}C spectra of disaccharides possessing a β-D-glycosidic linkage involving D-glucose, D-galactose, L-rhamnose, 2-acetamido-2-deoxy-D-galactose and 2-acetamido-2-deoxy-D-glucose (*261*). The compounds were studied as models of immunological polysaccharides. ^{13}C spectral data provided valuable information on the sequence of aglucone and glucose units in metabolites GI-3 (**858**), GI-6 and GI-7, isolated from *Fraxinus americana* (*262*).

The mutarotation of D-ribose (*263*), D-fructose and l-amino-l-deoxy-D-fructose (*264*) was studied quantitatively by ^{13}C n.m.r. spectrometry. The equilibrium mixture of mutarotated D-ribose was found to consist of 6.1% α-furanose, 11.6% β-furanose, 20.3% α-pyranose and 62% β-pyranose (*263*). D-fructose, after equilibration, yielded 46% β-D-pyranose, 7% α-D-pyranose, 30% β-D-furanose, 12% α-D-furanose and 5% keto form (*264*). Mutarotation equilibria and ionization of various hexoses (α- and β-glucopyranoses, α- and β-mannopyranoses and β-fructopyranose) have been studied by ^{13}C n.m.r. in aqueous alkaline solution (*265*).

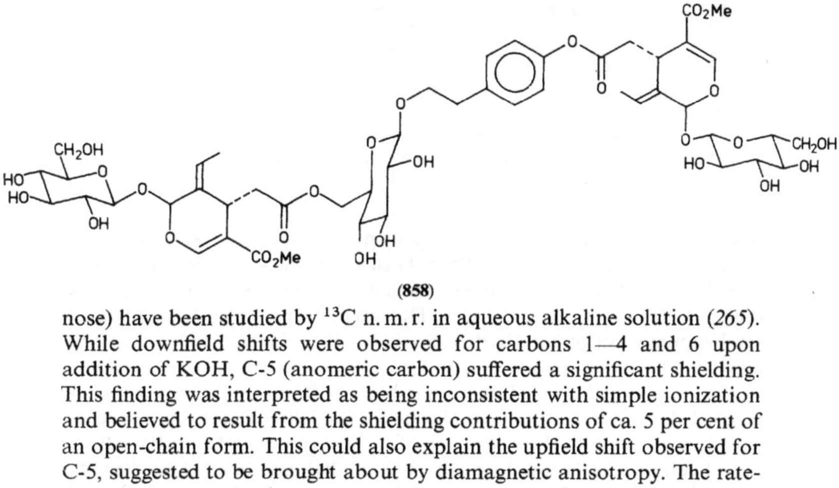

(**858**)

While downfield shifts were observed for carbons 1—4 and 6 upon addition of KOH, C-5 (anomeric carbon) suffered a significant shielding. This finding was interpreted as being inconsistent with simple ionization and believed to result from the shielding contributions of ca. 5 per cent of an open-chain form. This could also explain the upfield shift observed for C-5, suggested to be brought about by diamagnetic anisotropy. The rate-determining step in the mutarotation process was found to be rotation around the C-1—C-2 bond. This is corroborated by the fact that fructose mutarotates more slowly than glucose and mannose.

^{13}C spin-lattice relaxation data provided complementary criteria for the spectral assignment of paeoniflorin (**859**) and albiflorin (**860**), two

Table 55. ^{13}C *Chemical Shifts and Spin-Lattice Relaxation Times for Monoterpene Glucosides* **(859)** *and* **(860)** *(266)*

Carbon	**(859)**[1] T_1 (sec)	**(859)**[1] δ C	**(860)**[1] δ C
1	3.2	88.9	91.3
2	2.3	86.1	86.3
3	0.08	44.8	41.8
4	2.2	106.0	67.4
5	0.13	43.9	41.4
6	[2]	71.7	56.1
7	0.08	23.5	28.3
8	0.08	61.5	61.7
9	0.16	101.7	175.7
10	0.61	19.9	20.7
1′	0.21	100.5	100.3
2′	0.25	74.7	74.8
3′	0.23[3]	78.4	78.4
4′	[2]	71.7	71.6
5′	0.23[3]	78.4	78.4
6′	0.11	62.9	62.8
1″	>6	130.6	130.8
2″/6″	0.45	129.9	130.1
3″/5″	0.46	128.8	128.7
4″	0.22	133.3	133.2
7″	>6	166.6	166.6

[1] In pyridine-d_5.
[2] Not measured due to overlapping resonances.
[3] Average of T_1 (C-3) and T_1 (C-5) due to overlap of resonances.

pinane-type monoterpene glucosides from the roots of *Paeonia albiflora Pallas* (*266*). C-1 and C-2 in **(859)**, both quaternary carbons resonating at δ_C 88.9 and δ_C 86.1 resp., were designated on the basis of a significantly longer T_1 for C-1 (3.2 vs. 2.3 sec). This was interpreted as arising from a smaller number of geminal hydrogens for C-1 (2 vs. 5) (cf. p. 16, Section 2.5). Naively one would predict a $T_1(C\text{-}1)/T_1(C\text{-}2) \sim 5:2$ if all C-H distances were assumed to be equal and contributions from vicinal protons were ignored. However, it should be borne in mind that not all five protons equally contribute to the relaxation of C-2, since internal rotation causes the effective correlation time for the methyl protons to shorten by approximately an order of magnitude, relative to those of the backbone carbons. This fact, which was pointed out by the authors of Ref. 266, also provided an explanation for the T_1(C-4) being very similar to T_1(C-2) in spite of C-4 bearing only four protons. Unfortunately the exact concentrations were not given, so that the data cannot be reproduced,

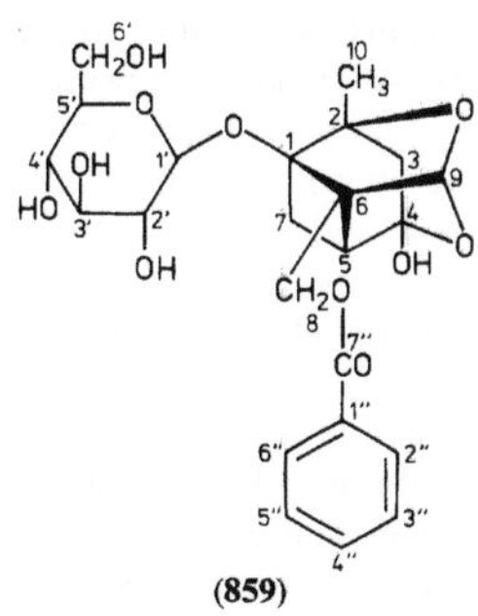

(859)

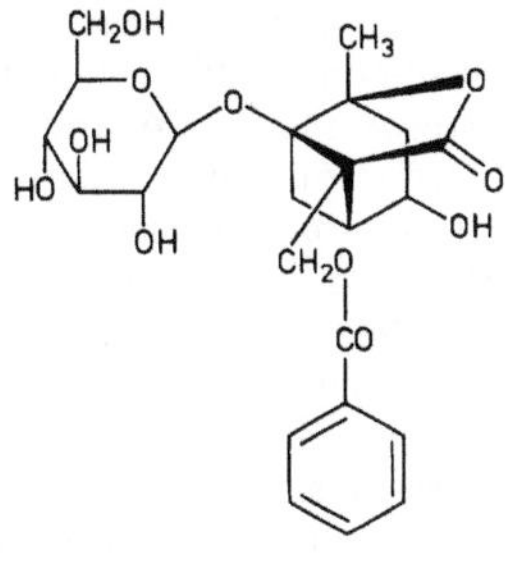

(860)

since T_1's for molecules which can associate are particularly concentration dependent (*267*).

^{13}C spin-lattice relaxation times were also reported for some D-glucopyranose derivatives (*268*). In this paper it was shown that in these molecules quaternary carbons were only partly (but dominantly) relaxed by the ^{13}C-^{1}H dipolar mechanism. This was evidenced by an approximately 20% increase of T_1 upon degassing the sample.

The utility of ^{13}C n. m. r. spectrometry as an alternative to common quantitative methods has recently been demonstrated by some Australian workers (*269*), who developed a fully automated technique for analyzing carbohydrates. In some experiments, aimed at analyzing carbohydrates from tissues of *Pinus radiata*, ^{13}C n. m. r. furnished data comparable to those obtained from GC except that it allowed in addition the identification of unknown constituents.

3.5. Antibiotics

Although antibiotics may be found in various categories of natural products which, from a structural viewpoint, have little or nothing in common, the similarity of their biological properties seems to warrant joint treatment. No data are reproduced in this section, however, the interested reader is referred to the pertinent primary literature.

^{13}C data have been reported for the following natural products possessing antibiotic activity: conocandin, a novel C_{18} fatty acid (*270*), junipal and related thiophenes (*271*), and the spirocyclic griseofulvin (*272*). Among natural products formally derived from quinone, ^{13}C spectral data have appeared for the following systems: stemphone (*273*), the bianthraquinonic biological pigments luteoskyrin and rugulosin (*274*) as well as some related polyhydroxyanthraquinone analogues, the naphtoquinone antibiotic juglone (*275*) isochromanquinone antibiotics griseusins A and B (*276*) and tetracycline (*277*). A ^{13}C n.m.r. analysis has been carried out on antitumor antibiotics daunorubicin and adriamycin (*278*). The spirocyclic bislactonic metabolite leudrin (*279*) was characterized with the aid of its ^{13}C spectrum.

Complete assignments have been reported for rifamycin S (*280*), sterigmatocystin (*31*, *281*), aflatoxins (*31*, *281*) and streptovaricins (*282*). The ^{13}C n.m.r. spectra lent support to the structures suggested for streptovaricins I—IV (*283*) and spectinabilin (*284*). A further class of important antibiotics examined by ^{13}C n.m.r. includes the aminoglucosides. Systems studied comprise: tobramycin, kanamycin B (*285*), apramycin (*286*), gentamycins (*287*, *288*), deoxystreptamine and its glycopyranosyl derivatives (*289*) and the nucleoside antibiotic hikizimycin and its constituents (*290*).

The ^{13}C n.m.r. spectra of a number of macrolide antibiotics have been assigned and the data evaluated in terms of conformational and substituent effects (*291*—*295*). ^{13}C spin-lattice relaxation experiments were carried out on 14-membered macrolide antibiotics with the objective of differentiating the equatorial and axial methyls attached to the macrocycle (*296*). It is well known that axial methyls have a lower barrier to internal motion and thus are less efficiently relaxed (cf. Section 2.5). However, no significant, T_1 differentials were observed, probably due to deviations of the methyl substituents from axial and equatorial disposition. Larger NT_1 values were observed for glycosidic sidechain carbons as compared to those of the aglucone moiety, reflecting higher mobility of the former. This observation is in accordance with earlier findings for 16-membered analogues (*297*). Information regarding conformational mobility was also derived from spin-lattice relaxation measurements in a ^{13}C n.m.r. study of the antibiotic lincomycin (*298*).

Structure and stereochemistry of a number of penam and cepham derivatives have been elucidated with the aid of their ^{13}C n.m.r. spectra (*299*—*302*).

Peptide antibiotics lend themselves to ^{13}C examination since ^{13}C shieldings of amino acids are well characterized and the incremental shifts induced upon peptide bond formation are known (*303*). ^{13}C chemical shift data have been furnished for a series of actinomycins and model peptides (*304*, *305*). A revised structure has been reported for echinomycin (*306*) as well as for griseoviridin, a cyclic peptide antibiotic (*307*). In both cases ^{13}C n.m.r. has provided the crucial argument. Structural information was obtained from the ^{13}C n.m.r. spectra of peptide antibiotics thiostrepton and siomycin A (*308*) as well as from those of novel althiomycin (*309*).

^{13}C n.m.r. spectrometry was also utilized for probing the interactions between tetracycline and metal ions (*310*, *311*). Similar studies were conducted on rifamycin S with respect to complexation of calcium and sodium ions (*312*). The studies afforded information regarding the binding sites in these antibiotics.

4. Biosynthetic Studies

Since the first incorporation experiment using ^{13}C-labelled precursors, and subsequent analysis of the labelled metabolite by ^{13}C n.m.r., in 1970 (*313*), the technique has enjoyed increasing popularity over the past years. Without exaggeration it may be stated that in this domain ^{13}C spectroscopy probably has celebrated its greatest triumph in that it has proven clearly superior to radioactive labelling. While the latter technique often provides information regarding the level of incorporation, the location of the label is often not accessible due to unavailability of a suitable degradation scheme. Moreover, the ^{13}C technique has the virtue of being non-destructive and far less time-consuming. A further useful by-product of ^{13}C labelling consists in its providing structural information too. Because of this the technique has in several instances permitted the correction of erroneous structures.

In view of the vast amount of available data and because the subject has been dealt with in several reviews (*4*—*9*), the present authors have decided to discuss only one example explicitly. For the remainder of systems, whose biosynthetic pathway has been elucidated on the basis of ^{13}C labelling experiments, the respective structures and appropriate references are provided.

While earlier labelling studies employed mainly ^{13}C-labelled acetate, other labelled precursors such as propionate, butyrate, mevalonate, amino acids etc., have more recently been used in ^{13}C feeding experiments. A new dimension was added by administering doubly labelled precursors (e.g. doubly labelled acetate) to the appropriate cultures. Double labelling has several advantages over the use of singly labelled compounds.

The following example may illustrate the potential of the technique. In a study of the sesquiterpenoidal stress metabolites of the *Solanaceae*, 90% enriched [1,2-^{13}C]-acetate was fed to potatoes, inoculated with the fungus *Glomerella cingulata*, furnishing a high yield of phytuberin (**861**)

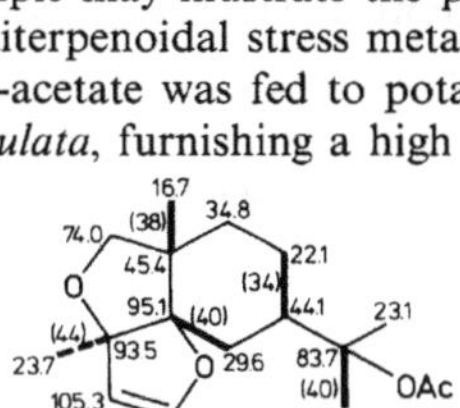

(**861**)

(*314*). Since the level of incorporation of labelled material is at best in the order of a few percent, one-bond ^{13}C-^{13}C spin-spin coupling will only be observed for carbons pertaining to the same acetate unit, i.e. for those carbons whose acetate precursor has been incorporated *intact*. Others, even though they originate from enriched material, will give rise to singlets only, since the probability for adjacent positions being labelled is much lower. This is exemplified by the spectrum of (**861**) in Fig. 9, showing five signals, which lack ^{13}C satellites, while the remainder of lines is flanked by the ^{13}C satellites forming pairs of AB doublets. Since the coupling constant appears twice each and the relative intensities are related to the $\Delta\delta_C/J_{CC}$ ratio, signal

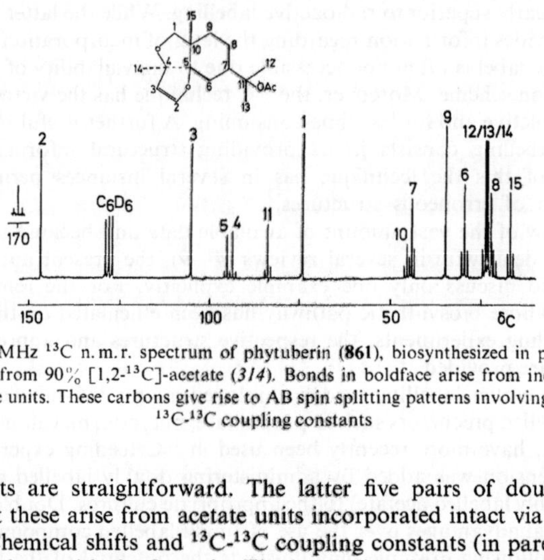

Fig. 9. 25.2 MHz ^{13}C n.m.r. spectrum of phytuberin (**861**), biosynthesized in potatoes by *G. cingulata* from 90% [1,2-^{13}C]-acetate (*314*). Bonds in boldface arise from incorporated intact acetate units. These carbons give rise to AB spin splitting patterns involving one-bond ^{13}C-^{13}C coupling constants

assignments are straightforward. The latter five pairs of doublets indicate that these arise from acetate units incorporated intact via mevalolactone. Chemical shifts and ^{13}C-^{13}C coupling constants (in parentheses) are given along with the formula. The heavy lines indicate bonds arising from the same acetate unit.

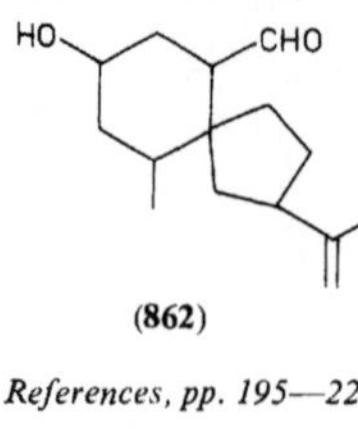

(**862**)

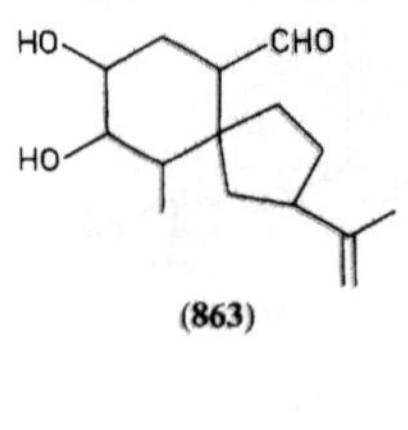

(**863**)

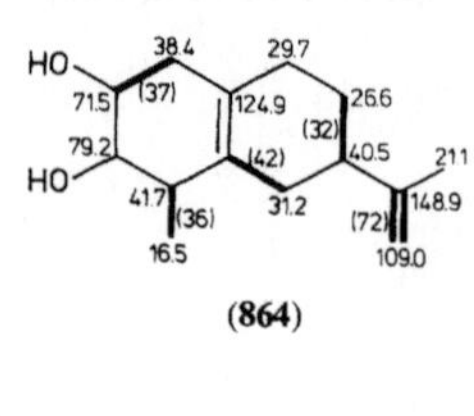

(**864**)

References, pp. 195—229

In the context of this work biogenetic data were also reported for related compounds lubimin (**862**), hydroxylubimin (**863**) and rishitin (**864**), all stress metabolites from potatoes (*314*). The labelling pattern for (**864**), together with chemical shifts and ^{13}C-^{13}C coupling constants, is given in the formula. This study was again conducted by feeding doubly enriched sodium acetate to potatoes, inoculated with a spore suspension of *Monilinia fructicola*. In the following a compilation of systems is provided, whose biosynthesis has been studied by ^{13}C n.m.r. labelling techniques.

Sesquiterpenoids

Ovalicin (**865**) (*315*, *316*); capsidiol (**866**) (*317*, *318*) cyclonerodiol (**867**) and cyclonerotriol (**868**) (*319*); trichothecolone (**869**) (*320*); several coriolins including 5-dihydrocoriolin C (**870**) (*321*); fomannosin (**871**) (*322*).

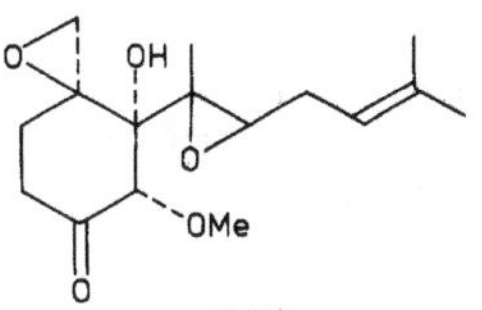

(**865**)

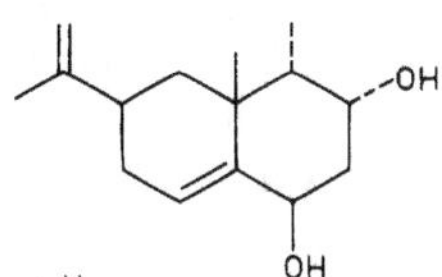

(**866**)

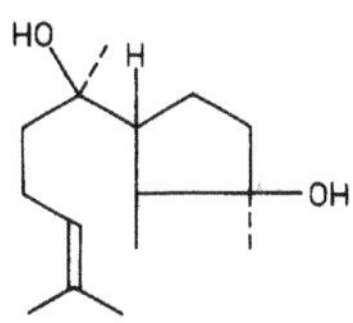

(**867**)

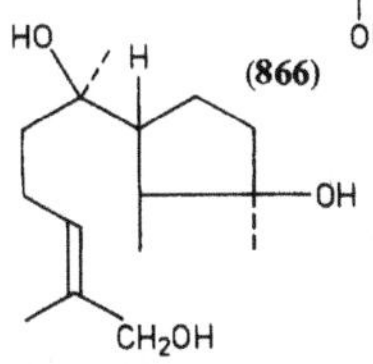

(**868**)

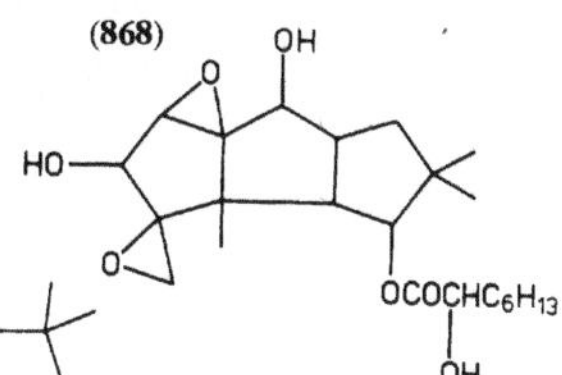

(**870**)

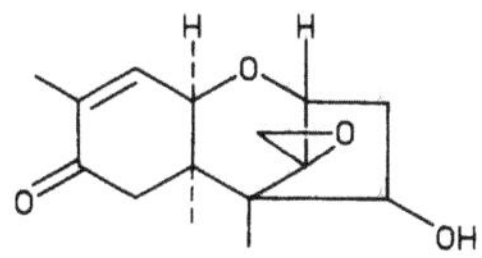

(**869**)

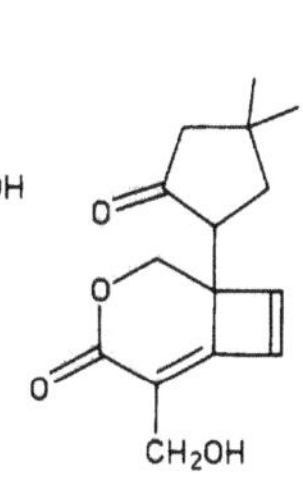

(**871**)

Diterpenoids

Virescenol A (**872**) and B (*323*); aphidicolin (**873**) (*324*); giberellins including methyl giberellate (**874**) (*122*); fusicoccin (**875**) (*325*).

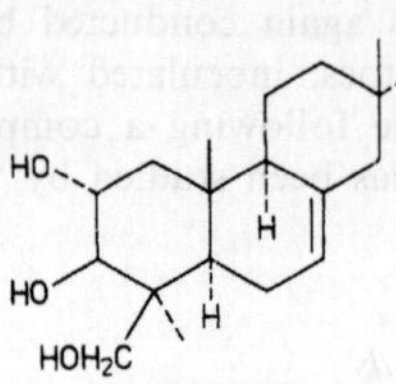

(**872**)

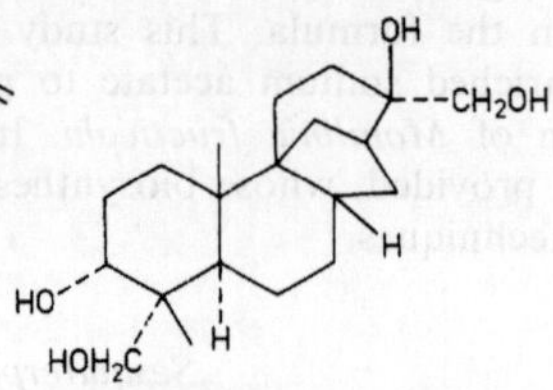

(**873**)

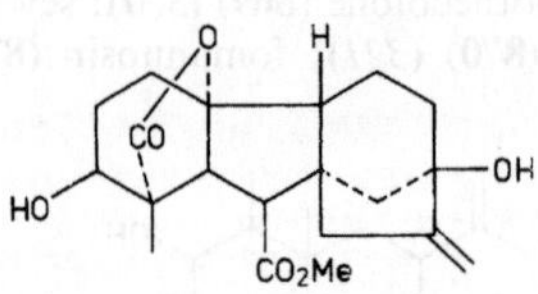

(**874**)

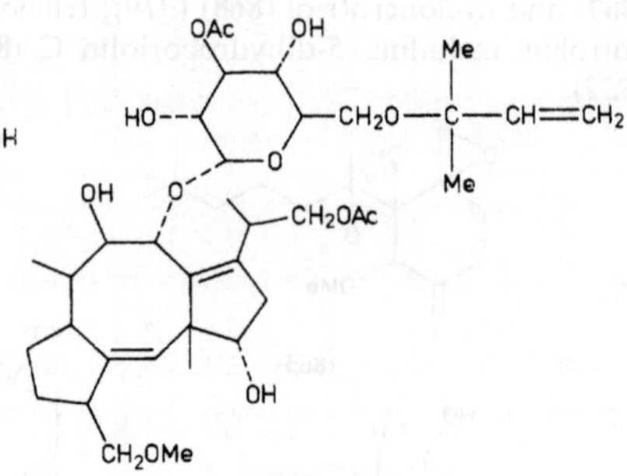

(**875**)

Triterpenoids

Oleanenes and ursenes (*326*, *141*); fusidic acid (**876**) (*327*).

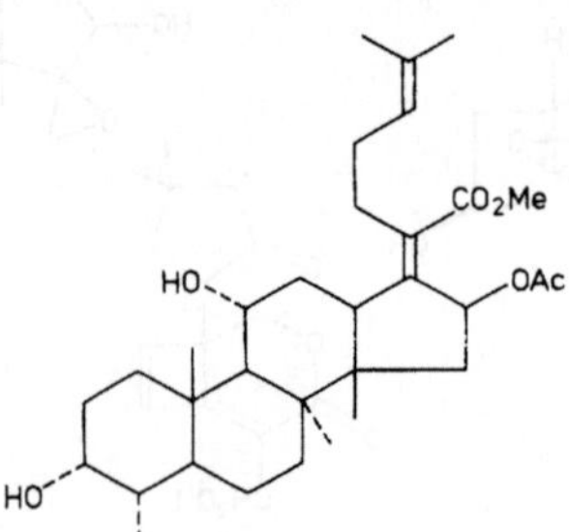

(**876**)

Steroids

Ergosterol (**877**) (*328*); cholesterol (**878**) (*329*).

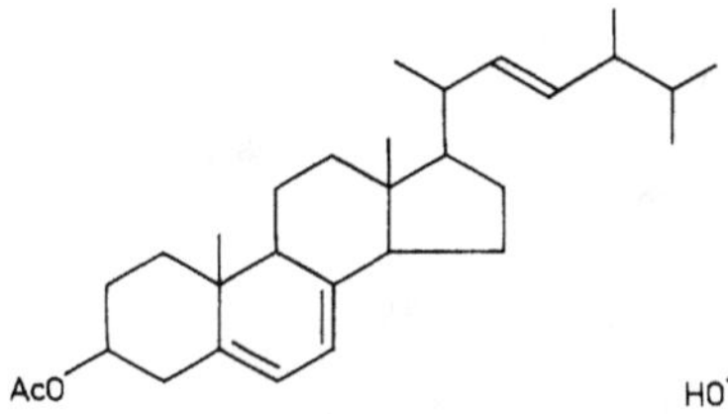

(**877**)

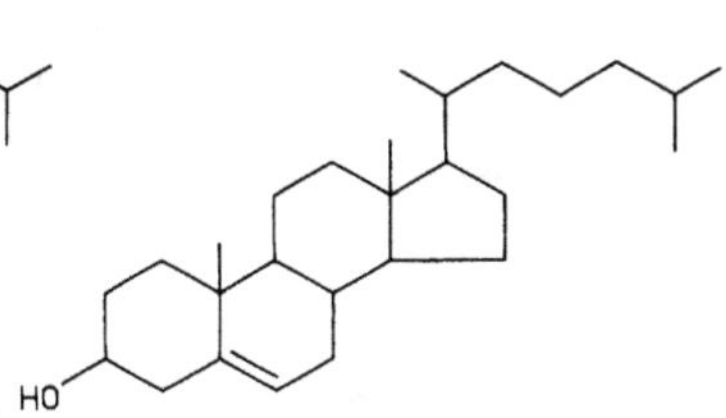

(**878**)

Others

Tetronic acid-type metabolites such as multicolic acid (**879**) and multicolosic acid (*330*); glauconic acid (**880**) (*331*); several pyrone metabolites (*332—334*) such as (**881**) (*332*) and (**882**) (rosellisin) (*334*); metabolites sclerin (**883**) (*335—337*); terrein (**884**) (*338*); a cyclopentene-type metabolite (**885**) (*339*); griseofulvin (**886**) (*340, 341*); ravenelin (**887**) (*342, 343*); mollicin (**888**) (*344, 345*); scytalone (**889**) (*346, 347*); bikaverin (**890**) (*348*), islandicin (**891**) (*349*); daunomycin (**892**) (*350*); tajixanthone (**893**) (*351*); chartreusin (**894**) (*352*); averufin (**895**) (*353*); sterigmatocystin (**896**) (*31, 354, 356*); aflatoxin B (**897**) (*31, 355—357*); narasin (**898**); nybomycin (**899**) (*360*); cryptoechinuline A (**900**) and echinuline (*361*); tenellin (**901**) (*362—364*); bassianin (*362*); cytochalasin B (**902**) and D (*365, 366*); anthramycin (**903**) (*367*); spectinomycin (**904**) (*368*); chloramphenicol (**905**) (*369*); geldanamycin (**906**) (*370*); 16-membered macrolides leucomycin A_3 (*371*) and tylosin (*372*).

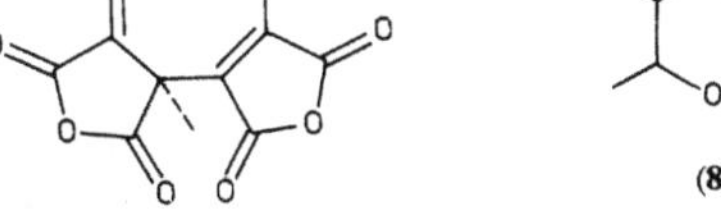

(**879**)

(**880**)

(**881**)

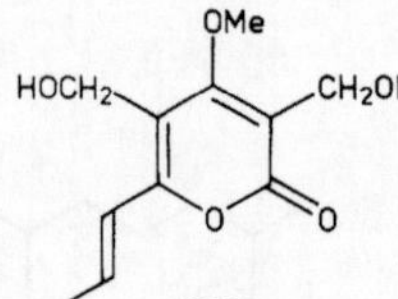

(882)

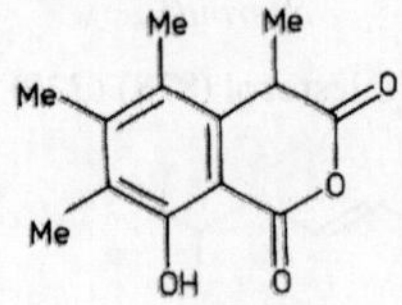

(883)

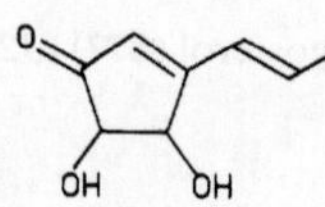

(884)

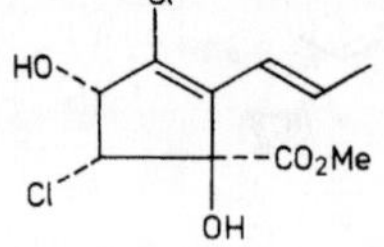

(885)

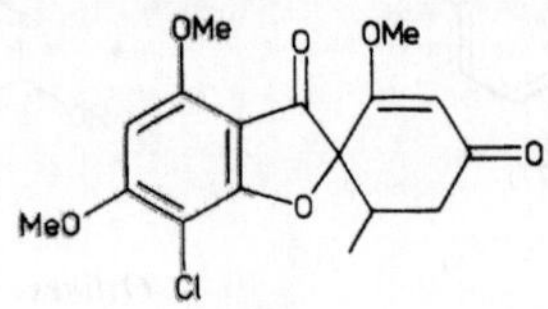

(886)

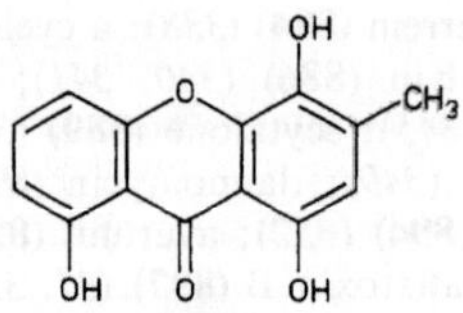

(887)

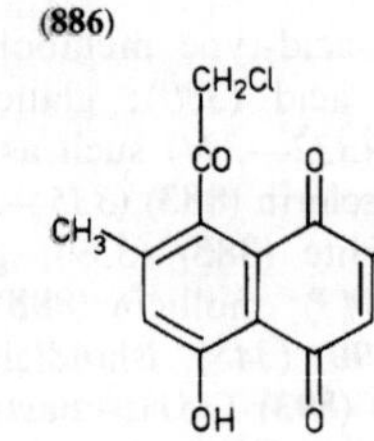

(888)

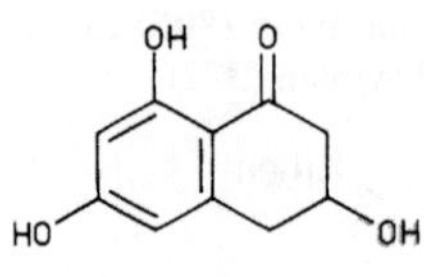

(889)

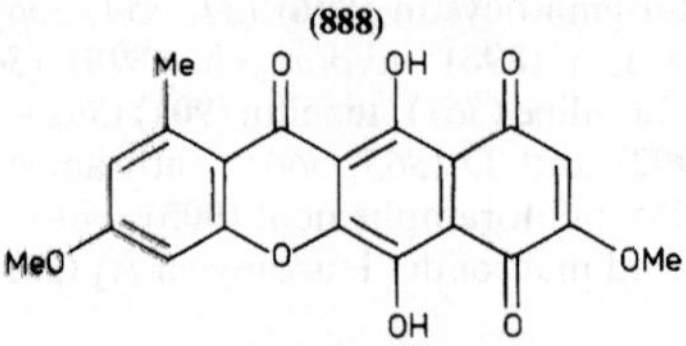

(890)

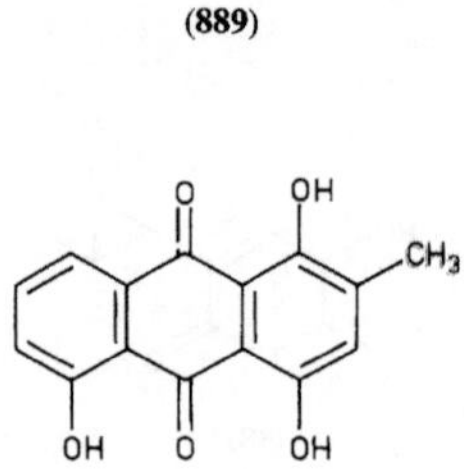

(891)

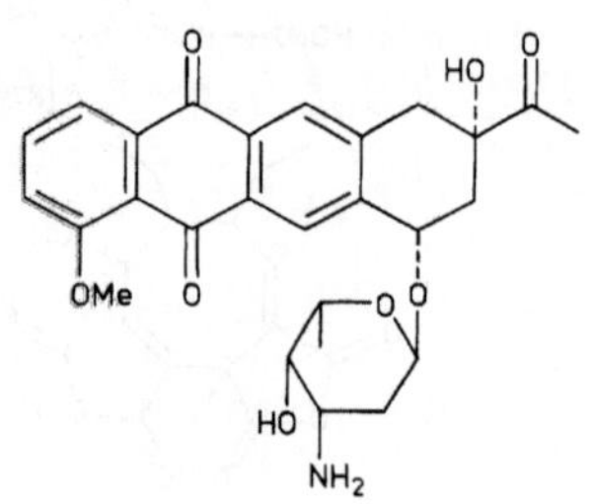

(892)

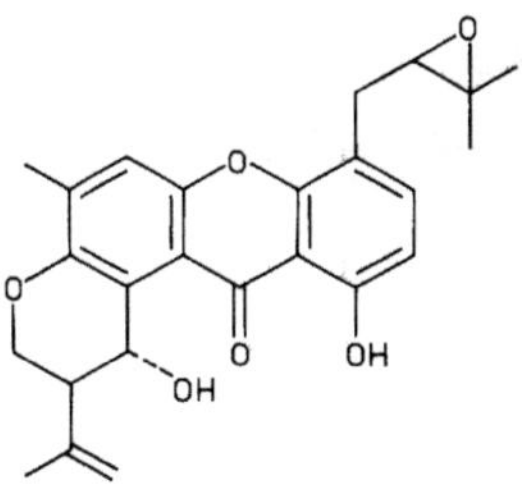

(893)

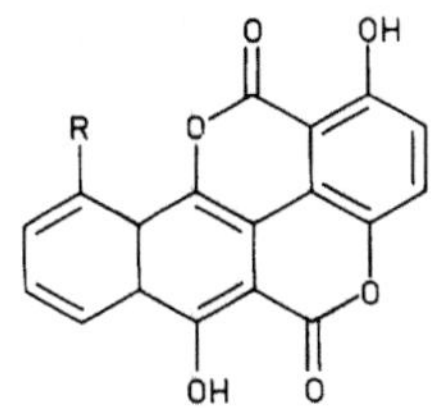

(894)

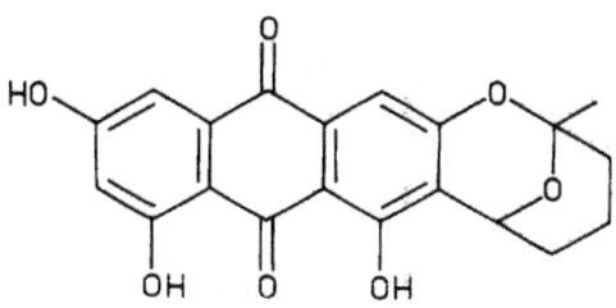

(895)

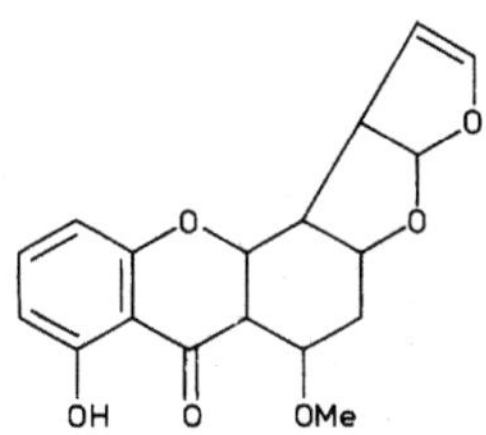

(896)

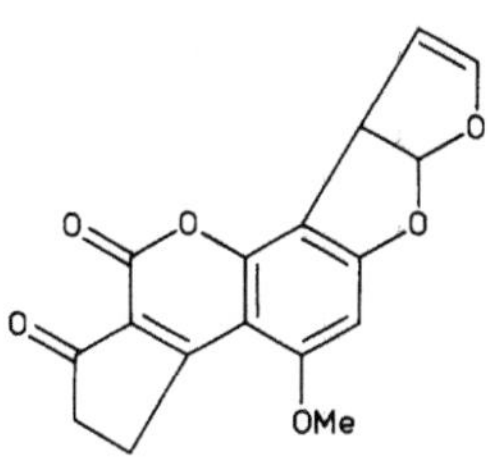

(897)

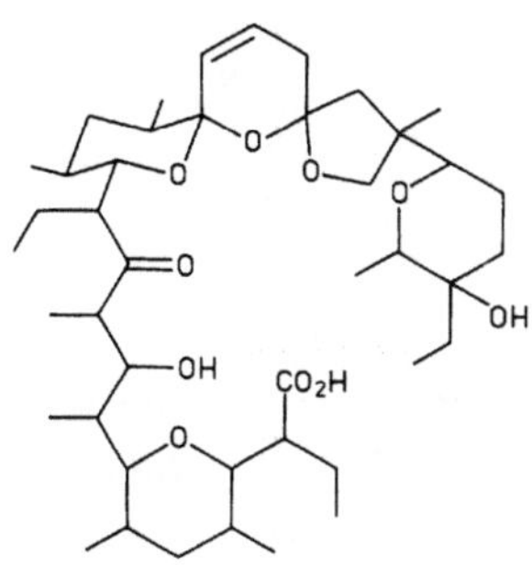

(898)

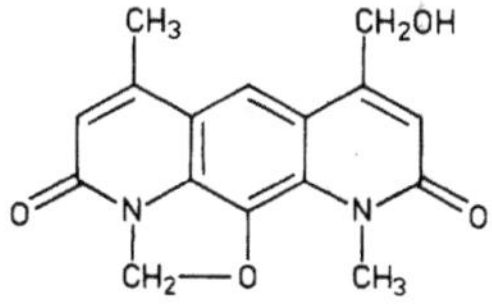

(899)

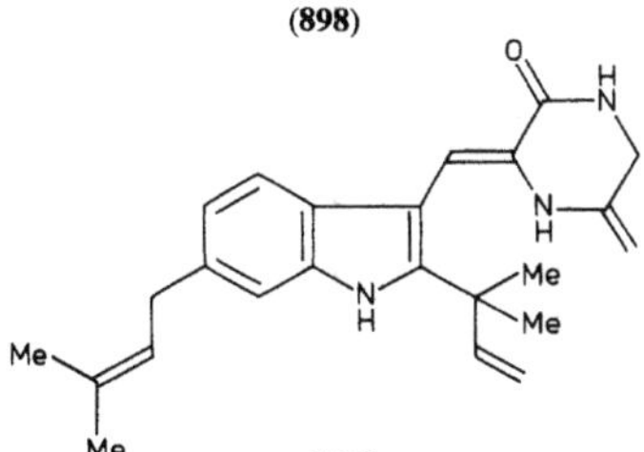

(900)

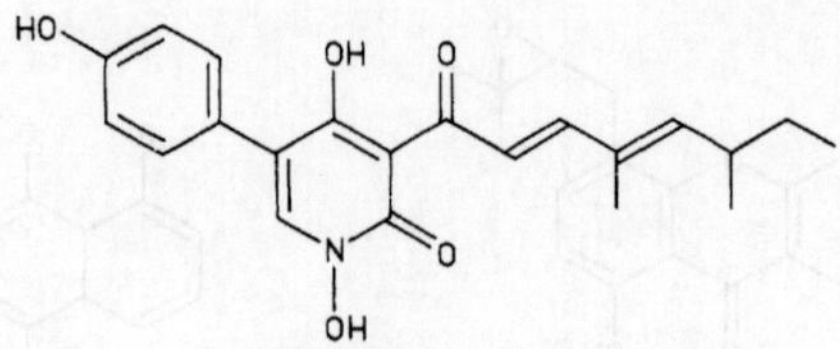

(901)

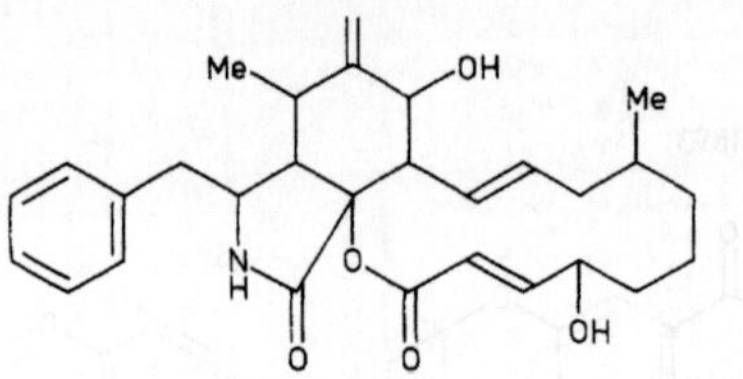

(902)

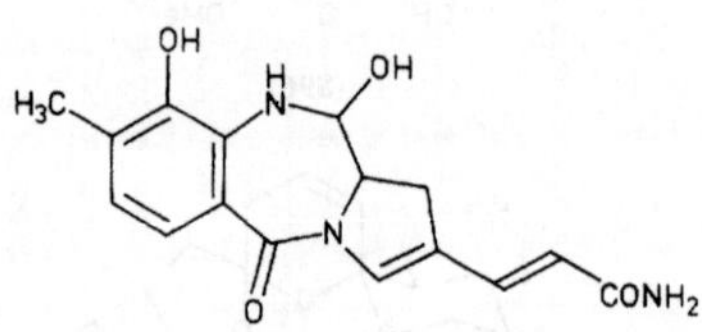

(903)

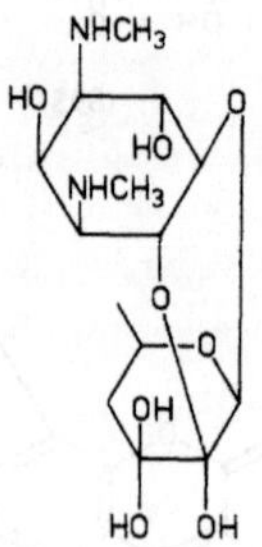

(904)

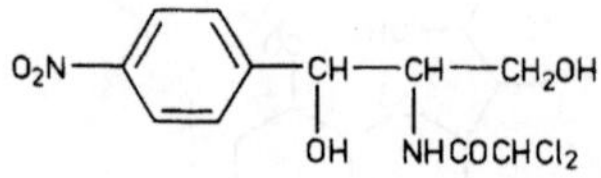

(905)

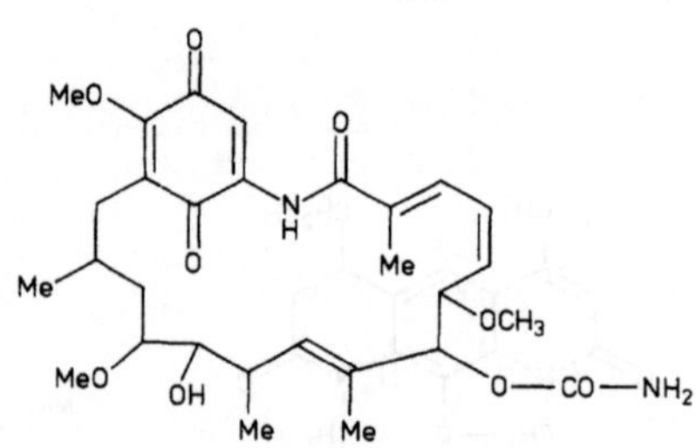

(906)

References, pp. 195—229

So far ^{13}C labelling methods have proved very successful in biosynthetic studies of microbial metabolites. However, the relatively low incorporation levels usually observed in higher plants have until very recently precluded the use of the technique in this domain. A successful experiment therefore demands a high incorporation or a small pool of substance of interest. Since either or both of these requirements are rarely met with higher plants, comparatively few alkaloid biosyntheses based on ^{13}C labelling have as yet been carried out. HUTCHINSON *et al.* (*373*) and BATTERSBY *et al.* (*374*) were the first to succeed in demonstrating the feasibility of ^{13}C n.m.r. spectrometry as a probe for the elucidation of the biosynthetic pathway of higher plant metabolites.

The biogenetic kinship between camptothecin (**908**) and (**907**) was proved by feeding [5-^{13}C]-(**907**) to *Catharantus roseus* G. Don (*373*). A possible biogenetic pathway leading from (**907**) to (**908**) is delineated by formulae (**907 a**) through (**907 e**). Comparison of the relative intensities in the spectra of the biosynthetic product with those of natural abundance ^{13}C camptothecin revealed a ca. 50% higher intensity for C-5 in the former, consistent with a specific incorporation of ca. 1.6%.

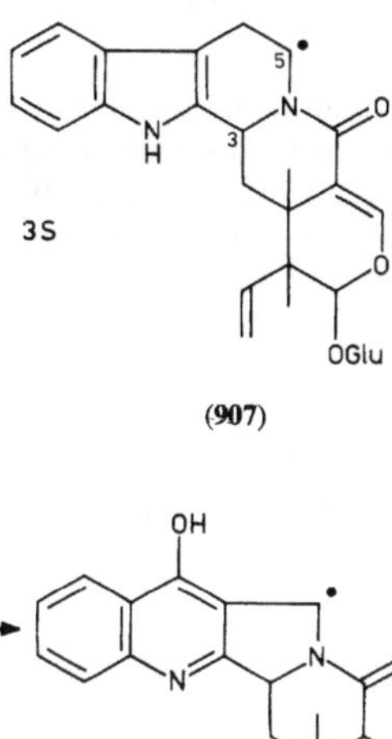

(**907**)

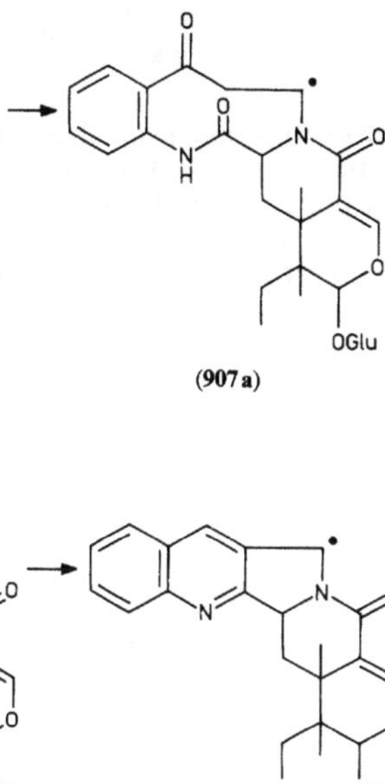

(**907 a**)

(**907 b**)

(**907 c**)

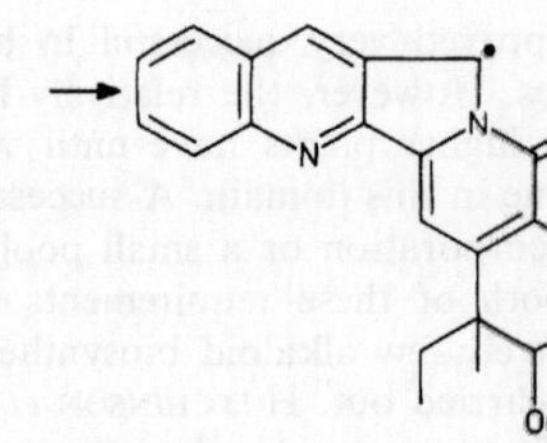

(907d)

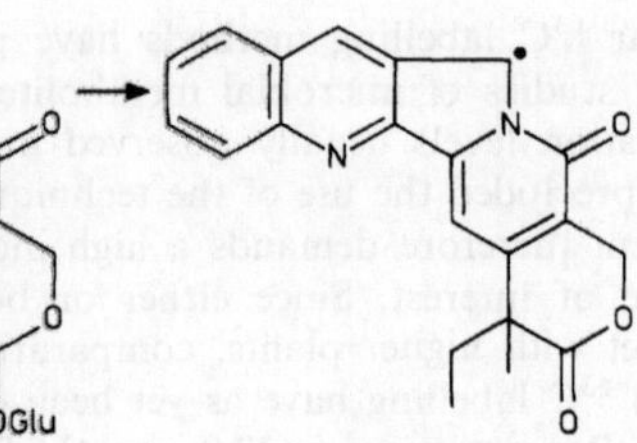

(907e)

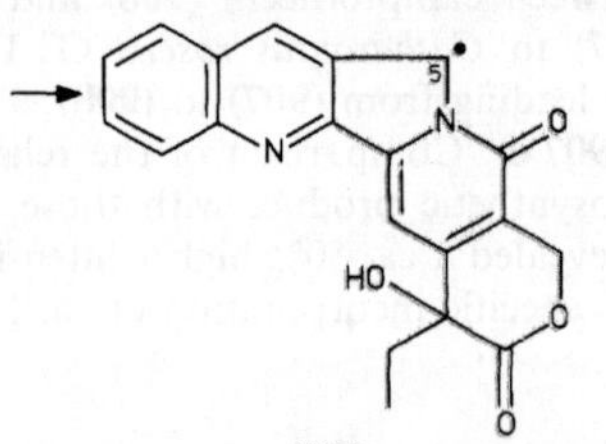

(908)

Shihunine (**910**), a unique phtalidopyrrolidine, has been shown not to be of polyketide origin (*375*). In order to prove o-succinylbenzoic acid (**909**) to be a precursor via biotransformation intermediates (**909 a—c**), [^{13}C-1]-(**909**) was fed to *D. pierardii*. The ^{13}C spectrum of the subsequently isolated product (**910**) showed exclusive incorporation of ^{13}C at position 5.

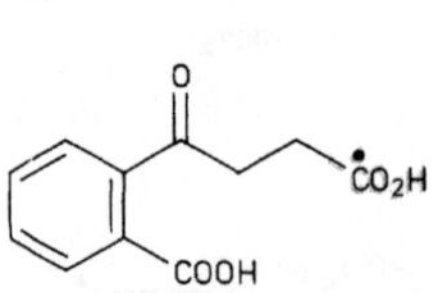

(909)

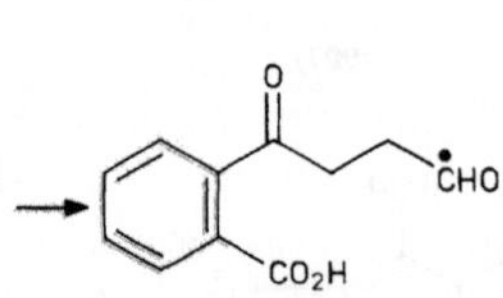

(909a)

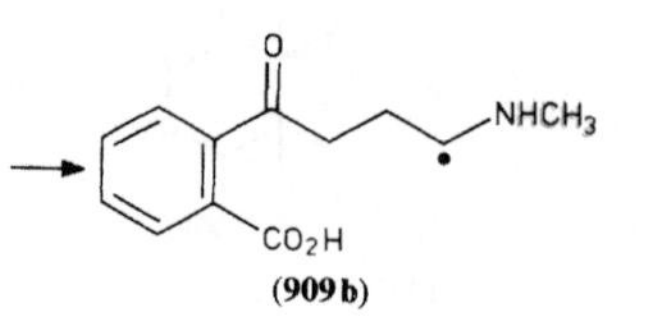

(909b)

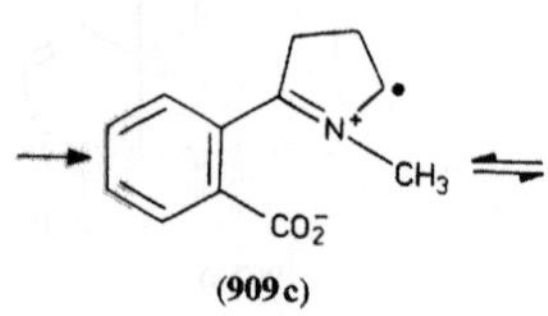

(909c)

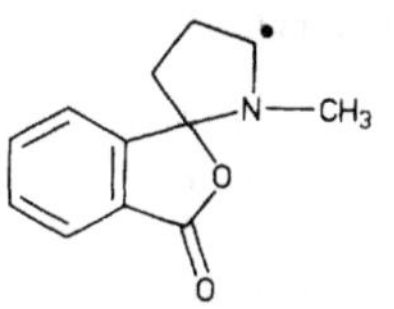

(910)

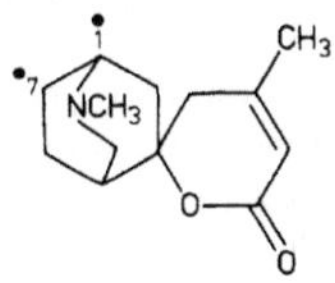

(911)

In a similar way it was shown that the biosynthesis of dioscorin (**911**) proceeded via nicotinic acid as an intermediate (*376*). For this purpose [5,6-$^{13}C_2$]-nicotinic acid was administered to *D. hispida* plant and the isolated dioscorin subjected to ^{13}C analysis. The spectrum of the biosynthetic (**911**) exhibited enhanced intensity for carbons pertaining to sites 1 and 7. Moreover an AB spin splitting pattern indicated intact incorporation of the nicotinic acid moiety.

In one case $^{13}CO_2$ was used in a study of secondary plant metabolite biosynthesis (*377*). It appeared that information similar to that obtained from doubly labelled acetate could be derived. In addition the experiment furnished insight into the dynamics of the biosynthesis.

Double labelling, previously outlined as a method suitable for proving *intact* precursor incorporation, lends itself as well for disclosing the contrary, namely that the precursor has undergone rearrangement. Tenellin (**913**), a yellow pigment produced by *Beauvaria tenella*, is well known to be derived from acetate, phenylalanine and methionine. It had been shown that the carbonyl carbon of phenylalanine (**912**) migrates,

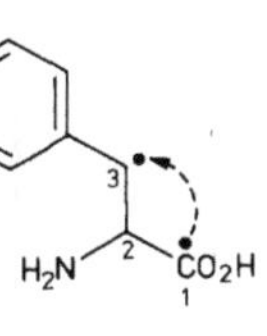

(912)

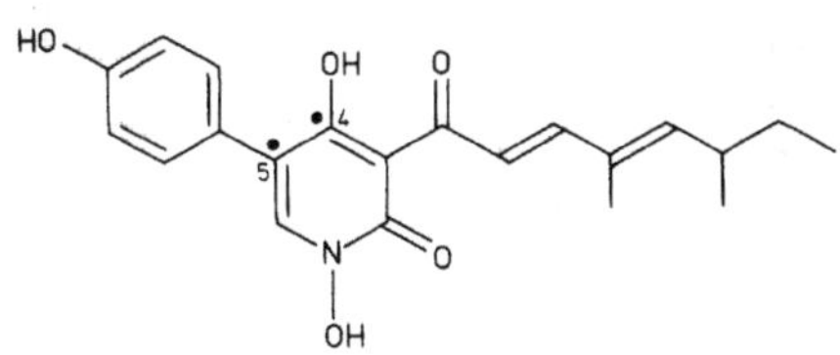

(913)

ultimately becoming C-4 of (**913**). In order to ascertain the nature of this rearrangement, [1,3-$^{13}C_2$]-phenylalanine (**912**) was added to a culture of *B. bassiana* (*364*). The spectrum of the isolated tenellin exhibited AX doublets at chemical shifts consistent with C-4 and C-5. This unequivocally demonstrates carboxyl migration to be of the intramolecular type as indicated in (**912**). In the case of an intermolecular rearrangement the enriched sites would pertain to different isotopomers hence resulting in enhanced singlets for C-4 and C-5.

5. Concluding Remarks

In his personal view a few years ago, DJERASSI (*378*) made the following prophetic statement regarding the future implications of instrumental analysis methods on the development of natural products chemistry:

"Physical methods will continue to play the dominant role in structural elucidation, and the major advances along these lines to be expected during the next few years are the much wider use by organic chemists themselves of X-ray crystallography and the routine use of ^{13}C n.m.r. spectroscopy. In fact, the most significant contribution that could probably be made to natural products methodology in the near future is an instrumental one — the availability of ^{13}C n.m.r. instrumentation that would permit the routine obtention of ^{13}C n.m.r. spectra on 1—5 mg of substance."

Today obtention of ^{13}C n.m.r. spectra on a milligram and even submilligram scale has become reality thanks to the joint application of state-of-the-art n.m.r. technology in conjunction with microsampling techniques. The latter have mainly been due to SHOOLERY (*379*), who first showed that in order to optimize the sensitivity for a given limited amount of sample, the volume of both the sample and the receiver coil have to be reduced. Reduction of the sample volume has a further benefit in that the frequently encountered problem of obscuring solvent resonances is

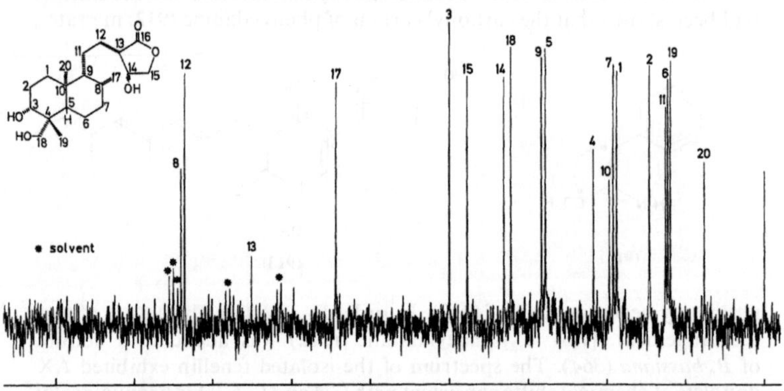

Fig. 10. 25.2 MHz ^{13}C n.m.r. spectrum of 2.0 mg andrographolide in 10 microliters of pyridine-d_5 using a 1.7 mm outer diameter tube in conjunction with a microinsert (Varian XL-100). The spectrum is the result of 8200 0.8 sec transients (1.8 hours) using a 45° pulse flip angle. All signals except the one pertaining to the carbonyl resonance are clearly visible

References, pp. 195—229

obviated. These demands culminated in the design of microprobes by utilizing capillary tubes of ca. 1 mm inner diameter as sample tubes and typical sample volumes of 5—15 microliters. With such an experimental array it has become possible to obtain ^{13}C spectra of 5 mg of a medium-molecular-weight compound routinely (i.e. in a few hours accumulation time). The spectrum of 2.0 mg of andrographolide (Fig. 10), recorded in 10 microliters of pyridine-d_5 may serve to illustrate this capability. At the present time the practical detection limit lies in the order of 100 micrograms requiring 2—3 days accumulation time.

A trend which has emerged only very recently, deals with the evaluation of the experimental data rather than with the experiment itself. Exploitation of digital computers, which carry out a comparison of the spectral features pertaining to the sample of interest, with those contained in a data bank (*380—382*) may become a valuable aid in structure identification and elucidation. The technique has been developed to the stage where joint processing of the available spectroscopic data results in a set of computer-plotted structures (*383*). Although it is still a curiosity at present, "artificial intelligence" will probably be used routinely in the not too distant future.

References

1. Highet, R. J., and E. A. Sokoloski: Structural Investigations of Natural Products by Newer Methods of NMR Spectroscopy. Fortschr. Chem. org. Naturstoffe **32**, 119 (1975).

2. Wenkert, E., J. S. Bindra, C.-J. Chang, D. W. Cochran, and F. M. Schell: Carbon-13 Nuclear Magnetic Resonance Spectroscopy of Naturally Occurring Substances. Alkaloids, Accounts Chem. Res. **7**, 46 (1974).

3. Wenkert, E., B. L. Buckwalter, I. R. Burfitt, M. J. Gašić, H. E. Gottlieb, E. W. Hagaman, F. M. Schell, and P. M. Wovkulich: Carbon-13 Nuclear Magnetic Resonance Spectroscopy of Naturally Occurring Substances. In: Topics in Carbon-13 N.M.R. Spectroscopy, Vol. 2, p. 81 (G. C. Levy, Ed.). New York: Wiley-Interscience. 1976.

4. Tanabe, M.: NMR with Stable Isotopes in Biosynthetic Studies. In "Specialist Periodical Reports, Biosynthesis". The Chemical Society (London) **2** (1973); **3**, 247 (1975); **4**, 204 (1976).

5. Scott, A. I.: Biosynthesis of Natural Products. Science **184**, 760 (1974).

6. Séquin, U., and A. I. Scott: Carbon-13 as a Label in Biosynthetic Studies. Science 101 (1974).

7. Mc Innes, A. G., and J. L. C. Wright: Use of Carbon-13 Magnetic Resonance Spectroscopy for Biosynthetic Investigations. Account Chem. Res. **8**, 313 (1975).

8. Simpson, T. J.: Carbon-13 Nuclear Magnetic Resonance in Biosynthetic Studies. Chem. Soc. Reviews **4**, 497 (1975).

9. McInnes, A. G., J. A. Walter, J. L. C. Wright, and L. C. Vining: ^{13}C NMR Biosynthetic Studies. In: Topics in Carbon-13 NMR Spectroscopy, Vol. 2 (G. C. Levy, Ed.). New York: Wiley-Interscience. 1976.

10. WEHRLI, F. W., and T. WIRTHLIN: "Interpretation of Carbon-13 NMR Spectra". London: Heyden. 1976.

11. FREEMAN, R., and H. D. W. HILL: Fourier Transform Study of NMR Spin-Lattice Relaxation by Progressive Saturation. J. Chem. Phys. **54,** 3367 (1971); BIRDSALL, B., N. J. M. BIRDSALL, and J. FEENEY: Simplified ^{13}C Spectral Assignments Using a Graphical Method to Present ^{13}C Spectra Recorded Under Conditions of Proton Off-resonance Spin Decoupling. J. C. S. Chem. Commun. (London) **1972,** 316.

12. HAGAMAN, E. W.: Pattern Recognition of Geminal Proton Nonequivalence and Second-Order Coupling in ^{13}C Single-frequency Off-resonance Decoupled Spectra. Assignment Criteria and Structure Elucidation. Org. Magn. Res. **8,** 389 (1976).

13. NEWMARK, R. A., and J. R. HILL: Virtual Coupling in Carbon-13 NMR Spectra with Off-resonance Continuous Wave Proton Spin Decoupling. J. Amer. Chem. Soc. **95,** 4435 (1973).

14. JIKELI, G., W. HERRIG, and H. GÜNTHER: Assignment of ^{13}C Nuclear Magnetic Resonance Signals. Fingerprints in Off-Resonance ^{13}C, {^{1}H} Nuclear Magnetic Double Resonance Spectra. J. Amer. Chem. Soc. **96,** 323 (1974).

15. FRITZ, H., and H. SAUTER: Off-Resonance Decoupling in ABX Spin Systems. J. Magn. Res. **18,** 527 (1975).

16. MÜLLER, L., A. KUMAR, and R. R. ERNST: Two-dimensional Carbon-13 NMR Spectroscopy. J. Chem. Phys. **63,** 5490 (1975).

17. BODENHAUSEN, G., R. FREEMAN, and G. A. MORRIS: A Simple Pulse Sequence for Selective Excitation in Fourier Transform NMR. J. Magn. Res. **23,** 171 (1976).

18. FREEMAN, R., G. A. MORRIS, and M. J. T. ROBINSON: Proton-coupled Carbon-13 Nuclear Magnetic Resonance Spectra from Individual Carbon Sites in a Molecule: The Rotameric Equilibrium in Menthone. J. C. S. Chem. Commun. (London) **1976,** 754.

19. VÖGELI, U., and W. VON PHILIPSBORN: Vicinal C, H Spin Coupling in Substituted Alkenes. Stereochemical Significance and Structural Effects. Org. Magn. Res. **7,** 617 (1975).

20. PHILIPSBORN, W. VON: Application of Double Resonance and Fourier Transform NMR Spectroscopy in Organic Chemistry. Pure and Applied Chemistry **40,** 159 (1974).

21. GÜNTHER, H., H. SCHMICKLER, and G. JIKELI: Applications of Carbon-13 Resonance Spectroscopy. V. "Fingerprints" for the Assignment of Carbon-13 Resonance Signals. J. Magn. Res. **11,** 344 (1973).

22. Nuclear Magnetic Resonance Spectroscopy. Benezene-^{13}C. WEIGERT, F. J., and J. D. ROBERTS. J. Amer. Chem. Soc. **89,** 2967 (1967).

23. Ref. *10,* Chap. 3.

24. MASSOW, F. VON, and M. A. R. SMITH: Indirect Coupling in Asymmetrically Trisubstituted Benzenes: a Carbon-13 Nuclear Magnetic Resonance. Study J. C. S. Perkin II. **1976,** 977.

25. WEHRLI, F. W.: Proton-Coupled ^{13}C Nuclear Magnetic Resonance Spectra Involving ^{13}C – ^{1}H Spin-Spin Coupling to Hydroxyl Protons, A Complementary Assignment Aid. J. C. S. Chem. Commun. (London) **1975,** 663.

26. KINGSBURY, C. A., and J. H. LOOKER: Carbon-13 Spectra of Methoxyflavones. J. Org. Chem. **40,** 1120 (1975).

27. CHANG, C.: Carbon-13-Proton Long-Range Couplings of Phenols. Hydrogen Bonding and Stereospecificity. J. Org. Chem. **41,** 1881 (1976).

28. PACHLER, K. G. R., and P. L. WESSELS: Selective Population Inversion (SPI). A Pulsed Double Resonance Method in FT NMR Spectroscopy Equivalent to INDOR. J. Magn. Res. **12,** 337 (1973).

29. SORENSEN, S., R. S. HANSEN, and H. JAKOBSEN: Assignments and Relative Signs of ^{13}C-X Coupling Constants in ^{13}C FT NMR from Selective Population Transfer (SPT). J. Magn. Res. **14,** 243 (1974).

30. Nagel, D. W., K. G. R. Pachler, P. S. Steyn, R. Vleggaar, and P. L. Wessels: The Chemistry and ^{13}C NMR Assignments of Oxaline, a Novel Alkaloid from *Penicillium oxalicum*. Tetrahedron **32**, 2625 (1976).
31. Pachler, K. G. R., P. S. Steyn, R. Vleggaar, P. L. Wessels, and D. B. Scott: Carbon-13 Nuclear Magnetic Resonance Assignments and Biosynthesis of Aflatoxin B_1 and Sterigmatocystin. J. S. C. Perkin I 1182 (1976).
32. Walker, T. E., R. E. London, T. W. Whaley, R. Barker, and N. A. Matwiyoff: Carbon-13 Nuclear Magnetic Resonance Spectroscopy of [1-^{13}C] Enriched Monosaccharides. Signal Assignments and Orientational Dependence of Geminal and Vicinal Carbon-Carbon and Carbon-Hydrogen Spin-Spin Coupling Constants. J. Amer. Chem. Soc. **98**, 5807 (1976).
33. Eggert, H., and C. Djerassi: Carbon-13 Nuclear Magnetic Resonance Spectra of Keto Steroids. J. Org. Chem. **38**, 3788 (1973).
34. Feeney, J., P. Partington, and G. C. K. Roberts: The Assignment of Carbon-13 Resonances from Carbonyl Groups in Peptides. J. Magn. Res. **13**, 268 (1974).
35. Gorin, P. A. J., and M. Mazurek: Further Studies on the Assignment of Signals in ^{13}C Magnetic Resonance Spectra of Aldoses and Derived Methyl Glycosides. Can. J. Chem. **53**, 1212 (1975).
36. Wahlberg, I., S.-O. Almqvist, T. Nishida, and C. R. Enzell: Carbon-13 Nuclear Magnetic Resonance Spectra of Some Podocarpane Derivatives. Acta Chem. Scand. **B29**, 1047 (1975).
37. Joseph-Nathan, P., J. Maves, M. C. Hernández, and J. N. Shoolery: Proton and Carbon-13 Nuclear Magnetic Resonance Studies of Flavone and Deuterated Analogs. J. Magn. Res. **16**, 447 (1974).
38. Wehrli, F. W.: ^{13}C Spectral Assignments with the Aid of Spin-Spin Coupling and Spin-Lattice Relaxation Data Involving Labile Protons. Paper presented at the 3rd International Symposium on Nuclear Magnetic Resonance, St. Andrews, Scotland, 1975.
39. Kobayashi, M., Y. Terui, K. Tori, and N. Tsuji: Carbon-13 NMR Spectra of Juglone, Naphthazarin and their Derivatives. Tetrahedron Letters **1976**, 619.
40. Nishida, T.: Unpublished results.
41. Nishida, T., I. Wahlberg, and C. R. Enzell: Carbon-13 Nuclear Magnetic Resonance Spectra of Some Aromatic Diterpenoids. Org. Magn. Res. **9**, 203 (1977).
42. Ref. *3*, p. 104.
43. Lenkinski, R. E., and M. R. Willcott, III: Contact Shifts in the Nuclear Magnetic Resonance Spectra of Isoquinoline and of *endo*-Norborneol. J. Amer. Chem. Soc. **98**, 4250 (1976).
44. Gansow, O. A., P. A. Loeffler, R. E. Davis, M. R. Willcott, and R. E. Lenkinski: Evidence for Europium-Induced Contact Shifts in Carbon-13 Nuclear Magnetic Resonance Spectra. J. Amer. Chem. Soc. **95**, 3390 (1973).
45. Chadwick, D. J., and D. H. Williams: The full Assignment of the Carbon-13 Nuclear Magnetic Resonance Spectrum of 5α-Cholestan-3β-ol with the Aid of the Lanthanide Shift Reagent $Yb(dpm)_3$. J. C. S. Perkin II, **1974**, 1903.
46. Almqvist, S.-O., C. R. Enzell, and F. W. Wehrli: Carbon-13 NMR Studies of Labdane Diterpenoids. Acta Chem. Scand. **B29**, 695 (1975).
47. Lyerla, J. R., Jr., and G. C. Levy: Carbon-13 Nuclear Spin Relaxation. In: Topics in Carbon-13 NMR Spectroscopy, Vol. 1, Chap. 3 (Levy, G. C., ed.). New York: Wiley-Interscience. 1974.
48. Lyerla, J. R. Jr., and D. M. Grant: Carbon-13 Nuclear Spin Relaxation. In: International Reviews in Science. Physical Chemistry Series, Vol. 4, Chap. 5 (McDowell, C. A., Ed.). Medical and Technical Publishing, 1972.
49. Wehrli, F. W.: Organic Structure Assignments Using ^{13}C-Spin-Relaxation Data.

In: Topics in Carbon-13 N.M.R. Spectroscopy, Vol. 2, p. 343 (LEVY, G. C., Ed.). New York: Wiley-Interscience. 1976.
50. APSIMON, J. W., H. BEIERBECK, and J. K. SAUNDERS: Carbon-13 Spin-Lattice Relaxation Times for Selected Steroids. Can. J. Chem. **53,** 338 (1975).
51. LEVY, G. C., J. D. CARGIOLI, and F. A. L. ANET: Carbon-13 Spin-Lattice Relaxation in Benzene and Substituted Aromatic Compounds. J. Amer. Chem. Soc. **95,** 1527 (1973).
52. BERGER, S., F. R. KREISSL, D. M. GRANT, and J. D. ROBERTS: Determination of Anisotropy of Molecular Motion with ^{13}C Spin-Lattice Relaxation Times. J. Amer. Chem. Soc. **97,** 1805 (1975).
53. BECKER, R. S., S. BERGER, D. K. DALLING, D. M. GRANT, and R. J. PUGMIRE: Carbon-13 Magnetic Resonance Investigation of Retinal Isomers and Related Compounds. J. Amer. Chem. Soc. **96,** 7008 (1974).
54. ROWAN III, R., and B. D. SYKES: A Carbon-13 Nuclear Magnetic Resonance Study of the Visual Chromophores and Model Compounds. J. Amer. Chem. Soc. **96,** 7000 (1974).
55. See e.g. Ref. *45*, p. 373ff.
56. NISHIDA, T., unpublished data.
57. CZARNIECKI, M. F., and E. R. THORNTON: ^{13}C Spin-Lattice Relaxation in the Neuraminic Acids. Evidence for an Unusual Intramolecular Hydrogen Bonding Network. J. Amer. Chem. Soc. **98,** 1023 (1976).
58. WEHRLI, F. W.: ^{13}C Spectral Assignment and Spin-Lattice Relaxation in Medium-Sized Molecules. Advances in Molecular Relaxation Processes **6,** 555 (1974).
59. ECHOLS, R. E., and G. C. LEVY: Carbon-13 Nuclear Magnetic Resonance Spectral Analysis Using Spin-Lattice Relaxation Data and Specific Deuteration. Thiamine Hydrochloride. J. Org. Chem. **39,** 1321 (1974).
60. GRIBBLE, G. W., R. B. NELSON, J. L. JOHNSON, and G. C. LEVY: Carbon-13 Fourier Transform Nuclear Magnetic Resonance Spectroscopy of Indolo [2,3-*a*] quinolizines. Specific Deuteration and Relaxation Methods in Structural Assignments. J. Org. Chem. **40,** 3720 (1975).
61. NORTON, R. S., and A. ALLERHAND: Effect of ^{13}C-^{14}N Dipolar Interactions on Spin-Lattice Relaxation Times and Intensities of Nonprotonated Carbon Resonances. J. Amer. Chem. Soc. **98,** 1007 (1976).
62. FALLER, J. W., M. A. ADAMS, and G. N. LA MAR: The Application of Lanthanide Relaxation Reagents in CMR. Tetrahedron Letters **1974,** 699.
63. LEVY, G. C., U. EDLUND, and C. E. HOLLOWAY: Mechanisms for Interaction between Organic Molecules and Paramagnetic Relaxation Reagents. J. Magn. Res. **24,** 375 (1976).
64. BOHLMANN, F., R. ZEISBERG und E. KLEIN: ^{13}C-NMR-Spektren von Monoterpenen. Org. Magn. Res. **7,** 426 (1975).
65. ABRAHAM, R. J., C. M. HOLDEN, P. LOFTUS, and D. WHITTAKER: The NMR Spectra and Conformations of Cyclic Compounds IX — Conformational Studies of Bicyclo[3.1.0]hexane Derivatives by ^{13}C NMR. Org. Magn. Res. **6,** 184 (1974).
66. Ref. *10*, p. 110ff.
67. NISHIDA, T.: Unpublished data.
68. WENKERT, E., D. W. COCHRAN, and F. M. SCHELL: CMR Spectral Analysis of Tetrahydrocannabinol and its Isomers. Experientia **28,** 250 (1972).
69. WEINER, B. Z., and A. T. MEYER: Unpublished data quoted in the review of MECHOULAM, R., N. K. MCCALLUM, and S. BURSTEIN: Recent Advances in the Chemistry and Biochemistry of Cannabis. Chem. Review **76,** 75 (1976).
70. NISHIDA, T.: To be published. The original assignments given in reference 71 are partly in error.

71. THOMAS, A. F., W. THOMMEN, B. WILLHALM, E. W. HAGAMAN, and E. WENKERT: Terpenoids Derived from Linalyl Oxide. Part I, The Stereochemistry of the Davanones. Helv. Chim. Acta **57**, 2055 (1974).
72. BOCK, K., S. R. JENSEN, and B. J. NIELSEN: Secogalioside, an Iridoid Glucoside from *Galium album* Mill, and ^{13}C NMR Spectra of some Seco-Iridoid Glucosides. Acta Chem. Scand. **B30**, 743 (1976).
73. KUBO, I., I. MIURA, and K. NAKANISHI: The Structure of Xylomollin, a Secoiridoid Hemiacetal Acetal. J. Amer. Chem. Soc. **98**, 6704 (1976).
74. THOMAS, A. F., M. OZAINNE, F. NÄF, and G. LUKACS: 10-Epijunenol, A New *cis*-Eudesmane Sesquiterpenoid. Tetrahedron **32**, 2261 (1976).
75. NAKANISHI, K., R. CROUCH, I. MIURA, X. DOMINGUEZ, A. ZAMUDIS, and R. VILLARREAL: Structure of a Sesquiterpene, Cuauthemone, and Its Derivative. Application of Partially Relaxed Fourier Transform ^{13}C Nuclear Magnetic Resonance. J. Amer. Chem. Soc. **96**, 609 (1974).
76. COXON, D. T., K. R. PRICE, B. HOWARD, S. F. OSMAN, E. B. KALAN, and R. M. ZACHARIUS: Two New Vetispirane Derivatives: Stress Metabolites from Potato *(Solanum tuberosum)* Tubers. Tetrahedron Letters 2921 (1974).
77. STOESSL, A., J. B. STOTHERS, and E. W. B. WARD: The Structures of Some Stress Metabolites from *Solanum melongena*. Can. J. Chem. **53**, 3351 (1975).
78. MAGNUSSEN, G., and S. THORÉN: Fungal Extractives. — VI. Structure of Lactaral, a New Sesquiterpene Furan-3-aldehyde from *Lactarius*, by Spectroscopic Methods. Tetrahedron **30**, 1431 (1974).
79. FLAMM, B. L., J. A. PETTUS, JR., J. J. SIMS, J. P. SPRINGER, and J. CLARDY: Isolation of Epoxydecompostin from *Lepidospartum squamatum* Gray and its Structure Revision. Tetrahedron Letters 2671 (1976).
80. SHEIKH, Y. M., G. SINGY, M. KAISIN, H. EGGERT, C. DJERASSI, B. TURSCH, D. DALOZE, and J. C. BRAEKMAN: Terpenoids — LXXI. Chemical Studies of Marine Intertebrates — XIV. Four Representatives of a Novel Sesquiterpene Class — The Capnellane Skeleton. Tetrahedron **32**, 1171 (1976).
81. WEI, R.-D., H. K. SCHNOES, P. A. HART, and F. M. STRONG: The Structure of PR Toxin, a Mycotoxin from *Penicillium roqueforti*. Tetrahedron **31**, 109 (1975).
82. MCGAHREN, W. J., G. A. ELLESTAD, G. O. MORTON, and M. P. KUNSTMANN: Spectral and Chemical Characterization of Fungal Metabolite LL-N313S. J. Org. Chem. **41**, 66 (1976).
83. BREITENSTEIN, W., and C. TAMM: ^{13}C NMR Spectroscopy of the Trichothecane Derivatives, Verrucarol, Verrucarins A, and B and Roridins A, D, and H. Helv. Chim. Acta **58**, 1172 (1975).
84. ELLISON, R. A., and F. N. KOTSONIS: Carbon-13 Nuclear Magnetic Resonance Assignments in the Trichothecene Mycotoxins. J. Org. Chem. **41**, 576 (1976).
85. BRAGA DE OLIVEIRA, A., M. DE L. M. FERNANDES, O. R. GOTTLIEB, E. W. HAGAMAN, and E. WENKERT: Aromatic Sesquiterpenoids from *Emmotum nitens*. Phytochemistry **13**, 1199 (1974).
86. THOMAS, A. F., and M. OZAINNE: The Stereochemistry of the Dihydroagarofurans. Tetrahedron Letters 1717 (1976).
87. DEN HERTOG, H. J., JR., C. KRUK, D. D. NAVANATI, and S. DEV: Stereochemistry of Malkanguniol and Stereostructures of Some Other Related Polyalcohols from *Celastrus paniculatus* Willd. Tetrahedron Letters 2219 (1974).
88. HERZ, W., and R. P. SHARMA: Complete Stereochemistry of Tenulin. Carbon-13 Nuclear Magnetic Resonance Spectra of Tenulin Derivatives. J. Org. Chem. **40**, 2557 (1975).
89. PREGOSIN, P. S., E. W. RANDALL, and T. B. H. MCMURRY: ^{13}C Fourier Studies. The Configurational Dependence of the Carbon-13 Chemical Shifts in Santonin Derivatives. J. C. S. Perkin I, 299 (1972).

90. MOSS, G. P., P. S. PREGOSIN, and E. W. RANDALL: Assignments in the Carbon-13 Fourier Spectra of Eudesmanolides. J. C. S. Perkin I, 1525 (1974).

91. VICHNEWSKI, W., I. K. SHUHAMA, R. C. ROSANSKE, and W. HERZ: Granilin and Ivasperin from *Ambrosia polystachia.* ^{13}C NMR Spectra of Hydroxylated Isoalantones. Phytochemistry **15,** 1531 (1976).

92. TORI, K., M. UEYAMA, I. HORIBE, Y. TAMURA, and K. TAKEDA: Carbon-13 N.M.R. Spectra of some Furano-Sesquiterpenes, Major Components of *Lindera Strychnifolia.* Tetrahedron Letters 4583 (1975).

93. TORI, K., I. HORIBE, Y. TAMURA, K. KURIYAMA, H. TADA, and K. TAKEDA: Reinvestigation of the Conformation of Laurenobiolide, a Ten-membered Ring Sesquiterpene Lactone, by Variable-Temperature Carbon-13 NMR Spectroscopy. Evidence for the Presence of Four Conformational Isomers in Solution. Tetrahedron Letters 387 (1976).

94. HERZ, W., and R. P. SHARMA: A *trans*-1,2-*cis*-4,5-Germacradienolide and Other New Germacranolides from *Tithonia* Species. J. Org. Chem. **40,** 3118 (1975).

95. HERZ, W., and R. P. SHARMA: Sesquiterpene Lactones of *Eupatorium hyssopifolium.* A Germacranolide with an Unusual Lipid Ester Side Chain. J. Org. Chem. **41,** 1015 (1976).

96. BHACCA, N. S., R. A. WILEY, N. H. FISCHER, and F. W. WEHRLI: Carbon-13 and Proton Magnetic Resonance Study of the Structure and Conformation of a New Germacranolide Sesquiterpene Dilactone. J. C. S. Chem. Commun. 614 (1973).

97. BHACCA, N. S., F. W. WEHRLI, and N. H. FISCHER: Carbon-13 Magnetic Resonance Study of Terpenoids, I. An Application of Heteronuclear Selective Decoupling Experiments to the Spectral Assignments of Non-proton-bearing Carbon-13 Resonances of a Germacranolide, Melampodin. J. Org. Chem. **38,** 3618 (1973).

98. HERZ, W., and P. S. KALYANARAMAN: Acanthospermal A and Acanthospermal B, Two New Melampolides from *Acanthospermun* Species. J. Org. Chem. **40,** 3486 (1975).

99. HERZ, W., and R. P. SHARMA: Pycnolide, a *seco*-Germacradienolide from *Liatris pycnostachya,* and Other Antitumor Constituents of *Liatris* Species. J. Org. Chem. **41,** 1248 (1976).

100. GARCIA, M., A. J. R. DA SILVA, P. M. BAKER, B. GILBERT, and J. A. RABI: Absolute Stereochemistry of Eremanthine. A Schistosomicidal Sesquiterpene Lactone from *Eremanthus elaeagnus.* Phytochemistry **15,** 331 (1976).

101. CROMBIE, L., R. W. KING, and D. A. WHITING: Carbon-13 Magnetic Resonance Spectra. Synthetic Presqualene Esters, Related Cyclopropanes, and Isoprenoids. J. C. S. Perkin II, 913 (1975).

102. CROMBIE, L., G. PETTENDEN, and D. J. SIMMONDS: Carbon-13 Nuclear Magnetic Resonance Spectra of the Natural Pyrethrins and Related Compounds. J. C. S. Perkin I, 1500 (1975).

103. UCHIO, Y., A. MATSUO, M. NAKAYAMA, and S. HAYASHI: Vulgarone, A Sesquiterpene Ketone with a New Carbon Skeleton from *Chrysanthemum vulgare.* Tetrahedron Letters 2963 (1976).

104. AASEN, A. J., T. NISHIDA, C. R. ENZELL, and M. DEVREUX: Tabacco Chemistry 37. The Absolute Configuration of Prenylsolanone, (9*S*)6,12-Dimethyl-9-isopropyltrideca-5*E*,10*E*,12-trien-2-one, a Nortunberganoid of *Nicotiana tabacum* L. Acta Chem. Scand. **B30,** 178 (1976).

105. ALMQVIST, S.-O., and C. R. ENZELL: Carbon-13 NMR Studies of Labdane Diterpenoids. Acta Chem. Scand. **B29,** 695 (1975).

106. BUCKWALTER, B. L., I. R. BURFITT, A. A. NAGEL, E. WENKERT, and F. NÄF: ^{13}C-NMR. Spectroscopy of Naturally Occurring Substances, XXXV. Labdanic Diterpenes. Helv. Chim. Acta **58,** 1567 (1975).

107. NISHIDA, T.: Unpublished results.

108. SAVONA, G., F. PIOZZI, J. R. HANSON, and M. SIVERNS: Structure of Ballotinone, a Diterpenoid from *Ballota nigra*. J. C. S. Perkin I, 1607 (1976).

109. SMITH, C. R., JR., R. V. MADRIGAL, D. WEISLEDER, K. L. MIKOLAJCZAK, and R. J. HIGHET: Potamogetonin, a New Furanoid Diterpene. Structural Assignment by Carbon-13 and Proton Magnetic Resonance. J. Org. Chem. **41**, 593 (1976).

110. CAMBIE, R. C., I. R. BURFITT, T. E. GOODWIN, and E. WENKERT: The Structure of Hallol. J. Org. Chem. **40**, 3789 (1975).

111. KATO, T., C. KABUTO, N. SASAKI, M. TSUNAGAWA, H. AIZAWA, K. FUJITA, Y. KATO, and Y. KITAHARA: Momilactones, Growth Inhibitors from Rice, *Oryza sativa* L. Tetrahedron Letters 3861 (1973).

112. MATSUO, A., S. UTO, M. NAKAYAMA, and S. HAYASHI: (−)-Thermarol, a New *Ent*-Pimarane-Class Diterpene Diol. Tetrahedron Letters 2451 (1976).

113. MANCHAND, P. S., and J. F. BLOUNT: X-Ray Structure and Absolute Stereochemistry of Stemolide. Tetrahedron Letters 2489 (1976).

114. KUBO, I., Y.-W. LEE, V. BALOGH-NAIR, K. NAKANISHI, and A. CHAPYA: Structure of Ajugarins. J. C. S. Chem. Commun. 949 (1976).

115. BILLET, D., M. DURGEAT, S. HEITZ et A. AHOND: Constituants d'*evodia floribunda* Baker. I. l'Acide Floribundique, Nouveau Diterpène de Type Clerodane. Tetrahedron Letters 3825 (1975).

116. HANSON, J. R., M. SIVERNS, F. PIOZZI, and G. SAVONA: The ^{13}C Nuclear Magnetic Resonance Spectra of Kauranoid Diterpenes. J. C. S. Perkin I, 114 (1976).

117. HANSON, J. R., G. SAVONA, and M. SIVERNS: ^{13}C Nuclear Magnetic Resonance Spectra and Microbiological Hydroxylation of 7α- and 7β-Hydroxy-kaurenolide. J. C. S. Perkin I, 2001 (1974).

118. ELLAMES, G., and J. R. HANSON: Microbiological Hydroxylation of 3β,7β-Dihydroxy-kaurenolide. J. C. S. Perkin I, 1666 (1976).

119. HERZ, W., and R. P. SHARMA: New Hydroxylated *ent*-kauranoic Acids from *Eupatorium album*. J. Org. Chem. **41**, 1021 (1976).

120. WAHLBERG, I., C. R. ENZELL, and J. W. ROWE: *Ent*-16-Kauren-19-ol from Coffee. Phytochemistry **14**, 1677 (1975).

121. YAMASAKI, K., H. KOHDA, T. KOBAYASHI, R. KASAI, and O. TANAKA: Structures of Stevia Diterpene-glucosides: Application of ^{13}C NMR. Tetrahedron Letters 1005 (1976).

122. CARSTEN-LICHTERFELDE, C. VON, C. PASCUAL, and J. PONS: Carbon-13 NMR Spectra of *ent*-Beyerane and *ent*-Beyerane Derivatives. Tetrahedron Letters 3569 (1975).

123. EVANS, R., J. R. HANSON, and M. SIVERNS: The ^{13}C Nuclear Magnetic Resonance Spectra of Some Gibberellins. J. C. S. Perkin I, 1514 (1975).

124. RADEGLIA, R., G. ADAM und P. D. HUNG: ^{13}C-NMR-Spektroskopie von Gibberellin-Verbindungen. Tetrahedron Letters 605 (1976).

125. YAMAGUCHI, I., M. MIYAMOTO, H. YAMANE, N. MUROFUSHI, N. TAKAHASHI, and K. FUJITA: Elucidation of the Structure of Gibberellin A_{40} from *Gibberella fujikuroi*: J. C. S. Perkin I, 996 (1975).

126. POLONSKY, J., Z. BASKÉVITCH, H. E. GOTTLIEB, E. W. HAGAMAN, and E. WENKERT: Carbon-13 Nuclear Magnetic Resonance Spectral Analysis of Quassinoid Bitter Principles. J. Org. Chem. **40**, 2499 (1975).

127. MINALE, L., and R. RICCIO: Constituents of the Digestive Gland of the Molluses of Genus *Aplysia* — I. Novel Diterpenes from *Aplysa depilans*. Tetrahedron Letters 2711 (1976).

128. TAYLOR, D. A. H.: ^{13}C Nuclear Magnetic Resonance Spectra of Some Limonoids. Part 1. The Structure of Procerin, an Extractive from *Carapa Procera*. J. C. S. Perkin I, 437 (1974).

129. NÄNNI, R., C. TAMM, V. GULLO, and K. NAKANISHI: Modification of the Structure of Bussein. J. C. S. Chem. Commun. 563 (1975).

130. HALSALL, T. G., and J. A. TROKE: The Structures of Three New Meliacins Isolated from *Khaya anthoteca* Heartwood. J. C. S. Perkin I, 1758 (1975).

131. OCHI, M., H. KOTSUKI, K. HIROTSU, and T. TOKOROYAMA: Sendanin, a New Limonoid from *Melia Azedarach* Linn. *var. japonica Makino*. Tetrahedron Letters 2877 (1976).

132. DREYER, D. L., R. D. BENNETT, and S. C. BASA: Limonoids from *Atalantia monophylla*, Isolation and Structure. Tetrahedron **32**, 2367 (1976).

133. ZANNO, P., I. MIURA, K. NAKANISHI, and P. L. ELDER: Structure of the Insect Phagorepellent Azadirachtin. Application of PRFT/CWD Carbon-13 Nuclear Magnetic Resonance. J. Amer. Chem. Soc. **97**, 1975 (1975).

134. BORREMANS, F., M. DE POTTER, and D. DE KEUKELEIRE: Carbon-13 NMR Spectroscopy of Hop Bitter Substances. Org. Magn. Res. **7**, 415 (1975).

135. RADICS, L., M. KAJTAR-PEREDY, S. NOZOE, and H. KOBAYASI: Carbon-13 Nuclear Magnetic Resonance Spectra of Ophibolins. Tetrahedron Letters 4414 (1975).

136. KNIGHT, S. A.: Carbon-13 NMR Spectra of Some Tetra- and Pentacyclic Triterpenoids. Org. Magn. Res. **6**, 603 (1974).

137. JONES, A. J., P. F. ALEWOOD, M. BENN, and J. WONG: Functionalization of the 4α-Methyl Group of Lanostanol via Azidoformate Thermolysis. Tetrahedron Letters 1655 (1976).

138. KHUONG-HUU, F., M. SANGARE, V. M. CHARI, A. BEKAERT, M. DEVYS, M. Barbier, and G. LUKACS: Carbon-13 Nuclear Magnetic Resonance Spectral Analysis of Cycloartanol and Related Compounds. Tetrahedron Letters 1789 (1975).

139. RADICS, L., M. KAJTAR-PEREDY, S. CORSANO, and L. STANDOLI: Carbon-13 NMR Spectra of Some Polycyclic Triterpenoids. Tetrahedron Letters 4287 (1975).

140. TORI, K., S. SEO, A. SHIMAOKA, and Y. TOMITA: Carbon-13 NMR Spectra of Olean-12-enes. Full Signal Assignments Including Quaternary Carbon Signals Assigned by Use of Indirect ^{13}C, ^{1}H Spin Couplings. Tetrahedron Letters 4227 (1974).

141. SEO, S., Y. TOMITA, and K. TORI: Carbon-13 NMR Spectra of Urs-12-enes and Application to Structural Assignments of Compounds of *Isodon japonicus* Hara Tissue Cultures. Tetrahedron Letters 7 (1975).

142. SEO, S., Y. TOMITA, and K. TORI: Biosynthesis of Ursene-type Triterpenes from Sodium[1,2-^{13}C]-Acetate in Tissue Cultures of *Isodon japonicus* Hara and Reassignments of ^{13}C N.m.r. Signals in Urs-12-enes. J. C. S. Chem. Commun. 954 (1975).

143. TORI, K., Y. YOSHIMURA, S. SEO, K. SAKURAWI, Y. TOMITA, and H. ISHII: Carbon-13 NMR Spectra of Saikogenins. Stereochemical Dependence of Hydroxylation Effects upon Carbon-13 Chemical Shifts of Oleanene-Type Triterpenoids. Tetrahedron Letters 4163 (1976).

144. TORI, K., S. SEO, Y. YOSHIMURA, M. NAKAMURA, Y. TOMITA, and H. ISHII: Carbon-13 NMR Spectra of Saikosaponins A, C, D and F, Isolated from *Buplearum falcatum* L. Tetrahedron Letters 4167 (1976).

145. NAKANISHI, K., V. P. GULLO, I. MIURA, T. R. GOVINDACHARI, and N. VISWANATHAN: Structure of two Triterpenes. Application of Partially Relaxed Fourier Transform ^{13}C Nuclear Magnetic Resonance. J. Amer. Chem. Soc. **95**, 6473 (1973).

146. EGGERT, H., C. L. VAN ANTWERP, N. S. BHACCA, and C. DJERASSI: Carbon-13 Nuclear Magnetic Resonance Spectra of Hydroxy Steroids. J. Org. Chem. **41**, 71 and 4051 (1976).

147. BLUNT, J. W.: Carbon-13 NMR Spectra of 6β-Substituted 5α-cholestan-3β.5-diols: γ- and δ-Effects. Aust. J. Chem. **28**, 1017 (1975).

148. CUSHLEY, R. J., and J. D. FILIPENKO: ^{13}C Fourier Transfor N.M.R. XIII. Reassignment of the ^{13}C Spectrum of Ergosterol. Org. Magn. Res. **8**, 308 (1976).

149. ABRAHAM, R. J., and J. R. MONASTERIOS: ^{13}C Nuclear Magnetic Resonance Spectra of some Ergosta-dienes and -trienes. J. C. S. Perkin II, 662 (1974).

150. HANSON, J. R., and M. SIVERNS: ^{13}C Nuclear Magnetic Resonance Spectra of Some Steroidal Unsaturated Ketones. J. C. S. Perkin I, 1956 (1975).

151. LETOURNEUX, Y., Q. KHUONG-HUU, M. GUT, and G. LUKACS: Identification of C-22 Epimers in Steroids by Carbon-13 Nuclear Magnetic Resonance Spectroscopy. J. Org. Chem. **40,** 1674 (1975).

152. ANDERSON, W. G., C. Y. BYON, M. GUT, and F. H. BISSETT: ^{13}C NMR Studies of the Four 20,22-Epoxy-cholesterols and the Two 20(22)-Dehydrocholesterols. Tetrahedron Letters 2193 (1976).

153. BANNAI, K., M. MORISAKI, and N. IKEKAWA: Studies on Steroids. Part. 37. Synthesis of the Four Stereoisomers of 20,22-Epoxycholesterol. J. C. S. Perkin I, 2116 (1976).

154. TORI, K., H. ISHII, Z. W. WOLKOWSKI, C. CHACHATY, M. SANGARÉ, F. PIRIOU, and G. LUKACS: Carbon-13 Nuclear Magnetic Resonance Spectra of Cardenolide. Tetrahedron Letters 1077 (1973).

155. WRAY, V., and S. LANG: Fourier Transform ^{13}C Nuclear Magnetic Resonance Studies of Steroids. II. Derivatives of 17β-(2,5-Dihydro-5-oxo-3-furyl)-3β,5α,6-triandrostane. Tetrahedron **31,** 2815 (1975).

156. EGGERT, H., and C. DJERASSI: ^{13}C NMR Spectra of Sapogenins. Tetrahedron Letters 3635 (1975).

157. GĂSIĆ, M., Z. DJARMATI, and S. W. PELLETIER: Carbon-13 Nuclear Magnetic Resonance Spectroscopy of Polycyclic δ-Lactones. J. Org. Chem. **41,** 1219 (1976).

158. SHRIVER, J., E. W. ABRAHAMSON, and G. D. MATEESCU: The Structure of Visual Pigments. I. Carbon-13 Nuclear Magnetic Resonance Spectroscopy of *N-all-trans*-Retinylidene propylimine and Its Protonated Species. J. Amer. Chem. Soc. **90,** 2407 (1976).

159. INOUE, Y., A. TAKAHASHI, Y. TOKITÔ, R. CHÛJÔ, and Y. MIYOSHI: Carbon-13 NMR Spectra of Retinal-1 and Its Related Compounds. Org. Magn. Res. **6,** 487 (1974).

160. ENGLERT, G.: A ^{13}C-NMR study of *cis-trans* Isomeric Vitamin A, Carotenoids and Related Compounds. Helv. Chim. Acta **58,** 2367 (1975).

161. FISCHLI, A., and H. MAYER: Carotinoidsynthesen über Sulfone; Synthese von β-Carotin. Helv. Chim. Acta **58,** 1584 (1975).

162. VÖGELI, U., W. ESCHENMOSER, and C. H. EUGSTER: Strukturbestimmung von O-Methylazafrinmethylester durch ^{13}C-NMR-Spektroskopie. Helv. Chim. Acta **58,** 2044 (1975).

163. BARANYAI, M., J. SZABOLCS, G. TOTH, and L. RADICS: Determination of the Geometrical Configuration of the Polyene Chain of Mono *cis* C_{40} Carotenoids — I. A ^{13}C NMR Study of Mono *cis* Zeaxantins and Mono *cis* Capsorubins. Tetrahedron **32,** 867 (1976).

164. BARLOW, L., and G. PATTENDEN: Synthesis of Poly-Z-isomers of 2,6,11,15-Tetramethylhexadeca-2,6,8,10,14-pentaene, a C_{20} Analogue of Phytoene. Re-examination of the Stereochemistry of a New Isomer of Phytoene from *Rhodospirillum rubrum*. J. C. S. Perkin I, 1029 (1976).

165. WENKERT, E., D. W. COCHRAN, E. W. HAGAMAN, F. M. SCHELL, N. NEUSS, A. S. KATNER, P. POTIER, C. KAN, M. PLAT, M. KOCH, H. MEHRI, J. POISSON, N. KUNESCH, and Y. ROLLAND: Carbon-13 Nuclear Magnetic Resonance Spectroscopy of Naturally Occurring Substances. XIX. *Aspidosperma* Alkaloids. J. Amer. Chem. Soc. **95,** 4990 (1973).

166. AHOND, A., M.-N. JANOT, N. LANGLOIS, G. LUKACS, P. POTIER, P. RASOANAIVO, M. SANGARÉ, N. NEUSS, M. PLAT, J. LE MEN, E. W. HAGAMAN, and E. WENKERT: On the Structure of Vindolinine. J. Amer. Chem. Soc. **96,** 633 (1974).

167. DAMAK, M., A. ABOND, and P. POTIER: Révision de la Structure de la Mélobaline. Tetrahedron Letters 167 (1976).
168. RASOANAIVO, P., and G. LUKACS: Revision of the Structure of the Bisindole Alkaloid 14',15'-Dihydro-Pycnanthine. A Carbon-13 Nuclear Magnetic Resonance Study. J. Org. Chem. **41,** 376 (1976).
169. WENKERT, E., C. J. CHANG, H. P. S. CHAWLA, D. W. COCHRAN, E. W. HAGAMAN, J. C. KING, and K. ORITO: General Methods of Synthesis of Indole Alkaloids. 14. Short Routes of Construction of Yohimboid and Ajmalicinoid Alkaloid Systems and Their ^{13}C Nuclear Magnetic Resonance Spectral Analysis. J. Amer. Chem. Soc. **98,** 3645 (1976).
170. KOCH, M. C., M. M. PLAT, N. PRÉAUX, H. E. GOTTLIEB, E. W. HAGAMAN, F. M. SCHELL, and E. WENKERT: Carbon-13 Nuclear Magnetic Resonance Spectroscopy of Naturally Occurring Substances. XXXIII. The Ochrofuanines and Emetine. J. Org. Chem. **40,** 2836 (1975).
171. MERLINI, L., R. MONDELLI, G. NASINI, F. W. WEHRLI, E. W. HAGAMAN, and E. WENKERT: ^{13}C NMR. Analysis of the Roxburghines. Helv. Chim. Acta **59,** 2254 (1976).
172. DAMAK, M., A. AHOND, P. POTIER, and M.-M. JANOT: Structure d'un Alcaloide Indolique Ancien: la Geissoschizine, Tetrahedron Letters 4731 (1976).
173. HUTCHINSON, C. R., M.-T. S. HSIA, A. H. HECKENDORF, and G. J. O'LOUGHLIN: A Biomimetic Synthesis of the Camptothecin Chromophore. J. Org. Chem. **41,** 3493 (1976).
174. HOFHEINZ, W., P. SCHÖNHOLZER, and K. BERNAUER: Überführung des (−)-Tabersonins in ein Indola[2,3-*b*]chinolizin, eine bemerkenswerte molekulare Umwandlung. Helv. Chim. Acta **59,** 1213 (1976).
175. LÉVY, J., C. PIERRON, G. LUKACS, G. MASSIOT, and J. LE MEN: Méthylène-Indolines, Indolénines et Indoléniums. XI. Nouveau Réarrangement de la Chloro-16 Déhydro-1 Tabersonine. Tetrahedron Letters 669 (1976).
176. KHUONG-HUU, F., M. CESARIO, J. GUILHEM, and R. GOUTAREL: Alcaloides Indoliques. CII. Deux Nouveaux Types d'Alcaloides Indoliques, l'Ibophyllidine, Dérivé du Nor-21(+)pandolane et l-Iboxyphylline, Dérivé de l'Abeo-21(20→19)(+)pandolane, Retirés des Feuilles de *Tabernanthe iboga* Baillon et de *T. subsessilis* Stapf. Tetrahedron **32,** 2539 (1976).
177. SRINIVASAN, P. R., and R. L. LICHTER: ^{13}C NMR Spectral Studies of Arecoline, Hordenine, Strychnine and Brucine. Org. Magn. Res. **8,** 198 (1976).
178. NEUSS, N., H. E. BOAZ, J. L. OCCOLOWITZ, E. WENKERT, F. M. SCHELL, P. POTIER, C. KAN, N. M. PLAT, and M. PLAT: The Structure of Vincarodine. Helv. Chim. Acta **56,** 2660 (1973).
179. BOMBARDELLI, E., A. BONATI, B. GABETTA, E. M. MARTINELLI, and G. MUSTICH: The Structure of Cuanzine. Tetrahedron **30,** 4141 (1974).
180. LE MEN, J., M. J. HOIZEY, G. LUKACS, L. LE MEN-OLIVIER, and J. LÉVY: Structure de la Pandine. Alcaloide Hexacyclique du Type Ψ-Vincadifformine. Tetrahedron Letters 3119 (1974).
181. KAN-KAN, C., G. MASSIOT, A. AHOND, B. C. DAS, H.-P. HUSSON, and P. POTIER: Structure and Biogenetic-type Synthesis of Andranginine: an Indole Alkaloid of a New Type. J. C. S. Chem. Commun. 164 (1974).
182. LANGLOIS, Y., F. GUÉRITTE, R. Z. ANDRIAMIALISOA, N. LANGLOIS, P. POTIER, A. CHIARONI, and C. RICHE: Réarrangement du Squelette de la Catharanthine. Tetrahedron **32,** 945 (1976).
183. AHOND, A., A.-M. BUI, P. POTIER, E. W. HAGAMAN, and E. WENKERT: Carbon-13 Nuclear Magnetic Resonance Analysis of Vobasine-like Indole Alkaloids. J. Org. Chem. **41,** 1878 (1976).
184. WENKERT, E., D. W. COCHRAN, H. E. GOTTLIEB, E. W. HAGAMAN, R. B. FILHO,

F. J. D. A. MATOS, and M. I. L. M. MADRUGA: ^{13}C-NMR Spectroscopy of Naturally Occurring Substances. XLV. Iboga Alkaloids. Helv. Chim. Acta **59**, 2437 (1976).

185. WENKERT, E., E. W. HAGAMAN, N. KUNESCH, N. WANG, and B. ZSADON: ^{13}C NMR Spectroscopy of Naturally Occurring Substances. XLII. Conformational Analysis of Quebrachamine-like Indole Alkaloids and Related Substances. Helv. Chim. Acta **59**, 2711 (1976).

186. BRUNETON, J., A. CAVÉ, E. W. HAGAMAN, N. KUNESCH, and E. WENKERT: The Carbon-20 Stereochemistry of Pandoline and *Epi*pandoline. Tetrahedron Letters 3567 (1976).

187. WENKERT, E., E. W. HAGAMAN, B. LAL, G. E. GUTOWSKI, A. S. KATNER, J. C. MILLER, and N. NEUSS: ^{13}C-NMR Spectroscopy of Naturally Occurring Substances. XXXII. Vincaleucoblastine and Related Alkaloids. Helv. Chim. Acta **58**, 1560 (1975).

188. DORMAN, D. E., and J. W. PASCHAL: ^{13}C NMR Spectroscopy: Comparison of the Spectra of Some Dimeric *Catharanthus* Alkaloids and their Derivatives. Org. Magn. Res. **8**, 413 (1976).

189. TAFUR, S. S., J. L. OCCOLOWITZ, T. K. ELZEY, J. W. PASCHAL, and D. E. DORMAN: Alkaloids of *Vinca rosea* L. (*Catharanthus roseas* G. Don). XXXVII. Structure of Vincathicine. J. Org. Chem. **41**, 1001 (1975).

190. CAVÉ, A., J. BRUNETON, A. AHOND, A.-M. BUI, H.-P. HUSSON, C. KAN, G. LUKACS, and P. POTIER: Structure Analysis by ^{13}C NMR Spectroscopy of Criophylline. A New Dimeric Indole Alkaloid. Tetrahedron Letters 5081 (1973).

191. DAS, B. C., J. P. COSSON, G. LUKACS, and P. POTIER: Structure Analysis by ^{13}C NMR Spectroscopy of Pleiocorine, a New Bisindole Alkaloid from *Alstonia deplanchei* van Heurck et Muell. Arg. Tetrahedron Letters 4299 (1974).

192. DAMAK, M., A. AHOND, H. DOUCERAIN, and C. RICHE: Bonafousine, a Novel Dimeric Indole Alkaloid: X-Ray Crystal Structure. J. C. S. Chem. Commun. 510 (1976).

193. DAMAK, M., C. POUPAT, and A. AHOND: Bis[hydroxy-11-coronaridinyl]-12, Nouvel Alcaloïde Dimère de Type Ibogane: Elucidation de la Structure par RM^{13}C. Tetrahedron Letters 3531 (1976), *ibid.* 3760 (1976).

194. BOMBARDELLI, E., A. BONATI, B. GABETTA, E. M. MARTINELLI, G. MUSTICH, and B. DANIELI: Structures of Tabernaelegantines A—D and Tabernaelegantinines A and B, New Indole Alkaloids from *Tabernaemontana elegans*. J. C. S. Perkin I, 1432 (1976).

195. ROLLAND, Y., N. KUNESCH, J. POISSON, E. W. HAGAMAN, F. M. SCHELL, and E. WENKERT: Carbon-13 Nuclear Magnetic Resonance Spectroscopy of Naturally Occurring Substances 43. Carbon-13 Nuclear Magnetic Resonance Analysis of Bis-Indoline Alkaloids of Two *Voacanga* Species. J. Org. Chem. **41**, 3270 (1976).

196. BACH, N. J., H. E. BOAZ, E. C. KORNFELD, C.-J. CHANG, H. G. FLOSS, E. W. HAGAMAN, and E. WENKERT: Nuclear Magnetic Resonance Spectral Analysis of the Ergot Alkaloids. J. Org. Chem. **39**, 1272 (1974).

197. ZETTA, L., and G. GATTI: ^{13}C NMR Spectra of Lysergic Acid Derivatives — I. 10-Methoxy-dihydrolysergic Acid Methyl Esters. Tetrahedron **31**, 1403 (1975).

198. MORELAND, C. G., A. PHILIP, and F. I. CARROLL: Carbon-13 Nuclear Magnetic Resonance Spectra of Cinchona Alkaloids. J. Org. Chem. **39**, 2413 (1974).

199. CARROLL, F. I., A. PHILLIPS, and M. C. COLEMAN: Synthesis and Stereochemistry of a Metabolite Resulting from the Biotransformation of Quinidine in Man. Tetrahedron Letters 1757 (1976).

200. DAUDON, M., M. H. MEHRI, M. M. PLAT, E. W. HAGAMAN, F. M. SCHELL, and E. WENKERT: Carbon-13 Nuclear Magnetic Resonance Spectroscopy of Naturally Occurring Substances, XXXIV. Monomeric Quinolinic *Melodinus* Alkaloids. J. Org. Chem. **40**, 2838 (1975).

201. DAUDON, M., M. M. MEHRI, M. M. PLAT, E. W. HAGAMAN, and E. WENKERT:

Carbon-13 Nuclear Magnetic Resonance Spectroscopy of Naturally Occurring Substances. 48. Dimeric Quinolinic *Melodinus* Alkaloids. J. Org. Chem. **41**, 3275 (1976).

202. HUGHES, D. W., H. L. HOLLAND, and D. B. MACLEAN: ^{13}C Magnetic Resonance Spectra of Some Isoquinoline Alkaloids and Related Model Compounds. Canad. J. Chem. **54**, 2252 (1976).

203. KAMETANI, T., K. FUKUMOTO, M. IHARA, A. UJIIE, and H. KOIZUMI: Conformational Analysis of the Dibenzo [*a. g.*]-quinolidines by Spectroscopic Methods. J. Org. Chem. **40**, 3280 (1975).

204. KAMETANI, T., A. UJIIE, M. IHARA, K. FUKUMOTO, and S.-T. LU: Studies on the Syntheses of Heterocyclic Compounds. Part DCL. Total Synthesis of Corytenchirine. J. C. S. Perkin I, 1218 (1976).

205. CARROLL, F. I., C. G. MORELAND, G. A. BRINE, and J. A. KEPLER: Carbon-13 Nuclear Magnetic Resonance Spectra of Morphine Alkaloids. J. Org. Chem. **41**, 996 (1976).

206. TERUI, Y., K. TORI, S. MAEDA, and Y. K. SAWA: C-13 NMR Spectra of Morphine Alkaloids, Codeine, Thebaine, and Sinomenine and Related Compounds. Tetrahedron Letters 2853 (1975).

207. BRINE, G. A., D. PRAKASH, C. K. HART, D. J. KOTCHMAR, C. G. MORELAND, and F. I. CARROLL: Ring-C Conformation of 6β-Naltrexol and 6α-Naltrexol. Evidence form Proton and Carbon-13 Nuclear Magnetic Resonance. J. Org. Chem. **41**, 3445 (1976).

208. ZETTA, L., G. GATTI, and C. FUGANTI: ^{13}C Nuclear Magnetic Resonance Spectra of Amaryllidaceae Alkaloids. J. C. S. Perkin II, 1180 (1973).

209. CROOKS, P. A., B. ROBINSON, and O. METH-COHN: The ^{13}C Nuclear Magnetic Resonance Spectra of Physostigmine and Related Compounds. Phytochemistry **15**, 1092 (1976).

210. SIMERAL, L., and G. E. MACIEL: Carbon-13 Chemical Shifts of Some Cholineric Neural Transmission Agents. Org. Magn. Res. **6**, 226 (1974).

211. LOUNASMAA, M., P. M. WOVKULICH, and E. WENKERT: Structures of Some *Knightia deplanchei* Alkaloids. J. Org. Chem. **40**, 3694 (1975).

212. NAKASHIMA, T. T., P. O. SINGER, L. M. BROWNE, and W. A. AYER: Carbon-13 Nuclear Magnetic Resonance Studies of Some *Lycopodium* Alkaloids. Canad. J. Chem. **53**, 1936 (1975).

213. VAN EIJK, J. L., M. H. RADEMA, and C. VERSLUIS: New Quinolizidine Alkaloids from *Cadia purpurea.* Tetrahedron Letters 2053 (1976).

214. BAXTER, R. L., L. CROMBIE, D. J. SIMMONDS, and D. A. WHITING: Structures of Cathedulin-2 and Catheduline-8. New Sesquiterpene Alkaloids from *Catha edulis.* J. C. S. Chem. Commun. 465 (1976).

215. LEBOEUF, M., M. HAMONNIÈRE, A. CAVÉ, H. E. GOTTLIEB, N. KUNESCH, and E. WENKERT: The Structure of Polyalthenol, an Indolosesquiterpene. Tetrahedron Letters 3559 (1976).

216. HART, N. K., S. R. JOHNS, J. A. LAMBERTON, H. SUARES, and R. I. WILLING: New Alkaloids of the *ent*-kaurene Type from *Anopterus* Species (*Escalloniaceae*). I. The Structure and Reactions of Anopterine. Aust. J. Chem. **29**, 1295 (1976).

217. HART, N. K., S. R. JOHNS, J. A. LAMBERTON, H. SUARES, and R. I. WILLING: New Alkaloids of the *ent*-kaurene Type from *Anopterus* Species. II. The Structure of the Minor Alkaloids. Aust. J. Chem. **29**, 1319 (1976).

218. PELLETIER, S. W., and Z. DJARMATI: Carbon-13 Nuclear Magnetic Resonance: Aconitine-Type Diterpenoid Alkaloids from *Aconitum* and *Delphinium* species. J. Amer. Chem. Soc. **98**, 2626 (1976).

219. PELLETIER, S. W., N. V. MODY, A. J. JONES, and M. H. BENN: The Structure of *Alkaloid A* from *Delphinium bicolor* Nutt. Tetrahedron Letters 3025 (1976).

220. PELLETIER, S. W., N. V. MODY, Z. DJARMATI, I. V. MIĆOVIĆ, and J. K. THAKKAR: The Structures of Staphidine, Staphinine and Staphimine, Three Novel Bis-Diterpene Alkaloids from *Delphinium staphisagria*. Tetrahedron Letters 1055 (1976).
221. PELLETIER, S. W., N. V. MODY, Z. DJARMATI, and S D. LAJŠIĆ: The Structures of Staphigine and Staphirine. Two Novel Bis-Diterpene Alkaloids from *Delphinium staphisagria*. J. Org. Chem. **41**, 3042 (1976).
222. PELLETIER, S. W., Z. DJARMATI, and N. V. MODY: The Structures of Staphisagnine and Staphisagrine, Two Novel Bis-Diterpene Alkaloids from *Delphinium staphisagria*. Tetrahedron Letters 1749 (1976).
223. BIRD, G. J., D. J. COLLINS, F. W. EASTWOOD, B. M. K. C. GATEHOUSE, A. J. JOZSA, and J. M. SWAN: New Steroidal Alkaloids from *Solanum callium*. Tetrahedron Letters 3653 (1976).
224. THORPE, M. C., W. C. COBURN, JR., and J. A. MONTGOMERY: The ^{13}C Nuclear Magnetic Resonance Spectra of Some 2-, 6- and 2,6-Substituted Purines. J. Magn. Res. **15**, 98 (1974).
225. BREITMAIER, E., and W. VOELTER Einflüsse von Substituenten in 6-Stellung auf die ^{13}C-chemischen Verschiebungen der Kohlenstoffatome des Purins. Tetrahedron **30**, 3941 (1974).
226. CHENON, M.-T., R. J. PUGMIRE, D. M. GRANT, R. P. PANZICA, and L. B. TOWNSEND: Carbon-13 Magnetic Resonance. XXV. A Basis Set of Parameters for the Investigation of Tautomerism in Purines Established from Carbon-13 Magnetic Resonance Studies Using Certain Purines and Pyrrolo[2,3-*d*]pyrimidines. J. Amer. Chem. Soc. **97**, 4627 (1975).
227. — — — — — Carbon-13 Magnetic Resonance. XXVI. A Quantitative Determination of the Tautomeric Populations of Certain Purines. J. Amer. Chem. Soc. **97**, 4636 (1975).
228. CHENON, M.-T., R. P. PANZICA, J. C. SMITH, R. J. PUGMIRE, D. M. GRANT, and L. B. TOWNSEND: Carbon-13 Magnetic Resonance Spectra of C-Nucleosides. 3. Tautomerism in Formycin and Formycin B and Certain Pyrazolo[4,3-*d*]-pyrimidines. J. Amer. Chem. Soc. **98**, 4736 (1976).
229. NICOLAU, C., and K. HILDENBRAND: 13C-Nuclear Magnetic Resonance Investigations of Xanthine and Some of its N-Methylated Derivatives. Z. Naturforsch. **29** C, 475 (1974).
230. BORDNER, J., W. E. THIESSEN, H. A. BATES, and H. RAPOPORT: The Structure of a Crystalline Derivative of Saxitoxin. The Structure of Saxitoxin. J. Amer. Chem. Soc. **97**, 6008 (1975).
231. SHIMIZU, Y., L. J. BUCKLEY, M. ALAM, Y. OSHIMA, W. E. FALLON, H. KASAI, I. MIURA, V. P. GULLO, and K. NAKANISHI: Structures of Gonyautoxin II and III from the East Coast Toxic Dinoflagellate *Gonyaulax tamareusis*. J. Amer. Chem. Soc. **98**, 5414 (1976).
232. MÜLLER, G., and W. v. PHILIPSBORN: ^{13}C NMR Spectra of Pteridines. Helv. Chim. Acta **56**, 2680 (1973).
233. FRICK, W. F., R. WEBER, and M. VISCONTINI: Über Pterinchemie, 50. Mitteil. ^{13}C-NMR-Spektren von 7,8-Dihydro- und 5,6,7,8-Tetrahydrofolsäure. Helv. Chim. Acta **57**, 2658 (1974).
234. LYON, J. A., R. B. DUNLAP, and P. D. ELLIS: Assignment of the Carbon-13 Nuclear Magnetic Resonance Spectra of Selected Folates. J. Magn. Res. **18**, 293 (1975).
235. CUSSANS, N. J., and T. N. HUCKERBY: Carbon-13 NMR Spectroscopy of Heterocyclic Compounds — II. A 20 MHz Study of Chemical Shifts and Carbon-Proton Coupling Constants for Coumarin and Some Bromacoumarins. Tetrahedron **31**, 2587 (1975).
236. — — Carbon-13 NMR Spectroscopy of Heterocyclic Compounds — III. A 20 MHz

Study of Chemical Shifts and Carbon-Proton Coupling Constants for the Methyl Coumarins. Tetrahedron **31**, 2591 (1975).

237. CUSSANS, N. J., and T. N. HUCKERBY: Carbon-13 NMR Spectroscopy of Heterocyclic Compounds — IV. A 20 MHz Study of Chemical Shifts and Carbon-Proton Coupling Constants in a Series of Hydroxy, Methoxy and Glucosyl Coumarins. Tetrahedron **31**, 2719 (1975).

238. SOJKA, S. A.: Carbon-13 Nuclear Magnetic Resonance Spectra of 2*H*-1-Benzopyran-2-one (Coumarins) in Chloroform and Sulfuric Acid. J. Org. Chem. **40**, 1175 (1975).

239. ERNST, L.: ^{13}C NMR Spectroscopy of Polycyclic Aromatics. VI. Coumarin and the Methylcoumarins. J. Magn. Res. **21**, 241 (1976).

240. LAPPER, R. D.: The Carbon-13 Nuclear Magnetic Resonance Spectrum of Siderin. Tetrahedron Letters 4293 (1974).

241. DREYER, D. L., K. P. MUNDERLOH, and W. E. THIESSEN: Extractives of *Dalea* Species (Leguminosae). Tetrahedron **31**, 287 (1975).

242. PELTER, A., R. S. WARD, and T. I. GRAY: The Carbon-13 Nuclear Magnetic Resonance Spectra of Flavonoids and Related Compounds. J. C. S. Perkin I, 2475 (1976).

243. WAGNER, H., V. M. CHARI, and J. SONNENBICHLER: ^{13}C-NMR-Spektren natürlich vorkommender Flavonoide. Tetrahedron Letters 1799 (1976).

244. TERNAI, B., and K. R. MARKHAM: Carbon-13 NMR Studies of Flavonoids — I. Flavones and Flavonols. Tetrahedron **32**, 565 (1976).

245. MARKHAM, K. R., and B. TERNAI: ^{13}C NMR of Flavonoids II. Flavonoids other than Flavone and Flavonol Aglycones. Tetrahedron **32**, 2607 (1976).

246. TORI, K., T. HIRATA, O. KOSHITANI, and T. SUGA: Carbon-13 NMR Spectral Studies of Aloenin and its Derivatives. Carbon-13 Signal Assignment Problem of 4-Methoxy-2-Pyrones. Tetrahedron Letters 1311 (1976).

247. BURROWS, B. F., W. B. TURNER, and E. R. H. WALKER: 8-Ethylidene-7,8-dihydro-4-methoxypyrano[4,3-*b*]-pyran-2,5-dione (Coarctatin), a Metabolite of *Chaetomium coarctatum*. J. C. S. Perkin I, 999 (1975).

248. KINGSBURY, C. A., M. CLIFFTON, and J. H. LOOKER: Carbon-13 Nuclear Magnetic Resonance Spectra of Kojic Acid and Other 4-Pyrone Derivatives.

249. CROMBIE, L., G. W. KILBEE, and D. A. WHITING: Carbon-13 Magnetic Resonance Spectra of Natural Rotenoids and their Relatives. J. C. S. Perkin I, 1497 (1975).

250. BRIMACOMBE, J. S.: NMR Spectroscopy and Conformational Features of Carbohydrates. ^{13}C NMR Spectroscopy. In: Specialist Periodical Report: Carbohydrates. The Chemical Society (London) **6**, 169 (1973), **7**, 182 (1975), **8**, 171 (1976).

251. RITCHIE, R. G. S., N. CYR, and A. S. PERLIN: Configurational Effects in ^{13}C Chemical Shifts of 1,6-Anhydrohexapyranoses and Related Compounds. Utility of ^{13}C-^{1}H Coupling Patterns for Signal Assignments. Can. J. Chem. **54**, 2301 (1976).

252. BOCK, K., and C. PEDERSEN: A Study of ^{13}CH Coupling Constants in Hexapyranoses. J. C. S. Perkin II, 293 (1974).

253. ABBAS, S. A., A. H. HAINES, and A. G. WELLS: Assignment of ^{13}C and ^{1}H Resonance of Methyl Groups in Me Tri-*O*-methyl Derivatives of Methyl Pentapyranosides; Some Observations on the Methoxy ^{13}C Chemical Shifts. J. C. S. Perkin I, 1351 (1976).

254. VIGNON, M. R., and PH. J. A. VOLTERO: RMN ^{13}C: Sur l'Utilisation des Esters pour l'Attribution des Carbones des Molecules Glucidiques. Tetrahedron Letters 2445 (1976).

255. SZAREK, W. A., D. M. VYAS, S. D. GERO, and G. LUKACS: Application of Carbon-13 Nuclear Magnetic Resonance Spectroscopy to the Structural Determination of Chlorodeoxy Sugars. Can. J. Chem. **52**, 3394 (1974).

256. CONWAY, E., R. D. GUTHRIE, S. D. GERO, G. LUKACS, and A.-M. SEPULCHRE: A ^{13}C Nuclear Magnetic Resonance Chemical Shift Study of *trans*-Fused Hexapyranoside Derivatives. J. C. S. Perkin II, 542 (1974).

257. MILJKOVIC, M., M. GLIGORIJEVIC, T. SATOH, D. GLISIN, and R. G. PITCHER: Carbon-

13 Nuclear Magnetic Resonance Spectra of Branched-chain Sugars. Configurational Assignment of the Branching Carbon Atom of Methyl Branched-chain Sugars. J. Org. Chem. **39**, 3847 (1974).

258. TORI, K., S. SEO, Y. YOSHIMURA, H. ARITA, and Y. TOMITA: Glycosidation Shifts in Carbon-13 NMR Spectroscopy: Carbon-13 Signal Shifts from Aglycone and Glucose to Glucoside. Tetrahedron Letters 179 (1977).

259. KASAI, R., M. SUZUO, J. ISAKAWA, and O. TANAKA: Carbon-13 Chemical Shifts of Isoprenoid-β-Glucopyranosides and -β-D-Mannopyranosides. Stereochemical Influences of Aglycone Alcohols. Tetrahedron Letters 175 (1977).

260. HOSTETTMANN, K., and A. JACOT-GUILLARMOD: Identification de Xanthones et de Nouveaux Arabinosides de C-Glucosides flavoniques dans *Swertia perennis* L. Helv. Chim. Acta **59**, 1584 (1976).

261. COLSON, P., and R. R. KING: The ^{13}C NMR Spectra of Disaccharides of D-Glucose, D-Galactose and L-Rhamnose as Models for Immunological Polysaccharides. Carbohydrate Research **47**, 1 (1976).

262. LALONDE, R. T., C. WONG, and A. I.-M. TSAI: Polyglucosidic Metabolites of Oleaceae. The Chain Sequence of Oleoside Aglucon. Tyrosol, and Glucose Units in Three Metabolites from *Fraxinus Americana*. J. Amer. Chem. Soc. **98**, 3007 (1976).

263. BREITMAIER, E., and V. HOLLSTEIN: Complete Assignment of the ^{13}C n. m. r. Spectrum of Mutarotated D-Ribose by Integration and Specific Deuteration. Org. Magn. Res. **8**, 573 (1976).

264. FUNCKE, W., and A. KLEMER: ^{13}C-NMR-Untersuchungen an Mutarotationsgleichgewichten von D-Fructose und 1-Amino-1-desoxy-D-fructose-derivaten (Amadori-Verbindungen). Carbohydrate Research **50**, 9 (1976).

265. DE WIT, G., A. D. G. KIEBOOM, and H. VAN BEKKUM: Ionization and Mutarotation of Hexoses in Aqueous Alkaline Solution as Studied by ^{13}C-NMR Spectroscopy. Tetrahedron Letters 3943 (1975).

266. YAMAZAKI, K., M. KANEDA, and O. TANAKA: Carbon-13 NMR Spectral Assignments of Paeoniflorin Homologues with the Aid of Spin-lattice Relaxation Times. Tetrahedron Letters 3965 (1976).

267. See for example p. 147 of Ref. *10*.

268. BOCK, K., and L. D. HALL: Carbon-13 Spin-lattice Relaxation Times of Some Carbohydrate Derivatives. Carbohydrate Research **40**, 63 (1975).

269. BLUNT, J. W., and M. H. G. MUNRO: An Automated Procedure for Qualitative and Quantitative Analysis of Mixtures by Means of Carbon Magnetic Resonance Spectroscopy, Applications to Carbohydrate Analysis. Aust. J. Chem. **29**, 975 (1976).

270. MÜLLER, J. M., H. FUHRER, J. GRUNER, und W. VOSER: Stoffwechselprodukte von Mikroorganismen, 160. Mitteil. Conocandin, ein fungistatisches Antibiotikum aus *Hormococcus conorum* (Sacc. et Roum.) Roback. Helv. Chim. Acta **59**, 2506 (1976).

271. HEARN, M. T. W.: The Carbon-13 Nuclear Magnetic Resonance Spectrum of Junipal. Tetrahedron Letters 2787 (1975).

272. LEVINE, S. G., R. E. HICKS, H. E. GOTTLIEB, and E. WENKERT: Carbon-13 Nuclear Magnetic Resonance Spectroscopy of Naturally Occurring Substances. XXX. Griseofulvin. J. Org. Chem. **40**, 2540 (1975).

273. HUBER, C., W. A. COURT, J. P. DEVLIN, and O. E. EDWARDS: Stemphone. Tetrahedron Letters 2545 (1974).

274. TOMA, F., F. C. BOUHET, P. PHAM VAN CHUONG, P. FROMAGEOT, W. HAAR, H. RÜTERJANS, and W. MAURER: Carbon-13 NMR Spectroscopy of the Biological Pigments Luteoskyrin and Rugulosin and Some Polyhydroxyanthraquinone Analogues. Org. Magn. Res. **7**, 496 (1975).

275. KOBAYASHI, M., Y. TERUI, K. TORI, and N. TSUJI: Carbon-13 NMR Spectra of Juglone, Naphthazarin and their Derivatives. Tetrahedron Letters 619 (1976).

276. TSUJI, N., M. KOBAYASHI, Y. TERUI, and K. TORI: The Structures of Griseusins A and B, New Isochromanquinone Antibiotics. Tetrahedron **32,** 2207 (1976).

277. ASLESON, G. L., and C. W. FRANK: Carbon-13 Nuclear Magnetic Resonance Spectral Analysis of Tetracycline Hydrochloride and Related Antibiotics. J. Amer. Chem. Soc. **97,** 6246 (1975).

278. ARNONE, A., G. FRONZA, R. MODELLI, and A. VIGEVANI: ^{13}C NMR Analysis of the Antitumor Antibiotics Daunorubicin and Adriamycin. Tetrahedron Letters 3349 (1976).

279. HIGHET, R. J., G. W. PEROLD, and E. A. SOKOLOSKI: Characterization of Spiro-Bislactonic Phenolic Metabolites of Proteaceae by ^{13}C Nuclear Magnetic Resonance. J. Org. Chem. **41,** 3860 (1976).

280. MARTINELLI, R., R. J. WHITE, G. G. GALLO, and P. J. BEYNON: Carbon-13 NMR Spectrum Rifamycin S: A Re-examination of the Assignments with Special Reference to their Biogenetic Implication. Tetrahedron Letters 1367 (1974).

281. COX, R. H., and R. J. COLE: Carbon-13 Nuclear Magnetic Resonance Studies of Fungal Metabolites, Aflatoxins and Sterigmatocystins. J. Org. Chem. **42,** 112 (1977).

282. KAKIMUMA, K., B. I. MILAVETZ, and K. L. RINEHART, JR.: Carbon-13 Nuclear Magnetic Resonance Spectra of the Streptovaricins and Related Compounds. J. Org. Chem. **41,** 1358 (1976).

283. DESHMUKH, P. V., K. KAKIMUMA, J. J. AMEEL, K. L. RINEHART, JR., P. F. WILEY, and L. H. LI: Protostreptovaricins I—V. J. Amer. Chem. Soc. **98,** 870 (1976).

284. KAKIMUMA, K., C. A. HANSON, and K. L. RINEHART, JR.: Spectinabilin, a New Nitrogen-Containing Metabolite Isolated from *Streptomyces spectabilis*. Tetrahedron **32,** 217 (1976).

285. KOCH, K. F., J. A. RHOADES, E. W. HAGAMAN, and E. WENKERT: Carbon-13 Nuclear Magnetic Resonance Spectral Analysis of Tobramycin and Related Antibiotics. J. Amer. Chem. Soc. **96,** 3300 (1974).

286. WENKERT, E., and E. W. HAGAMAN: Carbon-13 Nuclear Magnetic Resonance Spectroscopy of Naturally Occurring Substances. XXXIX. Apramycin — an Application of Amine Protonation Parameters. J. Org. Chem. **41,** 701 (1976).

287. NAGABHUSHAN, T. W., W. N. TURNER, P. J. L. DANIELS, and J. B. MORTON: The Gentamycin Antibiotics. 7. Structures of the Gentamicin Antibiotics A_1, A_3 and A_4. J. Org. Chem. **40,** 2830 (1975).

288. NAGABHUSHAN, T. L., P. J. L. DANIELS, R. S. JARET, and J. B. MORTON: The Gentamicin Antibiotics. 8. Structure of Gentamicin A_2. J. Org. Chem. **40,** 2835 (1975).

289. NAGABHUSHAN, T. L., and P. J. L. DANIELS: Carbon-13 Magnetic Resonance Spectroscopy and Absolute Configuration of Anomeric Center in Axially Linked 4-*O*- and/or 6-*O*-Glycopyranosyl Derivatives of Deoxystreptamine. Tetrahedron Letters 747 (1975).

290. UCHIDA, K., E. BREITMAIER, and W. A. KOENIG: ^{13}C NMR Investigations of the Nucleoside Antibiotic Hikizimycin and its Constituents. Tetrahedron **31,** 2315 (1975).

291. NOURSE, J. G., and J. D. ROBERTS: Nuclear Magnetic Resonance Spectroscopy. Carbon-13 Spectra of Some Macrolide Antibiotics and Derivatives. Substituent and Conformational Effects. J. Amer. Chem. Soc. **97,** 4584 (1975).

292. ŌMURA, S., A. NAKAGAWA, A. NESZMÉLYI, S. D. GERO, A.-M. SEPULCHRE, F. PIROU, and G. LUKACS: Carbon-13 Nuclear Magnetic Resonance Spectral Analysis of 16-Membered Macrolide Antibiotics. J. Amer. Chem. Soc. **97,** 4001 (1975).

293. ŌMURA, S., A. NESZMÉLYI, M. SANGARÉ, and G. LUKACS: Conformational Homogeneity in Solution of 14-Membered Macrolide Antibiotics as Evidenced by ^{13}C NMR Spectroscopy. Tetrahedron Letters 2939 (1975).

294. TERUI, Y., K. TORI, K. NAGASHIMA, and N. TSUJI: C-13 Nuclear Magnetic Resonance Spectra of Erythromycins. Tetrahedron Letters 2583 (1975).

295. MARTIN, J. R., R. S. EGAN, A. W. GOLDSTEIN, and P. COLLUM: Extension of the

Erythromycin Biosynthetic Pathway. Isolation and Structure of Erythromycin E. Tetrahedron **31**, 1985 (1975).

296. Neszmélyi, A., S. Ōmura, T. T. Thang, and G. Lukacs: A Carbon-13 Spin Lattice Relaxation Time Study of 14-Membered Macrolide Antibiotics. Tetrahedron Letters 725 (1977).

297. Neszmélyi, A., S. Ōmura, and G. Lukacs: Carbon-13 Spin-Lattice Relaxation Times and their Use for Spectral Analysis of 16-Membered Macrolide Antibiotics. J. C. S. Chem. Commun. 97 (1976).

298. Mizsak, S., G. Slomp, A. Neszmélyi, S. D. Gero, and G. Lukacs: Carbon-13 N.M.R. Spectral Analysis and Spin-Lattice Relaxation Times of the Antibiotic Lincomycin and Related Compounds. Tetrahedron Letters 721 (1977).

299. Kukoljy, S., N. D. Jones, M. D. Chaney, T. K. Elzey, M. R. Gleissner, J. W. Paschal, and D. E. Dorman: Structure and Stereochemistry of Isomeric Penam and Cepham Derivatives. J. Org. Chem. **40,** 2388 (1975).

300. Tori, K., T. Tsushima, Y. Tamura, H. Shigemoto, T. Tsuji, H. Ishitobi, and H. Tanida: Structure and Stereochemistry of 2-Chloromethylpenam and 3-chlorocepham Derivatives Studied by ^{13}C and ^{1}H NMR Spectroscopy. Tetrahedron Letters 3307 (1975).

301. Harrison, C. R., and P. Hodge: Determination of the Configuration of Some Penicillin *S*-Oxides by ^{13}C Nuclear Magnetic Resonance Spectroscopy. J. C. S. Perkin I, 1772 (1976).

302. Busson, R., H. Vanderhaeghe, and S. Toppet: Preparation and Stereochemical Analysis of 5-*Epi*benzylpenicillin (*S*)- and (*R*)-Sulfoxide Esters. J. Org. Chem. **41,** 3054 (1976).

303. Ref. *10,* Chap. 5, p. 227ff. and references cited therein.

304. Lackner, H.: ^{13}C-NMR-Spektren der Pentapeptidlactonringe von Actinomycinen. Tetrahedron Letters 1921 (1975).

305. Booth, H., A. B. Mauger, and W. J. Rzeszotarski: A ^{13}C NMR Study of Actinomycin D and Related Model Peptides. Org. Magn. Res. **8,** 219 (1976).

306. Dell, A., D. H. William, H. R. Morris, G. A. Smith, J. Feeney, and G. C. K. Roberts: Structure Revision of the Antibiotic Echinomycin. J. Amer. Chem. Soc. **97,** 2497 (1975).

307. Bycroft, B. W., and T. J. King: Revised Constitution, Absolute Configuration and Conformation of Griseoviridin, a Modified Cyclic Peptide Antibistic. J. C. S. Perkin I, 1966 (1976).

308. Tori, K., K. Tokura, K. Okabe, M. Ebata, H. Otsuka, and G. Lukacs: Carbon-13 NMR Studies of Peptide Antibiotics Thiostrepton and Siomycin A: The Structure Relationship. Tetrahedron Letters 185 (1976).

309. Bycroft, B. W., and R. Pinchin: Structure of Althiomycin, A Highly Modified Antibiotic. J. C. S. Chem. Commun. 121 (1975).

310. Gulbis, J., and G. W. Everett, Jr.: A ^{13}C Nuclear Magnetic Resonance Analysis of the Metal Binding Site in Tetracycline. J. Amer. Chem. Soc. **97,** 6248 (1975).

311. Gulbis, J., G. W. Everett, Jr., and C. W. Frank: Effect of Added Electrolyte on the Binding of Tetracycline to Paramagnetic Ion Probes. A ^{13}C and ^{1}H Nuclear Magnetic Resonance Study. J. Amer. Chem. Soc. **98,** 1280 (1976).

312. Leibfritz, D.: Nachweis spezifischer Calcium- und Natriumkomplexierung durch das Antibiotikum Rifamycin S mit Hilfe der C-13 NMR-Spektroskopie. Tetrahedron Letters 4125 (1974).

313. Tanabe, M., H. Seto, and L. F. Johnson: Biosynthetic Studies with Carbon-13. Carbon-13 Nuclear Magnetic Resonance Spectra of Radicinin. J. Amer. Chem. Soc. **92,** 2157 (1970).

314. Stoessl, A., E. W. B. Ward, and J. B. Stothers: Incorporation of Doubly Labelled

Sodium Acetate-$^{13}C_2$ into Phytuberin and Other Susquiterpenes in Potatoes: Experimental Confirmation of Postulated C-C Cleavages. Tetrahedron Letters 3271 (1976).
315. TANABE, M., and K. T. SUZUKI: Biosynthetic Studies with Carbon-13: Incorporation Pattern of 1,2-^{13}C-Acetate into the Fungal Sesquiterpene Ovalicin. Tetrahedron Letters 4417 (1974).
316. CANE, D. E., and R. H. LEVIN: Application of Carbon-13 Magnetic Resonance to Isoprenoid Biosynthesis. II. Ovalicin and the Use of Doubly Labelled Mevalonate. J. Amer. Chem. Soc. **98,** 1183 (1976).
317. BAKER, F. C., C. J. W. BROOKS, and S. A. HUTCHINSON: Biosynthesis of Capsidiol in Sweet Peppers (*Capsicum frutescens*) Infected with Fungi: Evidence for Methyl Group Migration from ^{13}C Nuclear Magnetic Resonance Spectroscopy. J. C. S. Chem. Commun. 293 (1975).
318. BAKER, F. C., and C. J. W. BROOKS: Biosynthesis of the Sesquiterpenoid, Capsidiol, in Sweet Pepper Fruits Inoculated with Fungal Spores. Phytochemistry **15,** 689 (1976).
319. EVANS, R., J. R. HANSON, and R. NYFELER: Studies in Terpenoid Biosynthesis, Part XVII. Biosynthesis of Sesquiterpenoids Cyclonerodiol and Cyclonerotriol. J. C. S. Perkin I, 1214 (1976).
320. HANSON, J. R., T. MARTEN, and M. SIVERNS: Studies in Terpenoid Biosynthesis, XII. Carbon-13 Nuclear Magnetic Resonance Spectra of the Trichothecanes and the Biosynthesis of Trichothecolone from [2-^{13}C]-Mevalonic Acid. J. C. S. Perkin I, 1033 (1974).
321. TANABE, M., K. T. SUZUKI, and W. C. JANKOWSKI: Biosynthetic Studies with Carbon-13: The FT-^{13}C NMR Spectra of the Sesquiterpenoid Coriolins. Tetrahedron Letters 2271 (1974).
322. CANE, D. E., and R. B. NACHBAR: Biosynthesis of Formannosin from [1,2-^{13}C]-Acetate. Tetrahedron Letters 2097 (1976).
323. POLONSKY, J., G. LUKACS, N. GAGNOLI-BELLAVITA, and R. CECCHERELLI: Application du Couplage ^{13}C-^{13}C à la Détermination de l'Origine Biogénétique des Substituants en C-4 des Virescenols A et B. Tetrahedron Letters 481 (1975).
324. ADAMS, M. R., and J. D. BU'LOCK: Biosynthesis of the Diterpene Antibiotic, Aphidicolin, by Radioisotope and ^{13}C Nuclear Magnetic Resonance Methods. J. C. S. Chem. Commun. 389 (1975).
325. BARROW, K. D., R. B. JONES, P. W. PEMBERTON, and L. PHILLIPS: Fusicoccin. V. The Biosynthesis of Fusicoccin from [1-^{13}C]- and [2-^{13}C]-Acetate. J. C. S. Perkin I, 1405 (1975).
326. SEO, S., Y. TOMITA, and K. TORI: Biosynthesis of Oleanene- and Ursene-type Triterpenes from [4-^{13}C]-Mevalonic Acid in Tissue Cultures of *Isodon japonicus* Hara. J. C. S. Chem. Commun. 270 (1975).
327. RIISOM, T., H. J. JAKOBSEN, N. RASTRUP ANDERSEN, and H. LORCK: Assignment of the ^{13}C NMR Spectra of Fusidic Acid Derivatives. Biosynthetic Incorporation of Sodium [1-^{13}C]-Acetate into Fusidic Acid. Tetrahedron Letters 2247 (1974).
328. CUSHLEY, R. J., and J. D. FILIPENKO: ^{13}C Fourier Transform n.m.r. XIII. Reassignment of the ^{13}C Spectrum of Ergosterol. Org. Magn. Res. **8,** 308 (1976).
329. POPJÁK, G., J. EDMOND, F. A. L. ANET, and N. R. EASTON, JR.: Carbon-13 NMR Studies on Cholesterol Biosynthesized from [^{13}C] Mevalonates. J. Amer. Chem. Soc. **99,** 931 (1977).
330. GUDGEON, J. A., J. S. E. HOLKER, and T. J. SIMPSON: Use of Singly and Doubly Labelled ^{13}C-Acetate in the Elucidation of the Structures and Biosyntheses of Multicolic and Multicolosic Acids, New Tetronic Acids from *Penicillium multicolor*. J. C. S. Chem. Commun. 636 (1974).
331. COX, R. E., and J. S. E. HOLKER: Biosynthesis of Glauconic Acid from [2,3-^{13}C]-Succinate. J. C. S. Chem. Commun. 583 (1976).

332. SIMPSON, T. J.: The ^{13}C NMR Spectrum of a Pyrone Metabolite of *Aspergillus melleus*. Biosynthetic Incorporation of Singly and Doubly Labelled [^{13}C]-Acetate. Tetrahedron Letters 175 (1975).
333. SIMPSON, T. J., and J. S. E. HOLKER: The Biosynthesis of a Pyrone Metabolite of *Aspergillus melleus*. An Application of Long-Range ^{13}C-^{13}C Coupling Constants. Tetrahedron Letters 4693 (1975).
334. NAIR, M. S. R.: Biosynthesis and Revised Structure of Rosellisin; Structure of Rosellisin Aldehyde. Phytochemistry **15**, 1090 (1976).
335. GARSON, M. J., and J. STAUNTON: Biosynthesis of Sclerin, Metabolite of *Sclerotinia Sclerotiorum*: Incorporation of [1-^{13}C]- and [1,2-^{13}C]-acetates. J. C. S. Chem. Commun. 928 (1976).
336. COX, R. E., and J. S. E. HOLKER: The Biosynthesis of Fungal Metabolites. IX. Sclerin: Feedings with [1,2-^{13}C]-Acetate and [Methyl-^{13}C]-Methionine. J. C. S. Perkin I, 2077 (1976).
337. YAMAZAKI, M., Y. MEBAYASHI, and T. TOKOROYAMA: Biosynthesis of Sclerin. Tetrahedron Letters 489 (1977).
338. HILL, R. A., R. H. CARTER (née RAYNER), and J. STAUNTON: Biosynthesis of Terrein, a Metabolite of *Aspergillus terreus* Thom. J. C. S. Chem. Commun. 380 (1975).
339. HOLKER, J. S. E., and K. YOUNG: Biosynthesis of Metabolites of *Perioconia macrospinosa* from [1-^{13}C]-, [2-^{13}C]- and [1,2-^{13}C]-Acetate. J. C. S. Chem. Commun. 525 (1975).
340. SATO, Y., T. ODA, and S. URANO: Griseofulvin Biosynthesis: New Evidence of Two Acetate Dispositions in the Ring A from ^{13}C Nuclear Magnetic Resonance Studies. Tetrahedron Letters 3971 (1976).
341. SIMPSON, T. J., and J. S. E. HOLKER: ^{13}C-NMR Studies on Griseofulvin Biosynthesis and Acetate Metabolism in *Penicillium patulum*. Phytochemistry **16**, 229 (1977).
342. BIRCH, A. J., T. J. SIMPSON, and P. W. WESTERMAN: Biosynthesis of Ravenelin from [1-^{13}C]- and [1,2-^{13}C]-Acetate. Tetrahedron Letters 4173 (1975).
343. BIRCH, A. J., J. BALDAS, J. R. HLUBUCEK, T. J. SIMPSON, and P. W. WESTERMAN: Biosynthesis of the Fungal Xanthone Ravenelin. J. C. S. Perkin I, 898 (1976).
344. SETO, H., L. W. CARY, and M. TANABE: Utilization of ^{13}C-^{13}C Coupling in Structural and Biosynthetic Studies. The Fourier Transform ^{13}C Nuclear Magnetic Resonance Spectrum of Mollisin. J. C. S. Chem. Commun. 867 (1973).
345. CASEY, M. L., R. C. PAULICK, and H. W. WHITLOCK, JR.: A Carbon-13 Nuclear Magnetic Resonance Study of Mollisin and Its Biosynthesis. J. Amer. Chem. Soc. **98**, 2636 (1976).
346. SETO, H., and H. YONEHARA: Utilization of ^{13}C-^{13}C Coupling in Structural and Biosynthetic Studies. VIII. The Cyclization Pattern of a Fungal Metabolite, Scytalone. Tetrahedron Letters 487 (1977).
347. SANKAWA, U., H. SHIMADA, T. SATO, T. KINOSHITA, and K. YAMASAKI: Biosynthesis of Scytalone. Tetrahedron Letters 483 (1977).
348. MCINNES, A. G., D. G. SMITH, J. A. WALTER, L. C. VINING, and J. L. C. WRIGHT: Homonuclear ^{13}C Decoupling in ^{13}C Nuclear Magnetic Resonance Studies of Biosynthesis Using Doubly Labelled Precursons. Assembly Pattern of the Acetate Units in Bikaverin. J. C. S. Chem. Commun. 66 (1975).
349. PAULICK, R. C., M. L. CASEY, D. F. HILLENBRAND, and H. W. WHITLOCK, JR.: A ^{13}C Nuclear Magnetic Resonance Study of the Biosynthesis of Islandicin from $^{13}CH_3$-$^{13}CO_2$-Na. J. Amer. Chem. Soc. **97**, 5303 (1975).
350. PAULICK, R. C., M. L. CASEY, and H. W. WHITLOCK, JR.: A ^{13}C Nuclear Magnetic Resonance Study of the Biosynthesis of Daunomycin from $^{13}CH_3$-$^{13}CO_2$-Na. J. Amer. Chem. Soc. **98**, 3370 (1976).
351. HOLKER, J. S. E., R. D. LAPPER, and T. J. SIMPSON: The Biosynthesis of Fungal

Metabolites. Part IV. Tajixanthone: ^{13}C Nuclear Magnetic Resonance Spectrum and Feeding with [1-^{13}C]- and [2-^{13}C]-Acetate. J. C. S. Perkin I, 2135 (1974).

352. CANHAM, P., L. C. VINING, A. G. MCINNES, J. A. WALTER, and J. L. C. WRIGHT: Pattern of Acetate Incorporation into the Aglycone of Chartreusin. Evidence from ^{13}C Nuclear Magnetic Resonance Studies for a Single-chain Polyketide Intermediate. J. C. S. Chem. Commun. 319 (1976).

353. GORST-ALLMAN, C. P., K. G. R. PACHLER, P. S. STEYN, P. L. WESSELS, and D. B. SCOTT: Biosynthesis of Averufin in *Aspergillus parasiticus* from [^{13}C]-acetate. J. C. S. Chem. Commun. 916 (1976).

354. SETO, H., L. W. CARY, and U. TANABE: Utilization of ^{13}C-^{13}C Coupling in Structural and Biosynthetic Studies. V. The ^{13}C FT NMR Spectrum of Sterigmatocystin. Tetrahedron Letters 4491 (1974).

355. STEYN, P. S., R. VLEGGAAR, P. W. WESSELS, and D. B. SCOTT: Biosynthesis of Aflatoxin B, from [2-^{13}C]- and [1,2-^{13}C]-Acetate. J. C. S. Chem. Commun. 193 (1975).

356. PACHLER, K. G. R., P. S. STEYN, R. VLEGGAAR, and P. L. WESSELS: Study of the Biosynthesis of Sterigmatocystin and Reassignment of ^{13}C Nuclear Magnetic Resonance Spectrum. J. C. S. Chem. Commun. 355 (1975).

357. HSIEH, D. P. H., J. N. SEIBER, C. A. REECE, D. L. FITZELL, S. L. YOUNG, J. I. DALEZIOS, G. N. LA MAR, D. L. BUDD, and E. MOTELL: ^{13}C Nuclear Magnetic Resonance Spectra of Aflatoxin B_1 Derived from Acetate. Tetrahedron **31**, 661 (1975).

358. HSIEH, D. P. H., R. C. YAO, D. L. FITZELL, and C. A. REECE: Origin of the Bisfuran Ring Structure in Aflatoxin Biosynthesis. J. Amer. Chem. Soc. **98**, 1021 (1976).

359. DORMAN, D. E., J. W. PASCHAL, W. M. NAKATSUKASA, L. L. HUCKSTEP, and N. NEUSS: The Use of ^{13}C-NMR Spectroscopy in Biosynthetic Studies. II. Biosynthesis of Narasin, a New Polyether Ionophore from Fermentation of *Streptomyces aureofaciens*. Helv. Chim. Acta **59**, 2625 (1976).

360. NADZAN, A. M., and K. L. RINEHART, JR.: Nybomycin. 8. Biosynthetic Origin of the Central Ring Carbons Studied by ^{13}C-Labelled Substrates. J. Amer. Chem. Soc. **98**, 5012 (1976).

361. CARDILLO, R., C. FUGANTI, G. GATTI, D. GHIRINGHELLI, and P. GRASSELLI: Molecular Structure of Cryptoechinuline A, a new Metabolite of *Aspergillus amstelodami*, Isolated during Investigations of Echinuline Biosynthesis. Tetrahedron Letters 3163 (1974).

362. MCINNES, A. G., D. G. SMITH, C.-K. WAT, L. C. VINING, and J. L. C. WRIGHT: Tenellin and Bassianin, Metabolites of *Beauveria* Species. Structure Elucidation with ^{15}N- and Doubly ^{13}C Enriched Compounds using ^{13}C Nuclear Magnetic Resonance Spectroscopy. J. C. S. Chem. Commun. 281 (1974).

363. MCINNES, A. G., D. G. SMITH, J. A. WALTER, L. C. VINING, and J. L. C. WRIGHT: New Techniques in Biosynthetic Studies Using ^{13}C Nuclear Magnetic Resonance Spectroscopy. The Biosynthesis of Tenellin Enriched from Singly and Doubly Labelled Precursors. J. C. S. Chem. Commun. 282 (1974).

364. LEETE, E., N. KOWANKO, R. A. NEWMARK, L. C. VINING, A. G. MCINNES, and J. L. C. WRIGHT: The Use of Carbon-13 Nuclear Magnetic Resonance to Establish that the Biosynthesis of Tenellin Involves an Intramolecular Rearrangement of Phenylalanine. Tetrahedron Letters 4103 (1975).

365. GRAF, W., J.-L. ROBERT, J. C. VEDERAS, C. TAMM, P. H. SOLOMON, I. MIURA, and K. NAKANISHI: Biosynthesis of the Cytochalasans. III. ^{13}C NMR of Cytochalasin B (Phomin) and Cytochalasin D. Incorporation of [1-^{13}C]- and [2-^{13}C]-Sodium Acetate. Helv. Chim. Acta **57**, 1801 (1974).

366. VEDERAS, J. C., W. GRAF, L. DAVID, and C. TAMM: Biosynthesis of Cytochalasans, 4. The Mode of Incorporation of Common Naturally-Occurring Carboxylic Acids into Cytochalasin D. Helv. Chim. Acta **58**, 1886 (1975).

367. HURLEY, L. H., M. ZMIJEWSKI, and C.-J. CHANG: Biosynthesis of Anthramycin.

Determination of the Labelling Pattern by the Use of Radioactive and Stable Isotope Techniques. J. Amer. Chem. Soc. **97,** 4372 (1975).

368. STROSHANE, R. M., M. TANIGUCHI, K. L. RINEHART, JR., J. P. POLLS, W. J. HAAK, and B. A. RUFF: Spectinomycin Biosynthesis Studied by Carbon Magnetic Resonance Spectroscopy. J. Amer. Chem. Soc. **98,** 3025 (1976).

369. MUNRO, M. H. G., M. TANIGUCHI, K. L. RINEHART, JR., and D. GOTTLIEB: A CMR Study of the Biosynthesis of Chloramphenicol. Tetrahedron Letters 2659 (1975).

370. JOHNSON, R. D., A. HABER, and K. L. RINEHART, JR.: Geldanamycin Biosynthesis and Carbon Magnetic Resonance. J. Amer. Chem. Soc. **96,** 3316 (1974).

371. ŌMURA, S., A. NAKAGAWA, H. TAKESHIMA, K. ATSUMI, J. MIYAZAWA, F. PIRIOU, and G. LUKACS: Biosynthetic Studies Using ^{13}C Enriched Precursors of the 16-Membered Macrolide Antibiotic Leucomycin A_3. J. Amer. Chem. Soc. **97,** 6600 (1975).

372. ŌMURA, S., A. NAKAGAWA, H. TAKESHIMA, J. MIYAZAWA, C. KITAO, F. PIRIOU, and G. LUKACS: A ^{13}C Nuclear Magnetic Resonance Study of the Biosynthesis of the 16-Membered Macrolide Antibiotic Tylosin. Tetrahedron Letters 4503 (1975).

373. HUTCHINSON, C. R., A. H. HECKENDORF, P. E. DADDONA, E. HAGAMAN, and E. WENKERT: Biosynthesis of Camptothecin. I. Definition of the Overall Pathway Assisted by Carbon-13 Nuclear Magnetic Resonance Analysis. J. Amer. Chem. Soc. **96,** 5609 (1974).

374. BATTERSBY, A. R., P. W. SHELDRAKE, and J. A. MILNER: Biosynthesis of Colchicine: Incorporation of a ^{13}C-Labelled Precursor in a Higher Plant. Tetrahedron Letters 3315 (1974).

375. LEETE, E., and G. B. BODEN: Biosynthesis of Shihunine in *Dendrobium pierardii.* J. Amer. Chem. Soc. **98,** 6321 (1976).

376. LEETE, B.: Biosynthesis of the Isoquinuclidine Moiety of Dioscorine. Incorporation of [5,6-$^{13}C_2$]-Nicotinic Acid Established by Means of ^{13}C Nuclear Magnetic Resonance. J. Amer. Chem. Soc. **99,** 648 (1977).

377. HUTCHINSON, C. R., M.-T. S. HSIA, and R. A. CARVER: Biosynthetic Studies with $^{13}CO_2$ of Secondary Plant Metabolites. *Nicotina* Alkaloids. 1. Initial Experiments. J. Amer. Chem. Soc. **98,** 6006 (1976).

378. DJERASSI, C.: Natural Products Chemistry 1950 to 1980 — A Personal View. Pure Appl. Chem. **41,** 113 (1975).

379. For an account on the experimental and methodological requirements see e. g. J. N. SHOOLERY, F. W. WEHRLI, and T. WIRTHLIN: Experimentelle Voraussetzungen zur Analyse kleinster Substanzmengen mittels ^{13}C- und ^{1}H-Kernresonanzspektroskopie. Chemie-Technik **6,** 55 (1977).

380. HELLER, S. R., G. W. A. MILNE, and R. J. FELDMANN: A Computer-Based Chemical Information System. Science **195,** 253 (1977).

381. SCHWARZENBACH, R., J. MEILI, H. KÖNITZER, and J. T. CLERC: A Computer System for Structural Identification in Organic Compounds from ^{13}C NMR Data. Org. Magn. Res. **8,** 11 (1976).

382. BREMSER, W., M. KLIER, and E. MEYER: Mutual Assignment of Subspectra and Substructures — A Way to Structure Elucidation by ^{13}C NMR Spectroscopy. Org. Magn. Res. **7,** 97 (1975).

383. CHEER, C. J., D. H. SMITH, C. DJERASSI, B. TURSCH, J. C. BRAEKMAN, and D. DALOZE: Application of Artificial Intelligence for Chemical Inference — XXI. Chemical Studies of Marine Invertebrates. XVII. The computer-assisted identification of [+]-Palustrol in the Marine Organism *Cespitularia* sp., aff. *subviridis.* Tetrahedron **32,** 1807 (1976).

Addendum

In a field growing as rapidly as ^{13}C n.m.r. spectroscopy of natural products, a time lag in excess of one year between completion and appearance of a review clearly impairs its value. In an effort to reduce obsolescence to the minimum, the authors complemented this work by adding additional 1977 as well as early 1978 references, grouped according to compound classes.

Terpenoids and Steroids

384. FRINGUELLI, F., H. E. GOTTLIEB, E. W. HAGAMAN, A. TATICCHI, E. WENKERT, and P. M. WOVKULICH: ^{13}C-NMR Spectroscopy of Carane Derivatives. Gazz. Chim. Ital. **105,** 1215 (1975).

385. HOLDEN, C. M., J. C. REES, S. P. SCOTT, and D. WHITTAKER: Stereochemistry and Reduction of Umbellulone (Thuj-3-en-2-one) and Isodihydroumbellulone (4β*H*-Thujan-2-one). J. C. S. Perkin II, 1342 (1976).

386. KREISER, W., L. JANITSCHKE, and L. ERNST: Configurational Assignment by NMR Spectroscopy of Stereoisomeric 2.6-Dimethyl-tricyclo[5,2,1,$0^{2,6}$]dec-3-enes and Decanes. Albene and Isoalbene. Tetrahedron **34,** 131 (1978).

387. BAILLEUL, F., P. DELAVEAU, A. RABARON, M. PLAT, and M. KOCH: Feretoside et Gardenoside du *Feretia apondanthera.* RMN du carbone-13 en Série Iridoïde. Phytochemistry **16,** 723 (1977).

388. MILES, D. H., U. KOKPOL, J. BHATTACHARYYA, J. L. ATWOOD, K. E. STONE, T. A. BRYSON, and C. WILSON: Structure of Sarracenin. An Unusual Enol Diacetate Monoterpene from Insectivorous Plant *Sarracenia Flava.* J. Am. Chem. Soc. **98,** 1569 (1976).

389. VAN ENGEN, D., J. CLARDY, E. KHO-WISEMAN, P. DREW, M. D. HIGGS, and D. J. FAULKNER: Violacenine, A. Reassignment of Structure. Tetrahedron Letters **29** (1978).

390. HIGGS, M. D., D. J. VANDERAH, and D. J. FAULKNER: Polyhalogenated Monoterpenes from *Plocamium cartilagineum* from the British Coast. Tetrahedron **33,** 2775 (1977).

391. WRATTEN, S. J., and D. J. FAULKNER: Carbonimidic Dichlorides from the Marine Sponge *Pseudaxinyssa pitys.* J. Am. Chem. Soc. **99,** 7367 (1977).

392. CREWS, P., and E. KHO-WISEMAN: Acylic Polyhalogenated Monoterpenes from the Red Alga *Plocamium violaceum.* J. Org. Chem. **42,** 2812 (1977).

393. NORTON, R. S., R. G. WARREN, and R. J. WELLS: Three New Polyhalogenated Monoterpenes from *Plocamium* Species. Tetrahedron Letters 3905 (1977).

394. SATO, T., M. TADA, T. TAKAHASHI, I. HORIBE, H. ISHII, and K. TORI: Carbon-13 and Hydrogen-1 NMR Studies of Conformations of Ligularol and 6-*Epi*ligularol, Naturally Occurring *cis*-Decalin Derivatives. Tetrahedron Letters 3895 (1977).

395. CUEVAS, O., and J. R. HANSON: Norbotryal Acetate. A Nor-sesquiterpenoid Aldehyde from *Botrytis cinerea.* Phytochemistry **16,** 1016 (1977).

396. KUBO, I., I. MIURA, M. J. PETTEI, Y.-W. LEE, F. PILKIEVICZ, and K. NAKANISHI: Muzigadial and Warburganal, Potent Antifungal Antiyeast, and African Army Worm Antifeedant Agents. Tetrahedron Letters 4553 (1977).

397. AASEN, A. J., T. NISHIDA, C. R. ENZELL, and H. H. APPEL: The Structure of (11ξ,12ξ)-11,12-Di(7-drimen-11-oxy)-11,12-epoxy-7-drimene. Acta Chem. Scand. **B 31,** 51 (1977).

398. ROSE, A. F., and J. J. SIMS: Marine Natural Products, XIV. 1-*S*-Bromo-4-*R*-hydroxy-selin-7-ene, A Metabolite of the Marine Alga *Laurencia sp.* Tetrahedron Letters 2935 (1977).

399. MURAI, A., N. KATSUI, F. YAGIHASHI, T. MASAMUNE, Y. ISHIGURI, and K. TOMIYAMA: Structure of Rishitin M-1 and M-2, Metabolites of Rishitin in Healthy Potato Tube Tissues. J. C. S. Chem. Commun. 670 (1977).

400. Ward, E. W. B., A. Stoessl, and J. B. Stothers: Metabolism of the Sesquiterpenoid Phytoalexins Capsidiol and Rishitin to their 13-Hydroxy Derivatives by Plant Cells. Phytochemistry **16,** 2024 (1977).
401. Coxon, D. T., K. R. Price, B. Howard, and R. F. Curtis: Metabolites from Microbially Infected Potato. Part I. Structure of Phytuberin. J. C. S. Perkin I, 53 (1977).
402. Southwell, I. A.: Biogenetically Significant Sesquiterpenoids from *Rubus rosifolius* Oil. Tetrahedron Letters 873 (1977).
403. Anderson, R. C., D. M. Gunn, J. Murray-Rust, P. Murray-Rust, and J. S. Roberts: Vetispirane Sesquiterpene Glucosides from Fluecured Virginia Tabacco: Structure, Absolute Stereochemistry, and Synthesis. X-Ray Structure of the *p*-Bromobenzenesulphonate of One of the Derived Aglycones. J. C. S. Chem. Commun. 27 (1977).
404. Matsuo, A., and S. Hayashi: Revised Structure and Absolute Configuration of the Sesquiterpene (+)-Bazzanene. J. C. S. Chem. Commun. 566 (1977).
405. Herz, W., P. S. Kalyanaraman, and G. Ramakrishnan: Sesquiterpene Lactones from *Eupatorium perfoliatum.* J. Org. Chem. **42,** 2264 (1977).
406. Doskotch, R. W., F. S. El-Feraly, E. H. Fairchild, and C.-T. Huang: Isolation and Characterization of Peroxyferolide, a Hydroperoxy Sesquiterpene Lactone from *Liriodendron tulipifera.* J. Org. Chem. **42,** 3614 (1977).
407. Herz, W., P. S. Subramaniam, R. Murai, N. Dennis, and J. F. Blount: Micordilin, a Complex Elemanolide from *Mikania cordifolia.* J. Org. Chem. **42,** 1720 (1977).
408. Vidari, G., M. DeBernardi, P. Vita-Finzi, and G. Fronza: Sesquiterpenes from *Lactarius blennius.* Phytochemistry **15,** 1953 (1976).
409. Zalkow, L. H., R. N. Harris, III., D. Van Derveer, and J. A. Bertrand: Isocomene, a Novel Sesquiterpene from *Isocoma wrightii,* X-Ray Crystal Structure of the Corresponding Diol. J. C. S. Chem. Commun. 456 (1977).
410. Beechan, C. M., C. Djerassi, J. S. Finer, and J. Clardy: Terpenoids, LXXIII. Sinularene. A Sesquiterpene Hydrocarbon Based on a Novel Skeleton from the Soft Coral, *Sinularia mayi.* Tetrahedron Letters 2395 (1977).
411. Matsuo, A., I. Terada, M. Nakayama, and S. Hayashi: Cuprenenol and Rosulantol. New Cuparane Class Sesquiterpene Alcohols from the Liverwort *Jungermannia rosulans.* Tetrahedron Letters 3821 (1977).
412. King, T. J.: Alliacolide, a New Bicyclic Sesquiterpene Expoxy-lactone with a Novel Carbon Skeleton from Cultures of the Fungus *Marasmius alliaceus* (Jacques ex Fr.) Fr, X-Ray Structure. J. C. S. Chem. Commun. 727 (1977).
413. Ranieri, R. L., and G. J. Calton: Quadrone, a New Antitumor Agent from *Aspergillus terreus.* Tetrahedron Letters 499 (1978).
414. Nozoe, S., H. Kobayashi, S. Urano, and J. Furukawa: Isolation of Δ^6-Protoilludene and the Related Alcohols. Tetrahedron Letters 1881 (1977).
415. Yuste, F., D. Diaz, F. Walls, and K. Jankowski: The Structure of Cacalone. J. Org. Chem. **41,** 4103 (1976).
416. Suzuki, T., A. Furusaki, N. Hashiba, and E. Kurosawa: Novel Skeletal Bromo Ether from the Marine Alga, *Laurencia nipponica* Yamada. Tetrahedron Letters 3731 (1977).
417. Coll, J. C., S. J. Mitchell, and G. J. Stokie: Studies of Australian Soft Corals V. A Novel Furano-sesquiterpene Acid from the Soft Coral *Sinularia gonatodes* (kolonko). Tetrahedron Letters 1539 (1977).
418. Sun, H. H., S. M. Waraszkiewicz, and K. L. Erickson: C_{15}-Halogenated Compounds from the Hawaiian Marine Alga *Laurencia nidifica,* VI. The Isomaneonenes. Tetrahedron Letters 4227 (1976).
419. Silva, M., A. Wiesenfeld, P. G. Sammes, and T. W. Tyler: New Sesquiterpenes from *Pleocarphus revolutus.* Phytochemistry **16,** 379 (1977).

420. EPPLEY, R. M., E. P. MAZZOLA, R. J. HIGHET, and W. J. BAILEY: Structure of Satratorin H, a Metabolite of *Stachybotrys atra*. Application of Proton and Carbon-13 Nuclear Magnetic Resonance. J. Org. Chem. **42**, 240 (1977).
421. BREITENSTEIN, W., and C. TAMM: Verrucarin K, the First Natural Trichothecene Derivative Lacking the 12,13-Epoxy Group. Helv. Chim. Acta **60**, 1522 (1977).
422. MATSUMOTO, M., H. MINATO, K. TORI, and M. UEYAMA: Structures of Isonoridin E, Epoxyisonoridin E, and Epoxy- and Diepoxy-nonidine H, New Metabolites Isolated from *Cylindrocarpon* Species Determined by Carbon-13 and Hydrogen-1 NMR Spectroscopy. Tetrahedron Letters 4093 (1977).
423. MANVILLE, J. F., K. BOCK, and E. VON RUDLOFF: Occurrence of Juvabione-Type and *Epi*juvabione-type Sesquiterpenoids in *Abies alba*. Phytochemistry **16**, 1967 (1977).
424. GERBER, N. N., and D. Z. DENNEY: The Carbon-13 Nuclear Magnetic Resonance Spectra of Four Eudesmane Sesquiterpenols. Phytochemistry **16**, 2025 (1977).
425. ARCHER, R. A., D. W. JOHNSON, E. W. HAGAMAN, L. N. MORENO, and E. WENKERT: Carbon 13 Nuclear Magnetic Resonance Spectroscopy of Naturally Occurring Substances. 47. Cannabinoid Compounds. J. Org. Chem. **42**, 490 (1977).
426. JONES, N. F.: The Pyrethrins and Related Compounds, Part 21. Carbon-13 Nuclear Magnetic Resonance Spectra of Synthetic Pyrethroids. J. C. S. Perkin I, 1878 (1977).
427. TAKEUCHI, S., J. UZAWA, H. SETO, and H. YONEHARA: New ^{13}C-NMR Techniques Applied to Pentalenolactone Structure. Tetrahedron Letters 2943 (1977).
428. CARSTEN-LICHTERFELDE, C. VON, C. PASCUAL, R. MA. RABANAL, B. RODRIGUEZ, and S. VALVERDE: ^{13}C NMR Substituent Effects in Tetracyclic Diterpenoids. Tetrahedron **33**, 1989 (1977).
429. CHALMERS, A. A., C. P. GORST-ALLMAN, and L. P. L. PIACENZA: Carbon-13 NMR Spectra of Stachenol and Related Derivatives. Tetrahedron Letters 1665 (1977).
430. IMAMURA, P. M., A. J. MARSAIOLI, L. E. S. BARATA, and E. A. RUVEDA: ^{13}C NMR Spectral Analysis of Eperuane Diterpenes. Phytochemistry **16**, 1842 (1977).
431. PELLETIER, S. W., N. V. MODY, J. BHATTACHARYYA, and D. H. MILES: Carbon-13 NMR Analysis of Naturally Occurring 9,10-Dihydrophenanthrenes and Their Derivatives. Tetrahedron Letters 425 (1978).
432. KATO, T., M. TSUNAKAWA, N. SASAKI, H. AIZAWA, K. FUJITA, Y. KITAHARA, and N. TAKAHASHI: Growth and Germination Inhibitors in Rice Husks. Phytochemistry **16**, 45 (1977).
433. HAYASHI, Y., T. MATSUMOTO, Y. YUKI, and T. SAKAN: New Congeners of Cytotoxic Nor-diterpenoid Dilactones in *Podocarpus nagi:* Three New Components of 7,8-Epoxy-Enolide Type. Tetrahedron Letters 4215 (1977).
434. KATO, T., H. AIZAWA, M. TSUNAKAWA, N. SASAKI, Y. KITAHARA, and N. TAKAHASHI: Chemical Transformations of the Diterpene Lactones, Momilactones A and B. J. C. S. Perkin I, 250 (1977).
435. SAVONA, G., F. PIOZZI, J. R. HANSON, and M. SIVERNS: Structures of Three New Diterpenoids from *Ballota* Species. J. C. S. Perkin I, 322 (1977).
436. CHATTERJEE, A., A. BANERJEE, and F. BOHLMANN: Crotocaudin, a Rearranged Labdane Type Norditerpene from *Croton caudatus* Geisel. Tetrahedron **33**, 2407 (1977).
437. MIYASE, T., P. RÜEDI, and C. H. EUGSTER: Unusual Rearranged Abietanoic Diterpenoids from *Solenostemon* Species. J. C. S. Chem. Commun. 859 (1977).
438. BLUNT, J. W., G. S. BOYD, M. P. HARTSHORN, M. H. G. MUNRO, and L. K. PANNELL: The Acid-Catalysed Dehydration of 13α-Substituted 13β-methyl-podocarpan-8β-ols. Aust. J. Chem. **30**, 2015 (1977).
439. MIYASE, T., P. RÜEDI, and C. H. EUGSTER: Diterpenoide Drüsenfarbstoffe aus Labiaten: Coleone U, V, W und 14-*O*-Formylcoleon-V sowie 2 Royleanone aus *Plectranthus myrianthus* Brig.; *cis*- und *trans*-A/B-6,7-Dioxoroyleanon. Helv. Chim. Acta **60**, 2770 (1977).

440. MIYASE, T., P. RÜEDI, and C. H. EUGSTER: Diterpenoide Drüsenfarbstoffe aus Labiaten: 3β-Acetoxyfuerstion, Nilgherron A und Nilgherron B, neue Chinomethane aus *Plectranthus nilgherricus* Benth.; absolute Konfiguration von Fuerstion. Helv. Chim. Acta **60,** 2789 (1977).
441. EKMAN, R., R. SJÖHOLM, and K. HANNUS: Isoabienol, the Principal Diterpene Alcohol in *Pinus sylvestris* Needles. Acta Chem. Scand. **B 31,** 921 (1977).
442. NAIR, A. G. R., S. S. SUBRAMANAN, F. BOHLMANN, S. SCHÖNEWEISS, and T. J. MABRY: A new Diterpene Galactoside from *Acanthospermum hispidum.* Phytochemistry **15,** 1776 (1976).
443. RODRÍGUEZ, B.: New Labdane Diterpenoids from *Sideritis chamaedryfolia.* Phytochemistry **17,** 281 (1978).
444. BRAUN, S., und H. BREITENBACH: Strukturaufklärung einer neuen Diterpensäure aus *Metasequoia glyptostroboides* mit Hilfe der ^{13}C-NMR-Spektroskopie. Tetrahedron **33,** 145 (1977).
445. GONZÁLEZ, A. G., J. M. ARTEAGA, J. L. BRETÓN, and B. M. FRAGA: Five new Labdane Diterpene Oxides from *Eupatorium jhanii.* Phytochemistry **16,** 107 (1977).
446. CECCHERELLI, P., M. CURINI, R. PELLICCIARI, M. S. RAJU, and E. WENKERT: Conversion of Virescenol A into Virescenol B. J. Org. Chem. **42,** 3438 (1977).
447. BHAT, S. V., B. S. BAJWA, H. DORNAUER, and N. J. DE SOUZA: Structures and Stereochemistry of New Labdane Diterpenoids from *Coleus forskohlii* Brig. Tetrahedron Letters 1669 (1977).
448. SAVONA, G., S. PASSANNANTI, M. P. PATERNOSTRO, F. PIOZZI, J. R. HANSON, and M. SIVERNS: 8β-Hydroxyfruticolone, a Diterpenoid from *Teucrium fruticans.* Phytochemistry **17,** 320 (1978).
449. BOHLMANN, F., und C. ZDERO: Ein neues Clerodan-Derivat sowie weitere Inhaltsstoffe aus der Gattung *Macowania.* Phytochemistry **16,** 1583 (1977).
450. KUBO, I., I. MIURA, K. NAKANISHI, T. KAMIKAWA, T. ISOBE, and T. KUBOTA: Structure of Isodomedin, a Novel *ent*-Kaurenoid Diterpene. J. C. S. Chem. Commun. 555 (1977).
451. VICHNEWSKI, W., H. DE FREITAS LEITÃO FILHO, R. MURARI, and W. HERZ: Cinnamoylgrandifloric Acid from *Mikania oblongifolia.* Phytochemistry **16,** 2028 (1977).
452. MANCHAND, P. S., and J. F. BLOUNT: Stereostructures of the Macrocyclic Diterpenoids Ovatodiolide and Isoovatodiolide. J. Org. Chem. **42,** 3824 (1977).
453. KASHMAN, Y., and A. GROWEISS: Lobolide: A New Epoxy Cembranolide from Marine Origin. Tetrahedron Letters 1159 (1977).
454. COLL, J. C., G. B. HAWES, N. LIYANAGE, W. OBERHÄNSLI, and R. J. WELLS: Studies of Australian Soft Corals. I. A New Cembrenoid Diterpene from a *Sacrophyton* Species. Aust. J. Chem. **30,** 1305 (1977).
455. COLL, J. C., S. J. MITCHELL, and G. J. STOKIE: Studies of Australian Soft Corals. II. A Novel Cembrenoid Diterpene from *Lobophytum michaelae.* Aust. J. Chem. **30,** 1859 (1977).
456. BOWDEN, B. F., J. A. BRITTLE, J. C. COLL, N. LIYANAGE, S. J. MITCHELL, and G. J. STOKIE: Studies of Austrial Soft Corals. VI. A New Cembrenoid Diterpene from the Soft Coral *Lobophytum crassum* (Coelenterata, Anthozoa, Octocorallia, Alcyonacea). Tetrahedron Letters 3661 (1977).
457. SPRINGER, J. P., J. CLARDY, R. H. COX, H. G. CUTLER, and R. J. COLE: The Structure of a New Type of Plant Growth Inhibitor Extracted from Immature Tobacco Leaves. Tetrahedron Letters 2737 (1975).
458. BEHR, D., I. WAHLBERG, A. J. AASEN, T. NISHIDA, C. R. ENZELL, A.-M. PILOTTI, and J.-E. BERG: (1*S*,2*E*,4*R*,6*E*,8*R*,11*S*,12*R*)- and (1*S*,2*E*,4*S*,6*E*,8*R*,11*S*,12*R*)-8,11-Epoxy-2,6-thunbergadiene-4,12-diol; Two New Diterpenoids of Greek Tobacco. Acta Chem. Scand., in press.

459. KAZLAUSKAS, R., P. T. MURPHY, R. J. WELLS, and P. SCHÖNHOLZER: Two New Diterpenes Related to Eunicellin from a *Cladiella* Species (Soft Coral). Tetrahedron Letters 4643 (1977).
460. IRELAND, C., and D. J. FAULKNER: Diterpenes from *Dolabella californica*. J. Org. Chem. **42,** 3157 (1977).
461. PRESTWICH, G. D., S. P. TANIS, F. G. PILKIEWICZ, I. MIURA, and K. NAKANISHI: Nasute Termite Soldier Frontal Gland secretions. 2. Structures of Trinervitene Congeners from *Trinervitermes* Soldiers. J. Am. Chem. Soc. **98,** 6062 (1976).
462. PRESTWICH, G. D., B. A. SOLHEIM, J. CLARDY, F. G. PILKIEWICZ, I. MIURA, S. P. TANIS, and K. NAKANISHI: Kempene-1 and -2, Unusual Tetracyclic Diterpenes from *Nasutitermes* Termite Soldiers. J. Am. Chem. Soc. **99,** 8082 (1977).
463. PASCARD, C., T. PRANGÉ, and J. POLONSKY: Crystal and Molecular Structure of the Quassinoid 6-Hydroxy-picrasin B. J. Chem. Research (S), 324 (1977); (M), 3636 (1977).
464. GONZÁLEZ, A. G., C. G. FRANCISCO, R. FREIRE, R. HERNÁNDEZ, J. A. SALAZAR, and E. SUÁREZ: C_{12} Stereochemistry of α- and β-Levantenolide. Carbon-13 NMR Spectra Labdanolic Diterpenes. Tetrahedron Letters 1897 (1976).
465. VALVERDE, S., and B. RODRÍGUEZ: The Use of ^{13}C-NMR in the Determination of Structures: A Correction of the Structure of Borjatriol. Phytochemistry **16,** 1841 (1977).
466. KASHMAN, Y., and A. RUDI: The ^{13}C-NMR Spectrum and Stereochemistry of Heteronemin. Tetrahedron **33,** 2997 (1977).
467. CIMINO, G., S. DESTEFANO, L. MINALE, and E. TRIVELLONE: 12-*epi*-Scalarin and 12-*epi*-Deoxoscalarin. Sesterterpenes from the Sponge *Spongia Nitens*. J. C. S. Perkin I, 1587 (1977).
468. STIPANOVIC, R. D., A. A. BELL, D. H. O'BRIEN, and M. J. LUKEFAHR: Heliocide H_3, an Insecticidal Terpenoid from *Gossypium hirsutum*. Phytochemistry **17,** 151 (1978).
469. CHEXAL, K. K., J. P. SPRINGER, J. CLARDY, R. J. COLE, J. W. KIRKSEY, J. W. DORNER, H. G. CUTLER, and B. J. STRAWTER: Austin, a Novel Polyisoprenoid Mycotoxin from *Aspergillus ustus*. J. Am. Chem. Soc. **98,** 6748 (1976).
470. MONDON, A., D. TRAUTMANN, B. EPE, and U. OELBERMANN: Zur Kenntnis der Bitterstoffe aus Cneoraceen, VI. Tetrahedron Letters 3291 (1976).
471. — — — — Zur Kenntnis der Bitterstoffe aus Cneoraceen, VII. Tetrahedron Letters 3295 (1976).
472. PURUSHOTHAMAN, K. K., S. CHANDRASEKHARAN, J. D. CONNOLLY, and D. S. RYCROFT: Tetranortriterpenoids and Related Substances. Part 18. Two New Tetranortriterpenoids with a Modified Furan Ring from the Bark of *Soymida febrifuga*. A. Juss (Meliaceae). J. C. S. Perkin I, 1873 (1977).
473. SIDDIQUI, S., S. FUCHS, J. LÜBCKE und W. VOELTER: Struktur eines neuen Naturstoffes aus *Melia azadirachta* Linn: 17-Hydroxyazadiradion. Tetrahedron Letters 611 (1978).
474. SABATA, B. J. D. CONNOLLY, G. LABBÉ, and D. S. RYCROFT: Tetranortriterpenoids and Related Substances. Part 19. Revised Structures of Atalantolide and Atalantin, Limonoids from the Root Bark of *Atalantia monophylla*. Correa (Rutaceae). J. C. S. Perkin I, 1875 (1977).
475. HALSALL, T. G., K. WRAGG, J. D. CONNOLLY, M. A. MCLELLAN, L. D. BREDELL, and D. A. H. TAYLOR: ^{13}C Nuclear Magnetic Resonance Spectra of Some Limonoids. Part III. The Spectra of Some Derivatives of Entandrophragmin and a Revised Structure for Condollein. J. Chem. Research (S), 154 (1977); (M), 1727 (1977).
476. ASAKAWA, J., R. KASAI, K. YAMASAKI, and O. TANAKA: ^{13}C NMR Study of Ginseng Sapogenins and their Related Dammarane Type Triterpenes. Tetrahedron **33,** 1935 (1977).
477. ISHII, H., S. SEO, K. TORI, T. TOZYO, and Y. YOSHIMURA: The Structures of Saikosaponin-E and Acetylsaikosaponins, Minor Components Isolated from *Bupleurum falcatum* L., Determined by C-13 NMR Spectroscopy. Tetrahedron Letters 1227 (1977).

478. YAMADA, Y., K. HAGIWARA, K. IGUCHI, and S. SUZUKI: Structures of Anvenin I and II, Bitter Principles from *Amagallis arvensis* L. (Primulaceae). New Cucurbbitacin Glucosides. Tetrahedron Letters 2099 (1977).

479. ZIEGLER, R., and C. TAMM: Isolation and Structure of Eucosterol and 16β-Hydroxy-eucosterol. Two Novel Spirocyclic Nortriterpenes, and of a New 24-Nor-5α-chola-8,16-diene-23-oic Acid from Bulbs of Several *Eucomis* Species. Helv. Chim. Acta **59,** 1997 (1976).

480. OKORIE, D. A., and D. A. H. TAYLOR: Triterpenes from the seed of *Entandrophragma* Species. Phytochemistry **16,** 2029 (1977).

481. DREYER, D. L., and E. K. TROUSDALE: Cucurbitacins in *Purshia tridentata.* Phytochemistry **17,** 325 (1976).

482. KAWAI, K., and S. SHIBATA: Pseudojujubogenin, a New Sapogenin from *Bacopa monniera.* Phytochemistry **17,** 287 (1978).

483. BLUNT, J. W., and J. B. STOTHERS: ^{13}C N.m.r. Spectra of Steroids — A Survey and Commentary. Org. Magn. Resonance **9,** 439 (1977).

484. VAN ANTWERP, C. L., H. EGGERT, G. D. MEAKINS, J. O. MINERS, and C. DJERASSI: Additivity Relationships in Carbon-13 Nuclear Magnetic Resonance Spectra of Dihydroxy Steroids. J. Org. Chem. **42,** 789 (1977).

485. BERMANN, E., Z. LUZ, Y. MAZUR, and M. SHEVES: Conformational Analysis of Vitamin D and Analogues. ^{13}C and ^{1}H Nuclear Magnetic Resonance Study. J. Org. Chem. **42,** 3325 (1977).

486. SMITH, W. B.: The Carbon-13 Spectra of Steroids on the Way to Ecdysone. Org. Magn. Resonance **9,** 644 (1977).

487. TAYLOR, E. J., and C. DJERASSI: Synthesis of Cholest-5-ene-3β-, 11α, 15β-triol-7-one, A Model for the Steroid Nucleus of Oogoniol, a Sex Hormone of the Water Mold *Achlya.* J. Org. Chem. **42,** 3571 (1977).

488. MCMORRIS, T. C., S. R. SHOW, and G. R. WEIHE: Evidence for C-29 Hydroxyl Group in Ooginiol from ^{13}C NMR Spectra of Model 3β,26- and 3β,29-Dihydroxy Stigmast-5-enes. Tetrahedron Letters 335 (1978).

489. HIKINO, H., K. MOHRI, Y. HIKINO, S. ARIHARA, T. TAKEMOTO, H. MORI, and K. SHIBATA: Inokosterone, an Insect Metamorphosing Substance from *Achyranthes faurier.* Absolute Configuration and Synthesis. Tetrahedron **32,** 3015 (1976).

490. KOREEDA, M., N. KOIZUMI, and B. A. TEICHER: Stereochemically Controlled Synthesis of 20,22-Epoxycholesterols. Tetrahedron Letters 4565 (1976).

491. MOSS, G. P.: Carbon-13 NMR Spectra of Carotenoids. Pure Appl. Chem. **47,** 97 (1976).

492. MATSUO, M., and S. URANO: ^{13}C NMR Spectra of Tocophenols and 2,2-Dimethyl-chromanols. Tetrahedron **32,** 229 (1976).

493. KAISER, R., A. KAPPELER, and D. LAMPARSKY: Inhaltsstoffe des *Osmanthus*-Absolues. 3. Mitteil. Derivate der Theaspirane. Helv. Chim. Acta **61,** 387 (1978).

494. FREI, B., H. EICHENBERGER, B. VON WARTBURG, H. R. WOLF, and O. JEGER: Photochemische Reaktionen, 94. Mitteil. Vinyloge β-Spaltung bei Epoxy-enonen der Jonon-Reihe. Helv. Chim. Acta **60,** 2968 (1977).

495. BLOUNT, J. F., R. L. HAN, B. A. PAWSON, R. G. PITCHER, and T. H. WILLIAMS: (*E*)- and (*Z*)-4-Methyl-5-[5-(2,6,6-trimethylcyclohexen-1-yl)-3-methyl-2(*E*),4(*E*)-pentadienylidene]-2(5*H*)-furanone. Synthesis and Spectral Properties. J. Org. Chem. **41,** 4108 (1976).

496. BEHR, D., I. WAHLBERG, T. NISHIDA, and C. R. ENZELL: Tobacco Chemistry. 41. Structure Determination and Synthesis of 5(13),7*E*-Megastigmadien-6,9-diol, a New Constituent of Greek Tobacco. Acta Chem. Scand. **B 31,** 609 (1977).

Alkaloids

497. HANISCH, P., A. J. JONES, A. F. CASEY, and J. E. COATES: Carbon-13 Magnetic Resonance Evidence for Non-Chair Conformations in Tropane Derivatives. J. C. S. Perkin II, 1202 (1977).

498. ZETTA, L., and G. GATTI: ^{13}C N.m.r. Spectra of Lysergic Acid Derivatives. II. Dihydrolysergamides. Org. Magn. Resonance **9**, 218 (1977).

499. RICCA, R. S., and C. CASAGRANDE: ^{13}C Magnetic Resonance Studies of Reduced Proaporphine Alkaloids. Org. Magn. Resonance **9**, 8 (1977).

500. VERPOORTE, R., and A. B. SVENDSEN: Carbon-13 Nuclear Magnetic Resonance of Some Strychnos Alkaloids. Pharm. Weekbl. **111**, 745 (1976).

501. VERPOORTE, R., P. J. HYLANDS, and N. G. BISSET: Carbon-13 N.m.r. Spectroscopy of Some *Strychnos* Alkaloids. Org. Magn. Resonance **9**, 567 (1977).

502. LEUNG, J., and A. J. JONES: Carbon-13 N.m.r. Analysis of Selected Strychnos Alkaloids. Org. Magn. Resonance **9**, 333 (1977).

503. HUGHES, D. W., B. C. NALLIAH, H. L. HOLLAND, and D. B. MACLEAN: ^{13}C Nuclear Magnetic Resonance Spectra of the Spirobenzylisoquinoline Alkaloids and Related Model Compounds. Can. J. Chem. **55**, 3304 (1977).

504. RONALDSON, J. W.: Sporidesmins. XV. The ^{13}C Nuclear Magnetic Resonance Spectra of Sporidesmin and Sporidesmin-D. The Evidence in the Spectra of Strain Imposed by an *Epi*dithio Bridge. Aust. J. Chem. **29**, 2307 (1976).

505. VAN BINST, G., D. TOURWÉ, and E. DECOCK: Benzo- and Indoquinolizines. X. Application of Carbon-13 N. M. R. to the Stereochemistry of the 4*b*,5,6,7,8*a*,10,11,16,16*b*-Decahydrobibenzo[f,h]indolo-[2,3-*a*]quinolizine Isomers. Org. Magn. Resonance **8**, 618 (1976).

506. WESTON, R. J., H. E. GOTTLIEB, E. W. HAGAMAN, and E. WENKERT: Carbon-13 Nuclear Magnetic Resonance Spectroscopy of Naturally Occurring Substances. LI. *Solanum* Glycoalkaloids. Aust. J. Chem. **30**, 917 (1977).

507. COLE, R. J., J. W. KIRKSEY, R. H. COX, and J. CLARDY: Structure of the Tremor-Producing Indole, TR-2. J. Agric. Food Chem. **23**, 1015 (1975).

508. COLE, R. J., J. W. PORNER, J. A. LANSDEN, R. H. COX, C. PAPE, G. CUNFER, S. S. NICHOLSON, and D. M. BEDELL: Paspalum Staggers: Isolation and Identification of Tremorgenic Metabolites from Sclerotia of *Clavicops paspali*. J. Agric. Food Chem. **25**, 1197 (1977).

509. JÖSSANG, A., H. JACQUEMIN, J.-L. POUSSET, A. CAVÉ, M. Damak, and C. RICHE: Structure de la Borreline, Nouvel Alcaloïde Indolique. Tetrahedron Letters 1219 (1977).

510. JÖSSANG, A., J.-L. POUSSET, H. JACQUEMIN, and A. CAVÉ: Structure de la Borrecarpine, Nouvel Alcaloïde Indolique. Tetrahedron Letters 4317 (1977).

511. LOUNASMAA, M., and C.-J. JOHANSSON: Synthetic Studies in the Alkaloid Field. IV. The Sodium Dithionite Reduction of 1-[2-(3-Indolyl)-ethyl]-3-methoxy-carbonyl Pyridinium Bromides. Tetrahedron **33**, 113 (1977).

512. BOMBARDELLI, E., A. BONATI, B. GABETTA, E. MARTINELLI, G. MUSTICH, and B. DANIELI: 17-*O*-Acetyl-19,20-dihydrovoachalotine, a New Alkaloid from *Voacanga chalotiana*. Phytochemistry **15**, 2021 (1976).

513. POUTA, C., A. AHOND, and T. SÉVENET: Alcaloides de *Acacia simplicifolia*. Phytochemistry **15**, 2019 (1976).

514. MELCHIO, J., A. BOUQUET, M. PAIS, and R. GOUTAREL: Alcaloïdes Indoliques. CVI. Identité de la Mayumbine et de l'*epi*-19 Ajmalicine. L'iso-3 Rauniticine, un Nouvel Alcaloïde Extrait du *Corynanthe mayumbensis* (R. Good) N. Hallé. Tetrahedron Letters 315 (1977).

515. SANTAMARIA, J., D. HERLEM, and F. KHUONG-HUU: Oxydation Photochimique d'Amines Tertiaires et d'Alcaloïdes — VIII. Oxydation Photochimique d'Alcaloides Indolinoin-dolizidiniques. Vincadifformine et *N*(a)-Acétyl-dihydro-2,16-tabersonine. Tetrahedron **33,** 2389 (1977).

516. Morita, Y., M. Hesse, H. Schmid, A. Banerji, J. Banerji, A. Chatterjee, and W. E. Oberhänsli: *Alstonia scholaris;* Struktur des Indolalkaloides Narelin. Helv. Chim. Acta **60,** 1419 (1977).

517. Das, B. C., J.-P. Cosson, and G. Lukacs: Structural Analysis by Carbon-13 Nuclear Magnetic Resonance Spectroscopy of Pleiocraline, a New Bisindole Alkaloid from *Alstonia dephanchei* van Heurck et Muell. Arg. J. Org. Chem. **42,** 2785 (1977).

518. Wagner, H., R. Brüning, H. Lotter, and A. Jones: Die Struktur von Cassinin, einem neuen Sesquiterpenalkaloid aus *Cassia metabolica* Loes. Tetrahedron Letters 125 (1977).

519. Pelletier, S. W., N. V. Mody, and N. Katsui: The Structures of Sachaconitine and Isodelphinine from *Aconitum miyabei* Nakei. Tetrahedron Letters 4027 (1977).

520. Pelletier, S. W., N. V. Mody, and H. S. Pur: Structures of Falaconitine and Mithaconitine. Two Novel Diterpenoid Alkaloids from *Aconitum falconeri* Stapf. J. C. S. Chem. Commun. 12 (1977).

521. Pelletier, S. W., and N. V. Mody: The Conformational Analysis of the E and F Rings of Atisine, Veatchine and Related Alkaloids. The Existence of C-20 Epimers. J. Am. Chem. Soc. **99,** 284 (1977).

522. — — The C-20 Epimers of Atisine and Behaviour of the Oxazolidine Ring. Tetrahedron Letters 1477 (1977).

523. Buzas, A., R. Cavier, F. Cossais, J.-P. Finet, J.-P. Jacquet, G. Lavielle, and N. Platzer: Synthese et Propriétés Amoebicides d'Analogues de l'Emétine. Analyse des Composés Nouveaux en ^{13}C-RMN. I. (±) Alkyl-1-déséthyl-3-émétine à Jonction B/C *cis* ou *trans.* Helv. Chim. Acta **60,** 2122 (1977).

524. Ahond, A., F. Picot, P. Potier, C. Poupat, and T. Sévenet: Alcaloïdes de *Melicope leratii.* Phytochemistry **17,** 166 (1978).

525. Stermitz, F. R., and I. A. Sharifi: Alkaloids of *Zanthoxylum monophyllum* and *Z. punctatum.* Phytochemistry **16,** 2003 (1977).

526. Zalkow, L. H., L. Gelbaum, and E. Keinan: Isolation of the Pyrrolizidine Alkaloid Europine *N*-oxide from *Heliotropium maris-mortui* and *H. rotundifolium.* Phytochemistry **17,** 172 (1978).

527. Joshi, B. S., N. Viswanathan, D. H. Gawad, V. Balakrishnan, and W. von Philipsborn: Piperaceae Alkaloids. Part IV. Structure and Synthesis of Cyclostachine A, Cyclostachine B and Cyclopiperstachine. Helv. Chim. Acta **58,** 2295 (1975).

528. Davidar, A. M., F. Winternitz, and S. R. Johns: Structure of Ipomine, a New Alkaloid from *Ipomoea muricata* Jacq. Tetrahedron **33,** 1733 (1977).

529. LaLonde, R. T., N. Muhammad, and C. F. Wong: A Stereocontrolled Synthesis of (±) Anhydronupharamine. The ^{1}H and ^{13}C Nuclear Magnetic Resonance of Piperidine Nuphar Alkaloids. J. Org. Chem. **42,** 2113 (1977).

530. Christofidis, I., A. Welter, and J. Jadot: Spectaline and Iso-6 Cassine. Two New Piperidin-3-ol Alkaloids from the Leaves of *Cassia spectabilis.* Tetrahedron **33,** 977 (1977).

531. Arpin, N., J. Favre-Bonvin, and S. Thivend: Structure de la Mycosorine 2. Nouvelle Molecule, Isolée de *Botrytis cinerea.* Tetrahedron Letters 819 (1977).

Purines, Flavonoids, Quinones, Carbohydrates, etc.

532. Coxon, B., A. J. Fatiadi, L. T. Sniegoski, H. S. Hertz, and R. Schaffer: A Novel Acylative Degradation of Uric Acid. Carbon-13 Nuclear Magnetic Resonance Studies of Uric Acid and its Degradation Products. J. Org. Chem. **42,** 3132 (1977).

533. KASAI, H., K. NAKANISHI, K. FRENKEL, and D. GRUNBERGER: Structures of 7,12-Dimethylbenz[*a*]-anthracene 5,6-Oxide Derivatives Linked to the Ribose Moiety of Guanosine. J. Am. Chem. Soc. **99,** 8500 (1977).

534. BAJURA, G. S., and W. G. BENTRUDE: Thymidine Nucleoside 3′,5′-Cyclic Phosphoramidites and Phosphites. Configuration at Phosphorus in Trivalent and Pentavent Cyclic Nucleotides by ^{31}P and ^{13}C NMR. Tetrahedron Letters 421 (1978).

535. WEGNER, M. M., and H. RAPOPORT: Perhydrogenation of 2,8-Diaminopurin. J. Org. Chem. **42,** 3065 (1977).

536. CHANG, C., H. G. FLOSS, and W. STECK: Carbon-13 Magnetic Resonance Spectroscopy of Coumarins. Carbon-13-Proton Long-Range Couplings. J. Org. Chem. **42,** 1337 (1977).

537. CHAN, K. K., D. D. GIANNINI, A. H. CAIN, J. D. ROBERTS, W. PORTER, and W. F. TRAGER: Carbon-13 Nuclear Magnetic Resonance Studies of Coumarin and Related Compounds. Tetrahedron **33,** 899 (1977).

538. SENDA, Y., J. ISHIYAMA, S. IMAIZUMI, and K. HANAYA: Carbon-13 Nuclear Magnetic Resonance Spectroscopy of 1-Tetralols and Chroman-4-ols. J. C. S. Perkin I, 217 (1977).

539. PELTER, A., R. S. WARD, and T. I. GRAY: The Carbon-13 Nuclear Magnetic Resonance Spectra of Flavonoids and Related Compounds. J. C. S. Perkin I, 2475 (1976).

540. WENKERT, E., and H. E. GOTTLIEB: Carbon-13 Nuclear Magnetic Resonance Spectroscopy of Flavonoid and Isoflavonoid Compounds. Phytochemistry **16,** 1811 (1977).

541. CHARI, V. M., M. ILYAS, H. WAGNER, A. NESZMÉLYI, F. CHEN, L. CHEN, Y. LIN, and Y. LIN: ^{13}C-NMR Spectroscopy of Biflavanoids. Phytochemistry **16,** 1273 (1977).

542. DESALBRES, X., M. DUTEIL, J. Y. LALLEMAND, W. DEGROOT-PFLEIDERER, and C. VERALOZA: Carbon-13 N. m. r. Studies of Flavanone 5,3′,4′-Trihydroxy-7β-Glucoside from *Juelia subterranea* (Balanophoraceae). Org. Magn. Resonance **9,** 659 (1977).

543. LALLEMAND, J. Y., and M. DUTEIL: ^{13}C N. m. r. Spectra of Quercetin and Rutin. Org. Magn. Resonance **9,** 179 (1977).

544. CASTELÃO, J. F., JR., O. R. GOTTLIEB, R. A. DELIMA, A. A. L. MESQUITA, H. E. GOTTLIEB, and E. WENKERT: Xanthonolignoids from *Kielmeyera* and Caraipa Species – ^{13}C NMR Spectroscopy of Xanthones. Phytochemistry **16,** 735 (1977).

545. IDRIS, M. S. H., A. JEFFERSON, and F. SCHEINMANN: Extractive from Guttiferae. Part 33. Synthesis of the Ozonolysis Product from Dimethylmangostin, 1-Hydroxy-3,6,7-trimethoxy-2,8-bis-(2-oxo ethyl)-xanthone; some ^{13}C Nuclear Magnetic Resonance Spectra of Xanthones. J. C. S. Perkin I, 2158 (1977).

546. WESTERMAN, P. W., S. P. GUNASEKERA, M. UVAIS, S. SULTANBAWA, and R. KAZLAUSKAS: Carbon-13 N. m. r. Study of Naturally Occurring Xanthones. Org. Magn. Resonance **9,** 631 (1977).

547. CHALMERS, A. A., G. J. H. RALL, and M. E. OBERHOLZER: ^{13}C NMR of Pterocarpans. Tetrahedron **33,** 1735 (1977).

548. PLATTNER, R. D., G. F. SPENCER, D. WEISLEDER, and R. KLEIMAN: Chromanone Acids in *Calophyllum brasiliense* Seed Oil. Phytochemistry **13,** 2597 (1974).

549. SCHILLING, E., K. WEINGES, D. MÜLLER, and W. MAYER: ^{13}C-NMR-Spektroskopische Konstitutionsermittlung der $C_{30}H_{24}O_{12}$-Procyanidine. Liebigs Ann. Chem. 1471 (1973).

550. POMILIO, A., B. ELLMANN, K. KÜNSTLER, G. SCHILLING, and K. WEINGES: Naturstoffe aus Arzneipflanzen. XXI. ^{13}C-NMR-spektroskopische Untersuchungen an Flavanoiden. Liebigs Ann. Chem. 588 (1977).

551. POMILIO, A., O. MÜLLER, G. SCHILLING, and K. WEINGES: Zur Kenntnis der Proanthocyanidine. XXII. Über die Konstitution der Kondensationsprodukte von Phenolen mit Flavyliumsalzen. Liebigs Ann. Chem. 597 (1977).

552. PELTER, A., R. HÄNSEL, and M. KALOGO: The Structure of Silychristin. Tetrahedron Letters 4547 (1977).

553. WAGNER, H., V. M. CHARI, M. SEITZ, and I. RIESS-MAURER: The Structure of Silychristin — A ^{13}C-NMR Study. Tetrahedron Letters 381 (1978).
554. DOMINGUEZ, X. A., C. MARTINEZ, A. CALERO, X. A. DOMINGUEZ, JR., M. HINOJOSA, and A. ZAMUDIO: Louisfieserone, an Unusual Flavanone Derivative from *Indigofera suffruticosa*, Mill. Tetrahedron Letters 429 (1978).
555. CHARI, V. M., N. JORDAN, H. WAGNER, and P. W. THIES: A ^{13}C-NMR Study of the Structure of an Acyl-linarin from *Valeriana wallichii*. Phytochemistry **16,** 1110 (1977).
556. BEUGELMANS, R., and C. MORIN: Action of Hydroxylamine on Chromone and Khellin. Oxime vs. Isoxazoles Structures. J. Org. Chem. **42,** 1356 (1977).
557. CHOPIN, J., M. L. BOUILLANT, A. G. R. NAIR, P. RAMESH, and T. J. MABRY: New *C*-Glycosylflavones from *Mollugo distica* T. Phytochemistry **17,** 299 (1978).
558. JENSEN, S. R., B. J. NIELSEN, and V. NORN: Dihydrochalcones from *Viburnum davidii* and *V. lantanoides*. Phytochemistry **16,** 2036 (1977).
559. KALDAS, M., I. MIURA, and K. HOSTETTMANN: Campestroside, a New Tetrahydro-xanthone Glucoside from *Gentiana campestris*. Phytochemistry **17,** 295 (1978).
560. MCDONALD, I. A., T. J. SIMPSON, and A. F. SIERAKOWSKI: ^{13}C NMR Spectral Studies of Some Naturally Occurring Quinones and Related Compounds. Aust. J. Chem. **30,** 1727 (1977).
561. HÖFLE, G.: ^{13}C-NMR-Spektroskopie Chinoider Verbindungen. II. Substituierte 1,4-Naphthochinone und Anthrachinone. Tetrahedron **33,** 1936 (1977).
562. HUNECK, S., W. STEGLICH, and G. HÖFLE: Canarion, ein neues Naphthochinon aus *Usnea canariensis*. Phytochemistry **16,** 121 (1977).
563. STIPANOVIC, R. D., A. A. BELL, D. H. O'BRIEN, and M. J. LUKEFAHR: Heliocide H_2: An Insecticidal Sesterterpenoid from Cotton *(Gossypium)*. Tetrahedron Letters 567 (1977).
564. NEVILLE, G. A., F. B. HASAN, and I. C. P. SMITH: Stereoselective Epimerization of Pilocarpine in Aqueous Solution as Determined by ^{13}C Nuclear Magnetic Resonance Spectroscopy. Can. J. Chem. **54,** 2094 (1976).
565. CRAIGIE, J. S., A. G. MCINNES, M. A. RAGAN, and J. A. WALTER: Chemical Constituents of the Physodes of Brown Algae. Characterization by ^{1}H and ^{13}C Nuclear Magnetic Resonance Spectroscopy of Oligomers of Phloroglucinol from *Fucus vesiculosus* (L). Can. J. Chem. **55,** 1575 (1977).
566. LOUNASMAA, M.: Dérivés Phloroglucinoliques d'*Hagenia abyssinica*. IV. Résonance Magnétique Nucléaire du ^{13}C de la Kasotoxine, du Pseudo-aspidinol et de l'α-Kosine. Acta Chem. Scand **B 31,** 77 (1977).
567. KELLEY, C. J., R. C. HARRUFF, and M. CARMACK: The Polyphenolic Acids of *Lithospermum ruderale*. II. Carbon-13 Nuclear Magnetic Resonance of Lithospermic and Rosmarinic Acid. J. Org. Chem. **41,** 449 (1976).
568. ARNONE, A., L. CAMARDA, L. MERLINI, and G. NASINI: ^{13}C Nuclear Magnetic Resonance Spectral Analysis of Santalin and Santarubin Permethyl Ethers. J. C. S. Perkin I, 2118 (1977).
569. HARRUFF, R. C., and W. T. JENKINS: A ^{13}C n. m. r. Study of the B_6 Vitamins and their Aldimine Derivatives. Org. Magn. Resonance **8,** 548 (1976).
570. GRANDE, H. J., R. GAST, C. G. VAN SCHLAGEN, W. J. H. VAN BERKEL, and F. MÜLLER: ^{13}C-NMR Study on Isoalloxazine and Alloxazine Derivatives. Helv. Chim. Acta **60,** 367 (1977).
571. MACDONALD, J. C., G. G. BISHOP, and M. MAZUREK: ^{13}C and Proton NMR Spectra of 2 (1H)Pyrazinones. Tetrahedron **32,** 655 (1976).
572. GLOMBITZA, K.-W., H.-U. RÖSENER, and M. KOCH: Polyhydroxyoligophenyle und Phenyläther aus *Bifurcaria bifurcata*. Phytochemistry **15,** 1279 (1976).
573. ANJANEYULU, A. S. R., A. M. RAO, V. K. RAO, L. R. ROW, A. PELTER, and R. S. WARD: The Isolation and Structure of 6″-Bromo-Isoarboreol. The First Bromine Containing Lignan. Tetrahedron Letters 4697 (1975).

574. ANJANEYULU, A. S. R., P. A. RAMAIAH, L. R. ROW, A. PELTER, and R. S. WARD: The Stucture of Wodeshiol. The First of a New Series of Lignans. Tetrahedron Letters 2961 (1975).

575. PELTER, A., R. S. WARD, E. V. RAO, and K. U. SASTRY: Revised Structures for Pluviatilol, Methyl Pluviatilol and Xanthoxylol. General Methods for the Assignment of Stereochemistry to 2,6-Diaryl-3,7-dioxabicyclo[3,3,0]octane Lignans. Tetrahedron **32,** 2783 (1976).

576. PELTER, A., R. S. WARD, and C. NISHINO: Revised Structures for Epiaschantin and Epimagnolin. Tetrahedron Letters 4137 (1977).

577. ANJANEYULU, A. S. R., A. M. RAO, V. K. RAO, L. R. ROW, A. PELTER, and R. S. WARD: Novel Hydroxy Lignans from the Heartwood of *Gmelina arborea.* Tetrahedron **33,** 133 (1977).

578. BAILEY, J. A., R. S. BURDEN, A. MYNETT, and C. BROWN: Metabolism of Phaseollin by *Septoria nodorum* and other Non-pathogens of *Phaseolus vulgaris.* Phytochemistry **16,** 1541 (1977).

579. LANGCAKE, P., and R. J. PRYCE: Oxidative Dimerisation of 4-Hydroxystilbenes *in vitro;* Production of a Grapevine Phytoalexin Mimic. J. C. S. Chem. Commun. 208 (1977).

580. KINNEL, R., A. J. DUGGAN, T. EISNER, and J. MEINWALD: Panacene: An Aromatic Bromoallene from a Sea Hare *(Aplysia brasiliana).* Tetrahedron Letters 3913 (1977).

581. NICOLLIER, G., and R. TABACCHI: Isolement et Identification de l'évernine dans la „Mousse de chêne" *(Evernia prunastri* (L.) Ach.). Helv. Chim. Acta **59,** 2979 (1976).

582. ITO, S., and Y. HIRATA: Isolation and Structure of a Mycosporine from the Zoanthid *palythoa* tuberculosa. Tetrahedron Letters 2429 (1977).

583. MUHLHEIRN, L. J., R. B. BEECHEY, D. P. LEWORTHY, and M. D. OSSELTON: Aurovertin B, a Metabolite of *Calcarisporium arbuscula.* J. C. S. Chem. Commun. 874 (1974).

584. CHEN, P. N., D. G. I. KINGSTON, and J. R. VERCELLOTTI: Reduction of Sterigmatocystin and Versicolorin A Hemiacetals with Sodium Borohydride. J. Org. Chem. **42,** 3599 (1977).

585. VOLKMANN, R. A., P. D. WEEKS, D. E. KUHLA, E. B. WHIPPLE, and G. N. CHMURNY: Reaction of Kojic Acid and Its Derivatives with Acrylonitrile. A New Look at an Old Problem. J. Org. Chem. **42,** 3976 (1977).

586. NIWA, M., M. IGUCHI, and S. YAMAMURA: The Isolation and Structure of Obtusilactone. Tetrahedron Letters 1539 (1975).

587. PIESTAP, H. A., J. O. BONAFEDE, and E. A. RUVEDA: Argentilactone, a Novel 5-Hydroxyacid Lactone from *Aristolochia argentina.* Phytochemistry **16,** 1579 (1977).

588. ENGSTROM, G. W., J. V. DELANCE, J. L. RICHARD, and A. L. BAETZ: Purification and Characterization of Roseotoxin B, a Toxic Cyclodepsipeptide from *Trichothecium roseum.* J. Agric. Food Chem. **23,** 244 (1975).

589. MARNER, F.-J., R. E. MOORE, K. HIROTSU, and J. CLARDY: Majusculamides A and B, Two Epimeric Lipodipeptides from *Lyngbya majuscula* Gomont. J. Org. Chem. **42,** 2815 (1977).

590. BERGER, S.: Vitamin C. A ^{13}C Magnetic Resonance Study. Tetrahedron **33,** 1587 (1977).

591. PERLIN, A. S.: Carbon-13 N.M.R. Spectroscopy of Carbohydrates. Int. Rev. Sci. Org. Chem. **7,** 1 (1976).

592. BURFITT, A. I. R., R. D. GUTHRIE, and R. W. IRVINE: A ^{13}C Nuclear Magnetic Resonance Study of Glycals (1,5-Anhydro-hox-1-enitols). Aust. J. Chem. **30,** 1037 (1977).

593. YAMASAKI, K., R. KASAI, Y. MASAKI, M. OKIHARA, O. TANAKA, H. OSHIO, S. TAKAGI, M. YAMAKI, K. MASUDA, E. NONAKI, M. TSUBOI, and I. NISHIOKA: Application of

C-13 NMR to the Structural Elucidation of Acylated Plant Glycosides. Tetrahedron Letters 1231 (1977).

Antibiotics

594. DOLAK, L.: The C-13 NMR Spectrum of Novobiocin. J. Antibiot. **29,** 710 (1976).
595. SETO, H., Y. MIYAZAKI, K. FUJITA, and N. ŌTAKE: Studies on the Ionophorous Antibiotics. X. The Assignment of ^{13}C-NMR Spectrum of Salinomycin. Tetrahedron Letters 2417 (1977).
596. SETO, H., T. YAHAGI, Y. MIYAZAKI, and N. ŌTAKE: Utilization of Carbon-13-Carbon-13 Coupling in Structural and Biosynthetic Studies. IX. Studies on the Ionophorous Antibiotics. IX. The Structure of 4-Methylsalinomycin (Narasin). J. Antibiot. **30,** 530 (1977).
597. CHAIN, E. B., and G. MELLOWS: Pseudomonic Acid. Part 3. Structure of Pseudomonic Acid B. J. C. S. Perkin I, 318 (1977).
598. CONNOR, D. T., R. C. GREENOUGH, and M. VON STRANDTMANN: W-7783, a Unique Antifungal Antibiotic. J. Org. Chem. **42,** 3664 (1977).
599. TROFAST, J., and B. WICKBERG: Mycorrhizin A and Chloromycorrhizin A, Two Antibiotics from a Mycorrhizal Fungus of *Monotropa hypopitys* L. Tetrahedron **33,** 875 (1977).
600. WILEY, P. F., R. B. KELLY, E. L. CARON, V. H. WILEY, J. H. JOHNSON, F. A. MACKELLAR, and S. A. MIZSAK: Structure of Nogalamycin. J. Am. Chem. Soc. **99,** 542 (1977).
601. KELLY, R. C., I. SCHLETTER, J. M. KOERT, K. A. MACKELLAR, and P. F. WILEY: Structures of Steffimycin and Steffimycin B. J. Org. Chem. **42,** 3591 (1977).
602. WRIGHT, L. H., J. A. CHAN, J. A. SCHROER, and A. A. ASZALES: A ^{13}C Nuclear Magnetic Resonance Study of N-Acetyldaunorubicinol. J. Org. Chem. **42,** 2344 (1977).
603. VUILHORGNE, M., S. ENNIFAR, B. C. DAS, J. W. PASCHAL, R. NAGARAJAN, E. W. HAGAMAN, and E. WENKERT: Structure Analysis of the Nucleoide Disaccharide Antibiotic Anthelmycin by Carbon-13 Nuclear Magnetic Resonance Spectroscopy. A Structural Revision of Hikizimycin and Its Identity with Anthelmycin. J. Org. Chem. **42,** 3289 (1977).
604. MARTIN, J. R., R. S. EGAN, A. W. GOLDSTEIN, R. S. STANASZEK, J. TADANIER, and W. KELLER-SCHIERLEIN: Minor Lankamycin — Related Antibiotics from *Streptomyces violaceoniger*. Helv. Chim. Acta **60,** 2559 (1977).
605. CORBETT, D. F., A. J. EGLINGTON, and T. T. HOWARTH: Structure Elucidation of MM 17880, a New Fused β-Lactam Antibiotic Isolated from *Streptomyces olivaceus:* a Mild β-Lactam Degradation Reaction. J. C. S. Chem. Commun. 953 (1977).
606. WILSON, W. L., H. W. AVDOVICH, D. W. HUGHES, and G. W. BUCHANAN: Application of Proton and ^{13}C-NMR Spectroscopy to Estimation of Diastereoisomer Ratio in Phenethicillin. J. Pharm. Sci. **66,** 1079 (1977).
607. BROWN, A. G., D. F. CORBETT, A. J. EGLINGTON, and T. T. HOWARTH: Structures of Olivanic Acid Derivatives MM 4550 and MM 13902; Two New Fused β-Lactams Isolated from *Streptomyces olivaceus*. J. C. S. Chem. Commun. 523 (1977).
608. DEPAIRE, H., J.-P. THOMAS, A. BRUN, and G. LUKACS: Acid and Alkaline Hydrolysis of the Antibiotic Nosiheptide. Tetrahedron Letters 1395 (1977).
609. WALKER, J., A. OLESKER, L. VALENTE, R. RABANAL, and G. LUKACS: Total Structure of the Polythiazole Containing Antibiotic Micrococcin P. The ^{13}C Nuclear Magnetic Resonance Study. J. C. S. Chem. Commun. 706 (1977).
610. PANDEY, R. C., J. C. COOK, JR., and K. L. RINEHART, JR.: High Resolution and Field Desorption Mass Spectrometry Studies and Revised Structures of Alamethicins I and II. J. Am. Chem. Soc. **99,** 8469 (1977).

611. BYCROFT, B. W.: Configurational and Conformational Studies on the Group A Peptide Antibiotics of the Mikamycin (Streptogramin, Virginiamycin) Family. J. C. S. Perkin I, 2464 (1977).
612. HASSALL, C. H., W. A. THOMAS, and M. C. MOSCHIDIS: Amino-acids and Peptides Part 19. Conformational Studies of the Monamycins, a Family of Cyclohexadepsipeptide Antibiotics. J. C. S. Perkin I, 2369 (1977).
613. MCGAHREN, W. J. G. O. MORTON, M. P. KUNSTMANN, and G. A. ELLESTAD: Carbon-13 Nuclear Magnetic Resonance Studies on a New Antitubercular Peptide Antibiotic LL-BM 547 β. J. Org. Chem. **42,** 1282 (1977).
614. MONDELLI, R., and P. VENTURA: ^{13}C Nuclear Magnetic Resonance of *N*-Heterocycles. Part 3. ^{13}C Chemical Shift Assignments of the Carbonyl Groups in Penicillins and Cephalosphorins. J. C. S. Perkin II, 1749 (1977).

Biosynthetic Studies

615. POPJAK, G., J. EDMOND, F. A. L. ANET, and N. R. EASTON, JR.: Carbon-13 NMR Studies on Cholesterol Biosynthesized from [^{13}C] Mevalonates. J. Am. Chem. Soc. **99,** 931 (1977).
616. BRADSHAW, A. P. W., J. R. HANSON, and M. SIVERNS: Use of Induced ^{13}C-^{13}C Coupling in Terpenoid Biosynthesis. J. C. S. Chem. Commun. 819 (1977).
617. KIMURA, Y., M. GOHBARA, and A. SUZUKI: Assignment of ^{13}C-NMR Spectrum and Biosynthesis of Colletotrichin. Tetrahedron Letters 4615 (1977).
618. DE ROSA, M., S. DE ROSA, and A. GAMBACORTA: ^{13}C-NMR Assignments and Biosynthetic Data for the Ether Lipids of *Caldariella.* Phytochemistry **16,** 1909 (1977).
619. SETO, H., T. SASAKI, and H. YONEHARA: Studies on the Biosynthesis of Radiclonic Acid. Tetrahedron Letters 4083 (1977).
620. YOSHIDA, S., S. SHIRAISHI, K. FUJITA, and N. TAKAHASHI: Biosynthetic Studies on Piericidin A and Its Structural Revision. Tetrahedron Letters 1863 (1975).
621. GARSON, M. J., R. A. HILL, and J. STAUNTON: Deuterium as a Tracer in Polyketide Biosynthesis: Incorporation of [2-^{13}C,2-$^{2}H_3$]Acetate into Terrein. J. C. S. Chem. Commun. 624 (1977).
622. EVANS, G. E., and J. STAUNTON: Biosynthesis of Citromycetin: Incorporation of [1-^{13}C]-, [2-^{13}C]-, and (1,2-$^{13}C_2$] Acetates. J. C. S. Chem. Commun. 760 (1976).
623. FELINE, T. C., R. B. JONES, G. MELLOWS, and L. PHILLIPS: Pseudomonic Acid. Part 2. Biosynthesis of Pseudomonic Acid A. J. C. S. Perkin I, 309 (1977).
624. CANHAM, P. L., L. J. VINING, A. G. MCINNES, J. A. WALTER, and J. L. C. WRIGHT: Use of ^{13}C in Biosynthetic Studies. Incorporation of ^{13}C-labelled Acetate into Chartreusin by *Streptomyces chartreusis.* Can. J. Chem. **55,** 2450 (1977).
625. HABER, A., R. D. JOHNSON, and K. L. RINEHART, JR.: Biosynthetic Origin of the C_2 Units of Geldanamycin and Distribution of Label from D-[6-^{13}C] Glucose. J. Am. Chem. Soc. **99,** 354 (1977).
626. COX, R. H., F. CHURCHILL, R. J. COLE, and J. W. DORNER: Carbon-13 Nuclear Magnetic Resonance Studies of the Structure and Biosynthesis of Versiconal Acetate. J. Am. Chem. Soc. **99,** 3159 (1977).
627. FITZELL, D. L., R. SINGH, D. P. HSIEH, and E. L. MOTELL: Nuclear Magnetic Resonance Identification of Versiconal Hemiacetal Acetate as an Intermediate in Aflatoxin Bioxynthesis. J. Agric. Food Chem. **25,** 1193 (1977).
628. GORST-ALLMAN, C. P., K. G. R. PACHLER, P. S. STEYN, and D. B. SCOTT: Carbon-13 Nuclear Magnetic Resonance Assignments of Some Fungal C_{20} Anthraquinones; their Biosynthesis in Relation to that of Aflatoxin B. J. C. S. Perkin I, 2181 (1977).
629. FITZELL, D. L., D. P. H. HSIEH, R. C. YAO, and G. N. LAMAR: Biosynthesis of Averufin from Acetate by *Aspergillus parasiticus.* J. Agric. Food Chem. **23,** 442 (1975).

630. Simpson, T. J.: ^{13}C Nuclear Magnetic Resonance Spectra and Biosynthetic Studies of Xanthomegnin and Related Pigments from *Aspergillus sulphureus* and *melleus*. J. C. S. Perkin I, 592 (1977).

631. Leete, E.: Biosynthesis of Dioscorine: Incorporation of Nicotinic Acid into the Isoquinuclidine Moiety. Phytochemistry **16**, 1705 (1977).

632. — The Incorporation of [5,6-$^{13}C_2$]-Nicotinic Acid into the Tabacco Alkaloids Examined by the Use of ^{13}C Nuclear Magnetic Resonance. Bioorg. Chem. **6**, 273 (1977).

633. Chang, C., H. G. Floss, L. H. Hurley, and M. Zmijewski: Application of Long-Range Spin-Spin Couplings in Biosynthetic Studies. J. Org. Chem. **41**, 2932 (1976).

634. Nadzan, A. M., and K. L. Rinehart, Jr.: Nybomycin. 9. Synthetic and Biosynthetic Incorporation of ^{15}N as a Means of Assigning the ^{13}C Nuclear Magnetic Resonance Spectrum of Nybomycin. J. Am. Chem. Soc. **99**, 4647 (1977).

635. Cardillo, R., C. Fuganti, D. Ghiringhelli, and P. Graselli: Pattern of Incorporation of Leucine Samples Asymmetrically Labelled with ^{13}C in the Isopropyl Unit into the C_5-Isoprenoid Units of Echinuline and Flavoglaucine. J. C. S. Chem. Commun. 474 (1977).

636. Ahrens, E. H., Jr., D. C. Williams, and A. R. Battersby: Biosynthesis of Porphyrins and Related Macrocycles. Part 11. Studies on Biosynthesis of the Phytyl Chain of Chlorophyll *a* by Use of Carbon-13. J. C. S. Perkin I, 2540 (1977).

637. Scott, A. I., A. J. Irwin, L. M. Siegel, and J. N. Shoolery: Sirohydrochlorin. Prosthetic Group of a Sulfite Reductase Enzyme and Its Role in the Biosynthesis of Vitamin B_{12}. J. Am. Chem. Soc. **100**, 3161 (1978).

(Received March 22, 1978)

The Role of Heteroatomic Substances in the Aroma Compounds of Foodstuffs

By G. OHLOFF and I. FLAMENT, Firmenich SA, Research Laboratories, Geneva, Switzerland

With 1 Figure

Contents

Acknowledgements. The authors are indebted to Drs. A. F. THOMAS, B. MAURER, and W. PICKENHAGEN for corrections of the text, stimulating discussion and valuable suggestions. The partial translation by Mrs. EDITH WENGER is gratefully acknowledged. Miss GERTRUD LINGESLEBEN is responsible for the lay-out of this paper, which she did with great skill and patience.

I. Introduction

Organic compounds containing heteroatoms are very widely distributed among aroma compounds. They mostly arise from odorless and tasteless precursors, either enzymatically or by the action of heat. In the present work, we do not propose to discuss the enzymatic route, which is responsible, for example, for the formation of a large number of sulfur-containing compounds occurring in various vegetables (*79, 160, 262*) but to restrict ourselves principally to a description of the aroma compounds resulting from thermal degradation of higher molecular weight precursors. Although the subject is quite different, we are following the methods adopted in the previous publication about oxygen-containing aliphatic and isoprenoid aroma compounds (*206*), so both articles are complementary to one another.

The aroma fraction of important foodstuffs is both qualitatively and quantitatively dominated by compounds containing heteroatoms. For example, 254 O-, N-, and S-containing heterocyclic compounds were found in coffee flavor alone (*314*). The 310 volatile compounds identified in cocoa by 1975 comprise 60 nitrogen-containing derivatives belonging to 6 different heterocyclic systems (*317*). So far 58 N-containing derivatives, 42 of which are pyrazines, have been detected in roasted filberts (*130*), while 34 aroma compounds of this type were identified in tea (*318*). Among the 2,200 compounds detected so far in tobacco and tobacco smoke, 600 (or 27% of the total) contain nitrogen (*246*). There have been 43 sulfur-containing compounds found in roasted coffee (*262*) and 46 in pressure-cooked beef (*338*). Onion flavor contains as many as 60 compounds of this type (*262*).

Heteroatomic compounds have an extraordinarily high odor strength and are often present in flavors in such minute traces that two decades ago their discovery was hindered because their concentration was lower than the analytical detection limit. Thus it is not surprising that certain thiols and pyrazines belong to those organic compounds which have the lowest detection threshold known today (Table 1). In

Table 1. *Threshold Values of O-, S-, and N-Containing Compounds*

Structure	Chemical Name	T*	Ref.
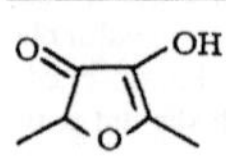	4-hydroxy-2,5-dimethyl-3(2*H*)-furanone (Furaneol®)	0.04	(*226*)
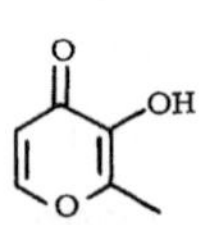	maltol	35 000	(*222*)
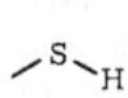	methanethiol	0.02	(*91*)

* T = flavor threshold value in µg (ppb) of water.

Table 1 *(continued)*

Structure	Chemical Name	T*	Ref.
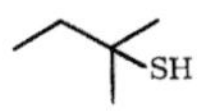	2-methyl-2-butanethiol	0.0007	(*162*)
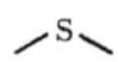	dimethyl sulfide	0.33	(*91*)
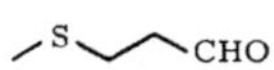	methional	0.2	(*32, 94*)
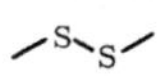	dimethyl disulfide	12	(*33*)
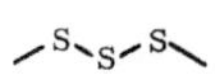	dimethyl trisulfide	0.01	(*33*)
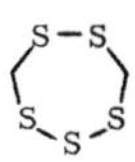	lenthionine	400	(*320*)
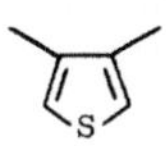	3,4-dimethylthiophene	1.3	(*18*)
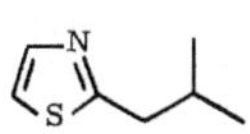	2-isobutylthiazole	2	(*34*)
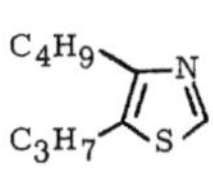	4-butyl-5-propylthiazole	0.003	(*126*)
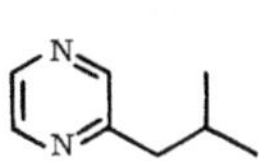	2-isobutylpyrazine	400	(*259*)
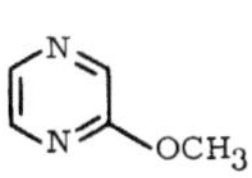	2-methoxypyrazine	700	(*259*)
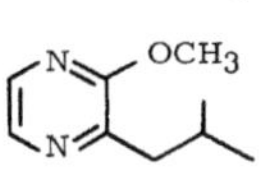	2-isobutyl-3-methoxypyrazine	0.002	(*259*)

* T = flavor threshold value in μg (ppb) of water.

many cases these compounds give a typical sensory impression and contribute to certain flavor types in which they partially constitute character impact compounds (*119*). Table 2 lists a selection of hetero-atom-containing derivatives which are considered to be typical food flavor components.

Table 2. *Character-Impact O-, S-, and N-Containing Compounds in Foods*

Structure	Chemical Name	Occurrence in Food	Ref.
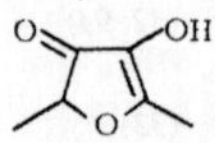	4-hydroxy-2,5-dimethyl-3(2*H*)-furanone (Furaneol®)	strawberries pineapple	(*229, 238, 336*)
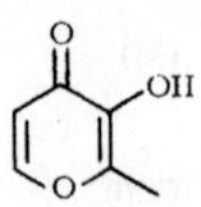	maltol	malt caramel coffee	(*284*)
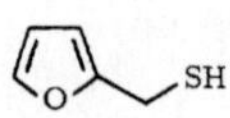	2-furylmethanethiol	coffee	(*6, 232*)
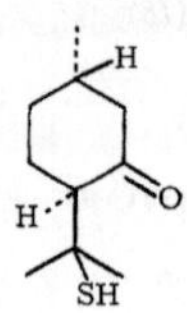	(+)-*trans*-8-mercapto-*p*-menthan-3-one	black currant principle in buchu leaf oil	(*140, 292*)
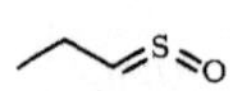	propanethial-*S*-oxide	lachrymatory factor in onions	(*23*)
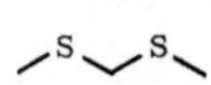	bis(methylthio)methane	white truffle	(*65*)
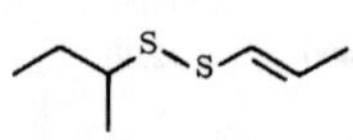	(*E* + *Z*)-(*R*)-2-butyl 1-propenyl disulfide	asafedita	(*191*)
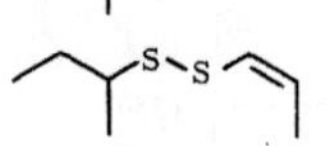			(*133*)
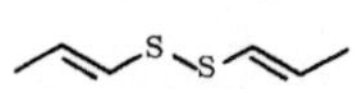	bis(1-propenyl) disulfide	boiled onion	(*18, 24*)
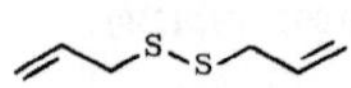	diallyl disulfide	garlic	(*160*)
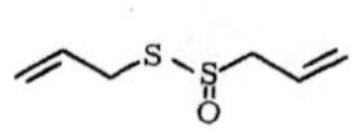	*S*-allyl 2-propene-1-thiosulfinate (allicin)		(*243*)

References, pp. 267—283

Table 2 *(continued)*

Structure	Chemical Name	Occurrence in Food	Ref.
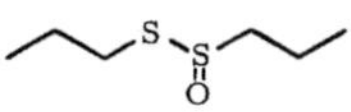	*S*-propyl propane thiosulfinate	fresh onion	(*18*)
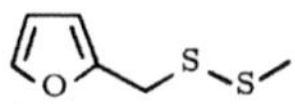	furfuryl methyl disulfide	fresh white bread	(*182*)
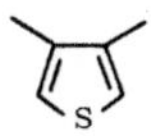	3,4-dimethylthiophene	fried onion	(*18*)
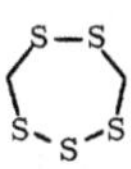	lenthionine	shiitake mushroom	(*173, 320*)
	allyl isothiocyanate	black mustard	(*263*)
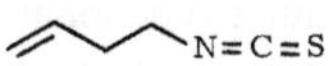	3-butenyl isothiocyanate	brown mustard	(*263*)
	4-pentenyl isothiocyanate	Hungarian horseradish	(*87*)
	4-(methylthio-3-butenyl isothiocyanate	pungent principle in radish	(*81*)
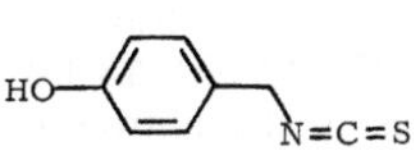	*p*-hydroxybenzyl isothiocyanate	white mustard	(*263*)
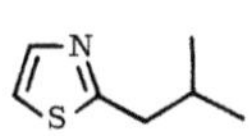	2-isobutylthiazole	green tomato leaves	(*312*)
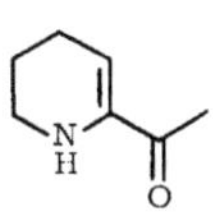	6-acetyl-1,2,3,4-tetrahydropyridine	crackers	(*114, 115*)
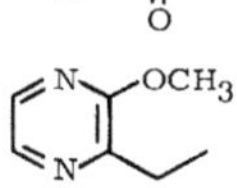	2-ethyl-3-methoxypyrazine	raw potato	(*93, 201*)
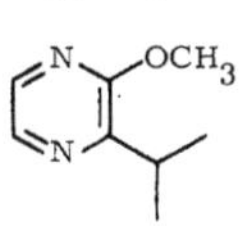	2-isopropyl-3-methoxypyrazine	earthy part of raw potato	(*32, 36, 40*)

Table 2 *(continued)*

Structure	Chemical Name	Occurrence in Food	Ref.
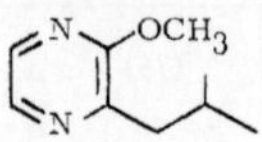	2-isobutyl-3-methoxy-pyrazine	bell peppers	*(38)*
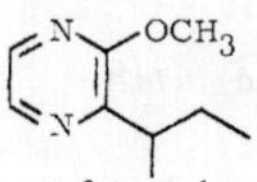 + 3-propyl and 3-isobutyl derivatives	2-sec. butyl-3-methoxy-pyrazine	green pea	*(185)*
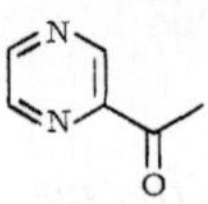	2-acetylpyrazine	popcorn	*(237)*

The structural diversity of this class of aroma compounds is particularly pronounced in the case of aliphatic sulfur compounds. Compounds of this type often have oxygen-containing functional groups and may also, but less frequently, have nitrogen-containing functional groups; a review of this topic was published recently *(72)*. The basic skeletons of the heterocyclic flavors found in foodstuffs up to now are shown in Fig. 1.

Due to their characteristic odor qualities as well as their extraordinarily high flavor values resulting from the low threshold concentration (*cf.* Table 1) the O-, S-, and N-containing aroma compounds are ideal compounding ingredients. In the USA the flavoring substances accepted for human consumption are listed in the FEMA GRAS list (Table 3). The use of their synthetic equivalents as food additives is limited to certain concentrations.

Table 3. *O-, S-, and N-Containing Groups of Food-Related Aroma Compounds on the FEMA GRAS* List*

60 furans	5 trisulfides
1 pyran	1 tetrasulfid
22 thioles	8 thiolesters
14 thiazoles	2 isothiocyanates
3 thiophenes	13 pyridines
23 monosulfides	33 pyrazines
8 disulfides	7 pyrroles

* FEMA = Flavoring Extract Manufacturers' Association
GRAS = Generally Recognized As Safe.

References, pp. 267—283

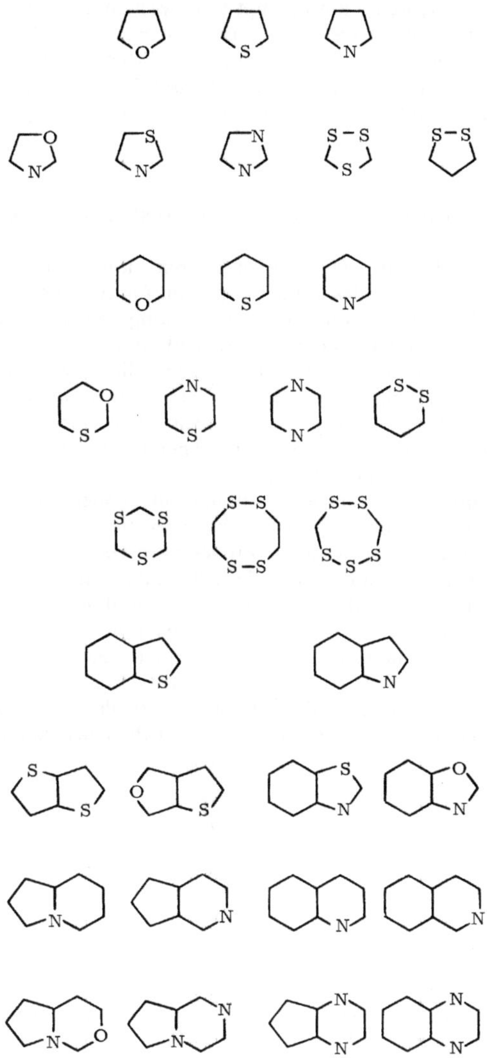

Fig. 1. Basic skeleton of heterocyclic aroma compounds

In this report, we do not attempt a complete literature coverage, but give a selection of interesting aroma substances with defined sensory properties in the hope that our examples will enable organic chemists to obtain some idea about a vital field of natural product chemistry which has undergone explosive development in the last fifteen years.

II. Pyranones, Furanones, and Related Aroma Compounds

"Browned flavors" (*112*) arising from non-enzymatic browning reactions of carbohydrates are developed during food processing mostly by caramelization, toasting, baking, roasting, concentration, or dehydration. The aroma compounds isolated from browned flavor complexes are mainly associated with a caramel-like, sweet, fruity, nutty and burnt odor impression. Several compounds of this kind have proved to be important flavoring chemicals in the food industry.

Structurally, these substances are generally cyclic ethers, furanoid compounds constituting the major part. They predominate both proportionally and in absolute amounts in condensates from carbohydrates that are subjected to browning reactions (*105, 106, 110*). In coffee aroma alone 70 furan derivatives have been found (*314*), while the volatile part of bread contains 25 (*149*) and caramel 37 (*68*) representatives of this class. Of the pyranoid and furanoid aroma substances, let us examine the group in which the oxygen functions on the ring are stabilized by intramolecular hydrogen bond formation.

It was found that a relationship exists between the molecular structure and the odor quality of these particular aroma compounds (*110, 112*). The common molecular feature of these heated food odorants

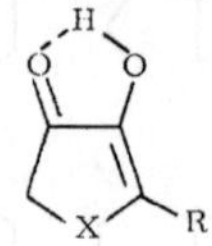

(A)

is a planar enol-carbonyl configuration of alkyl substituted five- or six-membered cyclic α-diketones, stabilized by hydrogen bonding and illustrated by the idealized model compound **A**. Small structural modifications of this model **A** can alter the odor within a restricted number of qualities, but on the other hand, they can cause a great change in potency.

References, pp. 267—283

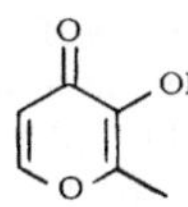
(1)

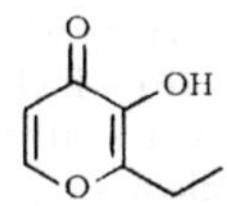
(2)

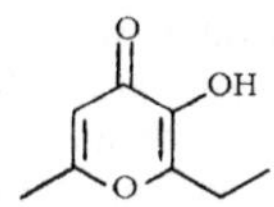
(3)

The pyranone derivative maltol (**1**), which smells like caramel, already discovered in 1894 as one of the character impact compounds in malt (*21, 217*) and identified since as an important flavor component in various processed foodstuffs (*284*), was the very first representative of this particular group of aroma compounds. 2-Ethyl-3-hydroxy-4(4*H*)-pyranone [=ethyl maltol, (**2**)], not yet found in nature but accepted for flavoring foodstuffs (*220*), causes an odor impression which is four to six times stronger than that of its lower homolog (**1**) (*244*). 2-Ethyl-3-hydroxy-6-methyl-4(4*H*)-pyranone (**3**), a methyl homolog of (**2**) which was isolated from refinery molasses, has an odor of preserved tang boiled down in soy sauce (*116*). Maltol (**1**) was found to be a sweetness synergist, 95 parts of sugar plus 0.015 part of (**1**) having the same sweetness as 103 parts of sugar (*19*) without imparting the taste of maltol. On the other hand, (**1**) and (**2**) are said to be capable of masking natural bitter principles such as those of wormwood, gentian, hop and cola (*210*).

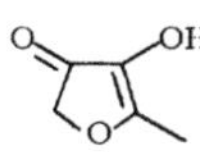
(4)

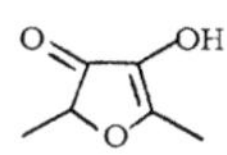
(5)

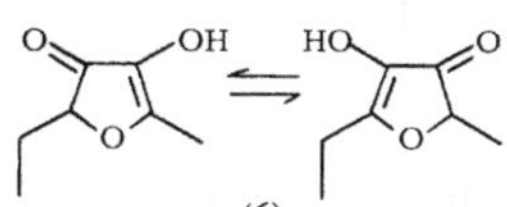
(6)

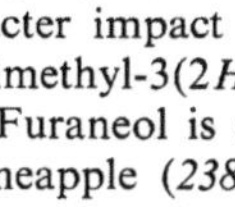
(7)

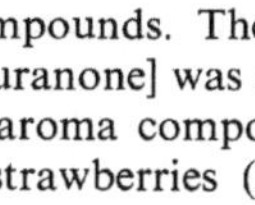
(8)

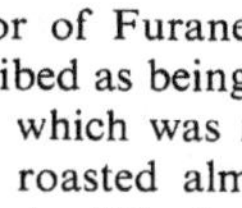
(9)

The 3(2*H*)-furanones (**4**) to (**6**) which have marked caramel-like odor properties were found to be present in various foodstuffs as character impact compounds. The odor of Furaneol® (**5**) [4-hydroxy-2,5-dimethyl-3(2*H*)-furanone] was described as being "burnt pineapple"-like. Furaneol is an aroma component which was found to be present in pineapple (*238*), strawberries (*229*), roasted almonds (*301*), and in the product mixture obtained by pyrolysis of **D**-glucose (*63, 106*). 4-Hydroxy-5-methyl-3(2*H*)-furanone (**4**), which is characterized by a roasted chicory root odor, was isolated together with (**5**) from beef broth (*398*). The homologous 3(2*H*)-furanone (**6**) which exists in tautomeric forms and which first became known as a synthetic product (*229*) was later found to be an important ingredient of soy sauce (*200*). The sensory impression of (**6**), which is stronger than that of (**5**) (*229*), was described as sweet and reminiscent of shortcake. 3-Hydroxy-4,5-dimethyl-2(5*H*)-furanone (**7**) following the molecular features of those browned flavoring compounds already mentioned imparts its characteristic and dominant burnt aroma to aged sake (*297*). The typical seasoning flavor of hydrolyzed vegetable proteins has been attributed to its homolog (**8**) (*291*). 5-Ethyl-3-hydroxy-4-methyl-2(5*H*)-furanone (**9**) (*14, 56*), which had not yet been found in foodstuffs, was described as having an intense caramel and curry-like odor (*14*). The addition of (**9**) to foods and beverages allows a fairly accurate reproduction of the flavor and taste of maple syrup (*57*).

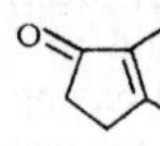

(10)

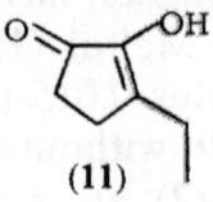

(11)

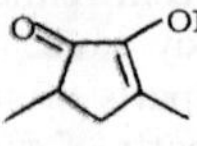

(12)

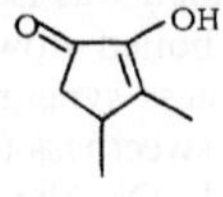

(13)

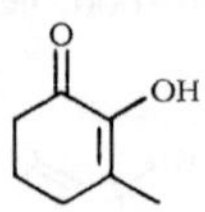

(14)

The five carbon analogs of furaneol and maltol (**10**)—(**14**) found among coffee aroma compounds (*86*) are characterized by a burnt sugar odor impression due to a planar contiguous alkyl-substituted enol-carbonyl function. Structure modification work in combination with a thorough flavor analysis showed that the caramel aroma of these carbo-analogs is closely related to the molecular features of model compound **A** ($X=CH_2$ and CH_2-CH_2). Cyclotene® (**10**), which is also known as "maple lactone" (*7*) and which was originally found in wood distillate (*164*), was also isolated from maple syrup (*66*) and roasted barley (*274*). 3-Ethyl-2-hydroxy-2-cyclopenten-1-one (**11**) is said to possess organoleptic qualities superior to those of (**10**) (*146, 222*) since it imparts a better and more natural flavor to food (*286*). 2-Hydroxy-3-methyl-2-cyclopenten-1-one (**10**) and its homolog (**11**) have been found in cigarette smoke (in quantities of 0.12 and 0.048 mg, respectively per cigarette) and are of some importance for tobacco flavor (*62*). Both compounds (**10**) and (**11**) are well-known flavor ingredients today and are used in caramel, nut, maple and butterscotch flavorings.

References, pp. 267—283

If enolization such as occurs in 3,4(2*H*, 5*H*)-furandione cannot take place (*128*), or if the acidic hydrogen atom of the 4-hydroxy-3(2*H*)-furanones is substituted by alkyl groups, as *e.g.* in 4-methoxy-2,5-dimethyl-3(2*H*)-furanone (**15**) (*205*), the caramel-like odor disappears (*203, 205*).

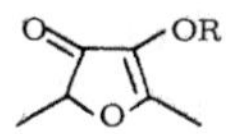

(**15**): R = CH_3
(**16**): R = Ac

The 3(2*H*)-furanone (**15**), first isolated from strawberry aroma (*336*) as a minor compound, was found to be a central metabolite of the aroma distillate of the arctic bramble (32%) (*122*) and a major volatile component of canned mango (*Mangifera indica* L.) (*113*). In spite of its odor which is reminiscent of sherry, (**15**) seems to contribute only little to the mango flavor (*113*). Compound (**16**), which is practically odorless, was found in coffee oil (*69*).

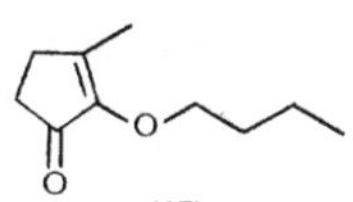

(**17**)

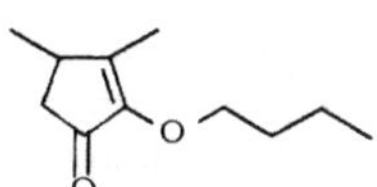

(**18**)

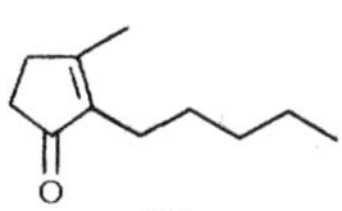

(**19**)

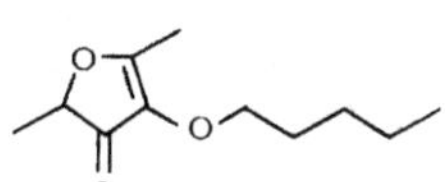

(**20**)

The *n*-butyl ether (**17**) (*339*) has a jasmin-like odor which is considerably more intense than that of the carbon analog dihydrojasmone (**19**). Further intensification of the odor is observed in the methylated derivative (**18**) (*339*) as well as in (**17**) (*83*) and the *n*-amyl ether homolog of (**18**) (*339*). The *n*-amyl ehter of furaneol (**20**) also had jasmin-like odor properties, whereas the *n*-butyl ether corresponding to (**20**) had a weaker and unspecific floral odor. A comparison of compounds (**17**) to (**20**) with regard to their molecular morphology (*3*) shows a certain concordance in the size and shape of the molecules. The correlation between the structure and the organoleptic qualities of the compounds included in this group should constitute an incentive for further investigations in the field of structure-activity relationships.

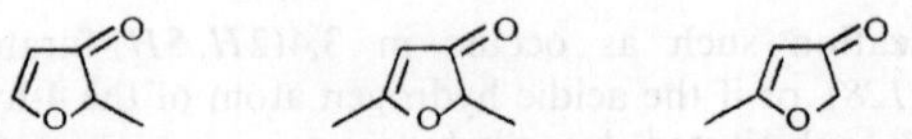

(21) (22) (23)

The group of aroma components which are formed in browning reactions and which are closely related to furaneol includes the 3(2*H*)-furanone derivatives (**21**) to (**23**). These compounds do not fulfill the molecular requirements necessary for inducing a caramel-like sensory impression (model **A**). 2,5-Dimethyl-3(2*H*)-furanone (**22**), which was the first one to be synthesized (*229*), was reported to carry a strong odor of freshly baked bread (*240*). Like 2-methyl-3(2*H*)-furanone (**21**), which was isolated from the aroma of white bread, the aroma of (**22**) is claimed not to be very reminiscent of bread aroma (*180, 183*). The sole character impact compound of bread aroma (*149*) is assumed to be 2-[(methyldithio)-methyl]-furan (**59**) (*182*). Compound (**22**) was also found to be present in coffee (*182*) and in mango flavor (*113*).

Many attempts have been made to show how browned cereal flavors are formed from individual sugar components in model reactions. Thermal decomposition and acid-catalyzed or base-induced reactions of sugars were extensively investigated. On heating, maltose was converted into a mixture containing maltol (**1**) (*54*), and thermal treatment of glucose resulted in the formation of aliphatic aldehydes, ketones, aromatic hydrocarbons and numerous furan derivatives, and, in addition, compounds (**5**), (**10**) and (**21**) (*63, 120*) which characterize the caramel flavor. The 3(2*H*)-furanone derivative (**4**) is formed by thermal decomposition of **D**-xylose (*260*). Mechanisms relating to the thermal formation of (**4**) and (**5**) were postulated (*204*). The alkaline degradation of sucrose (*264*) and especially of fructose yields (**5**) and the five cyclic diketones (**10**) to (**14**) (*264, 265*). Acid-catalyzed dehydration reactions of **D**-fructose and **D**-glucose in aqueous solution were also extensively studied (*64*). 2-Desoxy-**D**-ribose is dehydrated to 5-methyl-3(2*H*)-furanone (**23**) by an easily recognizable process (*261*), whereas (**4**) is formed by β-elimination of phosphoric acid from **D**-ribose-5-phosphate (*215*). Model sugar-amine reactions are interesting for organic chemistry since the formation under biomimetic conditions of aroma compounds smelling like caramel takes place in clearly defined partial steps according to known reaction mechanistic principles (*110, 112*). The conversions known as the Maillard reaction take place *via* the Amadori rearrangement of compounds derived from the condensation of amines or amino acids with reducing sugars resulting in the formation of 1-amino-1-desoxy-2-ketones. These non-volatile Amadori compounds are to be considered as the actual precursors which, on low-temperature thermolysis, are decomposed to form the browned flavor. For example, the reaction of **L**-rhamnose with piperidine acetate yielded the "burnt-pineapple" compound (**5**) (*111, 336*), whereas the action of diethylamine and acetic acid on **D**-(+)-xylose resulted in the formation of nor-derivative (**4**) (*216*). The pyrolysis of chemically homogeneous Amadori compounds is important for mechanistic studies. During distillation at 138° (1 Torr) 1-desoxy-1-(**L**-prolino)-**D**-fructose is converted into 2,3-dihydro-3,5-dihydroxy-6-methyl-4(4*H*)-pyranone which is probably the precursor in the formation of maltol (**1**) (*168*). On the other hand, 1-desoxy-1-piperidino-**D**-fructose is decomposed into isomaltol and the two furanone derivatives (**5**) and (**22**) (*167*). 1-Desoxy-1-dibenzylamino-**D**-fructuronic acid proved to be interesting as a model compound for studying the thermal formation of the furanone derivative (**4**) (*107, 108*).

III. Sulfur Compounds

Sulfur-containing flavor volatiles contribute to the pleasant as well as the unpleasant overall flavors in many foods. They are frequently formed during processing or storing of foodstuffs, *e.g.* by heat treatment, or, especially in vegetables, by enzymatic action. In a few cases sulfur compounds are formed as natural metabolic products.

Simple thiols have been detected in more than 60 different foodstuffs (*153*). Since some detailed reviews dealing with the occurrence, formation, properties and importance of thiols in foodstuffs have been published recently (*153, 254, 258, 262*) we limit ourselves to describe some pecularities of the compounds of this class.

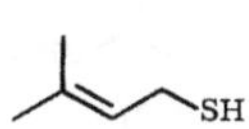

(**24**)

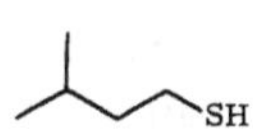

(**25**)

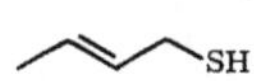

(**26**)

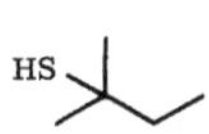

(**27**)

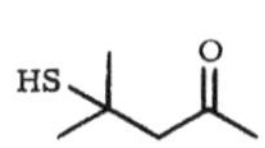

(**28**)

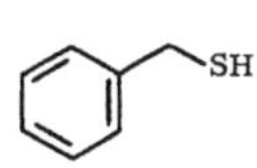

(**29**)

The "cooked" flavor of heated milk is attributed to compounds having sulfhydryl groups which are formed from β-lactoglobulin by thermal action (*255*). The so-called "sunlight" flavor, an off-flavor sometimes occurring in alcoholic beverages (*137, 138, 202, 228*), is caused by 3-methyl-2-butene-1-thiol (**24**), the odor of which is described as intense leek- or onion-like (*137*). The thiol (**24**) is responsible for the skunky odor of "sunstruck" beer (*137*), and is formed by the action of H_2S on a photolysis product of isohumulone or from *S*-(3-methyl-2-butenyl)-L-cysteine by the action of sunlight (*202*). The corresponding saturated compound (**25**), together with (*E*)-2-butene-1-thiol (**26**) and methyl(*E*)-2-butenyl disulfide (**54**), contributes to the offensive smell of the secretion of the North American skunk (*Mephitis mephitis*) (*4*). It may be recalled that 2-methyl-2-butanethiol (**27**) (*cf.* Table 1) belongs to the class of organic compounds having the lowest odor threshold (*162*).

H_2S is produced by pyrolysis of sulfur-containing aminoacids such as cysteine and cystine (*82, 143*) or by interaction of these with xylose (*145, 163*), ribose (*181*) or α-diketones (*134*). Methanethiol which is mainly formed by cleavage of methionine (*12, 82, 322*) has also been found in products of the model reaction glucose/hydrogen sulfide/ammonia (*270, 271*). Short-chain aliphatic thiols are produced in reaction of cysteine and cystine with ribose (*181*).

The action of H_2S on mesityl oxide results in the formation of 4-mercapto-4-methyl-2-pentanone (**28**) which has the extraordinarily unpleasant odor of cat urine (*211, 212, 214*). This odor was detected in canned beef (*285*) and undoubtedly constitutes a flavor defect (*295*). Phenylmethanethiol (**29**) and benzyl methyl sulfide (**30**) at levels of 1 ppb and 10 ppb, respectively may be responsible for the deterioration of the milk of cows which have consumed land cress (*325*). The sulfide (**30**) has been isolated from a model system (cysteine/xylose) approximating to cooked meat (*186*).

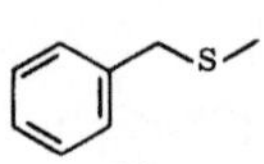

(30)

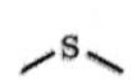

(31)

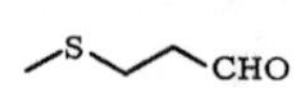

(32)

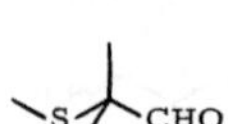

(33)

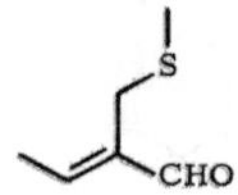

(34)

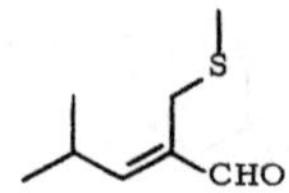

(35)

Dimethyl sulfide (**31**) and methional [3-(methylthio)propanal (**32**)] belong to the class of sulfur compounds which are most frequently found in foodstuffs. The former (**31**) was found in cocoa in amounts greatly exceeding its odor threshold (Table 1) and must, therefore, be contributing to chocolate flavor (*11*). Its precursors is an *S*-methylmethionine sulfonium salt which decomposes in aqueous medium according to the following pathway:

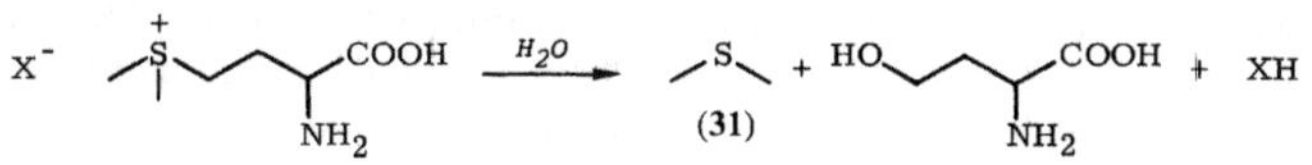

The formation of methionine which is also an important precursor of sulfur-containing compounds can be explained by the same mechanism (*148*):

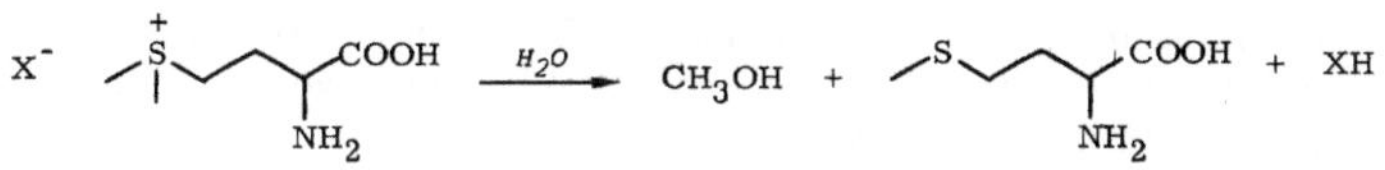

Dimethyl sulfide (**31**) is also formed on heating methionine in the presence of pectin, presumably in the same way as in autoclaved strawberries (*46, 282*). Both methylthio compounds (**31**) and (**32**) — although developing unpleasant odors — possess remarkable organoleptic properties resulting in odor addition or synergism (*95*). They are also known to be extraordinarily effective modifiers. The modification of the flavor quality depends on their concentration as well as on the chemical nature and organoleptic properties of the accompanying components. Compound (**31**) contributes to the character of high-quality mild coffee (*279*) and is also considered to have some influence on the overall flavor of tea (*332*) when present in quantities of ppb. The "laver-like" odor of green tea (*132*) and the "petroleum odor" of canned chum salmon (*176*) or frozen haddock (*157, 163*) are attributed to (**31**). The infection of wort by fermentation-resistent bacteria gives rise to the formation of (**31**) which develops in beer an astringent, sweet onion-like off-flavor (*16, 100, 197*). There is strong evidence that (**31**) is the principal volatile flavor component of processed tomato products (*165*). The feed-like flavor defect of milk is also attributable to (**31**) (*230, 231*) which is taken up from grass and corn silages (*171*). Compound (**32**) contributes significantly to the characteristic odor of fresh Pacific oysters (*239*), the aroma of canned sweet corn (*15*) and also the overall flavor of roasted barley (*328*). The degradation product (**32**) of methionine with its very low threshold (Table 1) was found to be responsible for the cheesy, brothy odor of cheddar and gruyere flavor (*60, 117, 127*) and the "sunlight" flavor of milk (*213*) (found to increase in the presence of riboflavin) and is considered to be an important component in the aroma of potato chips (*29, 35*). It is also present in baked potatoes (*32*), roasted filberts (*130*), soya protein hydrolysate (*267*), casein (*127*), cocoa (*225*) and tomatoes (*102, 242*), to mention only a few foodstuffs. Aliphatic sulfides have been found in some model reaction mixtures: dimethyl sulfide (**31**) and ethyl methyl sulfide are formed by interaction of cysteine or cystine with ribose (*181*) and diethyl sulfide was found in a glucose/hydrogen sulfide/ammonia reaction (*270, 271*). Ethyl methyl sulfide has also been isolated from a cysteine/xylose system (*186*). Some alkenyl sulfides result by interaction of aldehydes with hydrogen sulfide and ammonia, a model reaction producing a typical onion aroma (*17, 142*). A related compound, 2-methyl-2-(methylthio)propanal (**33**), is an important component of cocoa flavor (*323*). Other similar compounds include 2-(methylthiomethyl)-2-butenal (**34**) and 4-methyl-2-(methylthiomethyl)-2-pentenal (**35**). These compounds result from the aldol condensation of acetaldehyde and 2-methylpropanal with methional (**32**).

(36)

(37)

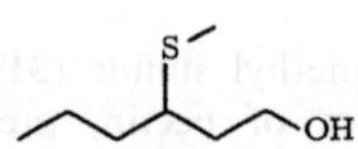

(38)

(39)

(40): R = CH_3
(41): R = C_2H_5
(42): R = H

(43)

3-(Methylthio)propan-1-ol (**36**), its acetate and 3-(ethylthio)propan-1-ol (**37**) were identified in flavor extracts of wines of various sorts and vintages (*184, 247, 249*). Thioalcohol (**36**) was also found in soy sauce (*2*) and tomato (*39*). It is the main metabolite of methionine formed in anaerobic model fermentations by *Saccharomyces cerevisiae* (*251*).

Recently, 3-(methylthio)hexan-1-ol (**38**) was identified in a flavor concentrate of the yellow passion fruit of Hawaiian origin. This compound has a green, fatty and sulfury note typical of certain exotic fruits and imparts the character of fresh fruit to the juice of passion fruit (*340*).

Some sulfur-containing esters were identified in pineapple flavor: methyl methylthioacetate (**39**) (*190*) and methyl and ethyl 3-(methylthio)-propionates (**40**) and (**41**) (*190, 342*), 1 g of (**41**) being present in 1000 kg of fruit pulp (*98*). This ester (**41**) and the corresponding acid (**42**) were also identified in wine (*59, 249*). Methyl 3-(mercapto)propionate and 3-(methylthio)propionate (**40**) have been found in the thermal

References, pp. 267—283

degradation products of cysteine methyl ester (*143*). 3-(Methylthio)-propionic acid (**42**) is presumably formed during fermentation by the addition of methanethiol to acrylic acid:

(**42**)

Methional (**32**) and the corresponding alcohol (**36**), ethyl ester (**41**) and its homolog (**43**) together with 2-methyl-4,5-dihydro-3(2*H*)-thiophenone (**110**) constitute the major portion of the volatile sulfur-containing compounds in beer (*250*).

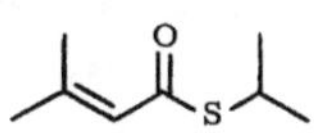

(**44**)

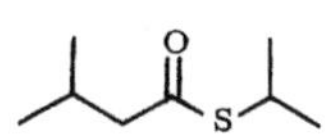

(**45**)

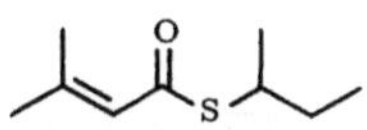

(**46**)

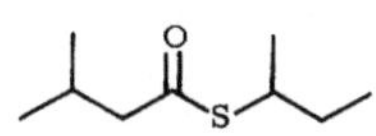

(**47**)

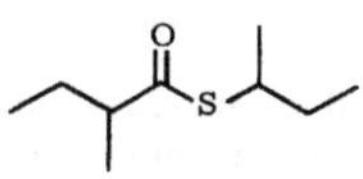

(**48**)

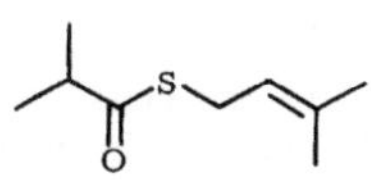

(**49**)

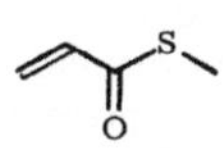

(**50**)

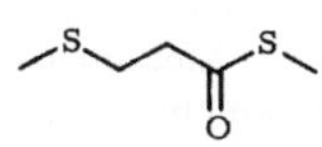

(**51**)

Thioesters are only rarely found in flavors. However, no less than 5 members of this class of compounds [(**44**) to (**48**) (*28*)] were found in galbanum. A close relative of these compounds is *S*-(3-methyl-2-butenyl)-thioisobutyrate (**49**) which was found in the essential oil of *Agathosma* species (*234*). *S*-Methyl thioacrylate (**50**) and *S*-methyl 3-(methylthio)-thiopropionate (**51**) were found to be the odor-developing agents in human urine after consumption of asparagus (*330*). A lower homolog of these thioesters, methyl thioacetate, has been identified in the products of a model reaction between cysteine and xylose (*186*).

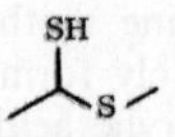

(52)

A new dithiohemiacetal, 1-methylthioethanethiol (**52**), was identified among the volatile components of beef broth (*22*). This compound possesses a strong, meaty, onion-like odor, and its formation can be easily explained by a reaction between acetaldehyde and methanethiol (*256*). The same mechanism explains the presence of such dithiohemiacetals in a model reaction involving aliphatic aldehydes, hydrogen sulfide and methanethiol (*17*).

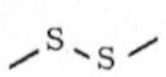

(53)

(54)

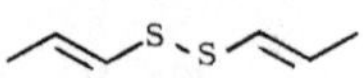

(55)

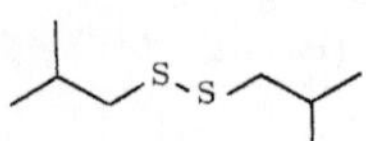

(56)

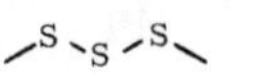

(57)

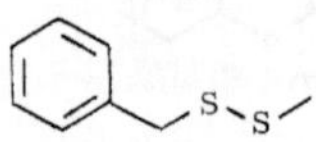

(58)

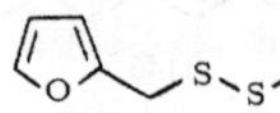

(59)

In addition to the class of symmetrical or mixed sulfides, a series of aliphatic di- and trisulfides has also acquired great importance as aroma components. These sulfides are frequently found in vegetables as character-impact compounds (Table 2) and in foodstuffs having undergone thermal treatment. Their importance is due to their low odor threshold (Table 1). Methanethiol and dimethyl disulfide (**53**) together make up 90% by weight of all sulfur compounds present in the flavor of cooked potatoes (*97*). Dimethyl trisulfide (**57**) in a concentration as low as 10 ppb (Table 1) significantly contributes to the overall flavor of roasted chicken leg meat (*294*). Furthermore, (**57**) is considered to be the most important aroma component of cooked cabbage, broccoli and cauliflower (*33*). Furfuryl methyl disulfide (**59**) (Table 2), the odoriferous principle of fresh white bread, was also found in pork liver (*187*). Benzyl methyl disulfide (**58**) was detected in cocoa (*323*) and roasted peanuts (*327*). Galbanum also contains two components of this class, bis-(1-propenyl)disulfide (**55**) (Table 2), known as a flavor-developing component of boiled onion, and the corresponding diisobutyl derivative (**56**) (*194*). The formation of aliphatic disulfides and of the trisulfide derivative (**57**) has been observed in the model reaction cysteine/xylose (*186*).

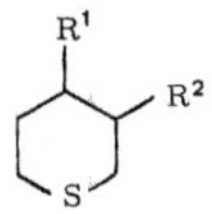

(60): $R^1, R^2 = H$
(61): $R^1 = H$; $R^2 = CH_3$
(62): $R^1 = CH_3$; $R^2 = H$

A series of alicyclic mono- and poly-sulfur compounds has been identified as flavor components within the last 10 years. Tetrahydrothiopyran (**60**) and its homologs (**61**) and (**62**) were tentatively identified in the flavor of roasted peanuts (*159*). These cyclic sulfides had already been found in crude Mexican petroleum and Middle Eastern oil distillates (*331*).

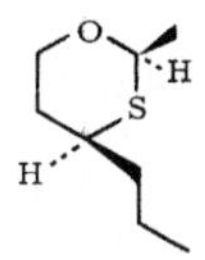

(**63**)

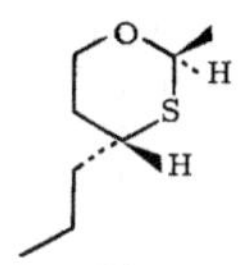

(**64**)

Recently, a mixture of diastereoisomeric 2-methyl-4-propyl-1,3-oxathianes (**63**) and (**64**) was identified in a flavor concentrate of the yellow passion fruit. Together with (**38**) these oxathianes belong to the few sulfur-containing compounds which have so far been found in fruit flavors. Both products, which were found in nature for the first time, have a strong and natural fruity odor with a green and slightly burnt note, particularly the *cis*-diastereomer (*340*).

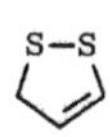

(**65**)

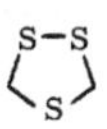

(**66**)

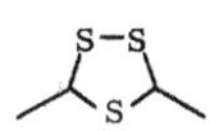

(**67**)

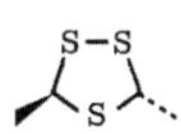

(**68**)

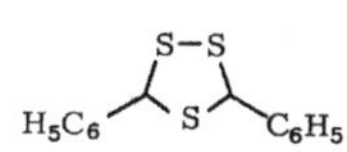

(**69**)

3*H*-1,2-Dithiole (**65**), 3-vinyl-3,4-dihydro-1,2-dithiin (**71**) and the homoconjugated isomer (**72**) have been found in the sulfur-containing flavor fraction of cooked asparagus (*309*), the characteristic odor of which is caused by (**65**).

In a model reaction, 1,2-dithiane (**70**) was formed by pyrolysis of cysteine (*143*) and by interaction of cysteine with xylose (*144*). Partially unsaturated compounds such as 3-(*E*)-propenyl-4-methyl-(4*H*)-1,2-dithiin (**73**) are isolated from the reaction of hydrogen sulfide on

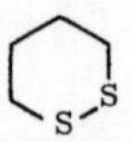
(70)

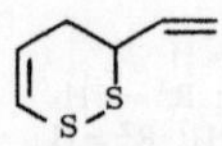
(71)

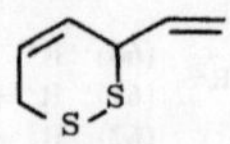
(72)

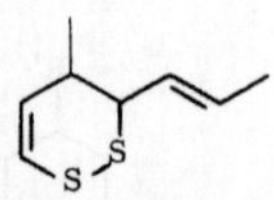
(73)

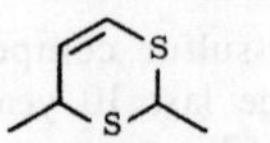
(74)

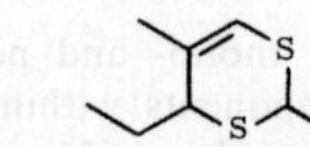
(75)

2-butenal (*347*); these compounds are described as being strongly odorous. 1,3-Dithiins of structures (**74**) and (**75**) have been detected in a synthetic onion aroma prepared by interaction of aliphatic aldehydes with hydrogen sulfide and ammonia (*142*).

The diastereoisomeric 3,5-dimethyl-1,2,4-trithiolanes (**67**) and (**68**) have been identified in the flavors of various food products prepared from beef (*22, 48, 103, 223*) and in the flavor of pork (*293*). They were also reported to occur in a volatile flavor of potato oil (*40*), in roasted filberts (*130*), in the mushroom *"Boletus edulis"* (*305*), in cheese, particularly when toasted (*60*), and in dry red beans (*41*). Their presence was also detected in the products of the thermal decomposition of a cysteine-xylose mixture (*144, 145, 172*). Trithiolaniacin [*cis*-3,5-diphenyl-1,2,4-trithiolane (**69**)] has been identified in guinea-heen weed (*Petiveria alliacea*) (*1*). Diastereoisomers (**67**) and (**68**) are easily synthesized by reaction of acetaldehyde with sulfur and hydrogen sulfide in the presence of a primary amine (*10*). This reaction is probably similar to that which takes place spontaneously in foodstuffs, since it wås recently shown that (**67**) and (**68**) result from the oxidation of bis-(1-mercaptoethyl)sulfide (**76**) which is a key product formed in the reaction of acetaldehyde with hydrogen sulfide (*17*):

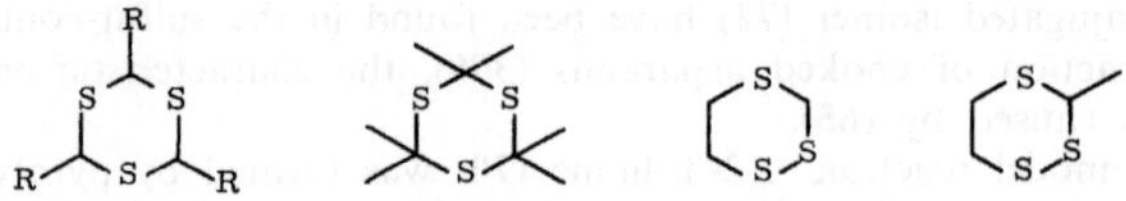

(76)

After finding isomers (**67**) and (**68**) in natural products, a synthesis was developed (*306*) in order to elucidate the stereochemistry of these compounds. When first discovered, the mixture (**67**)/(**68**) was described

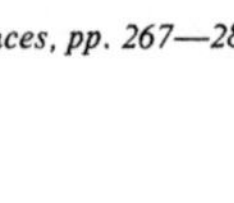

(**77**): R = H
(**78**): R = CH_3
(**79**): R = C_2H_5

(80)

(81)

(82)

as having a characteristic odor of boiled beef (*48*) although this opinion was later questioned (*109*).

2,4,6-Trimethyl-*s*-trithiane [trithioacetaldehyde (**78**)] and 2,2,4,4,6,6-hexamethyl-*s*-trithiane [trithioacetone (**80**)] were identified in the flavor of boiled beef (*338*). The formation of thialdine [2,4,6-trimethyl-5,6-dihydro-1,3,5-dithiazine (**83**)], which was also found in the same flavor (*338*), can be easily explained by the reaction between acetaldehyde, hydrogen sulfide and ammonia resulting from the decomposition of amino acids. *s*-Trithiane [trithioformaldehyde (**77**)] is also said to be a component of meat flavor (*337*). 1,2,4-Trithiane (**81**) and its 3-methyl derivative (**82**) have not yet been isolated from natural sources, but are formed by degradation of cysteine, cystine of thiazolidine derivatives (*143, 181*).

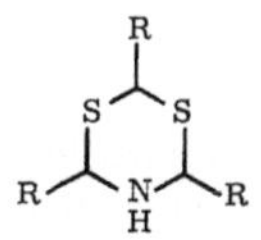

(**83**): $R = CH_3$
(**84**): $R = C_2H_5$

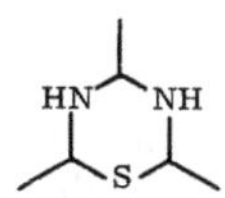

(**85**)

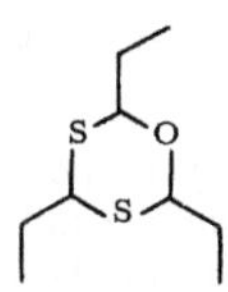

(**86**)

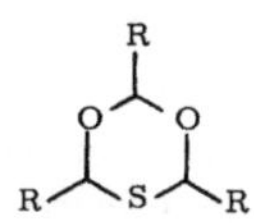

(**87**): $R = CH_3$
(**88**): $R = C_2H_5$

Various alkyl-substituted 1,3,5-dithiazines (**83**) and (**84**) have been identified in model reactions such as heat degradation of cysteine and cystine (*143*), reaction of cysteine with xylose (*144, 145*), interaction of acetaldehyde with hydrogen sulfide and ammonia (*17*) or from a mixture of acetaldehyde, propionaldehyde and ammonium sulfide (*142*). The following heterocyclic skeletons have also been identified in model reactions: 1,3,5-trithianes (**77**), (**78**) and (**79**) (*17, 142, 268*), perhydro-1,3,5-thiadiazine (**85**) (*17*), 1,3,5-oxadithiane (**86**) (*17, 142*), 1,3,5-dioxathianes (**87**) and (**88**) (*17, 142*) and 1,2,4,5-tetrathiane [(**89**), $R = C_2H_5$] (*142*). Some of these products have a typical flavor reminiscent of onion (*142*) and meat (*145*).

Lenthionine [1,2,3,5,6-pentathiepane (**92**)] was identified for the first time in "shiitake" *(Lentinus edodes)*, an edible mushroom highly prized in Asian countries, including Japan and China (*173, 320*). After having been subjected to various thorough physicochemical investigations (*9, 172, 198, 319*) this compound was synthesized in very low yields either from sodium sulfide, sulfur and formalin or by reaction of methylene iodide with sodium disulfide (*173, 174*). Lenthionine (**92**) has a very typical mushroom odor and taste (Table 2) (*175, 321*) and, in addition, possesses fungicidal properties (*298*). This compound has also found, to some extent, an application as a sensitizing agent in photographic emulsions (*311*).

The eight-membered ring system, 1,2,5,6-tetrathiocane (**91**), was recently identified in a steam-distillate of "Chenin blanc" grape leaf (*333*).

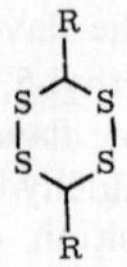

(**89**)

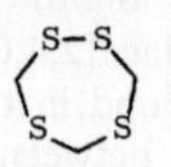

(**90**)

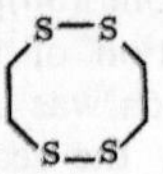

(**91**)

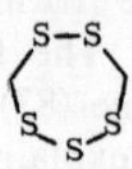

(**92**)

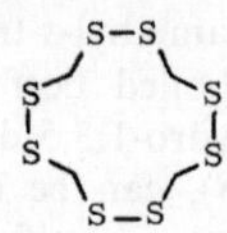

(**93**)

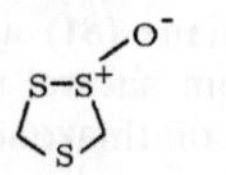

(**94**)

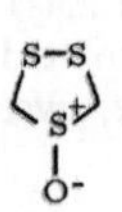

(**95**)

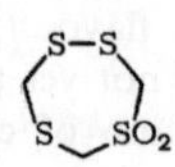

(**96**)

A series of related cyclic polysulfides was recently identified in the red alga *Chondria californica* (*345*), in which they and their oxidation products are responsible for the antibiotic activity. The mechanism of their biogenesis is probably similar to that occurring in *Lentinus edodes*, since in both the red alga and the mushroom 1,2,4-trithiolane (**66**) and lenthionine (**92**) are present. The 1-oxide (**94**), and 4-oxide (**95**), 1,2,4,6-tetrathiepane (**90**) and its 4,4-dioxide (**96**), and a twelve-membered heterocyclic compound (**93**) containing eight sulfur atoms are also found in the red alga (*345*). Compound (**90**) has recently been identified in the reaction products of furfural with hydrogen sulfide and ammonia (*268*).

IV. Thiophenes

Compounds of the thiophene class have been found only in the last ten years in flavors and, with the exception of American cranberries (*5*), exclusively in cooked or roasted products. Thiophenes have been identified in large numbers in the flavors of meat-related products (*187, 218, 338*) and coffee (*84, 288, 290*), and, in significantly lower numbers, in onion (*18*) (Table 2), roasted filberts (*130*), roasted peanuts (*327*), popcorn (*326*), bread (*183, 296*), cooked asparagus (*309*) and black tea (*47*). Some protein isolates contain alkylthiophenes (*227*). The thiophenes most frequently identified in these flavors are the alkylated derivatives (**97**) and the acylated derivatives (**98**) and (**99**) in which R = H, CH_3 or C_2H_5 and R^2, R^3, R^4, R^5 = H and/or alkyl.

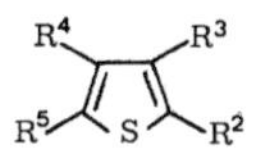

(97)

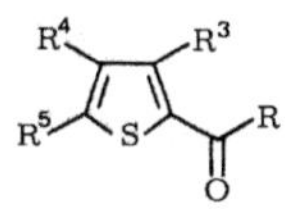

(98)

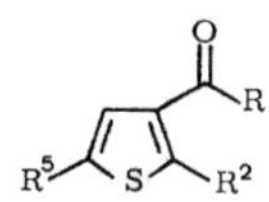

(99)

Numerous thiophenes have also been isolated from model reactions. Thiophene and some alkyl homologs result from cysteine and cystine degradation (*125*) which produces a typical sesame-like or popcorn-like aroma varying with temperature. Pyrolysis of the same amino acids in the presence of xylose (*144, 186*), glucose (*124, 245*), ribose (*181*), pyruvaldehyde (*124*) or from the interaction of glucose degradation products with hydrogen sulfide (*270, 271*) gave similar results. Thiophene aldehydes and ketones, 2-mercaptothiophene, 3-methylthiothiophene and 2-thiophenecarboxylic acid are formed under the same conditions.

The organoleptic properties of a series of substituted thiophenes have been described in detail (*342*). 2-Acetyl-3-methylthiophene [(**98**): R, R^3 = CH_3; R^4, R^5 = H], for example, in syrup at a concentration of 0.25 g/100 liter, produces a honey-like sensory impression, whereas the same compound at a concentration of 0.11 g/100 liter of coffee develops a nutty, starchy note.

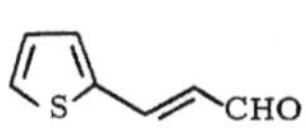

(100)

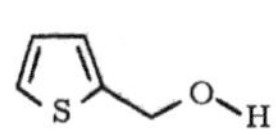

(101)

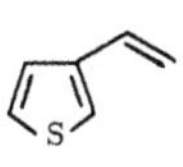

(102)

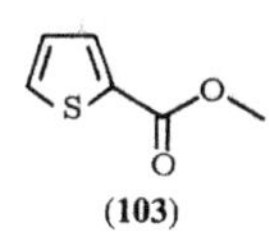

(103)

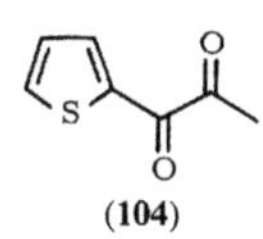

(104)

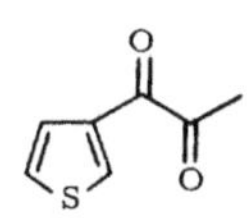

(105)

Apart from 3-(2-thienyl)propenal (**100**) (*187, 338*) and 2-thienylmethanol (**101**), which were identified in meat (*338*), compounds (**101**) to (**105**) were identified only in coffee flavor (*290*). Except for compound (**104**) which imparts a praline-like and woody note to coffee, none of these compounds seems to be really indispensable or characteristic.

Certain non-aromatic thiophene derivatives are more typical. Apart from the doubtful isolation of 2,3-dihydrothiophene (**106**) from milk caramel (*303*), the remaining members of this series are mainly dihydrothiophene-2(or -3)-ones (**107**) to (**110**).

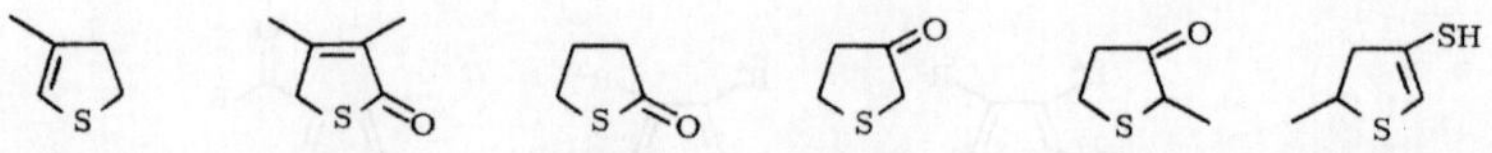

(106) (107) (108) (109) (110) (111)

(112) (113) (114) (115) (116) (117)

3,4-Dimethyl-2(5*H*)-thiophenone (**107**) was identified in the flavors of onion (*24*) and leek (*252*), 4,5-dihydro-2(3*H*)-thiophenone (**108**) is a component of the flavors of coffee (*287, 288*) and mushroom (*305*), and 4,5-dihydro-3(2*H*)-thiophenone (**109**) and its methyl homolog (**110**) were identified in roasted coffee (*290*), roasted peanuts (*327*), and roasted filberts (*130*). Thiophenone (**109**) and its methyl homolog (**110**) were also identified in cooked beef (*338*), beer (*251*) and wine (*248*). The latter two compounds were described as having respectively an onion, garlic taste and a green, burnt coffee taste (*343*). The formation of the carbonyl derivatives (**108**) to (**110**) in model reactions resulting from interaction of cysteine with xylose (*144, 186*), glucose (*245*) or pyruvaldehyde (*124*) has been demonstrated. The methylthiophenone derivative (**110**) is a fermentation product from methionine (*251*). The mercaptothiophenes (**111**) to (**116**) result from the reaction of H_2S on 4-hydroxy-5-methyl-3(2*H*)-furanone (**4**) (**111**) to (**115**) or on furaneol (**5**) (**116**) (*207*). Compound (**117**) has been described as flavoring agent (*129*) having a roasted and burnt meat aroma.

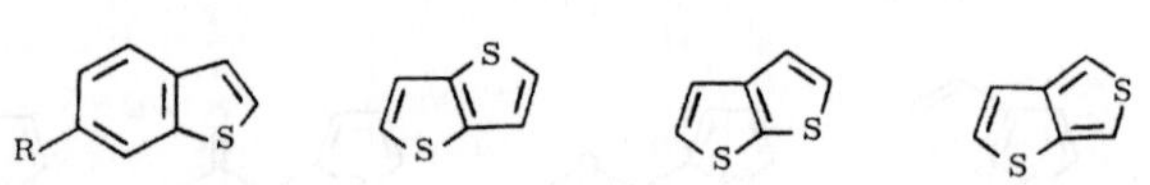

(**118**): R = H
(**119**): R = CH_3 (**120**) (**121**) (**122**)

Among the bicyclic thiophene compounds, benzo[*b*]thiophene (**118**) was identified in peas (**276**) and coffee (*290*), and 6-methylbenzothiophene (**119**) in milk caramel (*303*). Thieno[3,2-*b*]thiophene (**120**) or its isomer thieno[2,3-*b*]thiophene (**121**) was found to be present in coffee aroma (*290*). Isomers (**120**) and (**122**) are found among the products of the reactions between cysteine-cystine and ribose (*181*) and cysteine and glucose (*186, 245*), respectively.

6-Methyl-2,3-dihydro-thieno[2,3-*c*]furan (**123**), a bicyclic compound with a new structure (also known as kahweofuran), was isolated from

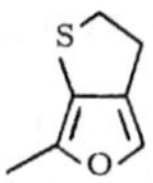

(123)

coffee aroma (*290*) and its structure confirmed by synthesis (*26*). Kahweofuran (**123**) in a pure state has a violent sulfury odor, but in high dilution it develops a pleasant roasted and smoky note.

A recently published review gives interesting details on the role of thiophenes in foodstuffs (*152*).

V. Thiazoles

After the discovery of the first member of this class in 1966 (*8*), numerous thiazoles were identified in food flavors. The highest number of thiazoles was found in the flavors of meat (*338*) and coffee (*313*). This series of thiazoles comprises mainly alkylated compounds (**124**) to (**127**).

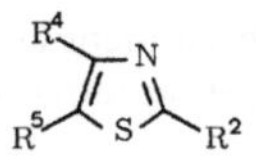

(124): $R^2, R^4, R^5 = CH_3$
(125): $R^2, R^4 = CH_3$; $R^5 = C_2H_5$
(126): $R^2 = H$; $R^4 = C_4H_9$; $R^5 = C_3H_7$
(127): $R^2 = i\text{-}C_4H_9$; $R^4, R^5 = H$

They are described as green, nutty and vegetable like (*221*). For instance, trimethylthiazole (**124**) and 5-ethyl-2,4-dimethylthiazole (**125**) are reported to have a cocoa, nutty character and a nutty, roasted and meaty character, respectively, whereas 4-butyl-5-propylthiazole (**126**), which has a low odor threshold (Table 1), reproduces the odor character of bell peppers (*34*). The most typical alkylthiazole, however, is probably 2-isobutylthiazole (**127**). This compound was isolated from tomato flavor (*312*) and was described as having a strong green odor resembling that of tomato leaf. Some authors (*126*) postulated that the structure of this compound was connected with the biogenesis of thiamine. They also reported that in water the pure compound has a spoiled wine-like, slightly horseradish-type flavor which is somewhat objectionable. However, when added to canned tomato puree or paste at levels of 20 to 50 ppb (Table 1), the compound develops a more intense, fresh tomato-like flavor.

Nearly 20 different alkylthiazoles have been isolated from model reactions involving either degradation of glucose in the presence of hydrogen sulfide and ammonia (*270, 271*), or, more frequently, fragmentation of cysteine or cystine (*125, 143*), or reaction of these with xylose (*144*), ribose (*181*), pyruvaldehyde (*124*) or glucose (*124*).

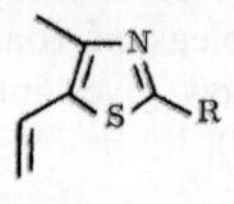

(**128**): R = H
(**129**): R = CH_3

Two vinylthiazoles have been identified in flavors: 4-Methyl-5-vinylthiazole (**128**) in cocoa (*289*), roasted filberts (*130*) and passion fruit (*344*) and 2,4-dimethyl-5-vinylthiazole (**129**) in meat flavor (*338*).

The homolog (**129**) has a strong, characteristic nut-like odor (*289*) and is used in flavor compositions (*96*). It is easily formed by dehydration of the corresponding alcohol 5-(2-hydroxyethyl)-2,4-dimethylthiazole (*289*).

Alkenylthiazoles have also been identified in model reactions such as degradation of cysteine and cystine alone (*143*) or in the presence of xylose (*144*).

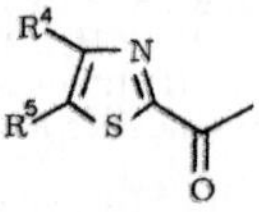

(**130**): R^4, R^5 = H
(**131**): R^4 = CH_3; R^5 = H
(**132**): R^4 = H; R^5 = CH_3

2-Acetylthiazole (**130**), which was identified in pressure-cooked beef (*338*), boiled potatoes (*37*), cooked pork liver (*187*), cooked asparagus (*309*), dry red beans (*41*) and canned beef (*219*) as well as in the products of the model reaction between cysteine-cystine and ribose (*181*), possesses a strong, characteristic nutty, cereal and popcorn flavor (*221*). 2-Acetyl-4-methylthiazole (**131**), which was detected in coffee flavor (*290, 313*), has an anthranilic burnt flavor (*343*). The addition of 0.034 g/liter of 2-acetyl-5-methylthiazole (**132**) to coffee results in a note like toasted coffee bean (*342*). A pathway for the formation of acylthiazoles has been proposed (*181*), and various model reactions starting from cysteine or cystine and glucose or pyruvaldehyde have been described (*124*). An interesting compound, 2-(2-furyl)thiazole, has also been isolated from the reaction of cysteine and cystine with ribose (*181*).

The utilization of 2-acetylthiazoles (**130**) to (**132**) as flavoring agents was also described (*224, 341*). They have organoleptic characters varying from green and fruity to burnt and nutty.

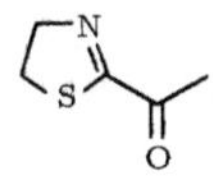

(**133**)

A hydrogenated form, 2-acetyl-2-thiazoline (**133**), was identified in beef broth (*50*). In a more detailed article (*307*) this compound is described as having an intense odor of freshly baked bread crust, although it is not at all the character impact flavor component of this foodstuff (*187*) (Table 2). Although the mechanism of formation of this product has not yet been elucidated, it may be assumed that it results either from the Strecker degradation of cysteine and subsequent addition of pyruvaldehyde, or from the addition of pyruvaldehyde to cysteamine and subsequent oxidation. Its utilization as flavoring agent has been described (*49*).

2-Thiazoline and its 2-methyl-, ethyl- and propyl-homologs have been identified in model reactions such as cysteine or cystine pyrolysis (*125, 143, 278*) or reaction of these aminoacids with xylose (*144*), glucose or pyruvaldehyde (*124*). 2,4-Dimethyl-3-thiazoline is formed by reaction of glucose with hydrogen sulfide and ammonia (*271*).

Thiazolidine and some 2-alkylsubstituted homologs have also been identified in cysteine and cystine degradation products (*82, 125, 143, 144, 278*). In this case, a scheme has been proposed to explain the formation of 2-methylthiazolidine by the reaction of acetaldehyde with mercaptoethylamine (*82*). Methylthiazolidine develops a sweet and fish-like aroma which also includes a sesame-like one and is thought to be essential to the development of the popcorn- and sesame-like aroma produced by pyrolysis of cystine and cysteine. Alkylthiazolidines have been patented as flavoring agents; used in combination with alkylpyrazines and alkylcyclohexenones they confer a meaty, nutty or cereal character to foods (*70*).

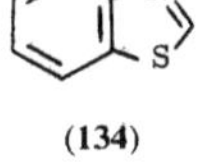

(**134**)

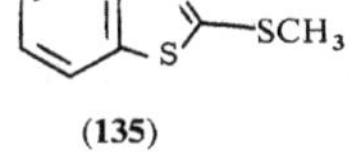

(**135**)

Benzothiazole (**134**) is abundant in flavors. Identified for the first time in a sterilized milk concentrate (*8*) in 1966, it was identified the following year in cocoa (*77*), butter (*78*), beer (*30*), cranberries (*5*) and Swiss cheese (*141*). Subsequently, it was detected in more than a dozen other flavors. It is described as having a quinoline-like, rubbery odor (*151*). 2-Methylthiobenzothiazole (**135**) has recently been identified in a meat flavor (*69*).

Reviews on the occurrence of thiazoles in foodstuffs have been published recently (*150, 151*). They also contain interesting details regarding the formation and the utilization of this new flavor class.

VI. Oxazoles and Oxazolines

In 1968 5-acetyl-2-methyloxazole (**136**), the first member of this class of heterocyclic compounds, was identified in coffee flavor (*287, 288*). Later, 27 additional oxazoles of various structures were found in the same flavor (*314*). This series includes 5-acetyl-2,4-dimethyloxazole (**137**) and 2-phenyloxazole (**138**) as well as the n-alkylated oxazoles (**139**) to (**142**). The last four compounds (**139**) to (**142**) are also found

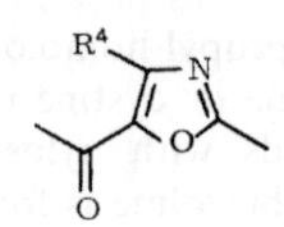

(**136**): $R^4 = H$
(**137**): $R^4 = CH_3$

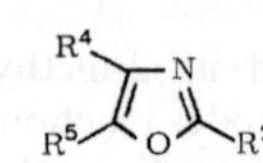

(**138**): $R^2 =$ 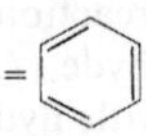 ; $R^4, R^5 = H$

(**139**): $R^2, R^5 = CH_3$; $R^4 = H$
(**140**): $R^4, R^5 = CH_3$; $R^2 = H$
(**141**): $R^2, R^4, R^5 = CH_3$
(**142**): $R^2 = C_3H_7$; $R^5 = CH_3$; $R^4 = H$

in cocoa flavor (*317*). The oxazoles might be formed by reactions between amino acids, reactions which are actually used for their preparation (*335*). Their formation might also be explained by the Strecker degradation of aminoketones which can result from the condensation of α-dicarbonyl compounds with amino acids (*314*).

2,4-Dimethyloxazole has recently been identified in the model reaction when furfural is treated with hydrogen sulfide and ammonia (*268*). Methyl- and dimethylbenzoxazoles (**143**) to (**147**) have been identified in roasted coffee (*314*), and 2-methyl-benzoxazole (**143**) in meat flavor (*69*).

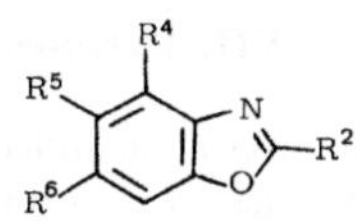

(**143**): $R^2 = CH_3$; $R^4, R^5, R^6 = H$
(**144**): $R^4 = CH_3$; $R^2, R^5, R^6 = H$
(**145**): $R^2, R^4 = CH_3$; $R^5, R^6 = H$
(**146**): $R^2, R^5 = CH_3$; $R^4, R^6 = H$
(**147**): $R^2, R^6 = CH_3$; $R^4, R^5 = H$

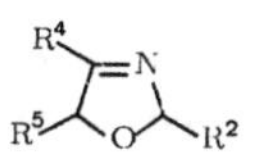

(**148**): $R^2, R^4 = CH_3$; $R^5 = H$
(**149**): $R^2, R^4, R^5 = CH_3$
(**150**): $R^2, R^4 = CH_3$; $R^5 = C_2H_5$
(**151**): $R^2, R^5 = CH_3$; $R^4 = C_2H_5$
(**152**): $R^2 = i\text{-}C_3H_7$; $R^4, R^5 = C_2H_5$
(**153**): $R^2, R^4, R^5 = C_2H_5$

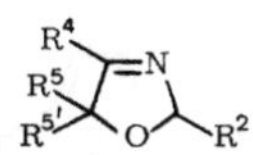

(**154**): $R^2 = i\text{-}C_3H_7$; $R^4, R^5, R^{5'} = CH_3$
(**155**): $R^2 = i\text{-}C_4H_9$; $R^4, R^5, R^{5'} = CH_3$

The four oxazolines (**148**) to (**152**) seem to have a certain importance in the development of the flavor of cooked beef (*48, 109, 189*). Reaction of acetaldehyde with acetoin in ammonia solution (*118*) provides an easy synthesis of (**149**). A study of the Strecker degradation shows that 2,4,5-trimethyl-3-oxazoline (**149**) might also be formed from alanine and 2,3-butanedione (*235*). By analogy, it can be expected that higher homologs are also present in food flavors in which they should be formed at the expense of the corresponding amino acid.

The interesting sensory properties of this class of compounds have been exploited for flavoring foodstuffs (*104*). Among these compounds are 2-isopropyl-4,5,5-trimethyl-3-oxazoline (**154**) which has a rum-like note, 2-isobutyl-4,5,5-trimethyl-3-oxazoline (**155**) which has an earthy (fungal) odor with a butter-like, green leaf-like note, and 2,4,5-triethyl-3-oxazoline (**153**) which has an odor strongly reminiscent of fresh carrots. Another particularly characteristic product is 2-isopropyl-4,5-diethyl-3-oxazoline (**152**) which possesses a typical cocoa aroma (*104*). A detailed report on the chemistry of oxazoles was published recently (*310*).

VII. Pyrroles

So far, 42 pyrroles have been identified in aromas. In addition to pyrrole itself, 11 alkylpyrroles, 15 formylpyrroles, 7 acetylpyrroles, 2 methyl esters, 4 N-furfurylpyrroles (of which one is formylated and one acetylated), and one pyrrole carrying a diketone chain. Products in which most representatives are found are roasted coffee (*85, 288, 290, 314, 316*), roasted peanuts (*327*) and filberts (*130, 266*), cocoa (*158, 323, 324*), mushrooms (*305*), tobacco (*52, 147*), and roasted green tea (*346*). A smaller number have been identified in rye bread (*296*), white bread (*180*), and cooked pork liver (*187*), and 14 other products contain from one to four representatives.

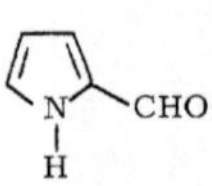

(156)

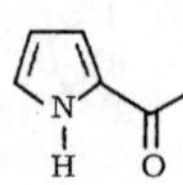

(157)

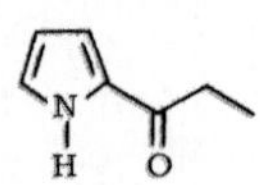

(158)

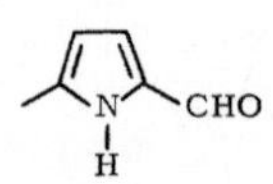

(159)

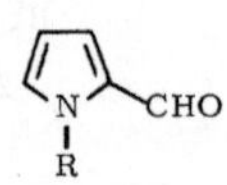

(160): R = CH_3
(161): R = C_2H_5

2-Formylpyrrole (**156**), found in 16 natural products, and 2-acetylpyrrole (**157**), found in 17, are the most widely distributed and the most abundant. 2-Propionylpyrrole (**158**), 2-formyl-5-methylpyrrole (**159**), 2-formyl-1-methylpyrrole (**160**) and 1-ethyl-2-formylpyrrole (**161**) are also frequently identified.

(162)

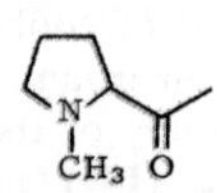

(163)

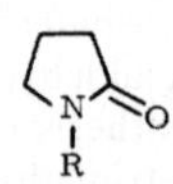

(164): R = CH_3
(165): R = C_2H_5
(166): R = C_3H_7

Pyrrolidine (**162**) has been found in caviar (*89*), fish (*121*), Russian (*88*) and Dutch (*90*) cheese, beer (*280, 281*) and wine (*58*). Its particularly powerful and characteristic amine smell accounts for the fishy and fermented notes of these foodstuffs. 2-Acetyl-1-methylpyrrolidine (**163**) has been found in bread (*115*), of which it possesses the typical odor. 1-Methyl-2-pyrrolidone (**164**) has been identified in roasted filberts (*130*), 1-ethyl-2-pyrrolidone (**165**) in cooked pork (*293*), and 1-propyl-2-pyrrolidone (**166**) in soya protein hydrolysate (*155*). Nearly 50 pyrroles have also been identified among the products of model reactions generally involving sugars and amino-acids, most of the pyrroles

References, pp. 267—283

found in foodstuffs being formed in this way. Certain more unusual substances possess a functional group in a chain attached to the nitrogen of the heterocycle.

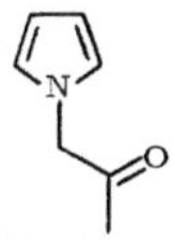

(167)

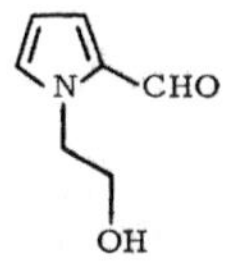

(168)

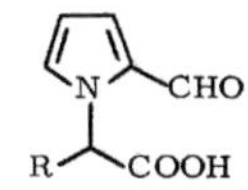

(169): R = H
(170): R = i-C_4H_9

1-Acetonylpyrrole (**167**) has been identified among the products of the reaction between hydroxyproline and glucose (*135*), a reaction which produces a cookie- or mushroom-like aroma. This pyrrole (**167**) may be formed by the reaction of hydroxyproline with the pyruvaldehyde derived from aldol cleavage of the 3-deoxyglucosone formed by dehydration of glucose (*135*). Product (**168**) is formed in a browning model reaction of glycolic aldehyde with aminoethanol (*241*). Products with structures (**169**) and (**170**) have been identified in the reaction products of xylose with glycine (**169**) or leucine (**170**) (*123*).

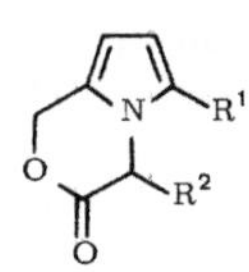

	R^1	R^2
(171):	H	CH_3
(172):	H	i-C_3H_7
(173):	H	C_4H_9
(174):	H	i-C_4H_9
(175):	CHO	H
(176):	CHO	C_2H_5
(177):	CHO	$CH_2-C_6H_5$

Some pyrrole lactones (**171**) to (**177**) are formed by reaction of alanine (**171**), valine (**172**), α-aminocaproic acid (**173**), leucine (**174**), glycine (**175**), and α-aminobutyric acid (**176**) with glucose (*272*). (**175**) and (**176**) improve the taste and/or smell of tobacco products by giving them a spicy, peppery flavor (*233*). Compounds (**175**) and (**177**) have been identified in flue-cured tobacco (*53*).

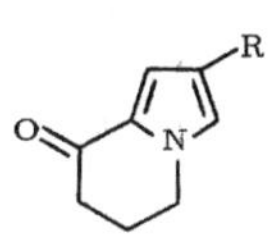

(178): R = H or CH_3

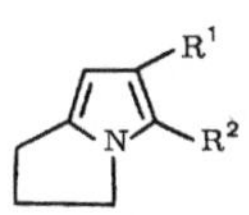

	R^1	R^2
(179):	H	$COCH_3$
(180):	CH_3	CHO
(181):	CH_3	$COCH_3$

Bicyclic compounds (**178**) to (**181**) have been isolated from a proline/glucose model reaction (*273*).

VIII. Pyrazines

The first isolation, if not the first identification, of an alkylpyrazine (**182**) in a product related to food flavor goes back to 1879, when two basic components having the formulas $C_8H_{12}N_2$ and $C_{10}H_{16}N_2$ were isolated

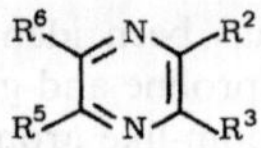

(**182**)

from fusel oil which had been obtained from molasses of sugar beet (*253*). It is reasonable to assume that these compounds were tetramethylpyrazine [(**182**): all R = CH_3] and a higher homolog. By the end of the last century and at the beginning of the present century, some alkylpyrazines were identified in alcoholic fermentation products (or by-products), such as fusel oils of various origins (*13, 24, 177, 199*) and whisky (*334*). In 1928, pyrazine [(**182**): all R = H], methylpyrazine [(**182**): R_2 = CH_3; R_3, R_5 = H] and 2,5- and 2,6-dimethylpyrazines [(**182**): R_2, R_5 = CH_3; R_3, R_6 = H and R_2, R_6 = CH_3; R_3, R_5 = H, respectively] were identified in a coffee flavor (*232*). No further pyrazines were found until 1964, when 2,5-dimethylpyrazine [(**182**): R_2, R_5 = CH_3; R_3, R_6 = H] was discovered in cocoa flavor (*55*). Since then nearly a hundred papers have been written on the identification of more than 50 different alkylpyrazines in about 30 natural or manufactured, cooked, roasted, or fermented products of vegetable or animal origin. Among the 43 pyrazines which have so far been identified in cocoa there are no less than 27 alkylpyrazines of different degrees of substitution (*317*). The alarm pheromones of the American *Odontomachus* workers which smell like chocolate consist mainly of alkylpyrazines. The mandibular gland secretion of *O. hastatus* contains 77% 2,5-dimethyl-3-isopentylpyrazine [(**182**): R_2, R_5 = CH_3; R_3 = i-C_5H_{12}], while the glandular exudates of *O. brunneus* consist of a mixture of 2,6-dimethyl-3-pentyl-, 3-butyl-, 3-propyl- and 3-ethyl-pyrazines (*329*).

It would take too much space to describe in detail the organoleptic properties of this interesting series of heterocyclic compounds which are the subject matter of various reviews (*45, 154, 302*). As a rule, they have a roasted nut-like note. Various studies have been devoted to the mechanism of formation (*235, 236, 269*) and to the metabolism of this important class of aroma compounds (*101*).

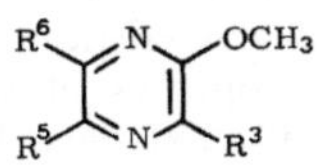

(183)

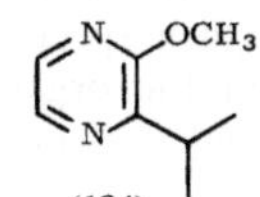

(184)

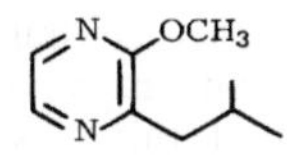

(185)

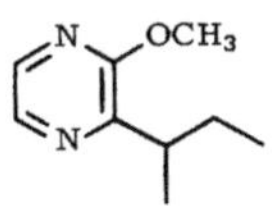

(186)

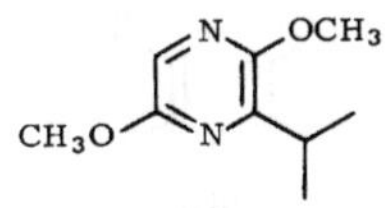

(187)

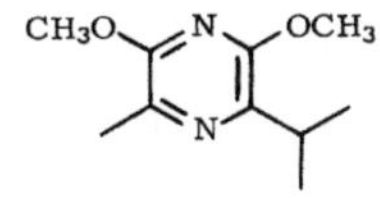

(188)

The organoleptic interest of the methoxypyrazines (**183**) was mentioned for the first time in a patent (*67*) which discloses the synthesis and the utilization of the three isomeric methoxymethylpyrazines (**183**) (R_3 = CH_3, R_5, R_6 = H; R_5 = CH_3, R_3, R_6 = H and R_6 = CH_3, R_3, R_5 = H) for flavoring foodstuffs. The first member of this class, 2-isobutyl-3-methoxypyrazine (**185**), was only identified in 1969 in an extract of bell pepper *(Caspicum annuum)* (*44*). In aqueous solution the odor threshold of this compound is 0.002 ppb (Table 1). Its odor is described as being characteristic of freshly cut green pepper (*42, 43*) (Table 2). The various odor thresholds of the main aroma components of pepper, in aqueous solution, are shown in a more detailed publication (*38*). After this discovery the same compound (**185**), 2-sec-butyl-3-methoxypyrazine (**186**) and the dimethoxy derivatives (**187**) and (**188**) were identified in galbanum oil (*20, 27*). In 1970, the methoxypyrazines substituted in position 3 with isopropyl, isobutyl and sec-butyl groups (**184**), (**185**) and (**186**) were in turn identified in peas *(Pisum sativum)* (*185*), and in 1971, 2-isobutyl-3-methoxypyrazine (**185**) was detected in the flavor of roasted coffee (*80*). It was also observed in 1971 that the 2-alkyl-3-methoxypyrazines (**183**) were very useful for flavoring mashed potatoes (*31, 93*), and indeed, in 1973, 2-isopropyl-3-methoxypyrazine (**184**) was identified in potatoes (*36*), to which it imparts a characteristic earthy note (Table 2). Soon after, a further 2-methoxypyrazine, 2-ethyl-3-methoxypyrazine [(**183**): R = C_2H_5; R_5, R_6 = H], was found in freeze-dried potato peel (*161*). This compound occurs as an off-flavor in "Gruyère de Comté" (*61*). 2-Isopropyl-3-methoxypyrazine (**184**) is responsible for the potato-like odor which appears during the first phase of the deterioration of chilled fish muscles caused by *Pseudomonas perolens* (*166*). The musty odor in the same kind of food has been attributed to the microbiological formation of 2-sec-butyl-3-methoxypyrazine (**186**) (*166*). Several patents relate to the utilization of alkoxypyrazines for flavoring coffee (*209*), pineapple (*136*) and potatoes (*92*). A method for the separation of methoxymethylpyrazines was developed to permit the determination of their spectroscopic characteristics (*193*). Several comparative studies on the organoleptic properties of some members of this class (*208*) and a detailed review (*154*) have been published.

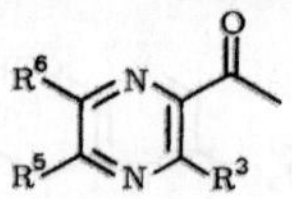

(**189**)

The acetylpyrazines (**189**) constitute a class of particularly valuable natural products. They have a surprisingly intense and characteristic roasted note reminiscent of popcorn (*237*). Acetylpyrazine [(**189**): R_3, R_5, R_6 = H] was identified for the first time in 1969 in an investigation on the components of sesame oil (*299*). After having been tentatively identified in popcorn (*326*), acetylpyrazine was isolated from roasted peanuts (*327*), tobacco (*52*), roasted filberts (*130*), boiled beef (*188*), roasted sesame seed (*156*), roasted coffee (*314*), cooked pork liver (*187*) and roasted almonds (*301*). It had already been synthesized in 1952 (*139, 275, 283*), but had exclusively been used as a nicotinamide antagonist (*51*), and its utility as a flavoring agent was described in a patent issued in 1968 (*237*).

The 3-, 5- and 6-methyl-homologs (*189*) were also identified in flavors. In addition to its occurrence in the flavors mentioned in the last paragraph 2-acetyl-3-methylpyrazine [(**189**): R_3 = CH_3; R_5, R_6 = H] was found in white bread (*179*) and among the products of the model reaction between cysteine and ribose (*181*). Syntheses of this isomer and of the 2,5-isomer [(**189**): $R_5 = CH_3$; R_3, $R_6 = H$] have been described

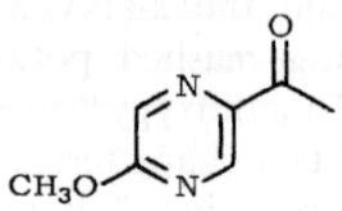

(**190**)

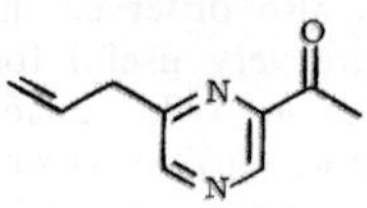

(**191**)

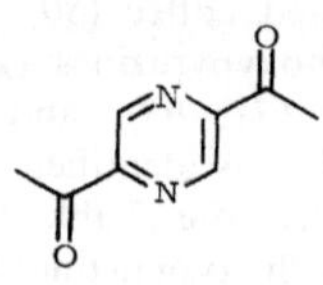

(**192**)

(*170, 224, 257*). The dimethyl homologs of acetylpyrazine have so far been identified only in roasted coffee (*314*). The acetylpyrazine homologs having an ethyl group in one of the positions 3, 5 or 6 were found in cooked pork liver (*187*), boiled beef (*188*), roasted almonds (*300, 301*), roasted peanuts (*25, 327*) and cocoa (*317*). 2-Acetyl-3-ethylpyrazine [(**189**): $R_3 = C_2H_5$; R_5, R_6 = H], with an odor and taste reminiscent of slightly roasted potatoes, is mentioned in a patent relating to the flavoring of foodstuffs (*169*).

Among other acetylpyrazines which were identified in foodstuffs, 2-acetyl-5-methoxypyrazine (**190**), 2-acetyl-6-allylpyrazine (**191**) and 2,5-diacetylpyrazine (**192**), have been identified in roasted almonds (*301*).

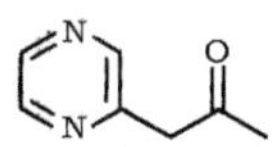

(193)

2-Acetonylpyrazine (**193**) was found in cooked beef (*188*) and among the products of a Maillard model reaction (*304*). When used in combination with thiazolidines and/or cyclohexenones, this pyrazine imparts a toasted and burnt note to foodstuffs (*70, 341*). Its utilization for modifying coffee flavor is patented (*341*).

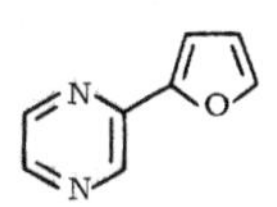

(194)

The identification of 2-(2-furyl)pyrazine (**194**) was mentioned for the first time in an article relating to the analysis of coffee flavor (*80*). It seems to be the most widely distributed compound of this series, having been found in cigarette smoke (*195, 196*) and roasted foodstuffs such as filberts (*130*), green tea (*346*), almonds (*300*), sesame seed (*156*), cocoa (*317*) and pressure-cooked pork liver (*187*). These identifications are based only on mass spectra of reference compounds given in the first article of the series (*178*). It has been suggested that 2-(2-furyl)-3-methylpyrazine (**196**) might be formed from 1(2-furyl)-1,2-propanedione (**195**) (*156*) and amino acids, these substances having themselves been identified in coffee flavors (*85, 288, 290*) and bread flavor (*180, 296*).

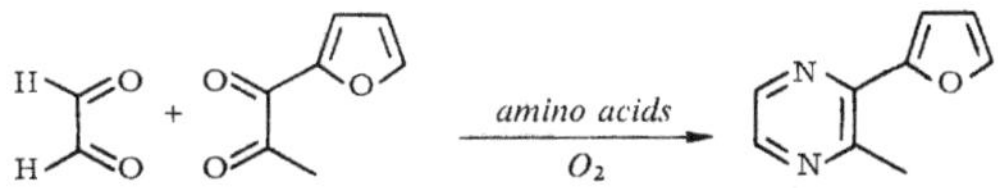

(195) **(196)**

A synthesis based on the condensation of (**195**) with ethylene diamine has also been described (*192*). A series of 8 isomers substituted with two methyl groups in various positions and a trimethyl homolog were also identified in coffee flavor (*314*).

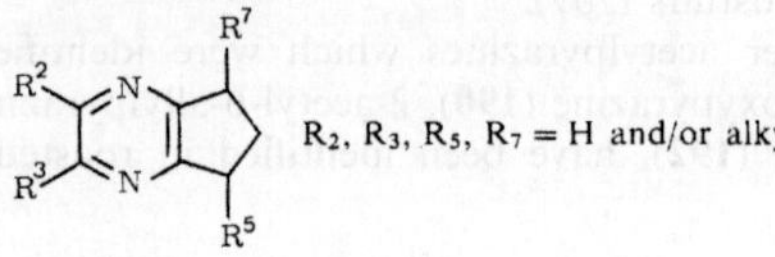

(**197**)

Since 1971, a dozen 6,7-dihydro-5*H*-cyclopenta[*b*] pyrazines (**197**) have been identified mainly in cooked or roasted products such as roasted peanuts (*277, 327*), roasted filberts (*130*), cooked beef (*188*), roasted green tea (*346*), roasted sesame seed (*131, 156*), cooked pork liver (*187*), roasted almonds (*300*), roasted cocoa (*317*), roasted coffee (*315*), roasted beef (*73*), cooked asparagus (*309*) and beer (*99*). The formation of these products in flavors can easily be explained by the condensation of 2-hydroxy-2-cyclopenten-1-ones with α-dicarbonyl compounds in the presence of ammonia (produced by degradation of amino acids) (*73*):

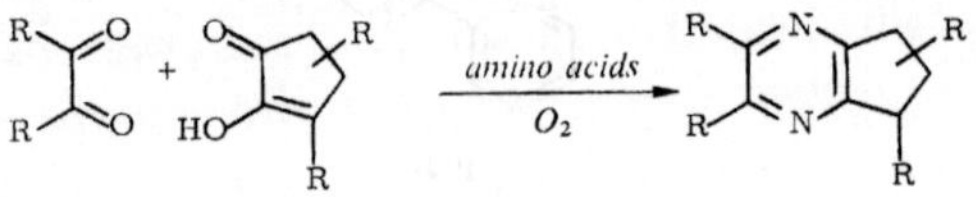

The organoleptic properties of some of these compounds were described in a patent prior to their identification in a natural flavor (*224*). They are prepared by condensation of 2-hydroxy-2-cyclopenten-1-ones with alkylenediamines, or of aliphatic α-diketones with 1,2-diaminocyclopentanes (*74*).

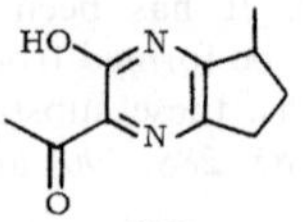

(**198**)

In addition to these alkylated compounds (**197**), 2-acetyl-3-hydroxy-5-methyl-6,7-dihydro-5*H*-cyclopenta[*b*] pyrazine (**198**) was identified in roasted almonds (*301*).

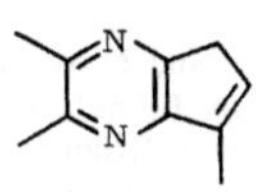

(**199**)

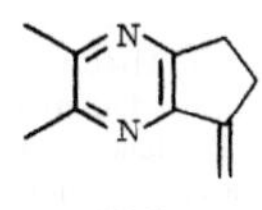

(**200**)

Among the model reaction products mentioned above (*73*) unsaturated bicyclic structures (**199**) and (**200**) have also been observed. Their synthesis (*76*) and their use as aromatizing ingredients (*71*) have been described. Such products have roasted, grilled, burnt and animal notes interesting for the aromatisation of meat and tobacco products.

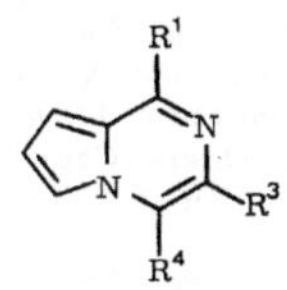

(**201**)

	R^1	R^3	R^4
a)	H	H	H
b)	CH_3	H	H
c)	H	CH_3	H
d)	H	H	CH_3
e)	CH_3	CH_3	H
f)	CH_3	H	CH_3
g)	H	CH_3	CH_3

Recently (*75*), a new bicyclic pyrazine skeleton was found in roasted meat flavor; seven alkylpyrrolo[1,2-*a*] pyrazines (**201**) were identified and the presence of higher homologs was suspected.

References

1. ADESOGAN, E. K.: Trithiolaniacin, a Novel Trithiolan from *Petiveria alliacea*. J. C. S. Chem. Comm. **1974,** 906.

2. AKABORI, S.: Aroma of Shoyu (Soy Sauce). I. J. Chem. Soc. Japan **57,** 828 (1936); Chem. Abstr. **31,** 1355/3 (1937).

3. AMOORE, J. E.: In Molecular Basis of Odor (I. NEWTON KUGELMASS, ed.). Springfield: Ch. C Thomas. 1970.

4. ANDERSEN, K., and D. T. BERNSTEIN: Noxious Chemicals in Skunk Scent Detailed. Chem. Eng. News **52** (37), 16 (1974).

5. ANJOU, K., and E. v. SYDOW: The Aroma of Cranberries. II. *Vaccinium macrocarpon* Ait. Acta Chem. Scand. **21,** 2076 (1967).

6. ANONYMUS: α-Furfurylmercaptan, einer der wichtigsten Duftstoffe des gerösteten Kaffees. Angew. Chem. **62,** 33 (1950).

7. ANONYMUS: Maple Lactone Process. Chem. Eng. News **50** (Febr. 7), 14 (1972).

8. ARNOLD, R. G., L. M. LIBBEY, and E. A. DAY: Identification of Components in the Stale Flavor Fraction of Sterilized Concentrated Milk. J. Food Sci. **31,** 566 (1966).

9. ASAHI, Y., K. TERADA, and M. ISHIO: Polarography of Cyclic Polysulfides and Trithiocarbonates. Rev. Polarogr. **14,** 382 (1969); Chem. Abstr. **71,** 21 598 g (1969).

10. ASINGER, F., M. THIEL und G. LIPPERT: Über die gemeinsame Einwirkung von elementarem Schwefel und gasförmigem Ammoniak auf Ketone. XXIII. Synthese von 1,2,4-Trithiolanen und 1,2,4,5-Tetrathianen. Liebigs Ann. Chem. **627,** 195 (1959).

11. BAILEY, S. D., D. G. MITCHELL, M. L. BAZINET, and C. WEURMAN: Studies on the Volatile Components of Different Varieties of Cocoa Beans. J. Food Sci. **27,** 165 (1962).
12. BALLANCE, P. E.: Production of Volatile Compounds Related to the Flavor of Foods from the STRECKER Degradation of **DL**-Methionine. J. Sci. Food Agric. **12,** 532 (1961).
13. BAMBERGER, E., und A. EINHORN: Zur Kenntnis der im Fuselöl enthaltenen Basen. Ber. dtsch. chem. Ges. **30,** 224 (1897).
14. BATCHELOR, J. F., D. W. H. CLARK und G. A. P. TUEY (MAY and BAKER Ltd.): Furanabkömmlinge enthaltende Kompositionen. DOS 1.955.390 (9. 7. 1970); Chem. Abstr. **73,** 77 034 d (1970).
15. BILLS, D. D., and T. W. KEENAN: Dimethyl Sulfide and Its Precursor in Sweet Corn. J. Agric. Food Chem. **16,** 643 (1968).
16. BISHOP, L. R.: The Evaluation of Beer Flavor. J. Inst. Brew. **77,** 389 (1971).
17. BOELENS, M., L. M. VAN DER LINDE, P. J. DE VALOIS, H. M. VAN DORT, and H. C. TAKKEN: Organic Sulfur Compounds from Fatty Aldehydes, Hydrogen Sulfide, Thiols, and Ammonia as Flavor Constituents. J. Agric. Food Chem. **22,** 1071 (1974).
18. BOELENS, M., P. J. DE VALOIS, H. J. WOBBEN, and A. VAN DER GEN: Volatile Flavor Compounds from Onion. J. Agric. Food Chem. **19,** 984 (1971).
19. BONCHARD, E. F., C. P. HETZEL, and R. D. OLSON (Chas. Pfizer and Co., Inc.): Sweetening Foods with Maltol and Sugar. US. Pat. 3.409.441 (5. 11. 1968); Chem. Abstr. **70,** 56 483 x (1969).
20. BRAMWELL, A. F., J. W. K. BURRELL, and G. RIEZEBOS: Characterisation of Pyrazines in Galbanum Oil. Tetrahedron Letters **1969,** 3215.
21. BRAND, J.: Über Maltol. Ber. dtsch. chem. Ges. **27,** 806 (1894).
22. BRINKMAN, H. W., H. COPIER, J. J. M. DE LEUW, and S. BOEN TJAN: Components Contributing to Beef Flavor. Analysis of the Headspace Volatiles of Beef Broth. J. Agric. Food Chem. **20,** 177 (1972).
23. BRODNITZ, M. H., and J. V. PASCALE: Thiopropanal S-Oxide — a Lachrymatory Factor in Onions. J. Agric. Food Chem. **19,** 269 (1971).
24. BRODNITZ, M. H., C. L. POLLOCK, and P. P. VALLON: Flavor Components of Onion Oil. J. Agric. Food Chem. **17,** 760 (1969).
25. BROWN, D. F., V. J. SENN, J. B. STANLEY, and F. G. DOLLEAR: Comparison of Carbonyl Compounds in Raw and Roasted Runner Peanuts. I. Major Qualitative and Some Quantitative Differences. J. Agric. Food Chem. **20,** 700 (1972).
26. BÜCHI, G., P. DEGEN, F. GAUTSCHI, and B. WILLHALM: Structure and Synthesis of Kahweofuran, a Constituent of Coffee Aroma. J. Organ. Chem. (USA) **36,** 199 (1971).
27. BURRELL, J. W. K., R. A. LUCAS, D. M. MICHALKIEWICZ, and G. RIEZEBOS: Characterisation of Pyrazines in Galbanum Oil. Chem. and Ind. **1970,** 1409.
28. — — — — Characterisation of Thiol Esters in Galbanum Oil. Tetrahedron Letters **1971,** 2837.
29. BUTTERY, R. G.: Some Unusual Volatile Carbonyl Components of Potato Chips. J. Agric. Food Chem. **21,** 31 (1973).
30. BUTTERY, R. G., D. R. BLACK, M. J. LEWIS, and L. LING: A Study of the Fate of Volatile Hop Constituents in Beer. J. Food Sci. **32,** 414 (1967).
31. BUTTERY, R. G., J. L. BOMBEN, D. G. GUADAGNI, and L. C. LING: Some Considerations of the Volatilities of Organic Flavor Compounds in Foods. J. Agric. Food Chem. **19,** 1045 (1971).
32. BUTTERY, R. G., D. G. GUADAGNI, and L. C. LING: Volatile Components of Baked Potatoes. J. Sci. Food Agric. **24,** 1125 (1973).
33. BUTTERY, R. G., D. G. GUADAGNI, L. C. LING, R. M. SEIFERT, and W. LIPTON: Additional Volatile Components of Cabbage, Broccoli, and Cauliflower. J. Agric. Food Chem. **24,** 829 (1976).

34. BUTTERY, R. G., D. G. GUADAGNI, and R. E. LUNDIN: Some 4,5-Dialkylthiazoles with Potent Bell Pepper-Like Aromas. J. Agric. Food Chem. **24,** 1 (1976).
35. BUTTERY, R. G., and L. C. LING: Characterisation of Non-Basic Steam-Volatile Components of Potato Chips. J. Agric. Food Chem. **20,** 698 (1972).
36. — — Earthy Aroma of Potatoes. J. Agric. Food Chem. **21,** 745 (1973).
37. — — Alkylthiazoles in Potato Products. J. Agric. Food Chem. **22,** 912 (1974).
38. BUTTERY, R. G., R. M. SEIFERT, D. G. GUADAGNI, and L. C. LING: Characterization of Some Volatile Constituents of Bell Peppers. J. Agric. Food Chem. **17,** 1322 (1969).
39. — — — — Characterization of Additional Volatile Components of Tomato. J. Agric. Food Chem. **19,** 524 (1971).
40. BUTTERY, R. G., R. M. SEIFERT, and L. C. LING: Characterization of Some Volatile Potato Components. J. Agric. Food Chem. **18,** 538 (1970).
41. — — — Characterization of Some Volatile Constituents of Dry Red Beans. J. Agric. Food Chem. **23,** 516 (1975).
42. BUTTERY, R. G., R. M. SEIFERT, R. E. LUNDIN, and D. G. GUADAGNI (United States Dept. of Agriculture): Food-Flavoring Compositions Containing 2-Methoxy-3-isobutylpyrazine. US. Pat. 3.630.750 (28. 12. 1971); Chem. Abstr. **76,** 139 279 k (1972).
43. — — — — 2-Methoxy-3-isobutylpyrazine. US. Pat. 3.720.672 (13. 3. 1973); Chem. Abstr. **79,** 5364 n (1973).
44. BUTTERY, R. G., R. M. SEIFERT, R. E. LUNDIN, D. G. GUADAGNI, and L. C. LING: Characterization of an Important Aroma Component of Bell Peppers. Chem. and Ind. **1969,** 490.
45. CALABRETTA, P. J.: Odor Characteristics and Thresholds of 2-Methylpyrazine and Derivatives. Cosm. Perf. **90,** 75 (1975); Chem. Abstr. **83,** 183 292 c (1975).
46. CASEY, J. C., R. SELF, and T. SWAIN: Origin of Methanol and Dimethyl Sulfide from Cooked Foods. Nature **200,** 885 (1963).
47. CAZENAVE, P., I. HORMANN, F. MÜGGLER-CHAVAN et R. VIANI: Sur la Composition de l'Arôme de Thé Noir. IV. Helv. Chim. Acta **57,** 206 (1974).
48. CHANG, S. S., C. HIRAI, B. R. REDDY, K. O. HERZ, A. KATO, and G. SIPMA: Isolation and Identification of 2,4,5-Trimethyl-3-oxazoline and 3,5-Dimethyl-1,2,4-trithiolane in the Volatile Flavour Compounds of Boiled Beef. Chem. and Ind. **1968,** 1639.
49. COPIER, H., C. H. T. TONSBEEK (Lever Brothers Co): Flavoring with Certain 2-Acyl-2-thiazolines. US. Pat. 3.778.518 (11. 12. 1973); Chem. Abstr. **80,** 69 405 e (1974).
50. COPIER, H., C. H. T. TONSBEEK, A. J. PLANCKEN, and J. A. LOSEKOOT: Identification of an Acylthiazoline as a Flavour Component of Beef Broth. 160th ACS National Meeting, Chicago, Sept. 1970 — AGFD 60.
51. COTE, L., J. J. OLESON, and J. H. WILLIAMS: Nicotinamide Antagonists. Exp. Med. Surg. **11,** 96 (1953); Chem. Abstr. **48,** 8354 c (1954).
52. DEMOLE, E., and D. BERTHET: A Chemical Study of Burley Tobacco Flavor (*Nicotiana tabacum* L.). I. Volatile to Medium-Volatile Constituents (b.p. ≦84°/0.001 Torr). Helv. Chim. Acta **55,** 1866 (1972).
53. DICKERSON, J. P., D. L. ROBERTS, C. W. MILLER, R. A. LLOYD, and C. E. RIX: Flue-Cured Tobacco Flavour. II. Constituents Arising from Amino Acid — Sugar Reactions. Tob. Sci. **20,** 71 (1976).
54. DIEMAIR, W., und H. HALA: Beitrag zur Bildung von 2-Methyl-3-hydroxypyron (Maltol). Z. Lebensm.-Unters. Forsch. **110,** 161 (1959).
55. DIETRICH, P., E. LEDERER, M. WINTER et M. STOLL: Recherches sur les Arômes. XI. Sur l'Arôme du Cacao. I. Helv. Chim. Acta **47,** 1581 (1964).
56. DIETRICH, P., and E. SUNDT (Firmenich et Cie): Compounds having Malt, Maple,

and Molasses-Type Flavors. Swiss Pat. 482.415 (30. 1. 1970); Chem. Abstr. **72,** 120 261 u (1970).
57. DIETRICH, P., and E. SUNDT (Firmenich SA): Tobacco Compositions Employing Flavoring Agents Comprising Unsaturated Butyrolactone Derivatives and Precursors Thereof. US. Pat. 3.884.247 (20. 5. 1975).
58. DRAWERT, F.: Components of Musts and Wines. V. Proof of Biogenic Amines in Wines and Their Significance. Vitis **5,** 127 (1965); Chem. Abstr. **64,** 1317c (1966).
59. DRAWERT, F., P. SCHREIER und W. SCHERER: Gaschromatographisch-massenspektrometische Untersuchung flüchtiger Inhaltsstoffe des Weines. III. Säuren des Weinaromas. Z. Lebensm.-Unters. Forsch. **155,** 342 (1974).
60. DUMONT, J. P., G. PRADEL, S. ROGER et J. ADDA: Etude des Composés Neutres Volatils Formés au Cours du Gratinage du Gruyère. Le Lait **1976,** No 551/552, 18.
61. DUMONT, J. P., S. ROGER, and J. ADDA: Identification of a Nitrogenous Heterocyclic Compound Responsible for a Potato-Like Off-Flavour in Gruyere de Comte. Le Lait **55,** 479 (1975).
62. ELMENHORST, H.: Nachweis von Methylcyclopentenolon und Aethylcyclopentenolen in Cigarettenrauch und Latakiatabak. Beitr. Tabakforsch. **6,** 205 (1972); Chem. Abstr. **78,** 121 572 g (1973).
63. FAGERSON, I. S.: Thermal Degradation of Carbohydrates. J. Agric. Food Chem. **17,** 747 (1969).
64. FEATHER, M. S., and J. F. HARRIS: Dehydration Reactions of Carbohydrates. Adv. Carbohydrate Chem. Biochem. **28,** 161 (1973) [recent summary].
65. FIECCHI, A., M. GALLI KIENLE, and A. SCALA: Bis-Methylthiomethane, an Odorous Substance from White Truffle, *Tuber magnatum* Pico. Tetrahedron Letters **1967,** 1681.
66. FILIPIC, V. J., J. C. UNDERWOOD, and C. O. WILLITS: The Identification of Methylcyclopentenolone and Other Compounds in Maple Sirup Flavor Extract. J Food Sci. **30,** 1008 (1965).
67. Firmenich & Co: Dérivés Pyraziniques (Methoxy- and Methylthiopyrazines). Fr. Pat. 1.391.212 (5. 3. 1965); Chem. Abstr. **62,** 16 270 f (1965).
68. FLAMENT, I. (Firmenich SA): Literatur-Compilation and Unpublished Results.
69. — Unpublished Results.
70. — (Firmenich et Cie): Compositions Aromatisantes et leur Application. Fr. Dem. 2.128.744 (8. 3. 1972); Chem. Abstr. **78,** 109 541 p (1973).
71. — (Firmenich SA): Utilisation de Dérivés de la Pyrazine comme Ingrédients Aromatisants. Swiss Pat. 559.518 (31. 1. 1975).
72. — Heterocyclics in Flavour Chemistry. Proc. int. Symp. Aroma Research, Zeist, p. 221. Wageningen: Pudoc. 1975.
73. FLAMENT, I., M. KOHLER et R. ASCHIERO: Sur l'Arôme de Viande de Bœuf Grillée. II. Dihydro-6,7-5*H*-cyclopenta[*b*]pyrazines; Identification et Mode de Formation. Helv. Chim. Acta **59,** 2308 (1976).
74. FLAMENT, I., P. SONNAY et G. OHLOFF: Pyrazines. II. Synthèse de Dihydro-6,7-5*H*-cyclopenta[*b*]pyrazines alkylées. Helv. Chim. Acta **56,** 610 (1973).
75. — — — Sur l'Arôme de Viande de Bœuf Grillée. III. Pyrrolo[1,2-*a*]pyrazines, Identification et Synthèse. Helv. Chim. Acta **60,** 1872 (1977).
76. FLAMENT, I., P. SONNAY, B. WILLHALM et G. OHLOFF: Pyrazines. III. Synthèse de 5*H*-cyclopenta[*b*]pyrazines alkylées. Helv. Chim. Acta **59,** 2314 (1976).
77. FLAMENT, I., B. WILLHALM et M. STOLL: Recherches sur les Arômes. 16. Sur l'Arôme du Cacao. III. Helv. Chim. Acta **50,** 2233 (1967).
78. FORSS, D. A., W. STARK, and G. URBACH: Volatile Compounds in Butter Oil. I. Lower-Boiling Compounds. J. Dairy Res. **34,** 131 (1967).
79. FREEMAN, G. G., and R. J. WHENHAM: Nature and Origin of Volatile Flavour Components of Onion and Related Species. Flavours **7,** 222 (1976).

80. Friedel, P., V. Krampl, T. Radford, J. A. Renner, F. W. Shephard, and M. A. Gianturco: Some Constituents of the Aroma Complex of Coffee. J. Agric. Food Chem. **19,** 530 (1971).
81. Friis, P., and A. Kjaer: 4-Methylthio-3-butenylisothiocyanate, the Pungent Principle of Radish Root. Acta Chem. Scand. **20,** 698 (1966).
82. Fujimaki, M., S. Kato, and T. Kurata: Pyrolysis of Sulfur-Containing Amino Acids. Agric. Biol. Chem. **33,** 1144 (1969).
83. Gen, A. van der: Corps Olfactifs à l'Odeur de Jasmin. Parf. Cosm. Sav. France **2,** 356 (1972).
84. Gianturco, M. A., A. S. Giammarino, and P. Friedel: Volatile Constituents of Coffee. Nature **210,** 1358 (1966).
85. Gianturco, M. A., A. S. Giammarino, P. Friedel, and V. Flanagan: The Volatile Constituents of Coffee. IV. Furanic and Pyrrolic Compounds. Tetrahedron **20,** 2951 (1964).
86. Gianturco, M. A., A. S. Giammarino, and R. G. Pitcher: The Structures of Five Cyclic Diketones Isolated from Coffee. Tetrahedron **19,** 2051 (1963).
87. Gilbert, J., and H. E. Nursten: Volatile Constituents of Horseradish Roots. J. Sci. Food Agric. **23,** 527 (1972).
88. Golovnya, R. V.: Zur gaschromatographischen Analyse von Aminen als Aromakomponenten von Lebensmitteln. Ernährungsforsch. **15,** 333 (1970).
89. Golovnya, R. V., G. A. Mironov, and I. L. Zhuraveleva: Gas-chromatographic Analysis of Amines in Volatile Components of Caviar. Zh. Anal. Khim. **22,** 612, 956 (1967); Chem. Abstr. **67,** 52 845 g (1967).
90. Golovnya, R. V., G. A. Mironov, I. L. Zhuravleva, and R. M. Abdullina: Organic Bases in Volatile Components of Cheese. Moloch. Prom. **29,** 11 (1968); Chem. Abstr. **68,** 113 416 n (1968).
91. Guadagni, D. G., R. G. Buttery, and S. Okano: Odour Thresholds of Some Organic Compounds Associated with Food Flavors. J. Sci. Food Agric. **14,** 761 (1963).
92. Guadagni, D. G., R. G. Buttery, and R. M. Seifert (United States Dept. of Agriculture): Imparting a Potato Flavor by Adding the Hydrochloric Acid Salt of 2-Methoxy-3-ethylpyrazine. US. Pat. 3.772.039 (13. 11. 1973); Chem. Abstr. **80,** 58 618 a (1974).
93. Guadagni, D. G., R. G. Buttery, R. M. Seifert, and D. Venstrom: Flavor Enhancement of Potato Products. J. Food Sci. **36,** 363 (1971).
94. Guadagni, D. G., R. Buttery, and J. G. Turnbaugh: Odour Thresholds and Similarity Ratings of Some Potato Chip Components. J. Sci. Food Agric. **23,** 1435 (1972).
95. Guadagni, D. G., J. C. Miers, and D. W. Venstrom: Concentration Effect on Odor Addition or Synergism in Mixtures of Methyl Sulfide and Tomato Juice. J. Food Sci. **34,** 630 (1969).
96. Gubler, B., et K. Jaeggi (Givaudan, L., et Cie. S.A.): 4-Methyl-5-vinylthiazole. Fr. Dem. 2.198.947 (5. 4. 1974); Chem. Abstr. **82,** 170 885 c (1975).
97. Gumbmann, M. R., and H. K. Burr: Volatile Sulfur Compounds in Potatoes. J. Agric. Food Chem. **12,** 404 (1964).
98. Haagen-Smit, A. J., J. G. Kirchner, C. L. Deasy, and A. N. Prater: Chemical Studies of Pineapple (*Ananas sativus* L.). II. Isolation and Identification of a Sulfur-Containing Ester in Pineapple. J. Amer. Chem. Soc. **67,** 1651 (1945).
99. Harding, R. J., H. E. Nursten, and J. J. Wren: Basic Compounds Contributing to Beer Flavour. J. Sci. Food Agric. **28,** 225 (1977).
100. Harrison, G. A. F.: The Flavor of Beer -- A Review. J. Inst. Brew. **76,** 486 (1970).

101. HAWKSWORTH, G., and R. R. SCHELINE: Metabolism in the Rat of Pyrazine Derivatives Having Flavor Importance in Foods. Xenobiotica **1975**, 389; Chem. Abstr. **83**, 188 835 s (1975).

102. HAZENIAC*, S. J., and R. M. HALL: Flavor Chemistry of Tomato Volatiles. J. Food Sci. **35**, 519 (1970).

103. HERZ, K. O.: A Study of the Nature of Boiled Beef Flavor. Thesis, Rutgers — The State University. Order No 68-14,248, University Microfilms, Ann Arbor/Mich. 1968.

104. HETZEL, D. S. (Pfizer Inc., USA): Alkylsubstituierte 3-Oxazoline als Geschmacksstoffe. DOS 2.120.577 (25. 11. 1971); Chem. Abstr. **76**, 72 505 k (1972).

105. HEYNS, K.: Lebensmittelchemische Probleme aus der Sicht des Naturstoffchemikers. Dtsch. Lebensm.-Rundsch. **73**, 111 (1977).

106. HEYNS, K., R. STUTE, and H. PAULSEN: Browning Reaction and Fragmentation of Carbohydrates. I. Volatile Products from the Thermal Degradation of **D**-Glucose. Carbohydrate Res. **2**, 132 (1966); Chem. Abstr. **65**, 7252 c (1966).

107. HICKS, K. B., and M. S. FEATHER: Studies on the Mechanism of Formation of 4-Hydroxy-5-methyl-3(2*H*)-furanone, a Component of Beef Flavor, from Amadori Products. J. Agric. Food Chem. **23**, 957 (1975).

108. HICKS, K. B., D. W. HARRIS, M. S. FEATHER, and R. N. LOEPPKY: Production of 4-Hydroxy-5-methyl-3(2*H*)-furanone, a Component of Beef Flavor, from a 1-Amino-1-deoxy-**D**-fructuronic Acid. J. Agric. Food Chem. **22**, 724 (1974).

109. HIRAI, C., K. O. HERZ, J. PODKORNY, and S. S. CHANG: Isolation and Identification of Volatile Flavor Compounds in Boiled Beef. J. Food Sci. **38**, 393 (1973); HIRAI, C., K. O. HERZ, B. R. REDDY, and S. S. CHANG: Isolation and Identification of Volatile Flavor Compounds in Boiled Beef. 28th Annual Meeting Inst. Food Technologists, Philadelphia, 1968.

110. HODGE, J. E.: Origin of Flavors in Foods. Nonenzymatic Browning Reactions. In: Symposium on Foods. The Chemistry and Physiology of Flavors. The Fourth in a Series of Symposia on Foods Held at Oregon State University, p. 465. Westport, Connecticut: The Avi Publ. Comp., Inc. 1967.

111. HODGE, J. E., B. E. FISHER, and E. C. NELSON: Dicarbonyls, Reductones and Heterocyclics Produced by the Reactions of Reducing Sugars with Secondary Amine Salts. Amer. Soc. Brewing Chemists Proc. **1963**, 84.

112. HODGE, J. E., F. D. MILLS, and B. E. FISHER: Compounds of Browned Flavor Derived from Sugar-Amine Reactions. Cereal Sci. Today **17**, 34 (1972).

113. HUNTER, G. L. K., W. A. BUCEK, and T. RADFORD: Volatile Components of Canned Alphonso Mango. J. Food Sci. **39**, 900 (1974).

114. HUNTER, I. R., M. K. WALDEN, W. H. MCFADDEN, and J. W. PENCE: Production of Bread-Like Aromas from Proline and Glycerol. Cereal Sci. Today **11**, 493 (1966).

115. HUNTER, I. R., M. K. WALDEN, J. R. SCHERER, and R. E. LUNDIN: Preparation and Properties of 1,4,5,6-Tetrahydro-2-acetoxypyridine, a Cracker-Odor Constituent of Bread Aroma. Cereal Chem. **46**, 189 (1969); Chem. Abstr. **71**, 111 609 r (1969).

116. ITO, H.: Isolation and Identification of the Characteristic Sweet-Aroma Compounds in Refinery Final Molasses. Agric. Biol. Chem. **40**, 827 (1976).

117. JACKSON, H. W.: Flavor Research on Cheese. Perf. Essent. Oil Rec. **49**, 256 (1958).

118. JASSMANN, E., und H. SCHULZ: Ringschlußreaktionen in der Ephedrinreihe. 2. Synthese neuer Δ^3-Oxazoline. Pharmazie **18**, 527 (1963); Chem. Abstr. **60**, 10 664 b (1964).

119. JENNINGS, W. G., and M. R. SEVENANTS: Volatile Esters of Bartlett Pear. III. J. Food Sci. **29**, 158 (1964).

* Ought to be: KAZENIAC.

120. JOHNSON, R. R., E. D. ALFORD, and G. W. KINZER: Formation of Sucrose Pyrolysis Products. J. Agric. Food Chem. **17**, 22 (1969).

121. JONES, N. R.: Fish Flavors. In: Symposium on Foods. The Chemistry and Physiology of Flavors. The Fourth in a Series of Symposia on Foods Held at Oregon State University, p. 267. Westport, Connecticut: The Avi Publ. Comp., Inc. 1967.

122. KALLIO, H.: Identification of Vacuum Steam-Distilled Aroma Compounds in the Press Juice of Arctic Bramble, *Rubus arcticus* L. J. Food Sci. **41**, 555 (1976);
— Development of Volatile Aroma Compounds in Artic Bramble, *Rubus arcticus* L., ibid. **41**, 563 (1976).

123. KATO, H., and M. FUJIMAKI: Formation of N-Substituted Pyrrole-2-aldehydes in the Browning Reaction Between D-Xylose and Amino Compounds. J. Food Sci. **33**, 445 (1968).

124. KATO, S., T. KURATA, and M. FUJIMAKI: Volatile Compounds Produced by the Reaction of L-Cysteine or L-Cystine with Carbonyl Compounds. Agric. Biol. Chem. **37**, 539 (1973).

125. KATO, S., T. KURATA, S. ISHIGURO, and M. FUJIMAKI: Additional Volatile Compounds Produced by Pyrolysis of Sulfur-Containing Amino Acids. Agric. Biol. Chem. **37**, 1759 (1973).

126. KAZENIAC, S. J., and R. M. HALL: Fresh Tomato Flavor Style. Food Technol. **23**, 58 (1969);
— — Flavor Chemistry of Tomato Volatiles. J. Food Sci. **35**, 519 (1970).

127. KEENEY, M., and E. A. DAY: Probable Role on the Strecker Degradation of Amino Acids in Development of Cheese Flavor. J. Dairy Sci. **40**, 874 (1957).

128. KENDALL, E. C., and Z. G. HAJOS: Tetrahydro-3,4-furandione. I. Preparation and Properties. J. Amer. Chem. Soc. **82**, 3219 (1960).

129. KING, B., E. DEMOLE, and A. F. THOMAS (Firmenich et Cie): 4-Hydroxy-2,5-dimethyl-2,3-dihydrothiophen-3-one, useful as flavoring. DOS 2.214.540 (5. 10. 1972); Chem. Abstr. **78**, 29 615a (1973).

130. KINLIN, T. E., R. MURALIDHARA, A. O. PITTET, A. SANDERSON, and J. P. WALRADT: Volatile Components of Roasted Filberts. J. Agric. Food Chem. **20**, 1021 (1972).

131. KINOSHITA, S., and T. YAMANISHI: Identification of Basic Aroma Components of Roasted Sesame Seeds. Nippon Nogei Kagaku Kaishi **47**, 737 (1973); Chem. Abstr. **80**, 106 964j (1974).

132. KIRIBUCHI, T., and T. YAMANISHI: Studies on the Flavor of Green Tea — Dimethyl Sulfide and Its Precursor. Agric. Biol. Chem. **27**, 56 (1963).

133. KJAER, A., M. SPONHOLTZ, K. O. ABRAHAM, M. L. SHANKARANARAYANA, R. RAGHAVAN, and C. P. NATARAJAN: 2-Butyl Propenyl Disulfids from Asafetida: Separation, Characterization, and Absolute Configuration. Acta Chem. Scand. B **30**, 137 (1976).

134. KOBAYASHI, N., and M. FUJIMAKI: On the Formation of Mercaptoacetaldehyde, Hydrogen Sulfide and Acetaldehyde on Boiling Cysteine with Carbonyl Compounds. Agric. Biol. Chem. **29**, 698 (1965).

135. KOBAYASHI, N., and M. FUJIMAKI: On the Formation of N-Acetonyl Pyrrole on Roasting Hydroxyproline with Carbonyl Compounds. Agric. Biol. Chem. **29**, 1059 (1965).

136. KUNG, J.-F., and F. MARTIN (General Foods Corp.): Use of 2-Methyl-6-ethoxy-pyrazine for Pineapple Flavor. US. Pat. 3.803.331 (9. 4. 1974); Chem. Abstr. **81**, 62 394n (1974).

137. KUROIWA, Y., and H. HASHIMOTO: Composition of Sun-Struck Flavor of Beer. Agric. Biol. Chem. **25**, 257 (1961).

138. KUROIWA, Y., H. HASHIMOTO, L. HASHIMOTO, E. KOKULO, and K. NAKAGAWA: Factors Essential for the Evolution of Sun-Struck Flavor. J. Inst. Brew. **78**, 78 (1967).

139. KUSHNER, S., H. DALALIAN, J. L. SANJURJO, F. L. BACH JR., S. R. SAFIR, V. K. SMITH JR., and J. H. WILLIAMS: Experimental Chemotherapy of Tuberculosis. II. The Synthesis of Pyrazinamides and Related Compounds. J. Amer. Chem. Soc. **74**, 3617 (1952).
140. LAMPARSKY, D., and P. SCHUDEL: *p*-Menthane-8-thiol-3-one, a New Component of Buchu Leaf Oil. Tetrahedron Letters **1971**, 3323.
141. LANGLER, J. E., L. M. LIBBEY, and E. A. DAY: Identification and Evaluation of Selected Compounds in Swiss Cheese Flavor. J. Agric. Food Chem. **15**, 386 (1967).
142. LEDL, F.: Analysis of a Synthetic Onion Aroma. Z. Lebensm.-Unters. Forsch. **157**, 28 (1975).
143. — Thermische Umwandlung von Cystein, Cystin, N-Acetylcystein, Thiazolidincarbonsäure und Cysteinmethylester in Sojaöl. Z. Lebensm.-Unters. Forsch. **161**, 125 (1976).
144. LEDL, F., und T. SEVERIN: Thermische Zersetzung von Cystein und Xylose in Tributyrin. Chem., Mikrobiol., Technol. Lebensm. **2**, 155 (1973); Chem. Abstr. **80**, 119 312 v (1974).
145. — — Schwefelhaltige Verbindungen aus Cystein und Xylose (Sulfur Containing Compounds from Cysteine and Xylose). Z. Lebensm.-Unters. Forsch. **154**, 29 (1974).
146. LEIR, C. M.: A New Synthesis of 3-Alkyl-2-cyclopenten-2-ol-1-ones. J. Organ. Chem. (USA) **35**, 3203 (1970).
147. LLOYD, R. A., C. W. MILLER, D. L. ROBERTS, J. A. GILES, J. P. DICKERSON, N. H. NELSON, C. E. RIX, and P. H. AYERS: Flue-Cured Tobacco Flavor. 1. Essence and Essential Oil Components. Tob. Sci. **20**, 43 (1976).
148. LOPEZ, A. S., and V. C. QUESNEL: Methyl-*S*-methionine Sulphonium Salt: A Precursor of Dimethyl Sulphide in Cacao. J. Sci. Food Agric. **27**, 85 (1976).
149. MAGA, J. A.: Bread Flavor. CRC Crit. Rev. Food Technol. **5**, 55 (1974) [recent summary of that subject].
150. — Thiazoles in Food. In: Fenaroli's Handbook of Flavor Ingredients (T. E. FURIA and N. BELLANCA, eds.), 2nd ed., Vol. I, p. 228. Cleveland, Ohio: CRC Press Inc. 1975.
151. — The Role of Sulfur Compounds in Food Flavor. I. Thiazoles. CRC Crit. Rev. Food Sci. Nutrition **6**, 153 (1975).
152. — The Role of Sulfur Compounds in Food Flavor. II. Thiophenes. CRC Crit. Rev. Food Sci. Nutrition **6**, 241 (1975).
153. — The Role of Sulfur Compounds in Food Flavor. III. Thiols. CRC Crit. Rev. Food Sci. Nutrition **7**, 147 (1976).
154. MAGA, J. A., and C. E. SIZER: Pyrazines in Foods. A Review. J. Agric. Food Chem. **21**, 22 (1973).
155. MANLEY, C. H., and I. S. FAGERSON: Major Volatile Neutral and Acid Compounds of Hydrolyzed Soy Protein. J. Food Sci. **35**, 286 (1970).
156. MANLEY, C. H., P. P. VALLON, and R. E. ERICKSON: Some Aroma Components of Roasted Sesame Seed (*Sesamum indicum* L.). J. Food Sci. **39**, 73 (1974).
157. MANNICH, C., and P. FRESENIUS: Chief Constituent of the Essential Oil from Asafoetida. Arch. Pharm. **274**, 461 (1936).
158. MARION, J. P., F. MÜGGLER-CHAVAN, R. VIANI, J. BRICOUT, D. REYMOND et R. H. EGLI: Sur la Composition de l'Arôme de Cacao. Helv. Chim. Acta **50**, 1509 (1967).
159. MASON, M. E., B. JOHNSON, and M. C. HAMMING: Volatile Components of Roasted Peanuts. The Major Monocarbonyls and some Noncarbonyl Components. J. Agric. Food Chem. **15**, 66 (1967).
160. MCLEOD, A. J.: The Chemistry of Vegetable Flavor. Flavour Ind. **1**, 665 (1970); Chem. Abstr. **74**, 63 202 r (1971).
161. MEIGH, D. F., A. A. FILMER, and R. SELF: Production of Growth Suppressing Volatile

Substances by Stored Potato Tubers. II. Growth-Inhibitory Volatile Compounds Produced by *Solanum tuberosum* Tubers. Phytochem. **12,** 987 (1973).
162. MEILGAARD, M. C.: Flavor Chemistry of Beer. II. Flavor Threshold of 239 Aroma Volatiles. Tech. Q. Master Brew. Assoc. Am. **12,** 151 (1975).
163. MERRITT, C., JR., and J. M. MENDELSON: Flavor in Fish. Abstr. Am. Chem. Soc. Natl. Meet. **142,** 14A (1962).
164. MEYERFELD, J.: Über einen neuen im Holzessig vorkommenden Körper. Chem. Ztg. **36,** 549 (1912); Chem. Zbl. **83,** II, 171 (1912).
165. MIERS, J. C.: Formation of Volatile Sulfur Compounds in Processes Tomato Products. J. Agric. Food Chem. **14,** 419 (1966).
166. MILLER III, A., R. A. SCANLAN, J. S. LEE, L. M. LIBBEY, and M. E. MORGAN: Volatile Compounds Produced in Sterile Fish Muscle *(Sebastes melanops)* by *Pseudomonas perolens.* Appl. Microbiol. **25,** 257 (1973); Chem. Abstr. **78,** 121 011 y (1973).
167. MILLS, F. D., B. G. BAKER, and J. E. HODGE: Amadori Compounds as Nonvolatile Flavor Precursors in Processed Food. J. Agric. Food Chem. **17,** 723 (1969).
168. MILLS, F. D., D. WEISLEDER, and J. E. HODGE: 2,3-Dihydro-3,5-dihydroxy-6-methyl-4*H*-pyran-4-one, a Novel Nonenzymatic Browning Product. Tetrahedron Letters **1970,** 1243.
169. MOOKHERJEE, B. D., M. G. BEETS, A. O. PITTET et al. (International Flavors and Fragrances, Inc.): Neue Pyrazinderivate, ihre Herstellung und Anwendung zum Würzen bzw. Aromatisieren. (Pyrazine Derivates for Seasoning or Flavouring). DOS 2.117.926 (14. 4. 1971); Chem. Abstr. **76,** 25 313 m (1972).
170. MOOKHERJEE, B. D., and E. M. KLAIBER: The Synthesis of 3-Alkyl-2-pyrazinyl Methyl Ketones and Related Compounds. J. Organ. Chem. (USA) **37,** 511 (1972).
171. MORGAN, M. E., and R. L. PEREIRA: Volatile Constituents of Grass and Corn Silage. II. Gas-Entrained Aroma. J. Dairy Sci. **45,** 467 (1962).
172. MORIARTY, R. M., N. ISHIBE, M. KAYSER, K. C. RAMEY, and H. J. GISLER, JR.: Conformational Studies on Seven-Membered Rings Containing the Disulfide and Trisulfide Groups. Tetrahedron Letters **1969,** 4883.
173. MORITA, K., and S. KOBAYASHI: Isolation and Synthesis of Lenthionine an Odorous Substance of *Shiitake,* an Edible Mushroom. Tetrahedron Letters **1966,** 573.
174. — — Isolation, Structure, and Synthesis of Lenthionine and Its Analogs. Chem. Pharm. Bull (Japan) **15,** 988 (1967); Chem. Abstr. **68,** 21 910 r (1968).
175. — — (Takeda Chem. Industries, Ltd.): Lenthionin. Jap. Pat. 67.20319 (11. 10. 1967); Chem. Abstr. **69,** 35 975 y (1968).
176. MOTOHIRO, T.: Studies on the Petroleum-Odor in Canned Chum Salmon. Mem. Fac. Fisheries Hokkaido **10** (1), 65 (1962); Chem. Abstr. **59,** 12 090e (1963).
177. MOZIN, E. C.: Sur les Bases Extraites des Liquides Ayant Subi la Fermentation Alcoolique. C. R. hebd. séances Acad. Sci. **106,** 360 (1888).
178. MÜGGLER-CHAVAN, F., et D. REYMOND: Constituants Aromatiques de Cacaos de Diverses Provenances. Mitt. Lebensmittelunters. Hyg. **58,** 466 (1967); Chem. Abstr. **69,** 95 245 v (1968).
179. MULDERS, E. J.: Odor of White Bread. Versl. Landbouwk. Onderz. **1973,** No 798, 19; Chem. Abstr. **80,** 13 787 p (1974).
180. — The Odour of White Bread. IV. Quantitative Determination of Constituents in the Vapor and Their Odour Values. Z. Lebensm.-Unters. Forsch. **151,** 310 (1973).
181. — Volatile Components from the Non-Enzymic Browning Reaction of the Cysteine/Cystine-Ribose System. Z. Lebensm.-Unters. Forsch. **152,** 193 (1973).
182. MULDERS, E. J., R. J. C. KLEIPOOL, and M. C. TEN NOEVER DE BRAUW: 2-[(Methyldithio)methyl]furan: A Characteristic Aroma Compound of White Bread. Chem. and Ind. **1976,** 613.

183. MULDERS, E. J., M. C. TEN NOEVER DE BRAUW, and S. VAN STRATEN: The Odour of White Bread. II. Identification of Components in Pentane-ether Extracts. Z. Lebensm.-Unters. Forsch. **150,** 306 (1973).
184. MULLER, C. J., R. E. KEPNER, and A. D. WEBB: Identification of 3-(Methylthio)propanol as an Aroma Constituent in Cabernet Sauvignon and Ruby Cabernet Wines. Amer. J. Enol. Viticult. **22,** 156 (1971); Chem. Abstr. **76,** 2516r (1972).
185. MURRAY, K. E., J. SHIPTON, and F. B. WHITFIELD: 2-Methoxypyrazines and the Flavour of Green Peas *(Pisum sativum).* Chem. and Ind. **1970,** 897.
186. MUSSINAN, C. J., and I. KATZ: Isolation and Identification of Some Sulfur Chemicals Present in two Model Systems Approximating Cooked Meat. J. Agric. Food Chem. **21,** 43 (1973).
187. MUSSINAN, C. J., and J. P. WALRADT: Volatile Constituents of Pressure Cooked Pork Liver. J. Agric. Food Chem. **22,** 827 (1974).
188. MUSSINAN, C. J., R. A. WILSON, and I. KATZ: Isolation and Identification of Pyrazines Present in Pressure-Cooked Beef. J. Agric. Food Chem. **21,** 871 (1973).
189. MUSSINAN, C. J., R. A. WILSON, I. KATZ, A. SANDERSON, and M. H. VOCK: Identification and Flavor Properties of Some 3-Oxazolines and 3-Thiazolines Isolated from Cooked Beef. Abstracts of papers, 170th ACS National Meeting, Aug. 1975, AGFD 22.
190. NÄF-MÜLLER, R., und B. WILLHALM: Über die flüchtigen Anteile der Ananas. Helv. Chim. Acta **54,** 1880 (1971).
191. NAIMIE, H., Z. SAMEK, L. DOLEJS, and Z. REHACEK: Volatile Components of Asafoetida. Two Novel Naturally Occurring Sulfur Compounds With Pesticidal Activity. Collect. Czech. Chem. Comm. **37,** 1166 (1972).
192. NAKATANI, Y., et Y. YANATORI: Synthèse Facile de Pyrazines Disubstituées en 2 et 3. Agric. Biol. Chem. **37,** 1509 (1973).
193. NAKEL, G. M., and L. V. HAYNES: Structural Identification of the Methoxymethylpyrazine Isomers. J. Agric. Food Chem. **20,** 682 (1972).
194. NAVES, Y. R.: Sur Quelques Constituants Inédits de la Gomme-Résine de Galbanum. Parf. Cosm. Savons **12,** 586 (1969).
195. NEURATH, G. B.: Recent Advances in Knowledge of the Chemical Composition of Tobac. Chem. Tob. Smoke, Proc. Symp. 1972, p. 77; Chem. Abstr. **77,** 111 705t (1972).
196. NEURATH, G., M. DÜNGER und I. KÜSTERMANN: Untersuchung der „Semi-Volatiles" des Cigarettenrauches. Beitr. Tabakforsch. **6** (1), 12 (1971); Chem. Abstr. **76,** 70 325c (1972).
197. NIEFIND, H. J.: Ph. D. Dissertation, Technical University, Berlin 1969; cited in Wallerstein Lab. Comm. **34,** 199 (1971).
198. NISHIKAWA, M., K. KAMIYA, S. KOBAYASHI, K. MORITA, and Y. TOMIIE: The X-Ray Analysis of Lenthionine, an Odorous Substance of *Shiitake* an Edible Mushroom. Chem. Pharm. Bull. (Japan) **15,** 756 (1967); Chem. Abstr. **67,** 94 826g (1967).
199. NISHIMURA, H., A. ASAHI, K. FUJIWARA, J. MIZUTANI, and Y. OBATA: Changes in Flavor Components of Onion by γ-Irradiation. Agric. Biol. Chem. **35,** 1831 (1971).
200. NUNOMURA, N., M. SASAKI, Y. ASAO, and T. YOKOTSUKA: Isolation and Identification of 4-Hydroxy-2(or 5)-ethyl-5(or 2)-methyl-3(2*H*)-furanone, as a Flavor Component in *Shoyu* (soy sauce). Agric. Biol. Chem. **40,** 491 (1976).
201. NURSTEN, H. E., and M. R. SHEEN: Volatile Flavour Components of Cooked Potato. J. Sci. Food Agric. **25,** 643 (1974).
202. OBATA, Y., and H. TANAKA: Studies on Sunlight Flavor of Beer. 9. Formation of Prenyl Mercaptan from *S*-Prenyl-**L**-cysteine by Sunlight. Agric. Biol. Chem. **29,** 104 (1965).
203. OHLOFF, G.: Die Chemie des Geruchssinns. Chemie in unserer Zeit **5,** 114 (1971).

204. — Classification and Genesis of Food Flavors. 3rd International Congress of Food Science and Technology, SOS, Washington 1970. Proceedings, p. 368. Chicago, Ill. 60 601: Inst. Food Technol. 1971; reprinted from Flavour Ind. **3,** 501 (1972).

205. — Les corps aromatiques dérivés de sucres. Parf. Cosm. Arômes **2,** 55 (1975).

206. — Recent Developments in the Field of Naturally-Occurring Aroma Compounds. In: Fortschritte der Chemie organischer Naturstoffe **35,** p. 431 (W. Herz, H. Grisebach, and G. W. Kirby, eds.). Wien-New York: Springer. 1978.

207. Ouweland, G. A. M., van den, and H. G. Peer (Unilever NV., Rotterdam): Furans and Thiophenes — Meat-Like Flavor Components. Germ. Offen. 1.932.800 (8. 1. 1970); Chem. Abstr. **72,** 90 265 (1970).

208. Parliment, T. H., and M. F. Epstein: Organoleptic Properties of Some Alkyl-Substituted Alkoxy- and Alkylthiopyrazines. J. Agric. Food Chem. **21,** 714 (1973).

209. Parliment, T. H., M. F. Epstein, W. P. Clinton, R. Scrapellino, and R. J. Soukup (General Foods Corp.): Enhancing Coffee Aroma with Substituted Pyrazines. Fr. Pat. 2.062.168 (30. 7. 1971); Chem. Abstr. **76,** 139 274e (1972).

210. Partipharm AG (Basel): Verwendung von 3-Hydroxy-2-alkylpyronen in Bitterstoff enthaltenden Zubereitungen. Swiss Pat. 441.543 (31. 10. 1969).

211. Patterson, R. L. S.: Catty Odours in Food: Their Production in Meat Stores from Mesityl Oxide in Paint Solvents. Chem. and Ind. **1968,** 548.

212. — Catty Odours in Food: Confirmation of the Identity of 4-Mercapto-4-methylpentan-2-one by GC-MS. Chem. and Ind. **1969,** 48.

213. Patton, S.: The Mechanism of Sunlight Flavor Formation in Milk with Special Reference to Methionine and Riboflavin. J. Dairy Sci. **37,** 446 (1954).

214. Pearce, T. J. P., J. M. Peacock, F. Aylward, and D. R. Haisman: Catty Odours in Food: Reactions between Hydrogen Sulfide and Unsaturated Ketones. Chem. and Ind. **1967,** 1562.

215. Peer, H. G., and G. A. M. van den Ouweland: Synthesis of 4-Hydroxy-5-methyl-2,3-dihydro-3-furanone from **D**-Ribose 5-phosphate. Rec. trav. chim. Pays-Bas **87,** 1017 (1968).

216. Peer, H. G., G. A. M. van den Ouweland, and C. N. de Groot: Reaction of Aldopentoses and Secondary Amine Salts, a Convenient Method of Preparing 4-Hydroxy-5-methyl-2,3-dihydro-3-furanone. Rec. trav. chim. Pays-Bas **87,** 1011 (1968).

217. Peratoner, A., und A. Tamburello: Über die Konstitution des Maltols. Giorn. di Scienze Nat. ed. Econ **25,** 272 (1905); Chem. Zbl. **76** (II), 680 (1905).

218. Persson, T., and E. von Sydow: Aroma and Canned Beef. Gaschromatographic and Mass Spectrometric Analysis of the Volatiles. J. Food Sci. **38,** 377 (1973).

219. Peterson, R. J., H. J. Izzo, E. Jungermann, and S. S. Chang: Changes in Volatile Flavor Compounds During the Retorting of Canned Beef Stew. J. Food Sci. **40,** 948 (1975).

220. Pfizer, Chas. and Co., Inc.: Potent Flavor Enhancer Under Development for Carbohydrate Foods. Strengthens Inherent Flavor and Aroma of Many Products. Food Processing Marketing **27** (5), 56 (1966).

221. Pittet, A. O., and D. E. Hruza: Comparative Study of Flavor Properties of Thiazole Derivatives. J. Agric. Food Chem. **22,** 264 (1974).

222. Pittet, A. O., P. Rittersbacher, and R. Muralidhara: Flavor Properties of Compounds Related to Maltol and Isomaltol. J. Agric. Food Chem. **18,** 929 (1970).

223. Pokorny, J.: Aroma of Beef and Its Changes During Processing. Prum. Potravin **21,** 262 (1970); Chem. Abstr. **73,** 119 365v (1970).

224. Polak's Frutal Works and Douwe Egberts Koninklijke Tabaksfabriek, Koffiebranderijen, Theehandel N. V. (Rotterdam): Coffee and Cocoa Flavoring Agents. Neth. Appl. 68 12 899 (13. 3. 1969); Chem. Abstr. **71,** 61 421a (1969).

225. PRAAG, M. VAN, H. S. STEIN, and M. S. TIBBETTS: Steam Volatile Aroma Constituents of Roasted Cocoa Beans. J. Agric. Food Chem. **16,** 1005 (1968).
226. PYYSALO, T.: Studies on the Volatile Compounds of Some Berries in Genus *Rubus*, specially Cloudberry (*Rubus chamaemorus* L.) and Hybrids Between Raspberry (*Rubus idaeus* L.) and arctic bramble (*Rubus arcticus* L.). Materials and Processing Technology, Technical Research Centre of Finland, Publication 14. Helsinki: 1976.
227. QVIST, I. H., and E. C. F. VON SYDOW: Unconventional Proteins as Aroma Precursors. Chemical Analysis of the Volatile Compounds in Heated Soy, Casein, and Fish Protein Model Systems. J. Agric. Food Chem. **22,** 1077 (1974).
228. RANKINE, B. C.: Nature, Origin, and Prevention of Hydrogen Sulfide Aroma in Wine. J. Sci. Food Agric. **14,** 79 (1963).
229. RE, L., B. MAURER und G. OHLOFF: Ein einfacher Zugang zu 4-Hydroxy-2,5-dimethyl-3(2*H*)-furanon (Furaneol), einem Aromabestandteil von Ananas und Erdbeere. Helv. Chim. Acta **56,** 1882 (1973).
230. REDDY, M. C., R. BASSETTE, G. WARD, and J. R. DUNHAM: Relationship of Methyl Sulfide and Flavor Score of Milk. J. Dairy Sci. **50,** 147 (1967).
231. REDDY, M. C., R. C. LINDSAY, and D. D. BILLS: Ester Production by *Pseudomonas fragi*. III. Synergistic Flavor Interaction of Esters at Subthreshold Concentrations. J. Dairy Sci. **52,** 1198 (1969).
232. REICHSTEIN, T., und H. STAUDINGER (Internationale Nahrungs- und Genussmittel-AG., Schaffhausen/Schweiz): Herstellung von Kaffeeersatzmitteln. Brit. P. 260.960 (1926); Chem. Zbl. **62,** 292 (1950); see also Angew. Chem. **62,** 292 (1950).
233. REYNOLDS, R. J. Ind. Inc. (Winston, Salem N. C.): Improving the Taste and/or Smell of Tobacco Products — Using Lactones of 2-(5-Hydroxymethyl-2-formyl-pyrrol-1-yl)alkanoic Acids. US. Pat. 3.996.941 (14. 12. 1976).
234. RIVETT, D. E. A.: S-Prenylthioisobutyrate from Some Agathosma Oils. Tetrahedron Letters **1974,** 1253.
235. RIZZI, G. P.: The Formation of Tetramethylpyrazine and 2-Isopropyl-4,5-dimethyl-3-oxazoline in the Strecker Degradation of **DL**-Valine with 2,3-Butanedione. J. Organ. Chem. (USA) **34,** 2002 (1969).
236. — A Mechanistic Study of Alkylpyrazine Formation in Model Systems. J. Agric. Food Chem. **20,** 1081 (1972).
237. ROBERTS, D. L. (Reynolds RJ Tobacco Co.): Process of Imparting a Popcorn-Like Flavor and Aroma to Food Stuffs and Tobacco by Incorporating Pyrazine Derivatives Therein and the Resulting Products. US. Pat. 3.402.051 (17. 9. 1968); Chem. Abstr. **70,** 27 784 r (1969).
238. RODIN, J. O., C. M. HIMEL, R. M. SILVERSTEIN, R. W. LEEPER, and W. A. GORTNER: Volatile Flavor and Aroma Components of Pineapple. I. Isolation and Tentative Identification of 2,5-Dimethyl-4-hydroxy-3(2*H*)-furanone. J. Food Sci. **30,** 280 (1965).
239. RONALD, A. P., and W. A. B. THOMSON: The Volatile Sulfur Compounds of Oysters. J. Fish. Res. Board Canada **21,** 1481 (1964).
240. ROSENKRANZ, R. E., K. ALLNER, R. GOOD, W. v. PHILIPSBORN und C. H. EUGSTER: Zur Kenntnis der Chemie einfacher Furenidone („β-Hydroxyfurane"). Helv. Chim. Acta **46,** 1259 (1963).
241. RUITER, A.: The Browning Reaction of Glycolic Aldehyde with Aminoethanol. III. Renewed Investigations on the Structure of a Reaction Product and Comparison with Aldose Aminoethanol Browning. Lebensm.-Wiss. Technol. **6,** 142 (1973).
242. RYDER, W. S.: Progress and Limitations in the Identification of Flavor Components. In: Flavor Chemistry. Advanc. Chem. Ser. **56,** 70 (R. F. GOULD, ed.). Washington: Amer. Chem. Soc. Publ. 1966.
243. SAGHIR, A. R., L. K. MANN, R. A. BERNHARD, and J. V. JACOBSON: Determination of Aliphatic Mono- and Disulfides in *Allium* by Gas Chromatography and Their

Distribution in the Common Food Species. Proc. Amer. Soc. Hortic. Sci. **84,** 386 (1964); Chem. Abstr. **62,** 4529a (1965).
244. Sanders, H. J.: Food Additives. Chem. Eng. News **44** (43), 108 (1966).
245. Scanlan, R. A., S. G. Kayser, L. M. Libbey, and M. E. Morgan: Identification of Volatile Compounds from Heated **L**-Cysteine HCl/**D**-Glucose. J. Agric. Food Chem. **21,** 673 (1973).
246. Schmeltz, J., and D. Hoffmann: Nitrogen-Containing Compounds in Tobacco and Tobacco Smoke. Chem. Rev. **77,** 295 (1977).
247. Schreier, P., und F. Drawert: 3-(Methylthio)-1-propanol, eine flüchtige Komponente des Weinaromas. Z. Lebensm.-Unters. Forsch. **154,** 27 (1974).
248. — — Gaschromatographisch-massenspektrometrische Untersuchung flüchtiger Inhaltsstoffe des Weines. V. Alkohole, Hydroxyester, Lactone und andere polare Komponenten des Weinaromas. Chem. Mikrobiol., Technol. Lebensm. **3,** 154 (1974); Chem. Abstr. **82,** 153 714d (1945).
249. Schreier, P., F. Drawert und A. Junker: Gaschromatographisch-massenspektrometrische Untersuchung flüchtiger Inhaltsstoffe des Weines. II. Thioäther-Verbindungen des Weinaromas. Z. Lebensm.-Unters. Forsch. **154,** 279 (1974).
250. — — — Flüchtige Schwefelverbindungen des Bieraromas. Brauwissenschaften **27,** 205 (1974); Chem. Abstr. **82,** 15 192f (1975).
251. Schreier, P., F. Drawert, A. Junker, H. Barton und G. Leupold: Über die Biosynthese von Aromastoffen durch Mikroorganismen. Z. Lebensm.-Unters. Forsch. **162,** 279 (1976).
252. Schreyen, L., P. Dirinck, F. van Wassenhove, and N. Schamp: Volatile Flavor Components of Leek. J. Agric. Food Chem. **24,** 336 (1976).
253. Schrötter, H.: Über eine im Fuselöl enthaltene Base. Chem. Ber. **12,** 1431 (1879).
254. Schutte, L.: Precursors of Sulfur-Containing Flavor Components. CRC Critical Rev. Food Techn. **4,** 457 (1974).
255. Schutte, L., and A. van der Heijden: Synthetic Precursors of Flavor Compounds with a Thiol Group. J. Agric. Food Chem. **21,** 226 (1973).
256. Schutte, L., and E. B. Koenders: Components Contributing to Beef Flavor. Natural Precursors of 1-Methylthio-ethanethiol. J. Agric. Food Chem. **20,** 181 (1972).
257. Schwaiger, W., and J. P. Ward: Reaction of Methylpyrazinecarbonitriles and 2-Methylquinoxaline with Phosphorus Oxychloride and Organomagnesium Compounds. Rec. trav. chim. Pay-Bas **91,** 1175 (1972).
258. Schwimmer, S., and M. Friedman: Genesis of Volatile Sulfur-Containing Food Flavors. Flavour Ind. **3,** 137 (1972).
259. Seifert, R. M., R. G. Buttery, D. G. Guadagni, D. R. Black, and J. G. Harris: Synthesis of Some 2-Methoxy-3-alkylpyrazines with Strong Bell Pepper-Like Odors. J. Agric. Food Chem. **18,** 246 (1970).
260. Severin, T., und W. Seilmeier: Studien zur Maillard-Reaktion. III. Umwandlung von Glucose unter dem Einfluß von Methylammoniumacetat. Z. Lebensm.-Unters. Forsch. **137,** 4 (1968).
261. Seydel, J. K., E. R. Garrett, W. Diller, and K. J. Schaper: 5-Methyl-3(2*H*)-furanone from Acid-Catalyzed Solvolysis of 2-Deoxy-**D**-ribose. J. Pharmac. Sci. **56,** 858 (1967).
262. Shankaranatayana, M. L., B. Raghavan, K. O. Abraham, and C. P. Natarajan: Volatile Sulfur Compounds in Food Flavors. CRC Critical Rev. Food Techn. **4,** 395 (1974).
263. Shankaranatayana, M. L., B. Raghavan, and C. P. Natarajan: Mustard. Varieties, Chemistry and Analysis. Z. Lebensm.-Wiss. Techn. **5,** 191 (1972).
264. Shaw, P. E., J. H. Tatum, and R. E. Berry: Base-Catalyzed Fructose Degradation and Its Relation to Nonenzymic Browning. J. Agric. Food Chem. **16,** 979 (1968).

265. SHAW, P. E., J. H. TATUM, and R. E. BERRY: Base-Catalyzed Sucrose Degradation Studies. J. Agric. Food Chem. **17,** 907 (1969).
266. SHELDON, R. M., R. C. LINDSAY, and L. M. LIBBEY: Identification of Volatile Flavor Compounds from Roasted Filberts. J. Food Sci. **37,** 313 (1972).
267. SHEMER, M., and E. G. PERKINS: Degradation of Methionine in Heated Soybean Protein and the Formation of β-Methylmercaptopropionaldehyde. J. Agric. Food Chem. **23,** 201 (1975).
268. SHIBAMOTO, T.: Formation of Sulfur- and Nitrogen-Containing Compound from the Reaction of Furfural with Hydrogen Sulfide and Ammonia. J. Agric. Food Chem. **25,** 206 (1977).
269. SHIBAMOTO, T., and R. A. BERNHARD: Investigation of Pyrazine Formation Pathways in Sugar-Ammonia Model Systems. J. Agric. Food Chem. **25,** 609 (1977).
270. SHIBAMOTO, T., and G. F. RUSSELL: Study of Meat Volatiles Associated with Aroma Generated in a D-Glucose-hydrogen Sulfide-ammonia Model System. J. Agric. Food Chem. **24,** 843 (1976).
271. — — A Study of the Volatiles Isolated from a D-Glucose-hydrogen Sulfide-ammonia Model System. J. Agric. Food Chem. **25,** 110 (1977).
272. SHIGEMATSU, H., T. KURATA, H. KATO, and M. FUJIMAKI: Formation of 2-(5-Hydroxymethyl-2-formylpyrrol-1-yl)alkyl Acid Lactones on Roasting Alkyl α-Amino Acid with D-Glucose. Agric. Biol. Chem. **35,** 2097 (1971).
273. SHIGEMATSU, H., S. SHIBATA, T. KURATA, H. KATO, and M. FUJIMAKI: 5-Acetyl-2,3-dihydro-1*H*-pyrrolizines and 5,6,7,8-Tetrahydroindilizin-8-ones, Odor Constituents Formed on Heating L-Proline with D-Glucose. J. Agric. Food Chem. **23,** 233 (1975).
274. SHIMIZU, Y., S. MATSUTO, Y. MIZUNUMA, and I. OKADA: Studies on the Flavor of Roast Barley (Mugi-cha). VI. Separation and Identification of 5-Hydroxy-maltol, Maltol, 5-Methylcyclopent-2-en-2-ol-1-one and Other Compounds. Agric. Biol. Chem. **34,** 845 (1970).
275. SHINDO, H.: Infrared Spectra of Heterocyclic Compounds. VIII. Infrared Spectra of Substituted Pyrazines and Their N-Oxides. Chem. Pharm. Bull. (Japan) **8,** 33 (1960); Chem. Abstr. **55,** 6139 d (1961).
276. SHIPTON, J., F. B. WHITFIELD, and J. H. LAST: Extraction of Volatile Compounds from Green Peas *(Pisum sativum)*. J. Agric. Food Chem. **17,** 1113 (1969).
277. SHU, C. K., and G. R. WALLER: Volatile Components of Roasted Peanuts: Comparative Analyses of the Basic Fraction. J. Food Sci. **36,** 579 (1971).
278. SIVERIN*, R., and F. LEDL: Thermal Decomposition of Cysteine in Tributyrin. Chem. Mikrobiol. Technol. Lebensm. **1,** 135 (1972); Chem. Abstr. **77,** 152 524 b (1972).
279. SIVETZ, M.: Coffee Processing Technology, Vol. II, p. 113. Westport, Conn.: AVI Publishing Co. 1963.
280. SLAUGHTER, J. C.: Volatile Amines in the Brewing Process. Identification and Occurrence of Pyrrolidine. J. Inst. Brew. **76,** 22 (1970); Chem. Abstr. **73,** 86 519 t (1970).
281. — Volatile Amines of Malt and Beer. J. Inst. Brew. **77,** 446 (1971); Chem. Abstr. **76,** 12 811 x (1972).
282. SLOAN, J. L., D. D. BILLS, and L. M. LIBBEY: Heat-Induced Compounds in Strawberries. J. Agric. Food Chem. **17,** 1370 (1969).
283. SMITH, V. K., JR., and S. KUSHNER (American Cyanamid Co): Acylpyrazines. US. Pat. 2.677.686 (4. 5. 1954); Chem. Abstr. **49,** 6322 a (1955).

* Ought to be: SEVERIN.

284. Solms, J.: Geschmacksverstärkende Verbindungen in Nahrungsmitteln. Chimia **21,** 169 (1967).

285. Spencer, R.: Catty Taints in Foods. Food Technol. **23,** 1372 (1969).

286. Stephens, C. R., Jr., A. Torres, and C. M. Leir (Chas. Pfizer and Co., Inc., New York): 2-Hydroxy-3-alkyl-2-cyclopentenones as Flavoring Compounds. DOS 2.005.160 (3. 9. 1970); Chem. Abstr. **73,** 98 467 n (1970).

287. Stoffelsma, J., and J. Pypker: Some New Constituents of Roasted Coffee. Rec. trav. chim. Pays-Bas **87,** 241 (1968).

288. Stoffelsma, J., G. Sipma, D. K. Kettenes, and J. Pypker: New Volatile Components of Roasted Coffee. J. Agric. Food Chem. **16,** 1000 (1968).

289. Stoll, M., P. Dietrich, E. Sundt et M. Winter: Recherches sur les Arômes. 15. Sur l'Arôme du Cacao. II. Helv. Chim. Acta **50,** 2065 (1967).

290. Stoll, M., M. Winter, F. Gautschi, I. Flament et B. Willhalm: Recherches sur les Arômes. 13. Sur l'Arôme de Café. I. Helv. Chim. Acta **50,** 628 (1967).

291. Sulser, H., J. DePizzol, and W. Büchi: A Probable Flavoring Principle in Vegetable-Protein Hydrolysates. J. Food Sci. **32,** 611 (1967).

292. Sundt, E., B. Willhalm, R. Chappaz und G. Ohloff: Das organoleptische Prinzip von Cassis-Flavor im Buccublätteröl. Helv. Chim. Acta **54,** 1801 (1971).

293. Swain, J. W.: Volatile Flavor Constituents of Pork Cured With and Without Nitrite. Thesis, Univ. Missouri 1972. Order No 73-21,836, University Microfilms, Ann Arbor/Mich. 1973; from Diss. Abstr. Int. B **34,** 1143 (1973).

294. Swoboda, P. A. T.: Dimethyl trisulfide: A Constituent of Roast Chicken Aroma. 160th ACS National Meeting, Chicago, Sept. 1970, AGFD 80.

295. Sydow, E. von: Flavor — A Chemical or Psychological Concept? Food Technol. **25,** 40 (1971).

296. Sydow, E. von, and K. Anjou: The Aroma of Rye Crispbread. Lebensm.-Wiss. Technol. **2,** 15 (1969).

297. Takahashi, K., M. Tadenuma, and S. Sato: 3-Hydroxy-4,5-dimethyl-2(5*H*)-furanone, a Burnt Flavoring Compound from Aged Sake. Agric. Biol. Chem. **40,** 325 (1976).

298. Takeda Chemical Industries, Ltd.: Fungicidal Cyclic Polysulfide. Fr. Pat. 1.502.924 (24. 11. 1967); Chem. Abstr. **69,** 96 795 m (1968).

299. Takei, Y., Y. Nakatani, A. Kobayashi, and T. Yamanishi: Aroma of Sesame Oil. II. Intermediate and High Boiling Compounds. Nippon Nogei Kagaki Kasshi **43,** 667 (1969); Chem. Abstr. **72,** 35 670 g (1970).

300. Takei, Y., K. Shimada, S. Watanabe, and T. Yamanishi: Volatile Components of Roasted Almonds: Basic Fraction. Agric. Biol. Chem. **38,** 645 (1974).

301. Takei, Y., and T. Yamanishi: Flavor Components of Roasted Almond. Agric. Biol. Chem. **38,** 2329 (1974).

302. Takken, H. J., L. M. van der Linde, M. Boelens, and J. M. van Dort: Olfactive Properties of a Number of Polysubstituted Pyrazines. J. Agric. Food Chem. **23,** 638 (1975).

303. Talapatra, K.: Studies on the Volatile Compounds in the Aroma Fraction of Milk Caramel. Thesis, Pennsylvania State Univ. 1974. Order No 74-21,031, University Microfilms, Ann Arbor/Mich. 1974.

304. Tas, A. C., und R. J. C. Kleipool: Produkte der Maillard-Reaktion. II. Allyl- und Propenylpyrazine. Riechst. Aromen Körperpflegem. **24,** 8 (1974); Chem. Abstr. **80,** 82 875 c (1974).

305. Thomas, A. F.: An Analysis of the Flavor of the Dried Mushroom, *Boletus edulis*. J. Agric. Food Chem. **21,** 955 (1973).

306. Tjan, S. B., J. C. Haakman, C. J. Teunis, and H. G. Peer: Synthesis of 3,5-Dialkyl-1,2,4-trithiolanes. Assignment of Configuration and Conformational Analysis by PMR. Tetrahedron **28,** 3489 (1972).

307. TONSBEEK, C. H. T., H. COPIER, and A. J. PLANCKEN: Components Contributing to Beef Flavor. Isolation of 2-Acetyl-2-thiazoline from Beef Broth. J. Agric. Food Chem. **19**, 1014 (1971).

308. TONSBEEK, C. H. T., A. J. PLANCKEN, and T. V. D. WERDHOF: Components Contributing to Beef Flavor. Isolation of 4-Hydroxy-5-methyl-3(2*H*)-furanone and Its 2,5-Homolog from Beef Broth. J. Agric. Food Chem. **16**, 1016 (1968).

309. TRESSL, R., D. BAHRI, M. HOLZNER, and T. KOSSA: Formation of Flavor Components in Asparagus. 2. Formation of Flavor Components in Cooked Asparagus. J. Agric. Food Chem. **25**, 459 (1977).

310. TURCHI, I. J., and M. J. S. DEWAR: The Chemistry of Oxazoles. Chem. Rev. **75**, 389 (1975).

311. USHIMARU, A., Y. NAKAJIMA, M. SUGIYAMA, H. NISHIO, and S. KOBAYASHI (Fuji Photo Film Co): 1,2,3,5,6-Pentathiacycloheptane Sensitizers for Photographic Silver Halide Emulsions. DOS 2.042.359 (18. 3. 1971); Chem. Abstr. **74**, 105 686j (1971).

312. VIANI, R., J. BRICOUT, J. P. MARION, F. MÜGGLER-CHAVAN, D. REYMOND et R. H. EGLI: Sur la Composition de l'Arôme de Tomate. Helv. Chim. Acta **52**, 887 (1969).

313. VITZTHUM, O. G., and P. WERKHOFF: Oxazoles and Thiazoles in Coffee Aroma. J. Food Sci. **39**, 1210 (1974).

314. — — Neuentdeckte Stickstoffheterocyclen im Kaffee-Aroma. Z. Lebensm.-Unters. Forsch. **156**, 300 (1974).

315. — — Cycloalkapyrazines in Coffee Aroma. J. Agric. Food Chem. **23**, 510 (1975).

316. — — Steam Volatile Aroma Constituents of Roasted Coffee: Neutral Fraction. Z. Lebensm.-Unters. Forsch. **160**, 277 (1976).

317. VITZTHUM, O. G., P. WERKHOFF, and P. HUBERT: Volatile Components of Roasted Cocoa: Basic Fraction. J. Food Sci. **40**, 911 (1975).

318. — — — New Volatile Constituents of Black Tea Aroma. J. Agric. Food Chem. **23**, 999 (1975).

319. WADA, S., H. NAKATANI, S. FUJINAWA, H. KIMURA, and M. HAGAYA: Lenthionine, a New Aroma-Bearing Substance from *Shiitake*. I. Physicochemical Properties of Lenthionine. Eiyo To Shokuryo **20**, 355 (1968); Chem. Abstr. **69**, 1836p (1968).

320. WADA, S., H. NAKATANI, and K. MORITA: A New Aroma-Bearing Substance from *Shiitake*, an Edible Mushroom. J. Food Sci. **32**, 559 (1967).

321. WADA, S., H. NAKATANI, J. TODA, and M. HAGAYA: Lenthionine, a New Aroma-Bearing Substance from *Shiitake*. II. Sensory Properties of Lenthionine and Its Application to Foods. Eiyo To Shokuryo **20**, 360 (1968); Chem. Abstr. **69**, 1837q (1968).

322. WAINWRIGHT, T., J. F. MCMAHON, and J. MCDOWELL: Formation of Methional and Methanethiol from Methionine. J. Sci. Food Agric. **23**, 911 (1972).

323. WAL, B. VAN DER, D. K. KETTENES, J. STOFFELSMA, G. SIPMA, and A. TH. J. SEMPER: New Volatile Components of Roasted Cocoa. J. Agric. Food Chem. **19**, 276 (1971).

324. WAL, B. VAN DER, G. SIPMA, D. K. KETTENES, and A. T. J. SEMPER: Some New Constituents of Roasted Cocoa. Rec. trav. chim. Pays-Bas **87**, 238 (1968).

325. WALKER, N. J., and I. K. GRAY: The Glucosinolate of Land Cress *(Coronopus didymus)* and Its Enzymatic Degradation Products as Precursors of Off-Flavor in Milk — A Review. J. Agric. Food Chem. **18**, 346 (1970). Pays-Bas **87**, 238 (1968).

326. WALRADT, J. P., R. C. LINDSAY, and L. M. LIBBEY: Popcorn Flavor: Identification of Volatile Compounds. J Agric. Food Chem. **18**, 926 (1970).

327. WALRADT, J. P., A. O. PITTET, T. E. KINLIN, R. MURALIDHARA, and A. SANDERSON: Volatile Components of Roasted Peanuts. J. Agric. Food Chem. **19**, 972 (1971).

328. WANG, P. S., H. KATO, and M. FUJIMAKI: Studies on Flavor Components of Roasted

Barley. IV. The Volatile Sulfur Compounds, Fatty Acids and Neutral Non-Carbonyl Oxygenated Compounds. Agric. Biol. Chem. **34,** 561 (1970).

329. WHEELER, J. W., and M. S. BLUM: Alkylpyrazine Alarm Pheromones in Ponerine Ants. Science **182,** 501 (1973).

330. WHITE, R. H.: Occurrence of S-Methyl Thioesters in Urines of Humans After They have Eaten Asparagus. Science **189,** 810 (1975).

331. WHITEHEAD, E. V., R. A. DEAN, and F. A. FIDLER: The Preparation and Physical Properties of Sulfur Compounds Related to Petroleum. II. Cyclic Sulfides. J. Amer. Chem. Soc. **73,** 3632 (1951).

332. WICKREMASINGHE, R. L., and T. SWAIN: Studies on the Quality and Flavor of Ceylon Tea. J. Sci. Food Agric. **16,** 57 (1965).

333. WILDENRATH, H. L., E. N. CHRISTENSEN, B. STACKLER, A. CAPUTI, JR., K. SLINKARD, and K. SCUTT: Volatile Constituents of Grape Leaves. I. *Vitis vinifera* variety Chenin blanc. Amer. J. Enol. Viticult. **26,** 148 (1975).

334. WILEY, H. W.: Beverages and Their Adulteration. Blakiston and Co, Philadelphia **1919,** 281; ref. mentioned by J. H. KAHN: Compounds Identified in Whisky, Wine and Beer: A Tabulation. J. Assoc. Offic. Anal. Chem. **52,** 1166 (1969).

335. WILEY, R. H.: The Conversion of α-Amino Acids to Oxazoles. J. Organ. Chem. (USA) **12,** 43 (1947).

336. WILLHALM, B., M. STOLL, and A. F. THOMAS: 2,5-Dimethyl-4-hydroxy-2,3-dihydrofuran-3-one. Chem. and Ind. **1965,** 1629; see also G. OHLOFF, Chemie der Geruchs- und Geschmacksstoffe. In: Fortschritte der chemischen Forschung (A. DAVISON et al., eds.), Vol. 12, No. 2, p. 219. Berlin-Heidelberg-New York: Springer. 1969.

337. WILSON, R. A., and I. KATZ: Review of Literature on Chicken Flavor and Report of Isolation of Several New Chicken Flavor Components from Aqueous Cooked Chicken Broth. J. Agric. Food Chem. **20,** 741 (1972).

338. WILSON, R. A., C. J. MUSSINAN, I. KATZ, and A. SANDERSON: Isolation and Identification of Some Sulfur Chemicals Present in Pressure-Cooked Beef. J. Agric. Food Chem. **21,** 873 (1973).

339. WINTER, M. (Firmenich SA, Geneva): Unpublished Results.

340. WINTER, M., A. FURRER, B. WILLHALM, and W. THOMMEN: Identification and Synthesis of Two New Organic Sulfur Compounds from the Yellow Passion Fruit *(Passiflora edulis* f. *flavicarpa).* Helv. Chim. Acta **59,** 1613 (1976).

341. WINTER, M., F. GAUTSCHI, I. FLAMENT, and M. STOLL (Firmenich et Cie): Coffee Flavoring Containing Pyrazines. Fr. Pat. 1.530.436 (28. 6. 1968); Chem. Abstr. **71,** 90 131 m (1969).

342. WINTER, M., F. GAUTSCHI, I. FLAMENT, M. STOLL, and I. M. GOLDMAN (Firmenich et Cie): Flavor Modified Soluble Coffee. US. Pat. 3.702.253 (7. 11. 1972).

343. WINTER, M., I. M. GOLDMAN, F. GAUTSCHI, I. FLAMENT, and M. STOLL (Firmenich et Cie): Flavoring Agents. US. Pat. 3.943.260 (9. 3. 1976).

344. WINTER, M., und R. KLÖTI: Über das Aroma der gelben Passionsfrucht *(Passiflora edulis* f. *flavicarpa).* Helv. Chim. Acta **55,** 1916 (1972).

345. WRATTEN, S. J., and D. J. FAULKNER: Cyclic Polysulfides from the Red Alga *Chandria californica.* J. Organ. Chem. (USA) **41,** 2465 (1976).

346. YAMANISHI, T., S. SHIMOJO, M. UKITA, K. KAWASHIMA, and Y. NAKATANI: Aroma of Roasted Green Tea (Hoji-cha). Agric. Biol. Chem. **37,** 2147 (1973).

347. BADINGS, H. T., H. MAARSE, R. J. C. KLEIPOOL, A. C. TAS, R. NEETER, and M. C. TEN NOEVER DE BRAUW: Formation of odorous compounds from hydrogen sulphide and 2-butenal. Z. Lebensm.-Unters. Forsch. **161,** 53 (1976).

(Received August 22, 1977)

Naturally Occurring Cembranes

By A. J. WEINHEIMER[1], C. W. J. CHANG[2], and J. A. MATSON[1]
Department of Chemistry, University of Oklahoma, U. S. A.

Contents

Present Address: [1] Department of Medicinal Chemistry and Pharmacognosy, University of Houston, Houston, TX 77004, U.S.A.
[2] Faculty of Chemistry, University of West Florida, Pensacola, FL 32504, U.S.A.

I. Introduction

The first studies which established the structures of members of the new group of diterpenes known as cembranes appeared in 1962 when DAUBEN, THIESSEN, and RESNICK (*1*), and KOBAYASHI and AKIYOSHI (*2*) simultaneously published the structure of the hydrocarbon cembrene (**1**) from pine oleoresins, and ROWLAND and ROBERTS (*3*) published the structure of two diastereomeric cembrenediols from tobacco. Since that time, many more members of this new class of diterpenes have been discovered. Those originating in terrestrial plants have been encountered primarily in pines and tobacco. More recently, growing numbers of cembranoids have been found in the coelenterate phylum of lower marine invertebrates.

The cembrane skeleton is isoprenoid and consists of a fourteen-membered carbocyclic ring substituted by an isopropyl residue at position 1, and by three symmetrically disposed methyl groups at positions 4, 8 and 12. Natural cembranoids range in functional complexity from hydrocarbons and alcohols to polyfunctional lactone derivatives. These compounds typically contain nuclear unsaturation at one or more of the sites of methyl substitution.

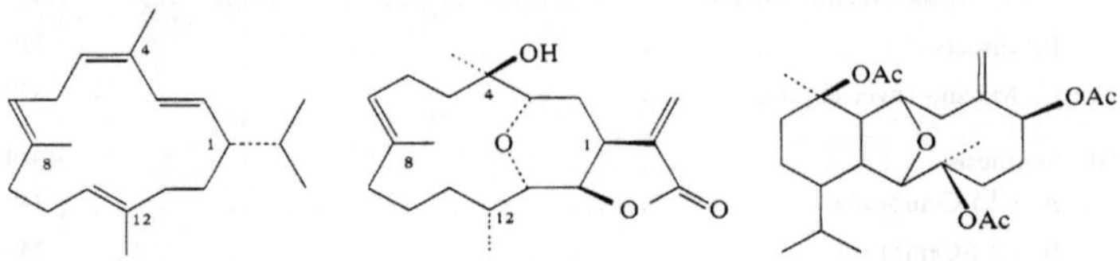

(1) (2) (3)

In this chapter, we have reviewed the literature through March 1977. Our coverage includes cembranoids containing transannular oxygen bridges, e.g., eunicin (**2**), but excludes bi- and polycarbocyclic compounds, e.g., eunicellin (**3**) which apparently are derived biogenetically from cembrane precursors by secondary ring closures.

A. Nomenclature

As many as four common names have been employed in the literature for this diterpene skeletal type and three of them are still in use. Of the three, the cembrane nomenclature appears to have the greatest validity since its use in the literature predates that of the other two by more than ten years. Therefore, cembrane nomenclature is used throughout this review.

The hydrocarbon cembrene occurs widely in pines. A compound with the properties of cembrene was first reported, but not named, as early as 1926 by WIENHAUS (*4*) who isolated it from an unnamed conifer. For a brief period it was referred to as the Wienhaus hydrocarbon. Cembrene was next reported in 1935 by SEBE (*5*), who isolated it from *Pinus formosana* and named it thumbelene. The custom of deriving the trivial name from the first species of origin was apparently disregarded in this instance. The name thumbelene was subsequently used only once, by AKIYOSHI (*6*) in 1937 who reported the isolation of cembrene from *P. thunbergii*. In their report on its structure in 1962, KOBAYASHI and AKIYOSHI (*2*) recommended the adoption of the name thunbergene to replace thumbelene, presumably to correct the nomenclatural oversight.

DAUBEN *et al.* (*1*), in their report on its structure in 1962, employed the name cembrene, which had first been used in 1951 by HAAGEN-SMIT, WANG and MIROV (*7*) who isolated the compound from *P. albicaulis*. The name cembrene was based on the taxonomic group of pines, *Cembrae*, which included the species *P. albicaulis*. MIROV subsequently reported cembrene in many other pine oleoresins in 1955 (*8*), 1956 (*9*) and 1961 (*10*). DAUBEN (*11*) clearly rejected the recommendation of the name thunbergene, possibly because it lacked the validity of a name more properly based on the first species of origin, and adopted the name cembrene, possibly because of its eleven years of use prior to the creation of the name thunbergene.

ROBERTS and ROWLAND (*3*), also in 1962, introduced the name duvane, when reporting on the first of the tobacco cembranoids. This name was coined from the Serbian word for tobacco and initially was used universally in the literature of the tobacco compounds. Recently, however, Scandinavian authors have dropped the duvane nomenclature in favor of thunberganoid nomenclature.

The cembrane nomenclature is the sole system used for the burgeoning numbers of this group of diterpenes being discovered in the marine kingdom.

Currently then, the literature of the cembranoid diterpenes reflects various authors' preferences for any one of three parent names, cembrane, thunbergane or duvane. In our own need for a single name for use in this review and in an attempt to remedy this intolerable confusion, we have chosen the name cembrane because of the logic of its considerably longer historical usage than either of the other two.

B. Structural Representation

The large size of the cembrane ring offers the opportunity for great diversity in modes of drawing or representing structures of individual compounds of this class. In fact, so many disparate representations of the ring system have appeared in the literature that detection of similarities or contrasts between structures is often quite difficult.

Since the object of our review is to focus attention upon precisely these features, we have adopted a single basic style for drawing the ring system. This drawing (**4**) has frequently been used in the literature without comment upon its merits. The advantages are made evident most readily by comparison with another frequently used drawing (**5**), which is a natural consequence of the organic chemist's predilection toward the use of the familiar six-membered ring outline.

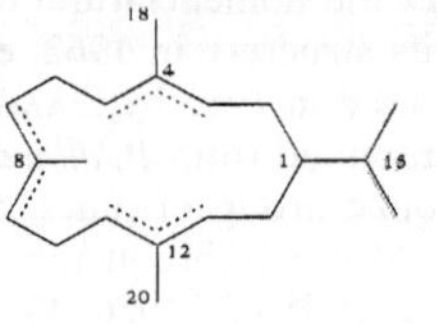

(**4**)

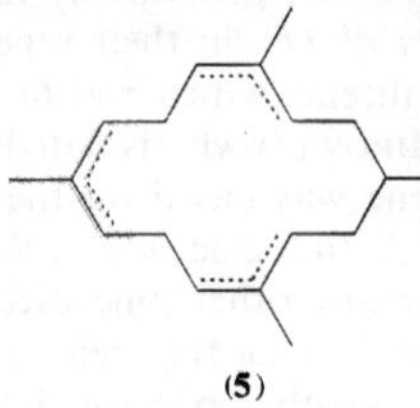

(**5**)

The superiority of (**4**) over (**5**) follows from the geometry of the nuclear double bonds which are normally located at the three sites of methyl substitution. In structure (**4**) the unsaturated sites have *trans* geometry which is now recognized as the norm for cembranoids. In fact, *cis* geometry is quite rare. Structure (**5**) and most other representations previously employed in the literature incorporate *cis* geometry at one or more of the three common sites of nuclear unsaturation.

Despite its accomodation of most cembranoids, (**4**) must be modified and adapted for occasional members of the system which contain uncommon structural and geometric features. The required changes are always minor, and, as in (**1**), are achieved by slight relocation of the position of one or two atoms within the ring. Aside from these variations, (**4**) will be used throughout the review, and will always be presented with position 1 at the right, and with the 1,8-axis of skeletal symmetry in the horizontal.

References, pp. 381—387

C. Configurations

Despite the symmetry of the cembrane carbon skeleton, all known natural cembranes are optically active because of the unsymmetrical disposition of nuclear unsaturations and/or oxygen containing functional groups. However, the symmetrical skeleton affords no point of reference, and configurational relationships among the cembranes are difficult to assess. For example, one may ask whether cembrene and eunicin, which have the absolute configurations shown in (**1**) and (**2**), are of the same or antipodal configurational series. In comparison with cembrene, eunicin could be regarded with equal justification either as (**2**), or as the "turned over" (**2′**), or even as (**2″**), which employs the alternate (clockwise) numbering sequence. It is obvious that in the absence of a skeletal marker, attempts at configurational comparisons are necessarily ambiguous.

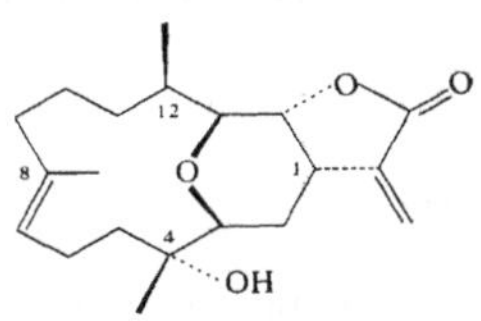

(**2′**)

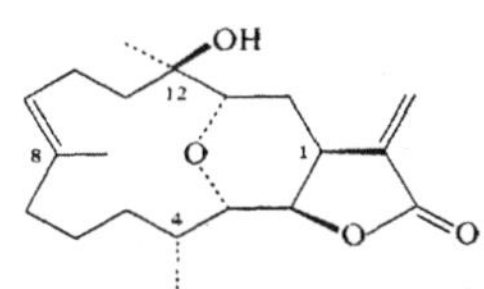

(**2″**)

Recently, WEINHEIMER *et al.* (*12*) proposed a convention for the designation of cembrane configurations. It was recommended that the nuclear unsaturation most remote from position 1, shown in (**1**) and (**2**) as Δ^7, be used as the required internal point of reference. This unsaturation, or some vestigal indication of its location, such as an epoxide, is present in all known natural cembranes. In fact, it is the only asymmetric marker which occurs universally in the cembranes known to date.

The convention requires that rings be numbered consistently so that the reference unsaturation is always located at position 7 (rather than at 8), thereby according it priority over "accidental" oxygen substituents. In applying the convention, structures are oriented (as they are drawn herein) so that a) position 1 is at the viewer's right, and b) the reference unsaturation, Δ^7, is situated above the 1,8-axis of skeletal symmetry. When oriented as described, compounds in which the three-carbon substituent at position 1 is disposed downward are designated as members of the α series, e. g., (**6**), and those having this substituent disposed upward, as members of the β-series, e. g., (**7**).

Returning to the comparison of cembrene and eunicin, it is seen that the eunicin structures (**2′**) and (**2″**) are oriented incorrectly and numbered incorrectly, respectively, whereas (**1**) and (**2**) incorporate the requirements of the convention. Thus cembrene (**1**) belongs to the α-series, and eunicin (**2**) to the β-series.

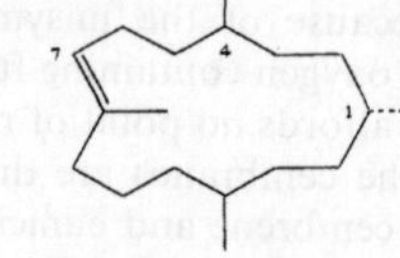

(6)

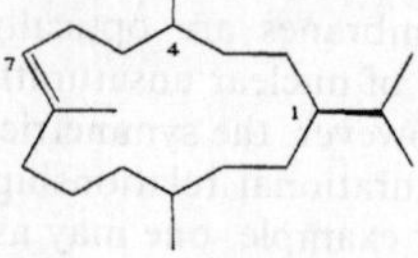

(7)

We will employ the above configuration convention in this review, as well as the usual anti-clockwise ring-numbering it stipulates. As a consequence, the numbering of certain of the compounds covered will conflict with that employed in the original papers. We feel, however, that this inconvenience is outweighed by the benefits of a systematic mode of presentation.

D. "Cembrane"

The parent saturated cycloalkane of the cembrane group of diterpenes is 1-isopropyl-4,8,12-trimethylcyclotetradecane (**8**), known commonly as cembrane or octahydrocembrene. Although it is not itself a natural product, this compound plays an important role in establishing the carbon skeleton of natural cembranoids, particularly those hydrocarbons and simpler hydroxylic derivatives which can be converted readily to cembrane by hydrogenation or combined dehydration-hydrogenation sequences.

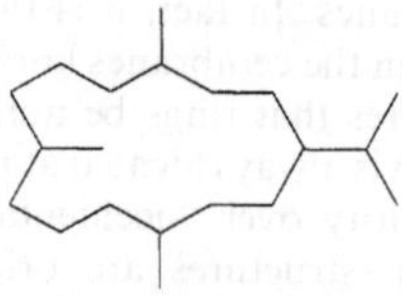

(8)

Even though these hydrogenations can, and presumably do, lead to mixtures of the several possible diastereomeric forms of the saturated hydrocarbon, the "cembrane" produced from various natural precursors is inseparable by glc methods and behaves as a single substance. The hydrocarbon mixture also exhibits the same infrared and mass spectra, irrespective of its source. These fortuitous circumstances offer a convenient and dependable method for characterizing the cembrane skeleton, which has been employed frequently in the structure elucidation of new members of the series.

E. Distribution of Cembranes

The natural cembranes have been encountered thus far in both the plant and animal kingdom, but only within narrowly limited phylogenetic ranges in each.

In the terrestrial plant kingdom, the oleoresins of conifers are common sources of hydrocarbons and simple monohydric alcohols in the cembrane series. The tobacco plant contains a chemically somewhat more elaborate group of di- and tri-oxygenated derivatives consisting of diols and hydroxy ethers, in which the ether functions bridge two atoms of the carbocyclic ring. Lactonic cembranoids have been found in two species of small plants of the *Anisomeles* genus.

In the animal kingdom, the hydrocarbon, cembrene-A, is the sole recorded cembranoid found in the insect phylum, and its occurrence in a species of termite is suggestive of an exogenous origin. The greatest number of cembranoids, and those of greatest chemical diversity, have been encountered in the coelenterates, a phylum of marine invertebrates. In fact, these diterpenes have been found thus far in only two groups of organisms within this phylum, the Caribbean gorgonians and the Pacific soft corals. The cembranoids from this source range in functional complexity from simple hydrocarbons and mono- and dihydric alcohols to, among others, epoxy alcohols, diepoxides, epoxy lactones, and polyoxygenated lactone derivates.

II. Natural Sources

A. Plants

1. Resinous Plants

Cembrene

The early literature dealing with this hydrocarbon refers primarily to its occurrence in various pine oleoresins, and has been reviewed in the Introduction in relation to the preferred nomenclature for the cembrane diterpenes. Two definitive chemical studies which established its structure appeared in 1962, one by DAUBEN, THIESSEN and RESNICK (*1*), and the other by KOBAYASHI and AKIYOSHI (*2*). These preliminary publications were followed by complete papers by each group (*11*, *13*).

DAUBEN *et al.* showed that cembrene (**1**), $C_{20}H_{32}$, formed an octahydro derivative, and thus contained four double bonds and one ring. Cembrene's ultraviolet spectrum showed absorption at 245 nm characteristic of a conjugated diene, which was judged to be *trans*oid from its failure to react with maleic anhydride. The infrared spectrum indicated a *trans*-disubstituted double bond, and also trisubstituted double bonds. The pmr spectrum indicated that the *trans* disubstituted double bond was part of the conjugated diene system. The other double bond of diene was judged to be methyl subsituted in the manner –CH=CMeCH=CH– from the triplet character of one of the vinyl methyl signals. The pmr spectrum also indicated an isopropyl group and two trisubsituted double bonds carrying methyl substituents.

Chemical reduction of cembrene with lithium in liquid ammonia gave a mixture of two dihydrocembrenes resulting from reduction of the diene system. The major component of the mixture (85%) was an isomer (**9**) which still contained the *trans* disubstituted double bond.

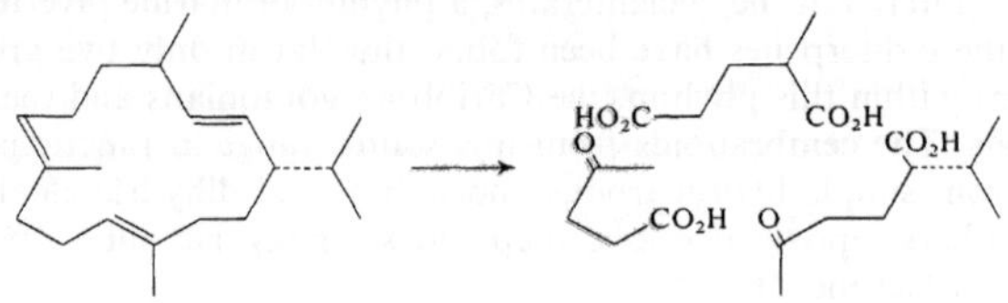

(**9**)

Ozonolysis of (**9**) followed by oxidative work-up gave three acids which were isolated as their methyl esters using preparative gas chromatography. They were identified as levulinic, 2-isopropyl-5-oxohexanoic and 2-methylglutaric acids. Since each was difunctional, and together accounted for all twenty carbon atoms of the molecule, a 14-membered ring was required.

Ozonolysis of cembrene itself followed by oxidative work-up, gave the first two of the above three acids, but acetic and malonic acids were formed in place of 2-methylglutaric acid. The two small acids were indicative of the location of the second double bond of the diene system, but one carbon atom was missing. This atom was located by reductive work-up of the ozonolysis products. This produced pyruvaldehyde which would be degraded to acetic acid under the oxidative conditions. These cleavage products could be pieced together as in (**10**).

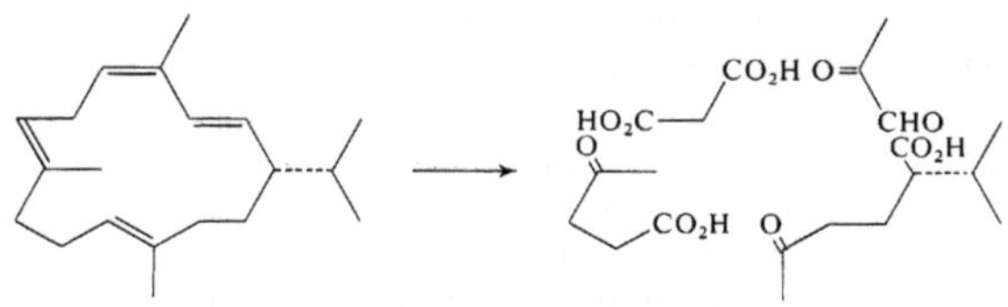

(1) (10)

Another dihydrocembrene (**11**), identical with the minor product from chemical reduction, was formed as the major product in the controlled catalytic reduction of cembrene over palladium on charcoal. Its structure was assigned from the absence of indications of the *trans* disubstituted double bond in its ir and pmr spectra. Ozonolysis and oxidative work-up of (**11**) gave malonic and levulinic acids as the major components of the acidic reaction products.

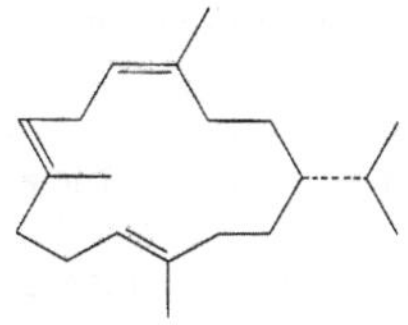

(11)

Only two structures for cembrene, (**1**) and (**12**), were compatible with the above ozonolysis results and the requirement of a *trans* disubstituted unsaturation as part of the conjugated diene system. The first of these (**1**) was isoprenoid, and the other was not.

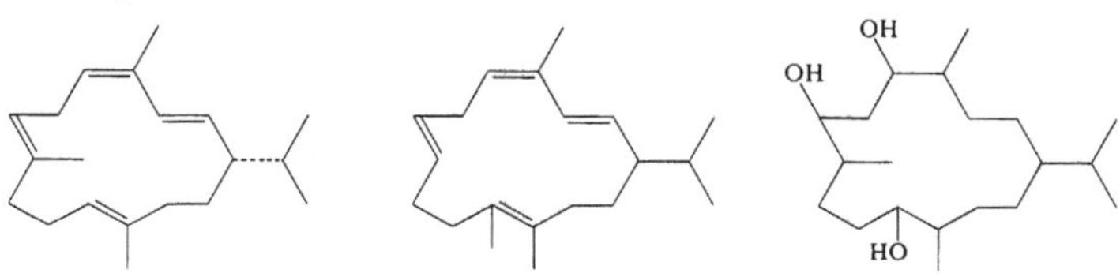

(1) (12) (13)

The choice between (**1**) and (**12**) was based on evidence from the hydroboration/oxidation of dihydrocembrene (**11**). The resulting triol (**13**) and its triacetate both showed pmr signals for three protons on carbon bearing oxygen. Thus structure (**1**) was established for cembrene, since the triol derived from the dihydro derivative of (**12**) would have contained only two such protons.

The configuration of the chiral center at position 1 was shown to be *S* from the known configuration of the corresponding center in the optically active 2-isopropyl-5-oxohexanoic acid obtained by ozonolysis.

In their structural studies on cembrene (which they called thunbergene), KOBAYASHI and AKIYOSHI (*2*, *13*) arrived at essentially the same spectral interpretations and assignments as DAUBEN *et al.* (*vide supra*). In their ozonolysis studies of cembrene (**1**), these authors isolated levulinic and 2-isopropyl-5-oxohexanoic-acids. They also noted the absence of formaldehyde and acetaldehyde which required that the unsaturations be part of the carbocycle.

Since the two acids from ozonolysis and the central three carbon atoms of the diene system (=CMeCH=) accounted for seventeen of the twenty carbons atoms, the remaining three carbon atoms and four hydrogen atoms necessarily consisted of a propanediylidene residue. Like the other residues, this was also bifunctional and necessarily part of the carbocycle, which was thus 14-membered.

They deduced that the carbon skeleton of cembrene was symmetrical because of the lack of optical activity of its octahydro derivative. Upon dehydrogenation with selenium, cembrene formed the phenanthrene (**14**). Its substitution pattern, the structures of the acid products of ozonolysis, and the requirement of a symmetrical skeleton, led to the proposal of structures (**1**) and (**15**) for cembrene, which differed only in the location of a double bond. The former was preferred on biogenetic grounds. These authors also assigned the *S* configuration to position 1.

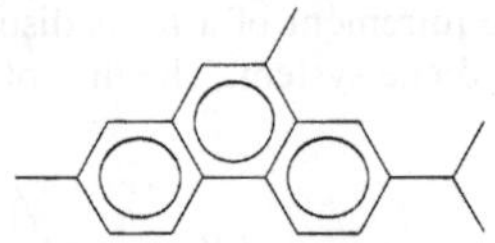

(**14**)

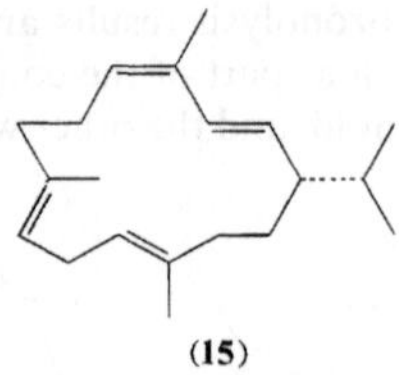

(**15**)

The identity of the hydrocarbons studied by the American and Japanese groups was verified by comparison of samples.

DREW, TEMPLETON, and ZALKIN (*14*) subsequently determined the structure of cembrene by X-ray diffraction and thereby confirmed structure (**1**) deduced by the chemical studies. Their work further showed that three of the unsaturations were *trans* and one was *cis*, as depicted in (**1**) and as had been postulated by DAUBEN *et al.* (*1*). Unfortunately, the structure of cembrene reproduced in this paper (*14*) incorrectly shows the configuration at position 1 as *R*.

Cembrene-A

Patil, Nayak, and Dev (*15*) isolated a new liquid hydrocarbon, which they named cembrene-A, during their investigation of the gum-resin exudate of *Commiphora mukul.* This resin was used in the ancient Ayurvedic system of Indian medicine; recent pharmacological studies provide some support for its possible efficacy in treatment of certain disorders.

Elemental analysis and catalytic hydrogenation revealed that cembrene-A was a monocyclic tetraene. Its pmr spectrum showed one terminal methylene, three other vinyl protons, and four vinyl methyl groups. The lack of ultraviolet absorption above 210 nm precluded olefin conjugation.

Recognizing the compound as a monocyclic diterpene, and considering only those structures which could be derived by a single cyclization of geranyl geranyl pyrophosphate, the normal biogenetic precursor of diterpenes, the authors arrived at three formulae resulting from 1→6, 1→10, or 1→14 cyclization of the open-chain precursor, each of which fulfilled the above spectral requirements of cembrene-A.

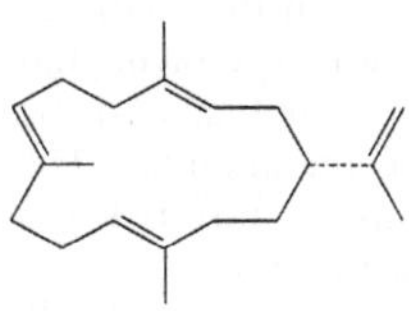

(**16**)

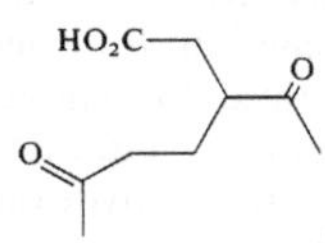

(**17**)

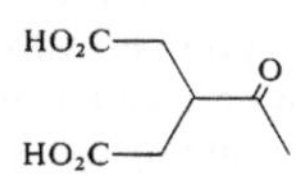

(**18**)

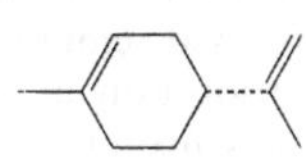

(**19**)

Of these, only structure (**16**) was compatible with either the presence of an isopropenyl substituent, established by double resonance studies, or the structures of cembrene-A's ozonolysis products. The ozonolysis reaction mixture was treated with Jones reagent to liberate three acids which were separated and identified as their methyl esters. They were levulinic, 3-acetyl-6-oxoheptanoic (**17**) and 3-acetylglutaric acids (**18**). Acid (**17**) could be obtained only from structure (**16**) for cembrene-A. This structure has been fully verified by a recent synthesis of cembrene-A (cf. Synthesis Section) which also established *trans* geometry for each of the nuclear unsaturations.

It is of interest that the degraded acid (**18**) appeared to be an artifact of the chromium trioxide work-up of the ozonolysis product. Similar ozonolysis of (+)-limomene (**19**) in methylene chloride/methanol, followed by chromium trioxide oxidation, afforded the same pair of acids (**17, 18**) in roughly the same proportions as obtained from cembrene-A.

Cembrene-A was also isolated from *Picea obovata* and *Pinus koraiensis* by Shmidt, Kashtanova, and Pentegova (*16*), who used the name neocembrene for this compound. Upon perhydrogenation, it formed octahydrocembrene, and upon controlled hydrogenation it formed the dihydrocembrene (**20**) resulting from saturation of the isopropenyl double bond. The latter triene was shown to be identical with the dihydro derivative obtained by controlled hydrogenation of cembrene (**1**) over palladium on calcium carbonate.

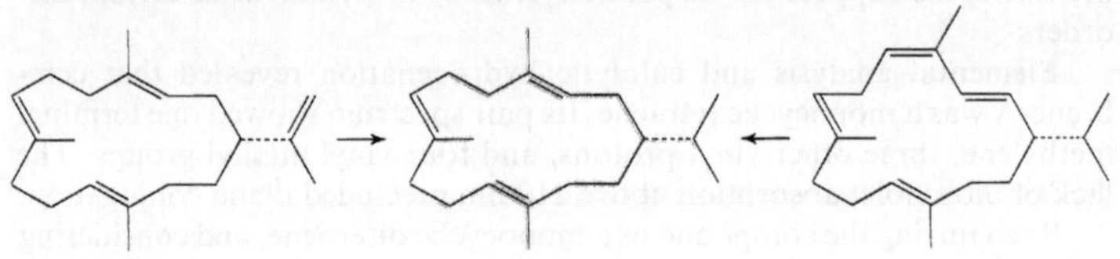

(**16**) (**20**) (**1**)

Isocembrene

A third cembrane hydrocarbon, isocembrene (**21**), was isolated as a liquid by KASHTANOVA, LISINA, and PENTEGOVA (*17*) from the oleoresin of *Pinus sibirica*. Like cembrene and cembrene-A, it was a tetraene and formed cembrane upon perhydrogenation. It displayed conjugated diene absorption at 238 nm, and isomerized to cembrene on alumina. The presence of an exo-methylene group was indicated in the infrared and was confirmed by the formation of formaldehyde on ozonolysis.

Under controlled conditions, the exo-methylene group was selectively hydrogenated, leading to the same dihydrocembrene (**9**) obtained by Dauben (*1*) by lithium-ammonia reduction of cembrene (**1**). This transformation established the location of the exo-methylene group, as well as structure (21) for isocembrene.

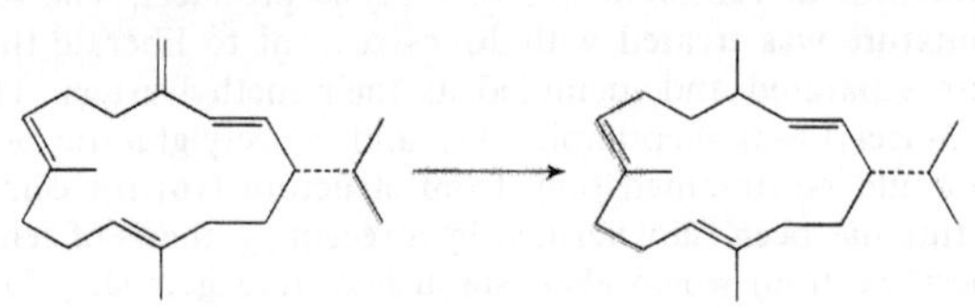

(**21**) (**9**)

Other Hydrocarbons

In their report on the occurrence of cembrene in *Pinus sibirica*, PENTEGOVA and KASHTANOVA (*18*) described three other cembranoid hydrocarbons. Their structures were not determined and no further reports on their identity seem to have appeared. All three compounds formed octahydrocembrene on catalytic hydrogenation and all three formed the phenanthrene (**14**) on dehydrogenation.

One of the hydrocarbons was isomeric with cembrene, $C_{20}H_{32}$, and showed a conjugated diene absorption at 244 nm in the ultraviolet. The other two hydrocarbons were tetrahydrocembrenes, $C_{20}H_{36}$. One of them showed a diene absorption at 253 nm, whereas the other showed no maximum above 200 nm.

Mukulol

RÜCKER (*19*), and PATIL, NAYAK, and DEV (*15*) independently isolated the same alcohol from the gum-resin of *Commiphora mukul*, which was also the source of the cembrene-A characterized by DEV's group. RÜCKER proposed the misleading name allylcembrol, suggestive of a molecule containing 23 carbon atoms. To avoid misinterpretation, the name coined by DEV's group, mukulol, is preferred.

RÜCKER observed that mukulol (**22**), $C_{20}H_{34}O$, a liquid, showed hydroxyl absorption in the ir, but no uv absorption higher than 207 nm. Its pmr spectrum indicated an isopropyl group, three vinyl methyls and three vinyl protons.

Mukulol readily underwent acid catalyzed dehydration to cembrene (**1**), and upon perhydrogenation formed a crystalline hexahydro alcohol.

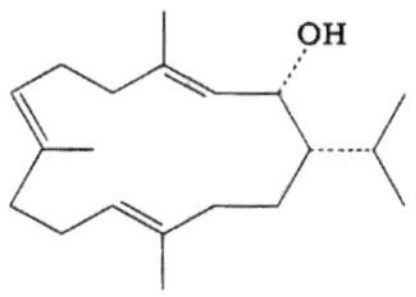

(**22**)

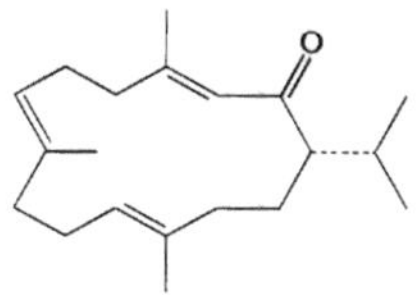

(**23**)

Mukulol was shown to be a secondary allylic alcohol by Jones oxidation to a conjugated ketone (**23**) which was substituted in the *alpha* position by a hydrogen, and in the *beta* position by a methyl group. Treatment of the ketone with lithium aluminium hydride regenerated mukulol as one of the reduction products.

The above results confirmed structure (**22**) for mukulol, and only the geometry of the 3,4-double bond remained in doubt. This question was settled by RALDUGIN *et al.* (*20*) who showed by the absence of a nuclear Overhauser effect between the 4-methyl and the 3-hydrogen of the ketone (**23**) that the double bond was *trans*.

In their study of mukulol, DEV's group (*15*) observed the same ir and uv characteristics and prepared the same unsaturated ketone (**23**). Decoupling experiments provided further evidence for the location of the hydroxyl group. Two of the vinyl protons were shown to be coupled to vicinal methylene groups, but the third was coupled to a single vicinal proton. This proved to be the carbinyl proton of the secondary alcohol function, establishing the partial structure (**24**). Since the carbinyl proton displayed no further coupling, it could not be vicinal to a methylene group, permitting expansion of the partial structure to (**25**).

−CMe=CH−CHOH− (**24**)

−CMe=CH−CHOH−C(C)−C (**25**)

Upon catalytic hydrogenation, mukulol suffered partial hydrogenolysis to the saturated hydrocarbon which was identical with octahydrocembrene-A. Mukulol also was smoothly dehydrated by heating in DMSO, forming cembrene with the same rotation as that obtained from pines. In view of the requirements of partial structure (**25**), structure (**22**) was established for mukulol.

In a subsequent study, Prasad and Dev (*21*) established the absolute stereochemistry shown in (**22**) for mukulol by chemical correlation of its methyl ether (**26**) with that of (+)-*cis*-piperitol (**27**). In each case, ozonolysis with oxidative work-up formed the same methoxy keto acid (**28**) which was characterized as its methyl ester.

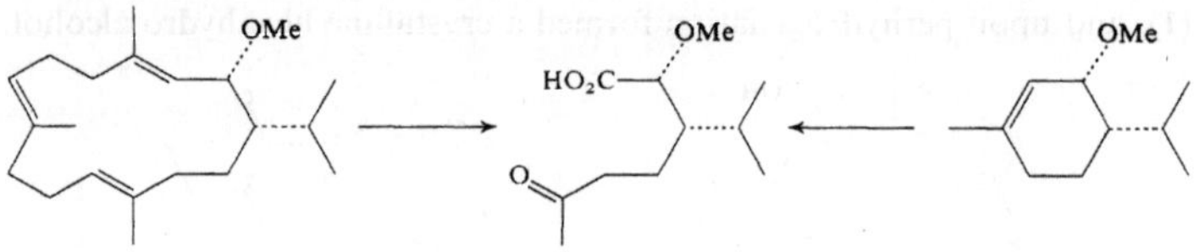

(**26**) (**28**) (**27**)

These authors also correlated mukulol (**22**) with cembrene-A (**16**) *via* the dihydrocembrene-A (**20**) formed by selective catalytic hydrogenation. The triene (**20**) was identical in all respects with the hydrocarbon obtained either by lithium-ammonia hydrogenolysis of mukulol acetate (**29**, R = Ac), or by lithium aluminum hydride reduction of mukulol tosylate (**29**, R = Ts), after separation from the expected allylic rearrangement products which were also formed in the reductions.

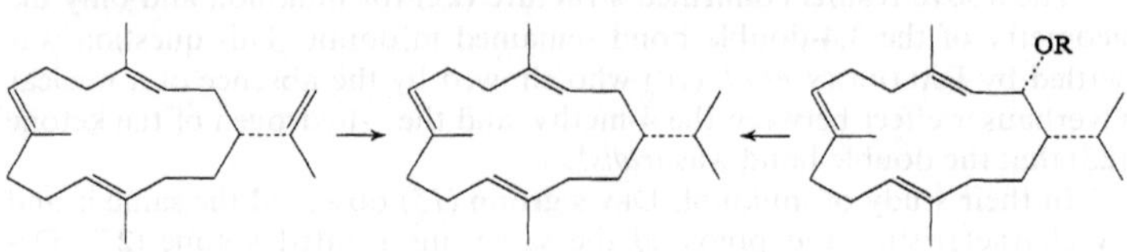

(**16**) (**20**) (**29**)

References, pp. 381—387

2,7,11-Cembratrien-4-ol (thunbergol)

ERDTMAN *et al.* (*22*), while studying the "pocket" resin of the North American Douglas fir *Pseudotsuga menziesii*, isolated an oily alcohol designated as thunbergol (=2,7,11-cembratrien-4-ol) (**30**). Its structure was elucidated by KIMLAND and NORIN (*23*). From elemental analysis and hydrogenation studies, it was apparent that this compound was a $C_{20}H_{34}O$ monocyclic triene. Its hexahydro derivative was nicely crystalline. Infrared and ultraviolet analysis of "thunbergol" disclosed the presence of an alcohol and the absence of olefinic conjugation. Feature by feature comparison of the pmr spectra of "thunbergol" and cembrene showed that the only significant difference was the replacement of the Δ^4 tri-substituted double bond of cembrene by a saturated tertiary methyl carbinol in "thunbergol". The proposal of structure (**30**) was supported by the ready dehydration of "thunbergol" by phosphoryl chloride-pyridine to cembrene (**1**).

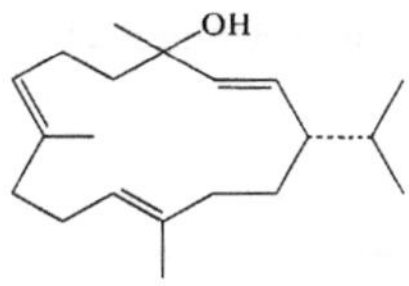

(**30**)

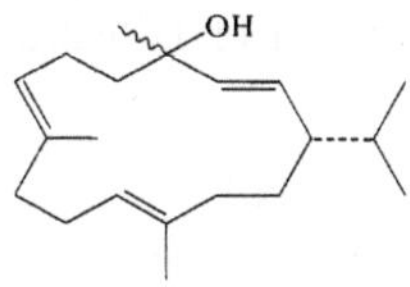

(**31**), (**32**)

The differences between "thunbergol" (**30**) and the next two alcohols to be discussed, isocembrol (**31**) and 4-epiisocembrol (**32**), are not clear.

Isocembrol (2,7,11-Cembratrien-4-ol)

Isocembrol has been isolated from three different pines, *Pinus sibirica* (*24, 25*) *Pinus koraiensis* (*26*) and *Picea obovata* (*27*) by PENTEGOVA's group, and has been assigned (*25*) structure (**31**). Its pmr spectrum shows a *trans* disubstituted double bond, two vinyl methyls and the methyl of a tertiary methyl carbinol. It undergoes dehydration to cembrene (**1**) with acid chlorides, thionyl chloride and acetic acid, but forms isocembrene (**21**) on dehydration with silver nitrate impregnated silica gel. Thus it is necessarily the 4-hydroxy compound (**31**).

By direct comparison, KIMLAND and NORIN (*23*) showed that hexa-hydrothunbergol was identical to a specimen of hexahydroisocembrol supplied by PENTEGOVA. Although this result did not necessarily establish the identity of the unsaturated alcohols, their pmr spectra and rotations were very similar, and the possibility of their identity exists. However, KIMLAND and NORIN did not comment on this point.

4-Epiisocembrol (2,7,11-Cembratrien-4-ol)

RALDUGIN and PENTAGOVA (*28*) found that minor amounts of 4-epiisocembrol (**32**) co-ocurred with isocembrol (**31**) in *Pinus koraiensis* and *Pinus sibirica*. The two isomers were separable by argentic chromatography on silica gel and showed very similar ir and pmr spectra. The location of the hydroxyl group at position 4 was established by the formation of a 2 : 1 mixture of cembrene (**1**) and isocembrene (**21**) upon dehydration with phosphoryl chloride-pyridine.

Cembrol (2,4,7-Cembratrien-12-ol)

LISINA, REZVUKHIN, and PENTEGOVA (*29*) isolated the liquid alcohol, cembrol (**33**) from the resin of *Pinus sibirica*. It showed a conjugated diene absorption at 244—246 nm, and a singlet methyl absorption at δ 1.32 for a tertiary methyl carbinol. Cembrol was dehydrated to cembrene (**1**) by acid chlorides, and formed octahydrocembrene on catalytic hydrogenation in acetic acid. Dehydrogenation with selenium gave the phenanthrene (**14**).

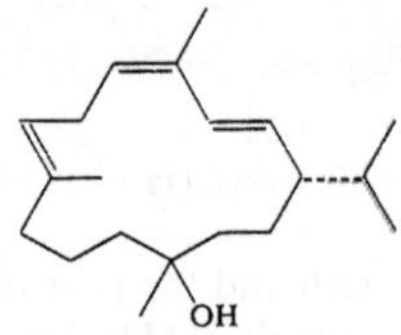

(**33**)

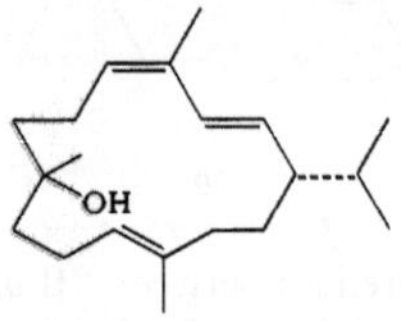

(**34**)

Of the two alternative structures for cembrol, (**33**) and (**34**), considered by these authors, (**33**) was favored because of a signal at δ 2.3 attributed to a doubly allylic methylene group. However, KIMLAND and NORIN (*23*) questioned this assignment because the pmr data reported by LISINA *et al.* (*29*) included signals for two vinyl methyl groups at δ 1.50 and 1.60 which are precisely the chemical shifts of the 8- and 12- methyls in both "thunbergol" (**30**) and cembrene (**1**). Both (**33**) and (**34**) should lack one of these signals. Since no clarification of this discrepancy has yet appeared in the literature, the structure of cembrol remains in doubt.

Artifacts

In the most recent paper from PENTEGOVA's group available to us, RALDUGIN and PENTEGOVA (*26*) reported the isolation of the nor-cembranoid ketone (**35**) from a *Pinus koraiensis* oleoresin which had been stored at room temperature for a year. It was characterized as a conjugated ketone having single hydrogens at both the α- and β-positions, and had other pmr features consistent with structure (**35**). The structure assignment was confirmed by synthesis of the ketone from isocembrene by permanganate oxidation.

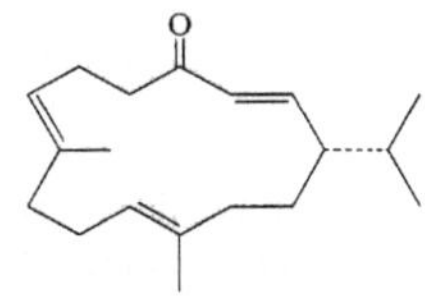

(**35**)

Upon comparing the cembranoid constituents of fresh and aged resin, these authors found that ketone (**35**) was not present in fresh resin, and presumably was formed by autooxidation of isocembrene during storage. They also noted that the initially high proportions of isocembrol and 4-epiisocembrol in the resin diminished with time, and demonstrated that several of the cembranoids they had previously reported from this pine were actually artifacts. These artifacts were the result of both aging of the resin and of the purification methods employed.

Of the three principal hydrocarbons, only cembrene-A was present in fresh resin. Cembrene and isocembrene were initially absent, but were formed during isolation if aluminia of high activity grade was employed for chromatographic separations. Their formation occurred at the expense of isocembrol and 4-epi-isocembrol.

Three isomeric hydrocarbons, the α-, β-, and γ-pinacenes (**36**), were similarly shown to be artifacts. These isomeric conjugated dienes were isolated from *P. koraiensis* by Raldugin, Kashtanova, and Pentegova (*30*) and shown to differ only in the geometry of the double bonds of the 1,3-diene system. Absent in fresh resin, the pinacenes were shown to be formed during vacuum distillation of the neutral resin fraction. This fraction contained traces of diterpene acids which were considered to be responsible for dehydration of the epimeric isocembrols at high temperatures, and isomerization of cembrene to the pinacenes.

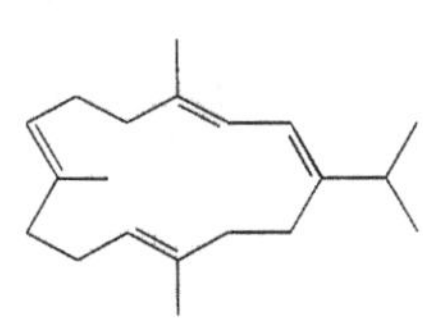

(**36**)

These observations of course lead to speculation that the hydrocarbon cembrene isolated from other sources might also be an artifact.

Incensole

NICOLLETI's group at the University of Rome has isolated and characterized three cembranoids from the frankincense of *Boswellia carteri*. Each of these, incensole (**37**), incensole oxide (**45**) and isoincensole oxide (**51**), contains a tetrahydrofuran unit within the cembrane carbon skeleton.

Incensole (**37**) was isolated (*31*) chromatographically as a low melting solid which could not be recrystallized. Its elemental analysis allowed for the assignment of a $C_{20}H_{34}O_2$ molecular formula. The presence of two unconjugated double bonds in its structure was established by hydrogenation and ultraviolet spectroscopy. Incensole was also shown to contain an hydroxyl group and not to contain a carbonyl group by infrared analysis. Intense absorption at 1050 cm^{-1} suggested an ether function. Structural elements deduced by interpretation of incensole's nmr spectrum included an isopropyl group, two vinylic methyls, two vinylic protons, a proton on carbon bearing hydroxyl and a methyl group on carbon bearing oxygen.

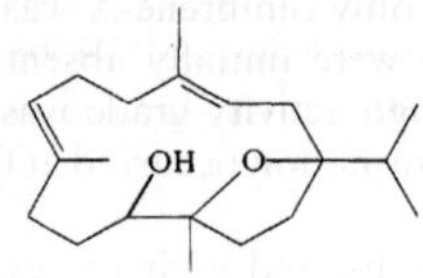

(**37**)

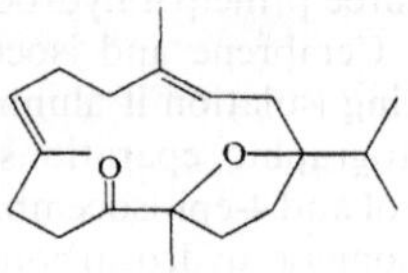

(**38**)

Oxidation of incensole to a monoketone (**38**), which lacked hydroxyl absorption in the infrared, confirmed that the second oxygen atom was present as an ether, presumably a ditertiary cyclic ether.

The presence of a cyclic ether and two double bonds required that the structure of incensole be monocarbocyclic, and the group of three methyl and one isopropyl substituents suggested a cembrane nucleus.

In the pmr spectrum of ketone (**38**), the signal of the methyl group at the ether terminus was shifted downfield relative to its location in incensole, indicating that it was *alpha* to the new carbonyl group. Postulating that the other tertiary terminus of the ether was at position 1, the authors considered trial structures (**39**) and (**40**). The latter was ruled out by the formation of levulinic acid in the Lemieux oxidation of incensole. The other cleavage product, the diketo acid (**41**), gave a pyrrole test showing it to be a 1,4-diketone. This refined structure (**39**) to (**37**).

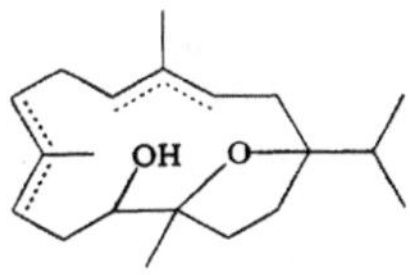

(39)

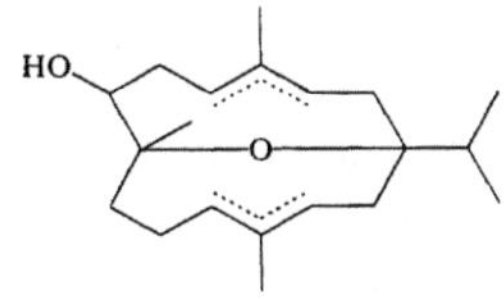

(40)

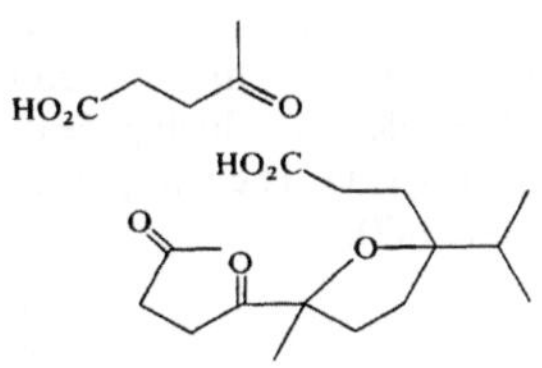

(41)

The tetrahydrofuran moiety, and also the cembrane skeleton, were confirmed by treatment of tetrahydroincensole (**42**) with acidic reagents. With sulfuric acid, (**42**) was converted to the unsaturated ketone (**44**) in a reaction involving pinacolic rearrangement to (**43**) followed by dehydration. When ketone (**44**) was subjected to Clemmensen reduction, both the ketone and the isolated double bond were reduced, forming cembrane and confirming structure (**37**) for incensole.

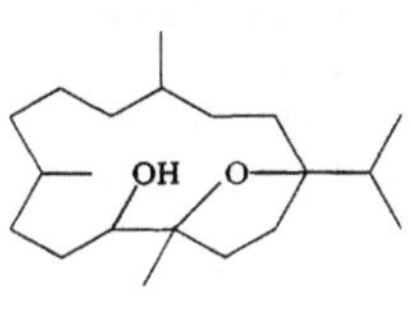

(42)

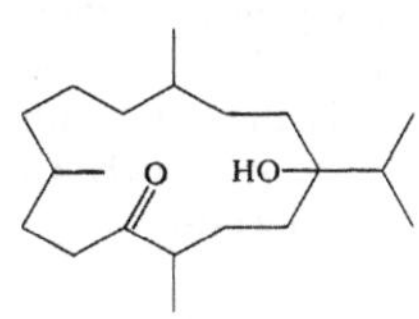

(43)

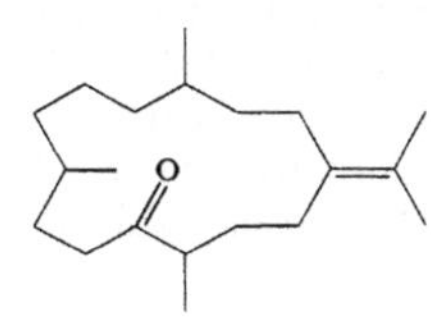

(44)

Incensole Oxide

The second cembranoid isolated from *B. carteri* by NICOLETTI's group (*32*) was incensole oxide (**45**), a crystalline solid of molecular formula $C_{20}H_{34}O_3$ which was spectrally very similar to incensole. The only significant difference between the two was observed in their pmr spectra, in which the signals for a vinyl methyl and a vinyl hydrogen in incensole were replaced by methyl and hydrogen signals typical of an oxirane in incensole oxide. The presence of an oxirane in (**45**) was confirmed by lithium aluminium hydride reduction to a tertiary methyl carbinol (**46**) which still contained one trisubstituted double bond.

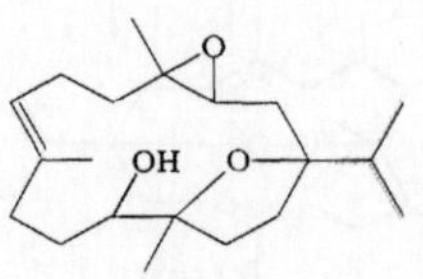
(45)

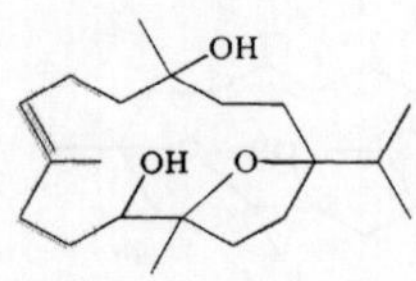
(46)

It was clear, therefore, that incensole oxide differed from incensole by epoxidation of a double bond. This was confirmed by reaction of incensole with *p*-nitroperbenzoic acid which gave a monoepoxide identical in all respects with incensole oxide. The two possible structures for incensole oxide thus were (**45**) and (**47**).

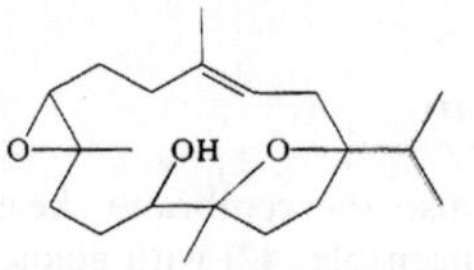
(47)

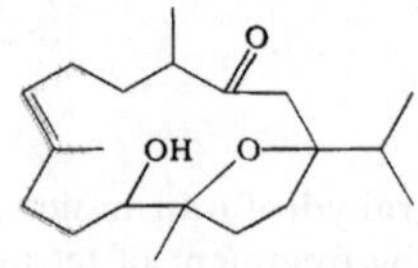
(48)

Selection between these two structures was guided by the properties of the ketone (**48**) obtained by the isomerization of incensole oxide with perchloric acid. The pmr spectrum of the ketone showed a doublet signal for the 4-methyl group in one *alpha* position, and a four-line AB pattern for the methylene group in the other *alpha* position. Since the latter signal showed no further splitting, it had to be flanked by a fully substituted carbon atom, as is the case with ketone (**48**) but not its alternative that would be formed from (**47**). Consequently, the structure of incensole oxide was proved to be (**45**).

Isoincensole Oxide

Isoincensole oxide was the third cembranoid isolated (*33*) from *B. carteri* by Nicoletti's group. After purification as its benzoate and hydrolysis, it was found to have the composition $C_{20}H_{34}O_3$. Spectral analysis showed that it was an alcohol, that it contained a trisubstituted double bond and that it was in general very similar to incensole oxide.

Isoincensole oxide was at first incorrectly regarded as simply the positional isomer (**47**) of incensole oxide (**45**), primarily because it was identical with one of the minor products obtained in the reaction between incensole and *p*-nitroperbenzoic acid. However, the supposed oxirane function of isoincensole oxide failed to be cleaved by either lithium aluminum hydride or catalytic hydrogenation, both under forcing conditions or by the generally more effective lithium-ethylamine reagent. In addition, the pmr signal for the supposed oxirane proton fell at δ 4.23, outside the usual range. The authors recognized these difficulties, but attributed the unusual stability of the supposed oxirane to transannular steric hindrance.

The authors verified the 3,4-location of the double bond in isoincensole oxide by cleaving it to the keto aldehyde (**49**) using osmium tetroxide followed by lead tetraacetate as shown in the following sequence. Its pmr spectrum demonstrated that the methylene *alpha* to the aldehyde was flanked by a fully substituted carbon atom.

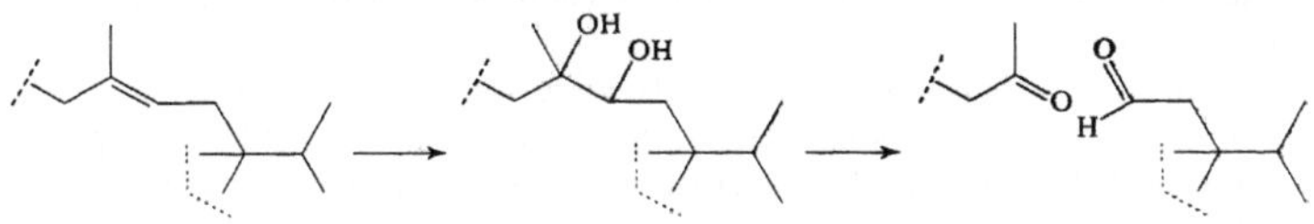

(**49**)

On the basis of further evidence, FORCELLESE *et al.* (*34*) subsequently revised the structure of isoincensole oxide to (**51**). The formation of this structure by reaction of incensole with peracid was interpreted as the result of transannular ring-opening of the initially formed oxirane (**47**) by the 11-hydroxyl group, leading to the formation of a second tetrahydrofuran ring.

Evidence for the new structure assignment was obtained by epoxidation of incensole acetate in which participation reactions were inhibited. This reaction afforded the expected oxirane acetate (**50**), which was different from the acetate prepared from isoincensole oxide. Careful alkaline hydrolysis of (**50**) formed the hydroxy oxirane (**47**) which was unstable and slowly converted to isoincensole oxide (**51**).

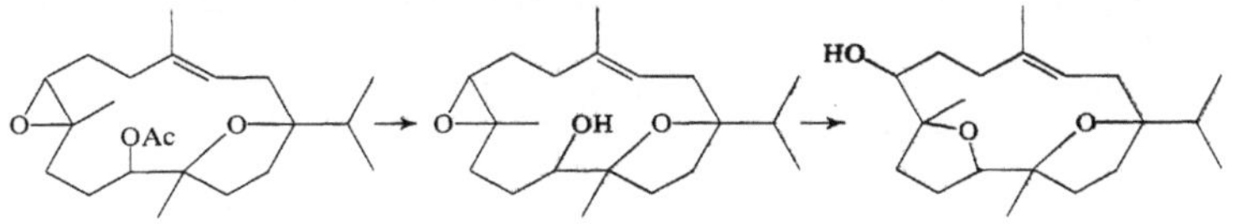

(**50**) (**47**) (**51**)

The structure (**51**) thus established for isoincensole oxide was compatible with all other evidence adduced for its structure, and accommodates the anomalous properties for the initially proposed "oxirane".

2. *Tobacco*

The tobacco cembranoids have been isolated primarily from tobacco leaf after commercial curing, but also from tobacco smoke and fresh plant material of *Nicotiana tabacum*. These compounds consist of a closely related group of hydroxy derivatives of cembrane, which contain at least one further oxygen function.

All of the natural tobacco cembranoids contain a 4-hydroxy substituent and six of the nine compounds discovered to date represent three epimeric pairs differing in the configuration of this function. Members of the pairs were chemically correlated, and were designated empirically as the α- and β-forms. This usage is convenient and still persists, but has no configurational implication and should not be confused with the designations of the two series of cembrane absolute configurations (cf. I. C.).

Although stereochemical details were known only sketchily at the time of the original publications, recent work by ENZELL's group (*35*) has established absolute configuration for most of the tobacco compounds. All are members of the cembrane α-series. In addition, the α- and β-forms were shown to have the hydroxyl configurations shown in (**52**) and (**53**). The structures presented for the natural tobacco cembranoids in the following discussion incorporate all the stereochemical detail currently known for each of them.

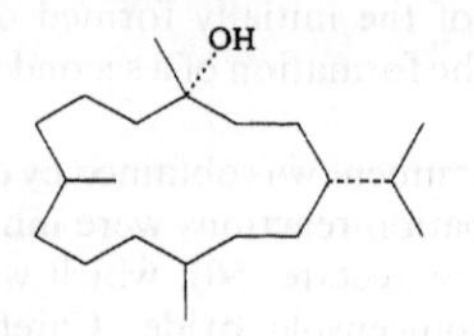

(**52**)

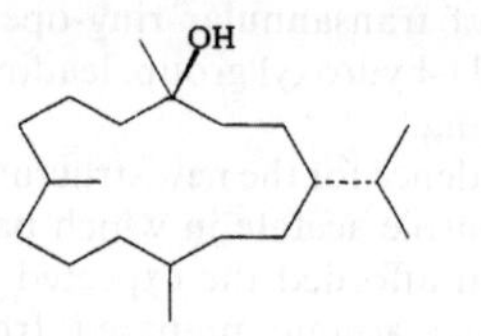

(**53**)

Including those of ROBERTS and ROWLAND's first report (*3*) the following cembrane diterpenes have been encountered in *Nicotiana tabacum*.

α- and β-2,7,11-cembratriene-4,6-diol (**54**) and (**55**)
α- and β-2,6,11-cembratriene-4,8-diol (**68**) and (**69**)
α-8,11-oxido-2,6,12-cembratriene-4-ol (**71**)
α- and β-8,11-oxido-2,6,12(20)-cembratriene-4-ol (**74**) and (**75**)
2,7,11-cembratriene-4-ol-6-one (**78**)
α-8,12-oxido-2,6-cembradiene-5,11-diol (**77**)
(cembrene) (**1**)

The first seven of these compounds were isolated from aged burley tobacco in a total yield of less than 0.02% (*36*). Compound (**78**) was isolated from natural fermented leaves (*37*), compound (**77**) was isolated from a Greek tobacco (*38*), and compound (**1**) was detected in extracts of tobacco leaf slices in the presence of enzyme inhibitors (*39*). Recently, the 4,6-diol diastereoisomers (**54**) and (**55**) have been found to represent the major components in cuticular wax of *N. tabacum* (*40*). The concentration of these isomers decreased with the age of the leaves while the total wax content remained constant (*41*).

α- and β- 2,7,11-Cembratriene-4,6-diol

In their first paper on the tobacco diterpenes which appeared in the same year, 1962, as the structure elucidations of cembrene, ROBERTS and ROWLAND of the R. J. Reynolds Tobacco Company reported (*3*) the characterization of two diastereomers of 2,7,11-cembratriene-4,6-diol. Thorough chemical studies and spectral interpretations led to the assignment of structures (**54**) and (**55**) to these isomers, which were designated as the α- and β-forms, respectively. The two isomers were shown to differ in configuration at position 4, and possibly also at position 6, but to have the same configuration at position 1.

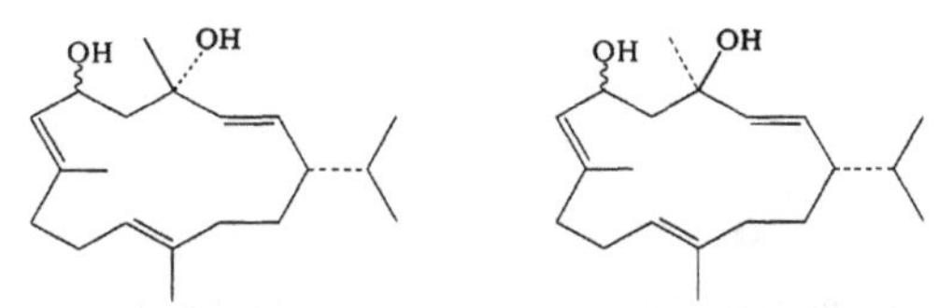

(**54**) (**55**)

Preliminary characterization showed the gross structural features of the two isomers, each $C_{20}H_{34}O_2$, to be very similar. Hydrogen uptake and active hydrogen determinations on the α- and β-isomers required that each be a monocyclic trienediol. Their pmr spectra demonstrated the presence in each of a secondary hydroxyl, a tertiary hydroxyl *alpha* to a methyl group, a *trans* disubstituted double bond, two vinyl methyls, four vinyl hydrogens and an isopropyl group attached to a saturated carbon atom.

The relative position of the two hydroxyl groups in these isomers and their hexahydro derivatives was shown to be 1,3 by chemical means. Upon hydrogenation, the α-diol (**54**) formed the hexahydroderivative (**56**) which was separated into two pure epimeric forms differing in configurations at either position 8 or position 12. Neither of these saturated epimers nor the unsaturated diol, reacted with lead tetraacetate or sodium periodate, thus precluding a 1,2-relationship for the hydroxyl groups. However, a 1,3-relationship of the hydroxyl groups was demonstrated by CrO_3/pyr oxidation of the two saturated diols (**56**) to two epimeric hydroxy ketones (**57**) which were clearly β-hydroxyketones.

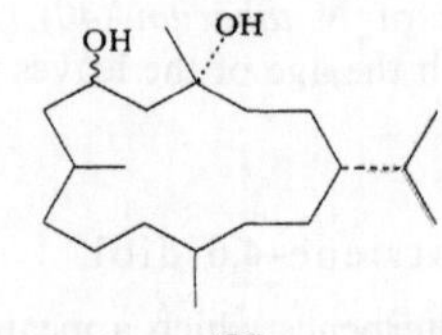

(56)

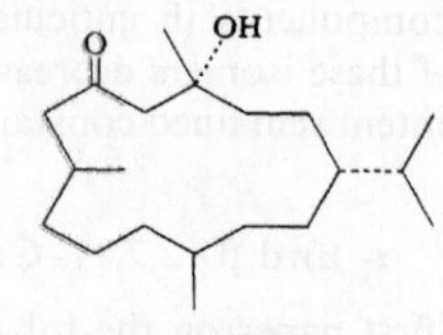

(57)

Upon treatment with refluxing aqueous base, each of the hydroxyketones (57) underwent retroaldol reaction to a different epimeric form of the ring-opened diketone (58). One of these diketones subsequently proved useful in establishing the epimeric relationship between (54) and (55) at position 4 and their configurational identity at position 1.

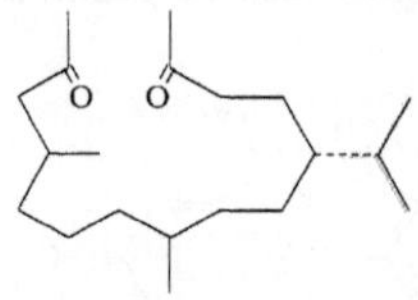

(58)

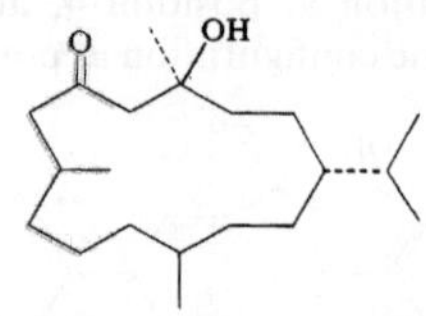

(59)

In a parallel series of chemical transformations, the natural β-diol (55) was converted to a similar pair of saturated epimeric diols, each different from the pair derived from the α-isomer. Separate oxidation of these again generated an epimeric pair of hydroxyketones (59). One of these underwent the retroaldol reaction to form a ring-opened diketone (58) which was identical to one of the open chain diketones obtained from the α-isomer, as indicated by mixture melting point of their bis-2,4-dinitrophenylhydrazones. Since the α- and β-series saturated hydroxyketones were non-identical, but were converted to the same openchain diketone in a reaction which merely eliminated one chiral center at position 4, it was shown that one of the differences between the two series was their epimeric relationship at position 4. It also demonstrated that the configuration at position 1 was identical in each series.

The secondary hydroxyl group of the unsaturated diols (54) and (55) could be oxidized by manganese dioxide, showing they were allylic alcohols. The pmr and uv spectra of the resulting conjugated ketones (60) and (61) showed that the conjugated double bond was trisubstituted, and bore a methyl group in the *beta* position.

These unsaturated hydroxyketones underwent interesting retroaldol reactions which were useful in locating the second of the three double bonds. Upon treatment with aqueous carbonate or with acidic alumina. the α-series hydroxyketone (60) formed an oily ring-opened diketone

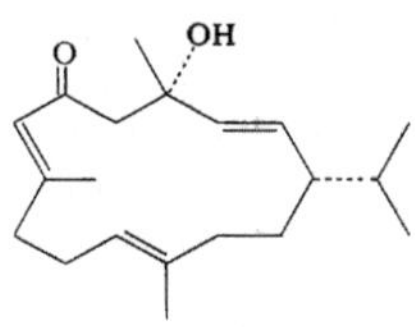

(60)

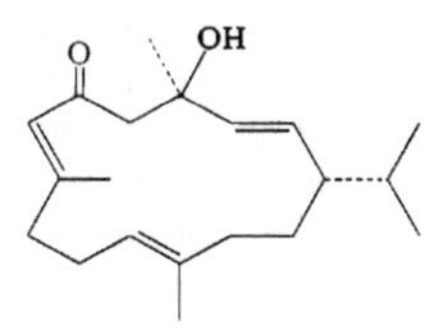

(61)

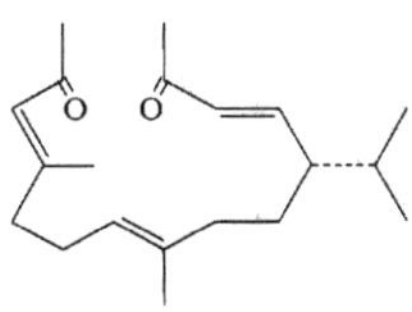

(62)

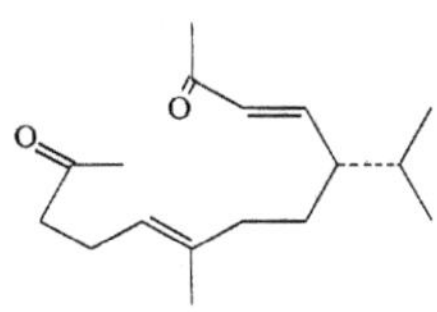

(63)

(62). Similar treatment of the β-series hydroxyketone afforded the same diketone (62). This result confirmed the conclusion reached in the saturated series that the α- and β-series were epimeric at position 4. However, it also showed that the balance of the molecule was identical in each series.

The diketone (62) contained, in addition to the expected conjugated tri-substituted double bond, a second conjugated double bond which had *trans* geometry, was disubstituted, and was flanked by a methine carbon atom. The diketone (62) underwent further retroaldol cleavage with the loss of acetone upon treatment with refluxing aqueous base. The product (63) contained a new saturated methyl ketone, but retained the conjugated disubstituted double bond. These results established the location of the Δ^2-unsaturation and provided chemical verification of the methyl substituent at position 8.

To complete the structure determination, there remained the need to establish the size of ring, the position of the isopropyl group, and the location of the remaining methyl substituted double bond. All of these features were demonstrated by the isolation of levulinic acid and 5-oxo-2-isopropyl-hexanoic acid (64) from the Lemieux-von Rudloff oxidation of the natural α-4,6-diol, thus firmly securing gross structures (54) and (55) for the two tobacco cembranoids.

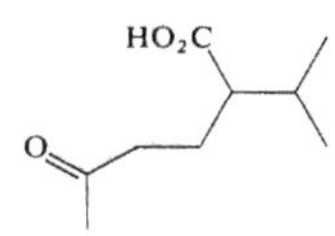

(64)

ROBERTS and ROWLAND showed by spectral means that the 2,3- and 7,8-unsaturations of (**54**) and (**55**) were both *trans*, but came to no conclusion regarding the geometry of the remaining 11,12-trisubstituted double bond. The latter problem was clarified by COURTNEY and MCDONALD (*42*) who subjected the β-diol (**55**) to fragmentation with p-toluenesulfonic acid. This reaction produced the acyclic aldehyde (**65**), itself a constituent of freshly picked tobacco flowers (*42*). The high field chemical shift (δ 1.60) of vinyl methyl group corresponding to the 12-methyl of the diol clearly established the *trans* geometry of this double bond.

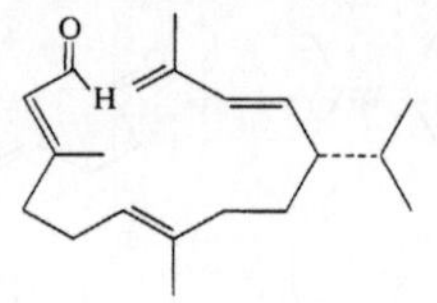

(**65**)

A closely related pair of 4,6-diols was reported recently by SPRINGER *et al.* (*43*). To test their hypothesis that tobacco plant development is controlled by natural growth inhibitors, they employed a bioassay for wheat coleoptile inhibition to guide the fractionation of a leaf extract. Two growth inhibitors were isolated. The structure and the relative configuration of the more active inhibitor was shown to be the 4,6-diol (**66**) by X-ray diffraction analysis. Its melting point and nmr spectral properties were similar to, but not identical with ROBERTS and ROWLAND's (*3*) β-4,6-diol (**55**). The structure of the less active inhibitor was not established, but it showed spectral similarities to the α-4,6-diol (**54**). Although, as inferred by the authors, this pair of inhibitors may be identical with the known pair of 4,6-diols, direct comparisons with authentic samples were not reported. Furthermore, SPRINGER *et al.* (*43*) observed pmr and cmr spectral differences between the two inhibitors which suggested that they differed in configurations at both hydroxylated positions. On the other hand, AASEN *et al.* (*35*) stated that ROBERTS and RHODE (*44*) have dehydrated the known α- and β-4,6-diols to the same cembratetraene-6-ol (**67**), indicating that the configuration at position 6 is identical in (**54**) and (**55**).

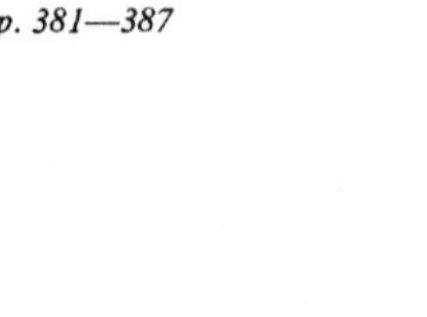

(**66**)

(**67**)

α- and β-2,6,11-Cembratriene-4,8-diols

In their second paper on tobacco constituents, ROWLAND and ROBERTS (*45*) characterized a second pair of diastereomeric diols, the α- and β-4,8 diols (**68**) and (**69**). These were shown to be allylic isomers of the 4,6-diols. The relationship between the two pairs of epimers was most simply demonstrated by the conversion of the α-4,6-diol (**54**) to the α-4,8-diol (**68**), and of the β-4,6-diol (**55**) to the β-4,8-diol (**69**) during slow chromatography on acidic alumina. The authors established that individual isomers of the set were stable to the isolation processes employed and thus were not artifacts.

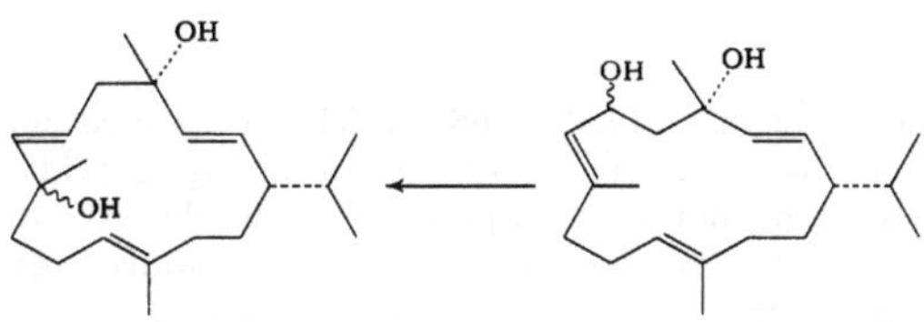

(**68**) (**54**)

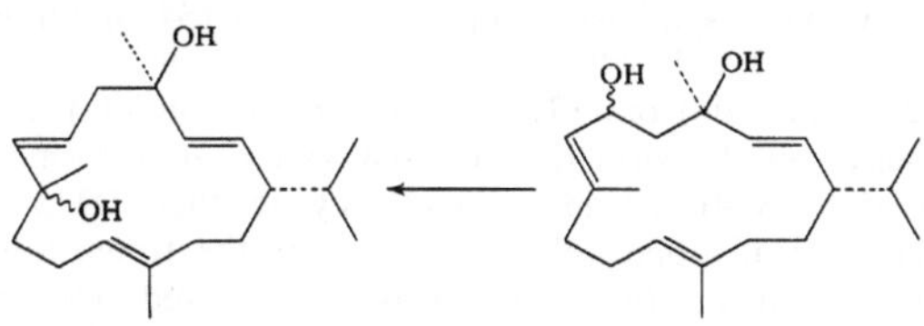

(**69**) (**55**)

The β-4,8-diol (**69**) was available in greater quantity, and was used for most of the characterization study. Hydrogenation-hydrogenolysis in acetic acid yielded the parent hydrocarbon, cembrane. This hydrocarbon exhibited an infrared spectrum identical with that of the hydrocarbon obtained similarly from the β-4,6-diol (**55**), thus establishing the common carbon skeleton.

The saturated hexahydro derivative of the β-4,8-diol was resistant to oxidation by chromium trioxide in pyridine or in aqueous acetic acid, thus characterizing both hydroxyl groups as tertiary. However, the unsaturated diol (**69**) itself underwent an interesting oxidation with a large excess of chromium trioxide in pyridine. The major product of this reaction was the hydroxyketone (**70**), which proved to be identical with the hydroxyketone obtained previously by oxidation of the β-4,6-diol (**55**). This anomalous reaction was interpreted in terms of a prior allylic isomerization of the ditertiary 4,8-diol to the oxidizable 4,6-diol system. The conjugated 7,8-

unsaturation in **(70)** was shown to have *trans* geometry by pmr comparison with the *cis* hydroxyketone which was also formed in the oxidation. Two other minor oxidation products were the 7,8-epoxy derivative of the hydroxyketone **(70)**, and the 6,7-epoxy derivative of the 4,8-diol **(69)** itself.

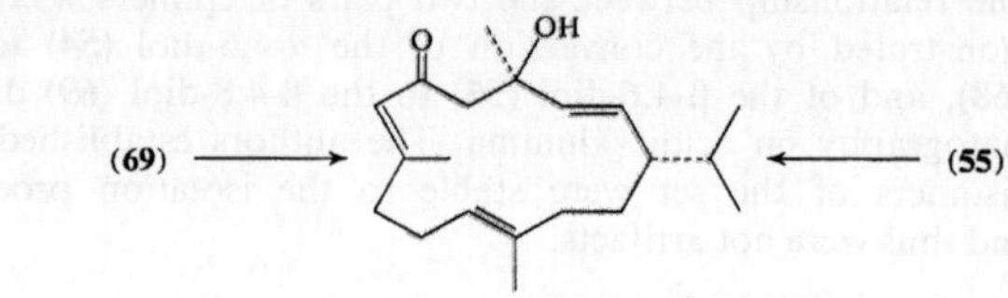

(70)

The structure of the α-4,8-diol **(68)**, which was isolated in only trace amounts from tobacco leaf, was established by analogous oxidation to the same hydroxyketone **(60)** obtained previously from the α-4,6-diol **(54)**.

ROWLAND *et al.* (*36*) also isolated three ether-bridged cembranoids containing tetrahydrofuran moieties from tobacco leaves and cigarette smoke. The structure suggested for the first of these compounds **(71)** rested in part upon its synthesis from the α-4,6-diol **(54)** and in part upon the presence of a doubly allylic methylene group in the dehydration product **(72)** obtained from **(71)**. Peracid oxidation of the diol **(54)** proceeded selectively to form the 11,12-epoxide **(73)**. Treatment of this epoxide with dilute sulfuric acid formed the cyclic ether **(71)** in 50% yield. Of the mechanistically plausible structures for the transannular cyclization product of this reaction, only **(71)** was considered to be compatible with the requirement of a double allylic methylene in its dehydration pro-

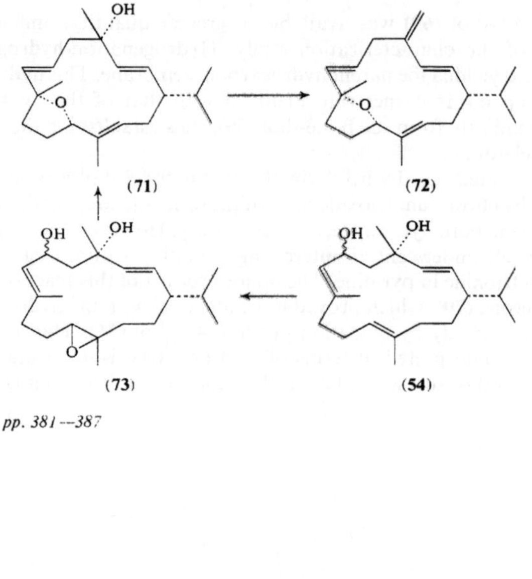

duct. Since the configuration of the 4-hydroxyl of the α-4,6-diol was not affected during this synthesis, the ether (**71**) was established as a member of the α-series.

The other two tetrahydrofurano cembranoids contained exocyclic unsaturations as shown in (**74**) and (**75**) for the α- and β-isomers, respectively. The gross structure of the α-isomer was established by hydrogenation to the hexahydro derivative which was identical by mixture melting point with the hexahydro derivative of the ether (**71**). The location of the double bonds in (**74**) was established by comparison of its nmr spectrum with that of (**71**). The structure of the β-isomer (**75**), differing only in the configuration at position 4, followed from its conversion to the same (mmp) tetraene (**76**) as obtained by dehydration of the α-isomer.

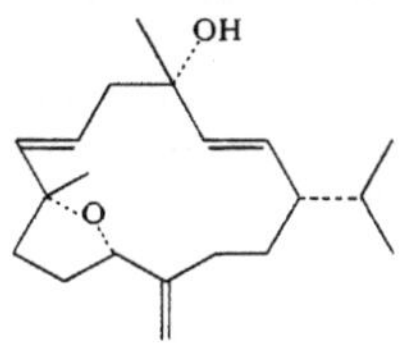

(**74**)

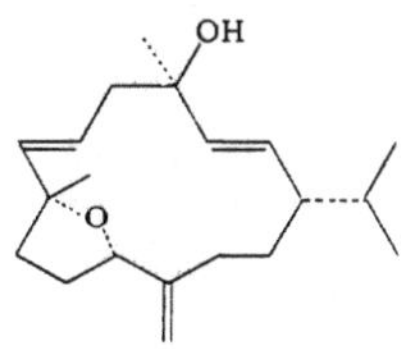

(**75**)

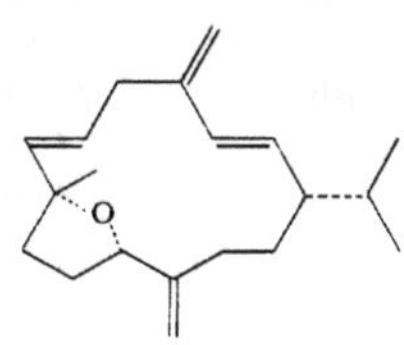

(**76**)

Recently, ENZELL's group at the Swedish Tobacco Company (*35*) has fully confirmed the structure (**74**) for the α-isomer by an X-ray diffraction study. In addition, the absolute configuration shown in (**74**) was established by the isolation of (2S)-5-oxo-2-isopropylhexanal from ozonolytic degradation of the β-4,6-diol (**55**) which had been correlated with (**74**) through an indirect series of transformations.

8,12-Oxido-2,6-cembradiene-4,11-diol

ENZELL's group (*37*) has also recently isolated an ether bridged cembranoid (**77**) containing a tetrahydropyran moiety. It was isolated from Greek tobacco at the level of 12 parts per billion. Its structure and

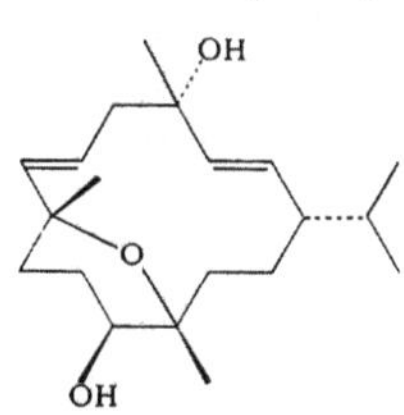

(**77**)

relative configuration were obtained by X-ray diffraction analysis of the 11-mono-acetate, and its absolute configuration was deduced from application of the benzoate sector rule to the 11-benzoate and the 11-*p*-nitrobenzoate. Although spectral data were presented in this paper, it is unfortunate that no melting points or rotations were recorded for (**77**) or any of its derivatives.

ZANE's isolation (*38*) of the hydroxyketone (**78**) from dark-fired tobacco was the first report of the occurrence of this compound in natural sources. Obtained as an oil, its structure was assigned by direct comparisons with authentic specimens of the individual hydroxyketones (**60**) and (**61**), prepared from the α- and β-forms of the 4,6-diols. However, since a mixture of the synthetic isomers could not be separated by glc or column chromatography, the isomeric purity of the natural material remained doubtful. Its description resembles that of ROBERTS and ROWLAND's (*3*) β-isomer (**55**) which was also an oil, whereas the α-isomer was crystalline. Although the optical rotation of the natural hydroxyketone was not reported, the distinctive rotations of the known α- and β-isomers may have allowed an assignment and/or an assessment of isomeric purity.

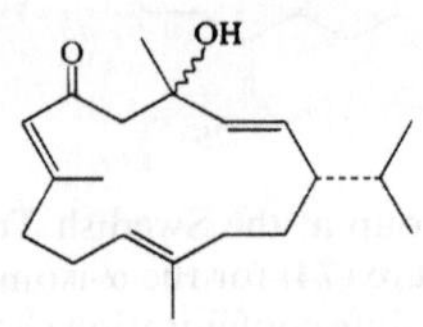

(**78**)

In a study of mevalonate uptake in tobacco leaf slices in the presence of inhibitors of mammalian sterol biosynthesis, REID (*39*) has claimed the occurrence and accumulation of the hydrocarbon, cembrene (**1**). However, no experimental basis for the identification was provided. This hydrocarbon is presumably the progenitor of the tobacco cembranoids, but has not otherwise been encountered in tobacco.

Configurational Interrelationships

In their paper which reported the X-ray structure determination and absolute configuration of the tetrahydrofurano cembranoid (**74**), ENZELL *et al.* (*35*) painstakingly summarized the scattered literature bearing upon the configurations and olefinic geometries in the known tobacco cembranoids and many of their chemical transformation products. In addition, they collected similar correlations for a number of other tobacco constituents, e. g. aldehyde (**65**), which are viewed as degradation products of the cembranoids. Several correlation gaps were bridged by the complete characterization of the ether (**74**). Other gaps were filled by the degradation of the β-4,6-diol (**55**) and the α-series hydroxyketone (**60**) to (2*S*)-5-oxo-2-isopropylhexanal.

All of the tobacco cembranoids were thereby shown to have the same configuration at position 1, and thus to belong to the α-series of cembranoids. Yet to be established are the configuration(s) at position 6 in the 4,6-diols, at position 8 in the 4,8-diols and the 6,7-olefinic geometries in the 4,8-diols.

As indicated at the beginning of the description of tobacco cembranoids, presently-known configurational features have been incorporated into the structures which have been presented.

3. Miscellaneous Plants

Ovatodiolide

The bis-lactone ovatodiolide (**79**) was one of the earliest cembrane derivatives to be reported from the plant kingdom, and still retains the distinction, by a wide margin, of being the chemically most elaborate member of the plant cembranoids.

In their preliminary communication, An, Toubiana, and Lederer (*46*) reported the partial characterization of ovatodiolide, $C_{20}H_{24}O_4$, which had been isolated from the Indochinese plant *Anisomeles ovata*. In the infrared, the compound showed very intense absorption at 1772 cm^{-1}, but no absorptions for hydroxyl or other oxygen containing functions, suggesting that it contained two γ-lactone functions. It showed no ultraviolet absorption above 220 nm.

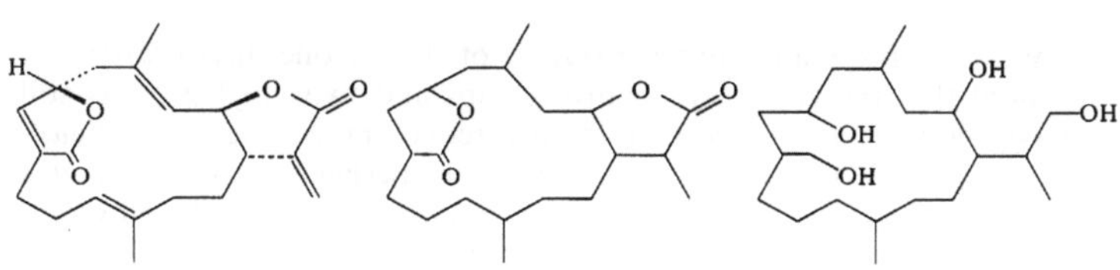

(**79**) (**80**) (**81**)

The presence of four double bonds was demonstrated by conversion to the octahydro derivative (**80**) by catalytic hydrogenation. The infrared spectrum of (**80**) showed intense γ-lactone absorption at 1782 cm^{-1}, and no hydroxyl absorption. Reduction of (**80**) with lithium aluminium hydride gave a crystalline tetrol (**81**), $C_{20}H_{40}O_4$, which clearly possessed only one carbocyclic ring. The pmr spectrum of its tetraacetate established that the tetrol consisted of two secondary and two primary alcohol functions, and the simple doublet character of the overlapping signal for the primary methylenes showed that each was vicinal to a methine position. Thus, in the octahydro derivative (**80**), the *gamma* position of each lactone was secondary, and the *alpha* position of each was monosubstituted.

In the second paper from this group, IMMER, POLONSKY, TOUBIANA, and AN (*47*) provided a complete chemical argument for structure (**79**) for ovatodiolide. A conjugated exo-methylene substituent was recognized in one of the lactones by the characteristic pair of downfield pmr signals which appeared as doublets because of allylic coupling with the *beta* hydrogen at position 1. Formaldehyde was produced by ozonolysis of ovatodiolide, but not of (**82**), its dihydro derivative formed by selective catalytic hydrogenation. The pmr spectrum of the dihydro derivative showed a new secondary methyl signal in place of the exo-methylene signals.

Since the ultraviolet spectrum of (**82**) still displayed the characteristic 216 nm absorption, it was concluded that the second lactone function of ovatodiolide also was α,β-unsaturated. In this case, the unsaturation was recognized as being endocyclic from the one proton absorption at δ 7.05 which is typical of the β-proton of α,β-butenolides.

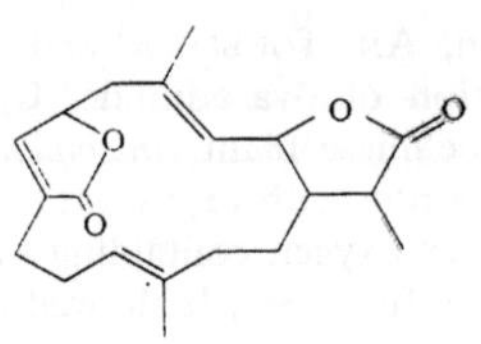

(**82**)

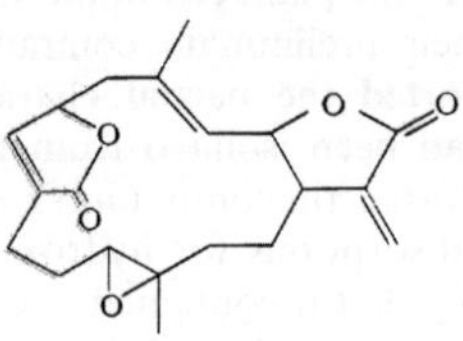

(**83**)

With the preceding characterization of the lactone unsaturations, it followed that the two vinyl methyl groups and two vinyl protons still unassigned were associated with the two remaining double bonds. Selective epoxidation of ovatodiolide with *p*-nitroperbenzoic acid formed a mono epoxide (**83**) whose nmr spectrum still showed one vinyl methyl and one vinyl proton, but also displayed methyl and proton signals characteristic of epoxides. Thus both double bonds were trisubstituted, and each bore a methyl group. Further, since no acetaldehyde was formed on ozonolysis of ovatodiolide, in neither double bond were both the methyl and hydrogen on the same carbon.

Two ketonic derivatives were prepared from ovatodiolide in an attempt to establish the size of its carbocycle from their carbonyl absorption frequencies. The acid (**84**), formed in minor amounts together with (**80**) upon perhydrogenation, was decarboxylated with lead tetracetate. The

References, pp. 381—387

resulting mixed acetates were hydrolyzed to the epimeric alcohols which were oxidized to a single compound, the nor-ketone (**85**) whose carbonyl absorption occurred at 1700 cm^{-1}. The other ketone was obtained from the saturated epoxide (**86**), prepared by hydrogenation of (**83**). Isomerization of (**86**) with boron trifluoride formed the ketone (**87**) which had a carbonyl absorption at 1710 cm^{-1}. The absorption frequency of the former ketone definitely indicated a ring size of greater than six atoms, and the latter was not incompatible with this conclusion.

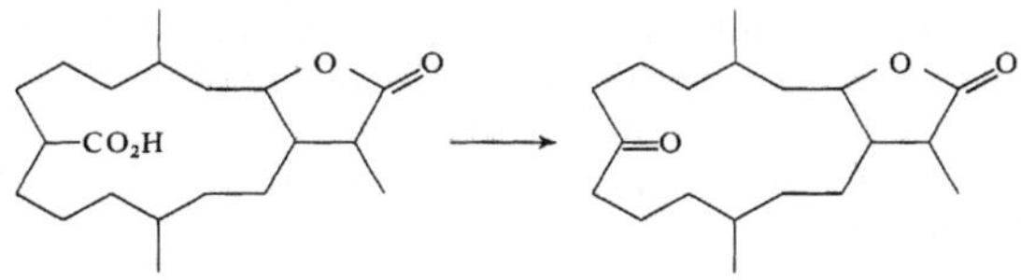

(**84**) (**85**)

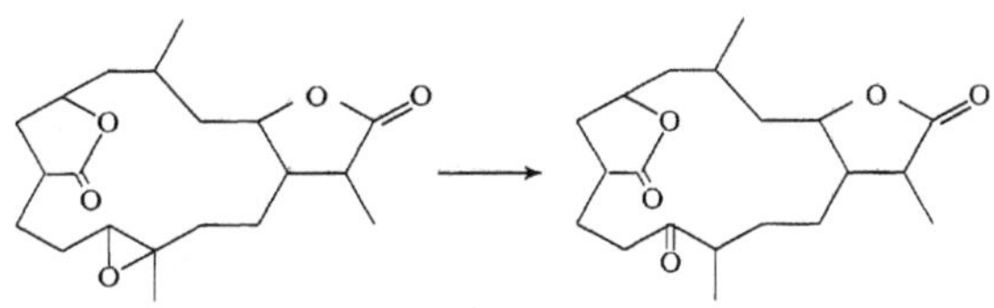

(**86**) (**87**)

The structural features established thus far accounted for six carbon atoms that could not be part of the one carbocycle required for ovatodiolide. The authors felt that ovatodiolide could therefore contain a 14-membered ring, more specifically a cembrane ring because the extracyclic structural elements were equivalent to the three methyl and one isopropyl substituents of that system.

These speculations were supported by establishment of the sequence (**88**) in dihydro-ovatodiolide (**82**) through decoupling experiments which showed that protons at position 5 were coupled both to H_3 and to H_6. Together, sequence (**88**) and sequence (**89**), for the isolated double bond, account for eleven of the fourteen members of a cembrane nucleus.

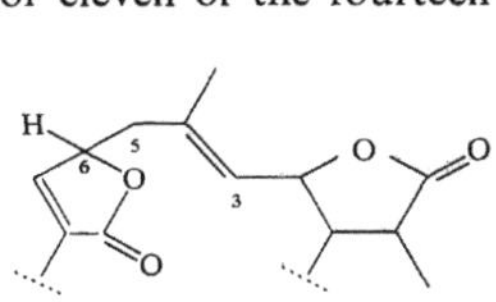

(**88**)

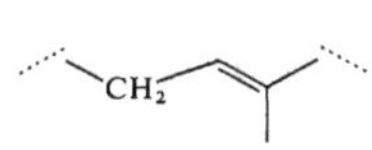

(**89**)

The location of the remaining three methylene groups, as well as the relationship between the two sequences (**88**) and (**89**), was established by mass spectrometry of the open-chain derivative (**90**). This compound was prepared from the monoepoxide (**83**), in which sequence (**88**) was also demonstrated. After conversion to the hexahydro derivative, the epoxide was hydrolysed ($HClO_4$) to the diol (**91**). Jones oxidation, and esterification of the cleavage product, produced keto ester (**90**). Its mass spectrum showed ions resulting from fragmentations at points (a) and (b) in (**90**). These results definitely located the methyl group at position 12, and established the cembrane nucleus for ovatodiolide. However, they did not distinguish between the 11,12 and the 12,13 locations for the original double bond, since mass spectral ions of identical m/e values would be expected from the open-chain compound derived similarly from the latter alternative.

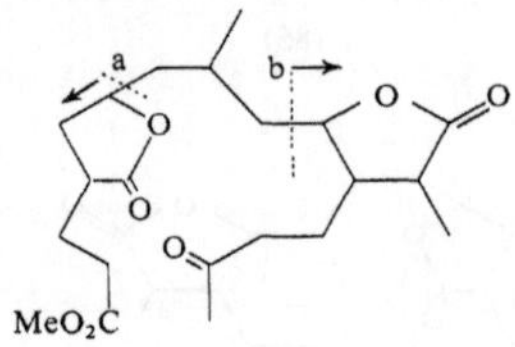

(**90**)

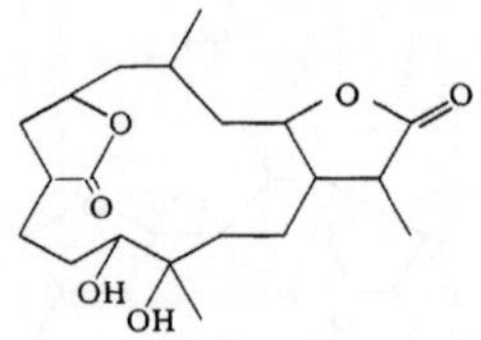

(**91**)

The 11,12 location for the double bond was confirmed by the formation of levulinic acid from the Lemieux oxidation of ovatodiolide. The formation of this acid was considered to arise from cleavage between positions 11 and 12 and positions 1 and 2. The latter cleavage was postulated to arise from prior allylic isomerization in the basic medium of the lactone exo-methylene group to the endocyclic position. In confirmation of this interpretation, dihydroovatodiolide (**82**) did not form levulinic acid under the same conditions.

The above evidence firmly established the gross structure (**79**) for ovatodiolide. Recently, TOUBIANA *et al.* (*48*) confirmed this structure and provided the relative stereochemistry shown in (**79**) by an X-ray diffraction study. They were also able to deduce the same stereochemical relationships from a detailed pmr spectral study using aromatic solvent induced shifts ($CDCl_3$ vs C_6D_6) and nuclear Overhauser effects.

Anisomelic Acid

The closely related lactonic acid, anisomelic acid (**92**) was recently reported by PURUSHOTHAMAN, RAO, and KALYANI (*49*) to occur together with ovatodiolide in the South Indian medicinal plant *Anisomeles mala-*

References, pp. 381—387

barica. The acid, $C_{20}H_{26}O_4$, showed extensive pmr similarities to ovatodiolide, including the α-methylenic lactone. However, the doublet at 7.05 ppm due to the proton in the β-position of the butenolide of ovatodiolide was replaced by a triplet at δ 5.85, characteristic of a proton in the β-position of an α,β-unsaturated acid.

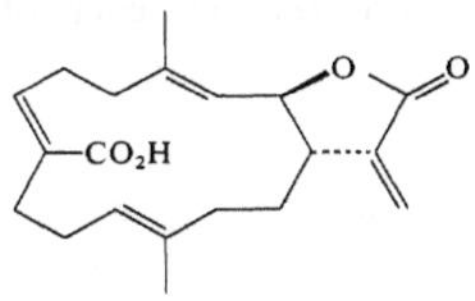

(92)

The relationship of the acid to ovatodiolide was firmly established by the identity of octahydroanisomelic acid with the saturated acid (**84**) previously prepared by AN *et al.* (*47*). The locations of the double bonds in anisomelic acid were not established.

B. Insects

Cembrene-A (Neocemrene-A)

The existence of a highly active scent-trail pheromone in the termite *Nasutitermes exitisosis* was first observed by MOORE (*50*) in 1966. Its identity with cembrene-A (**16**) was shown by BIRCH, BROWN, CORRIE, and MOORE (*51*) who employed chemical and spectral studies to establish its structure.

The pheromone typically was isolated in 2—3 mg quantities from batches of 12 kg (live weight) of termites, thus necessitating the use of micro techniques. The compound, $C_{20}H_{32}$, formed an octahydro derivative, $C_{20}H_{40}$, establishing that it was monocyclic. Upon ozonolysis, it formed two equivalents of levulinaldehyde and one of formaldehyde. Its pmr spectrum showed the presence of an isopropenyl group, and three trisubstituted double bonds, each containing a methyl group.

BIRCH *et al.* first considered biogenetically reasonable cyclohexanes with polyprenyl side chains which were compatible with the above evidence, and synthesized several stereoisomers of a likely candidate. They were, however, unlike the pheromone in activity, and in spectral and physical properties.

It was then recognized that the octahydro derivative of the pheromone was identical, by glc and ms, with octahydrocembrene provided by DAUBEN (*1*). The only cembrene structure compatible with the spectral and ozonolysis evidence was (**16**). To confirm this possibility, the pheromone was treated with the anion of DMSO which could be predicted to isomerize the double bond of the isopropenyl group to the conjugated endocyclic position (**36**). This hydrocarbon was indeed formed as one of the isomerization products, as shown by its ozonolysis which formed two equivalents of levulinaldehyde and one each of glyoxal and 6-methylheptane-2,5-dione.

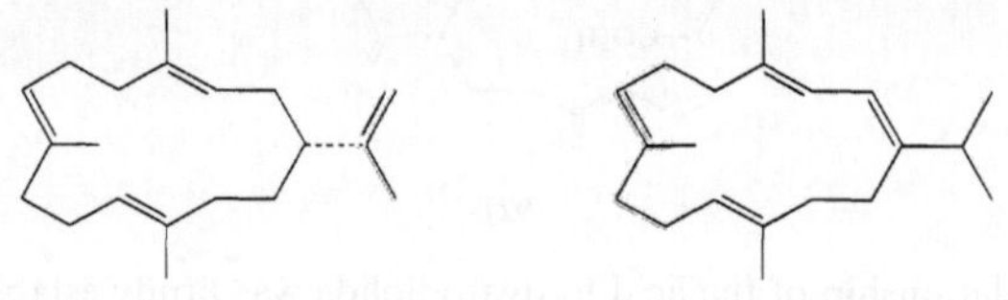

(**16**) (**36**)

These products accounted for all the carbon atoms in the molecule, and firmly established the structure of the trail pheromone as (**16**). The pmr spectrum of the pheromone differed slightly from that of PENTEGOVA's (*16*) cembrene-A (neocembrene), but was identical to that of DEV's (*15*) cembrene-A. Until the identity was established, BIRCH *et al.* had provisionally named the pheromone neocembrene-A.

C. Marine Invertebrates

(−)-Cembrene-A

Among the numerous cembranoids found by HERIN and TURSCH (*52*) of the Free University of Brussels in their studies of the Indonesian soft coral *Sinularia flexibilis* was the hydrocarbon (−)-cembrene-A (**16**). It was identified by comparison of its pmr, ir and mass spectra with those of the resin-derived cembrene-A provided by DEV (*15*), and displayed the same optical rotation. It also showed the same glc retention time as the cembrene-A prepared by dehydration of the side-chain alcohol, nephthenol (*vide infra*).

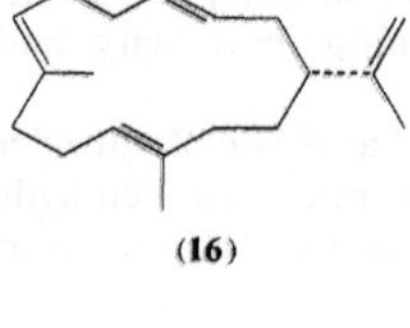

(**16**)

Flexibilene

HERIN, COLIN, and TURSCH (*53*) isolated a second diterpene hydrocarbon from *Sinularia flexibilis* which accompanied the major hydrocarbon, cembrene A. Although not itself a cembranoid, it is worthy of mention because of its very close relationship. The new hydrocarbon, named flexibilene (**93**), appears to possess the exceptional feature of a fifteen-membered ring.

Flexibilene, $C_{20}H_{32}$, is isomeric with cembrene (**1**) and has five degrees of unsaturation. It showed no conjugated diene absorption in the uv, and the ir indicated a *trans* disubstituted double bond. Its pmr spectrum showed a gem-dimethyl grouping, three vinyl methyls, five vinyl hydrogens and a doubly allylic methylene. It contained no indication of a terminal or exo double bond.

The authors argued that an acyclic diterpene having five double bonds, and devoid of a terminal olefin, would have to contain six methyl groups. This requirement was incompatible with the pmr data for flexibilene. On the other hand, a monocyclic structure having four double bonds and five methyl groups was a reasonable alternative accounting for the unsaturations, and was compatible with flexibilene. Taking into account the presence of one doubly allylic methylene, the absence of conjugation, and the absence of any indications of terminal groups of side chains, the authors proposed (**93**) as the only reasonable structure for flexibilene. This structure possesses no asymmetric center, compatible with the fact that flexibilene is optically inactive.

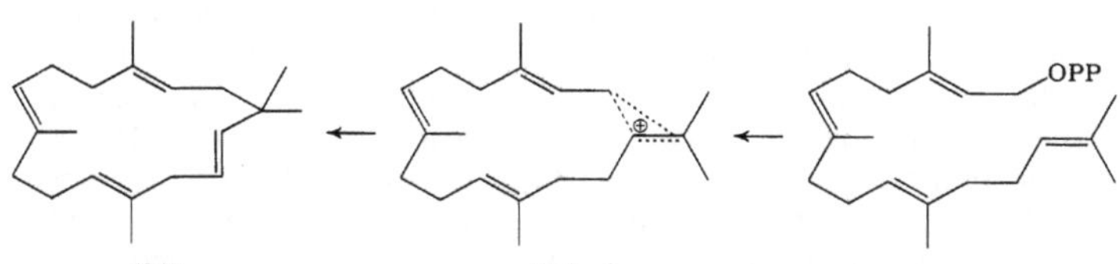

(**93**) (**94**)

HERIN and TURSCH observed that flexibilene has the same relationship to cembrene as the sesquiterpene humulene has to germacrene, and that flexibilene may be regarded as an isoprenolog of humulene. They speculated that flexibilene and cembrene may be formed biogenetically from geranylgeranyl pyrophosphate *via* the same bridged cationic intermediate (**94**).

It may be noted that ROBERTS and ROWLAND (*3*) predicted the existence of the flexibilene skeleton in their earliest paper on the tobacco cembranoids.

Nephthenol and (−)-Nephthenol

SCHMITZ, VANDERAH, and CIERESZKO (*54*) of the University of Oklahoma isolated nephthenol (**95**) as a minor constituent of an undetermined *Nephthea* species of soft coral from Enewetak, Marshall Islands. Its ir spectrum showed hydroxyl absorption, and the pmr showed three vinyl methyls, three vinyl protons, twelve allylic hydrogens, and a six proton singlet for a 2-hydroxy-2-propyl group. Degradation with potassium permanganate-sodium periodate formed levulinic acid and homoterpenyl ketone (**96**), permitting assignment of structure (**95**). Ketone (**96**) was identical with an authentic sample by glc comparison, but not enough material was available to obtain its optical rotation. The *trans* olefin geometries shown in (**95**) were assigned tentatively by analogy to the structure of epoxynephthenol acetate (*vide infra*) which occurred in the same organism.

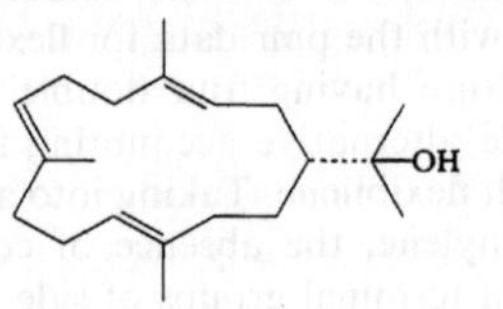

(**95**)

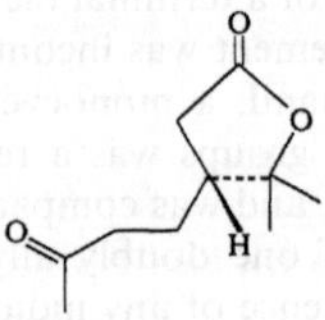

(**96**)

TURSCH, BRAEKMAN, and DALOZE (*55*) isolated larger quantities of nephthenol, of the levorotatory form, from the soft coral *Litophyton viridis* from Indonesia. It was identified by spectral comparison with the nephthenol reported by SCHMITZ *et al.* It formed two moles of levulinaldehyde upon ozonolysis, and was converted to cembrane by a dehydration-hydrogenation sequence. Lemieux oxidation of the Indonesian nepthenol formed (−)-homoterpenyl ketone (**96**) which confirmed structure (**95**) and also established the configuration of the asymmetric center at position 1. Phosphoryl chloride-pyridine dehydration of nephthenol afforded a 73 : 27 mixture of cembrene-A (**16**) and its isopropylidene isomer, thereby establishing *trans* geometry for the three double bonds in (**95**).

10-Hydroxycembrene

One of the minor constituents isolated by KASHMAN *et al.* (*56*) of Tel-Aviv University in their work on the Red Sea soft coral *Sarcophytum glaucum* was the alcohol (**97**). This structural proposal was based solely upon interpretations of the compound's ir and pmr spectra. The cembrane carbon skeleton was assumed.

Absorptions due to five vinyl hydrogens and three vinyl methyls indicated the presence of four nuclear double bonds, since an isopropyl group was also evident. Deshielding of one of the vinyl methyls (δ 1.77) and one of the vinyl hydrogens (δ 5.97, J = 15 Hz) suggested the presence of a conjugated diene. Thus the $\Delta^{2,4,7,11}$ tetraene system of cembrene (**1**) itself was assumed, although the ultraviolet spectrum was not reported.

Assignment of the secondary hydroxyl group to C-10 was based upon the interpretation that the chemical shift of its carbinyl proton, δ 4.51, required its attachment to an allylic position and upon the fact that it was coupled to three vicinal protons. Only position 10 fulfilled these requirements.

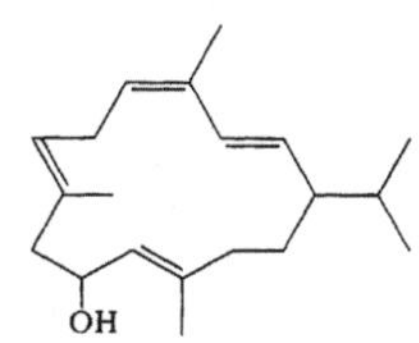

(**97**)

Structure (**97**) has been modified from that presented by KASHMAN *et al.* to reflect the normal geometry of the cembrene double bonds. The Israeli authors depicted the Δ^2 and Δ^4 unsaturations as *cis* and *trans*, respectively, whereas these bonds in cembrene are *trans* and *cis*, respectively. The large coupling constant of the low field diene proton described above supports the *trans* geometry of the disubstituted double bond (Δ^2) in this compound.

2-Hydroxynephthenol

In addition to nephthenol (**95**), TURSCH, BRAEKMAN and DALOZE (*55*) found the diol, 2-hydroxynephthenol (**98**), in the soft coral *Litophyton viridis*. Comparison of the diol's ir and pmr spectra with those of nephthenol suggested a close similarity and indicated that in addition to the tertiary alcohol function of nephthenol the diol contained a secondary allylic alcohol.

Like nephthenol, the diol formed two moles of levulinaldehyde upon ozonolysis and formed cembrane by dehydration-hydrogenation.

Confirmation of the secondary allylic alcohol was provided by careful oxidation to the conjugated ketone (**99**) which had α-hydrogen and β-methyl substituents on the conjugated double bond. Retroaldol cleavage of ketone (**99**) by methanolic base occurred with loss of acetone and formation of an equilibrium mixture (3 : 1) of the α,β- and β,γ-unsaturated ketones (**100**) and (**101**). This result provided firm evidence for the relationship of the two hydroxyl groups in the diol, and permitted assignment of gross structure (**98**) to 2-hydroxynephthenol. Its nuclear double bonds were presumed to be *trans* by analogy to other marine cembranoids.

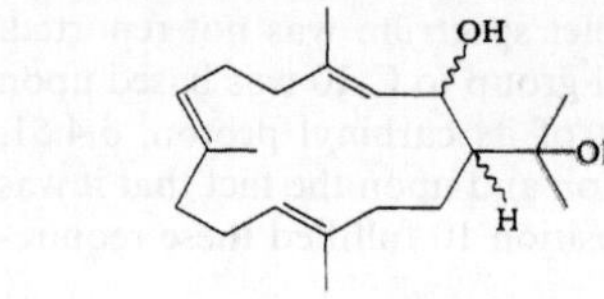

(**98**)

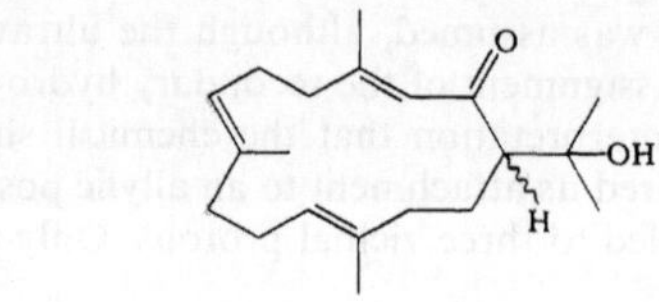

(**99**)

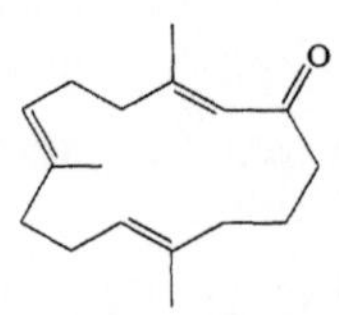

(**100**)

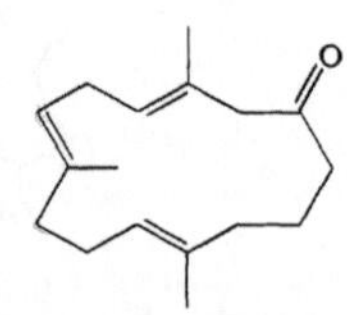

(**101**)

2,7,11-Cembratriene-4,10-diol

This diol, like 10-hydroxy cembrene (**97**), was isolated in minute amounts from *Sarcophytum glaucum* by KASHMAN *et al.* (*56*). As with the mono alcohol, the skeleton was assumed to be cembranoid. Structure (**102**) was tentatively proposed on the basis of the pmr spectrum, which showed the presence of an isopropyl group, a tertiary methyl carbinol, two vinyl methyls, four vinyl protons and the carbinyl proton of a secondary allylic alcohol coupled to three vicinal hydrogens.

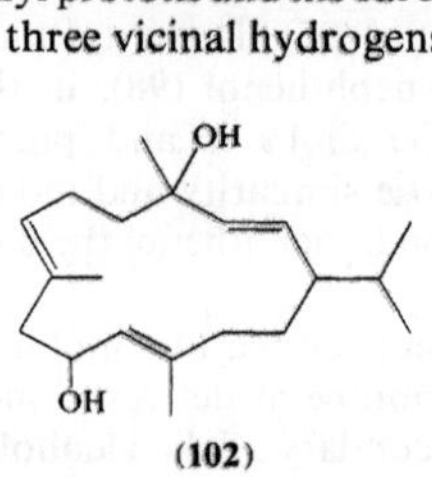

(**102**)

Position 4 was chosen as the site of the tertiary alcohol function because of the apparent absence of the conjugated diene present in (**97**). However, as was true of (**97**), no ultraviolet spectral data was presented. Of the two positions compatible with the coupling requirements of the secondary alcohol's carbinyl proton, position 10 was chosen over position 6 because the new diol differed from the 4,6-diol isolated from tobacco. In dismissing the 4,6-diol structure, it was not clear to which of the tobacco diols, (**54**) or (**55**) the authors referred, nor whether any consideration was given to their as yet unreported epimers.

Epoxynephthenol Acetate

In addition to nephthenol, Schmitz *et al.* (*54*) also isolated epoxynephthenol acetate, $C_{22}H_{36}O_3$, (**103**) from the Enewetak *Nephthea* species. The acetate residue was evident from the ir and pmr spectra, and the third oxygen atom was shown to be an epoxide by a positive test with acidic silver periodate. The pmr spectrum showed a six proton singlet for the 2-acetoxy-2-propyl group, two vinyl methyls, two vinyl hydrogens, and the methyl and hydrogen substituents of the epoxide, which was shown to have *trans* geometry by the absence of an nOe effect. The preceding structural features indicated that (**103**) contained a single carbocyclic ring.

Structure (**103**) was established by oxidative degradation of the free alcohol obtained by mild reduction with lithium aluminium hydride. Oxidation with osmium tetroxide-sodium periodate, followed by treatment with Jones reagent, gave the two ten carbon lactones (**104**) and (**96**). These resulted from cleavage at the two double bonds, and subsequent hydrolysis and oxidation of the epoxide to the *alpha*-hydroxyketone.

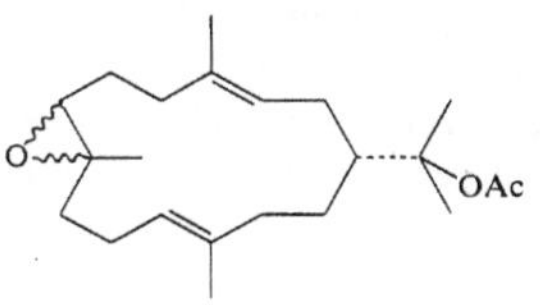

(**103**)

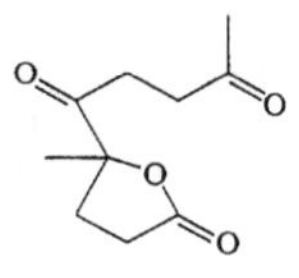

(**104**)

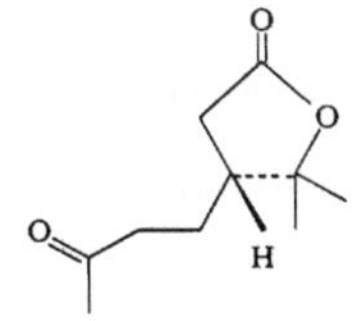

(**96**)

Lactone (**104**) was identified by pmr and ms analysis, and lactone (**96**) was shown to be identical with (*R*) (−)-homoterpenyl ketone. Thus epoxynephthenol acetate was shown to have structure (**103**) and to be a member of the α-series of cembranoids.

Despite the obvious structural relationship between the co-occurring epoxy-nephthenol acetate (**103**) and nephthenol (**95**), no direct correlation was made between the two substances because of the limited quantities of nephthenol available.

3,4,11,12-Diepoxycembrene-A

Another cembranoid isolated by HERIN and TURSCH (*52*) from Indonesian *Sinularia flexibilis* was the cembrene-A diepoxide (**105**). The diepoxide, $C_{20}H_{32}O_2$, showed no hydroxyl or carbonyl absorptions in the ir. The pmr spectrum indicated that the two oxygen atoms were present as methyl substituted epoxides from appropriate signals for two methyl groups and two epoxide protons. The pmr spectrum also exhibited signals for two vinyl methyl groups, and three vinyl protons, one of which was on a trisubstituted double bond and two of which indicated the presence of a terminal methylene group.

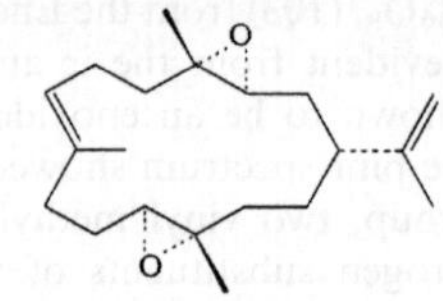

(**105**)

These features suggested that the compound was a diepoxide of cembrene-A; this was confirmed by deoxygenation of the two epoxide functions with zinc-copper couple to form exclusively the all-*trans* hydrocarbon, cembrene-A. Consequently, the carbon skeleton and the configuration at position 1 were established, as well as the fact that the compound was one of the three positional diepoxides derivable from the nucleus of cembrene-A.

Although the zinc-copper reduction of epoxides is generally not stereospecific, the authors avoided known stereospecific methods because of the sensitivity of both the diepoxide and the product, cembrene-A, to the acidic conditions required for their use. In a study of model epoxides, they showed that the geometry of the major alkene formed in the zinc-copper couple reduction of *cis* and *trans* pairs of epoxides invariably corresponded to the geometry of the starting epoxide. They concluded that the epoxides in cembrene-A diepoxide possessed *trans* geometry, as of course did the trisubstituted double bond.

The authors employed an ingenious assessment of the lithium aluminium hydride reduction products of the diepoxide to complete the structure determination and the assignment of its absolute configuration. The reduction gave a mixture of four isomeric diols consisting of a small amount of the di-secondary diol, the pair of isomeric secondary, tertiary diols (**106**) and (**107**) which were not separated, and the ditertiary diol (**108**).

The ditertiary diol upon hydrogenation gave the saturated diol (**109**) which was optically inactive at several wave lengths and thus possessed

References, pp. 381—387

a plane of symmetry. Only (**109**) or its 4,12-bis-epimer (**110**) could fulfil this requirement, thereby eliminating the remaining two positional diepoxides of cembrene-A as possibilities for the natural product.

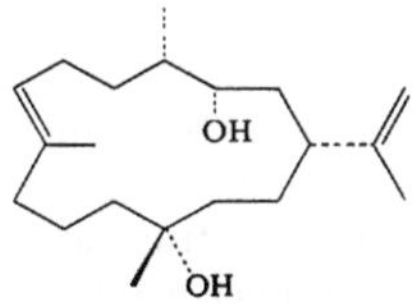

(**106**)

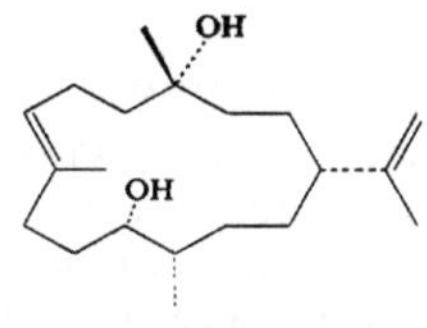

(**107**)

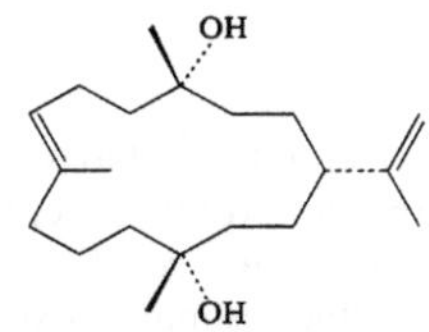

(**108**)

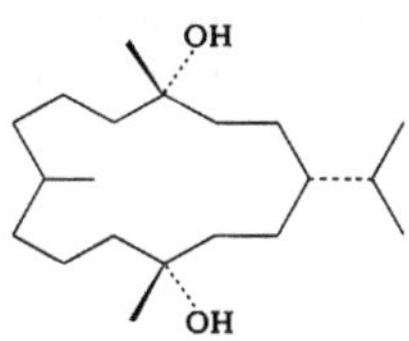

(**109**)

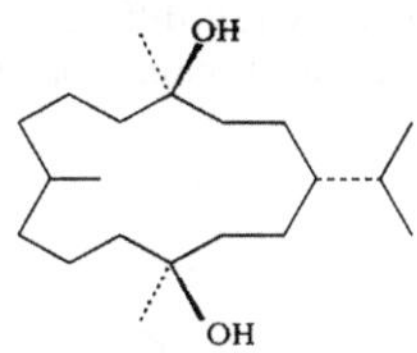

(**110**)

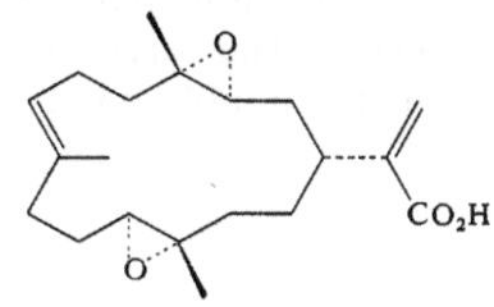

(**111**)

Evidence that the diepoxide had structure (**105**), which would lead to the ditertiary diol (**109**), and not the alternative *trans, trans* geometry which would lead to ditertiary diol (**110**), was provided by an examination of the configurations of the secondary alcohol centers in the pair of secondary, tertiary diols (**106**) and (**107**). The authors reasoned that the configurations of the secondary alcohol centers in the mixed diols which would be formed from the same *trans, trans*-diepoxide, that leads to symmetrical diol (**109**), would both necessarily be (*S*). Conversely, the configurations of these secondary centers in the mixed diols which would be formed from the *trans, trans*-diepoxide that leads to diol (**110**), would both be (*R*).

The configurations of the secondary alcohols were shown to be (*S*) by application of Horeau's rule to the reaction of the mixture of mixed diols (**106**) and (**107**) with (+)-α-phenylbutyric anhydride, which produced an excess of (−)-α-phenylbutyric acid. This result demonstrated the configurations of the diols (**106**), (**107**) and (**108**), and permitted assignment of the absolute configuration shown in (**105**) to the natural diepoxide.

Herin and Tursch speculated that the diepoxide (**105**) might lie on the biogenetic pathway to the formation of both sinulariolide (**157**), which occurs in *S. flexibilis* from Indonesian waters, and sinularin (**155**) which is present in this species from Australian waters. Allylic oxidation of the isopropenyl residue could form the hypothetical intermediate (**111**), which, by inversion opening of the epoxides at positions 3 or 12, would generate the lactones (**155**) or (**157**), respectively.

Asperdiol

Guided by bioassays for cytotoxicity, WEINHEIMER *et al.* (*57*) isolated the epoxydiol, asperdiol (**112**), from two Caribbean gorgonians of the genus *Eunicea, E. asperula* and *E. tourneforti.* Its pmr spectrum and double resonance experiments revealed an isopropenyl residue, a substituted 2-buten-1,4-diol function, a trisubstituted double bond, and a trisubstituted epoxide. The molecular structure and absolute configuration of asperdiol were deduced by X-ray diffraction and are represented in (**112**). The structure is unusual because of the *cis* geometry of one of the double bonds.

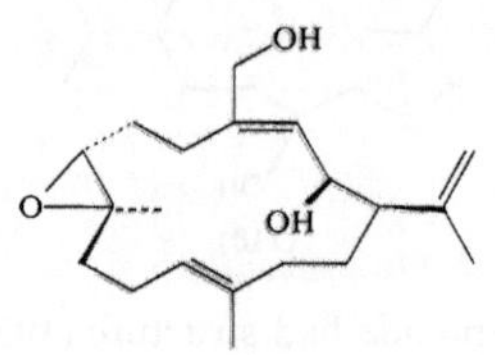

(**112**)

Deoxosarcophine and 2-Epideoxosarcophine

Two further compounds found by KASHMAN *et al.* (*56*) in *Sarcophytum glaucum* were the closely similar dihydrofurans, $C_{20}H_{30}O_2$, (**113**) and (**114**), which appeared to differ from sarcophine (**117**) by replacement of the sarcophine lactone carbonyl by a methylene group. Since the authors failed to name the compounds, the reviewers suggest the names deoxosarcophine and 2-epideoxosarcophine.

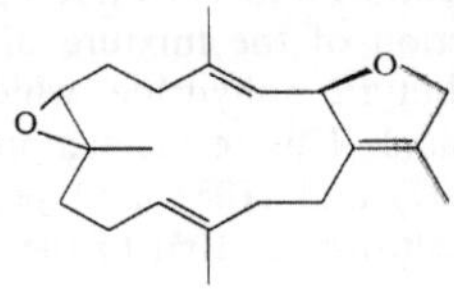

(**113**)

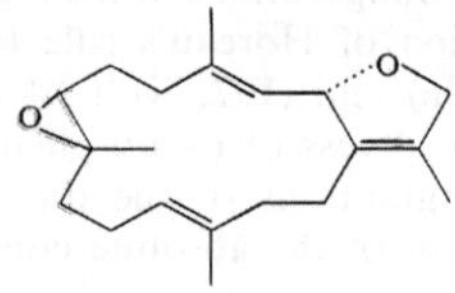

(**114**)

The absence of carbonyl and hydroxyl absorptions in deoxosarcophine (**113**) suggested that both oxygens were ethereal. The pmr spectrum showed a methyl substituted epoxide and two methyl-bearing trisubstituted double bonds. Extensive decoupling work established the dihydrofuran moiety, and its placement in relation to the 3,4-double bond. Much of the pmr spectrum was similar to that of sarcophine, and the authors assigned the balance of structure (**113**) by analogy.

The isomer, 2-epideoxosarcophine (**114**), displayed spectral properties very similar to (**113**). It was assumed to be epimeric at position 2 because of differences in pmr spectral features for the dihydrofuran unit.

Lobophytolide

TURSCH *et al.* (*58*) isolated the epoxycembranolide lobophytolide (**115**), $C_{20}H_{28}O_3$, from the soft coral *Lobophytum cristagalli* collected in Indonesian waters. Its ir spectrum showed carbonyl absorption due to a γ-lactone, but no hydroxyl absorption. The uv spectrum showed a maximum at 209 nm indicative of conjugation in the lactone, which was confirmed by the presence of the downfield doublets of an α-methylene group in the pmr spectrum. This spectrum also showed the two vinyl methyls and two vinyl protons of a pair of trisubstituted double bonds, and the proton and singlet methyl characteristic of an oxirane.

The β-proton of the lactone was shown by double resonance to be coupled to the two exo-methylene protons, and also to the proton at the lactone alkyl terminus, confirming the ring size of the lactone.

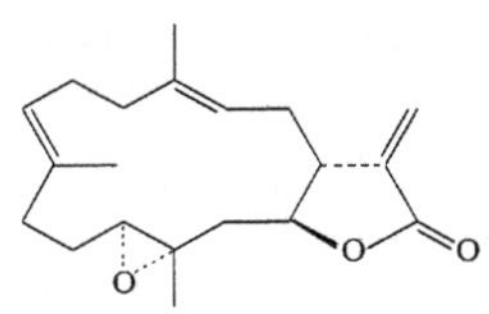

(**115**)

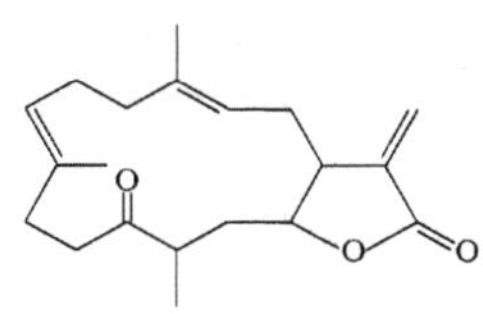

(**116**)

Ozonolysis of (**115**) formed levulinaldehyde, and toluenesulfonic acid isomerization of the epoxide formed ketone (**116**) which showed no downfield signals below 3 ppm for methylene groups that were both allylic and *alpha* to the carbonyl. No coupling could be detected between the β-lactone proton at position 2 and either the epoxy proton or the vinyl protons. The data available favored structure (**115**) but did not rigorously rule out one other structure. The structure and relative stereochemistry shown in (**115**) was determined by an X-ray diffraction study (*58*); KARLSSON (*59*) has recently established the absolute configuration shown therein.

Sarcophine

The epoxycembranolide, sarcophine (**117**), $C_{20}H_{28}O_3$, was isolated from the Red Sea soft coral *Sarcophytum glaucum* by BERNSTEIN *et al.*

(*60*). Structural elements were deduced from ir and pmr spectra and the complete structure was determined by X-ray diffraction.

The infrared spectrum of sarcophine indicated an α,β-unsaturated γ-lactone, and also suggested an ether function. The nmr spectrum confirmed the presence of a methyl substituted oxirane and two isolated double bonds carrying methyl substituents. The lactone was recognized as an α-methyl butenolide, the *gamma* position of which was vicinal to one of the isolated double bonds. These structural elements accounted for six of the seven degrees of unsaturation, thus requiring one carbocyclic ring.

Sarcophine selectively formed the 11,12-monoepoxide with peracid and the 11,12-dihydro derivative on hydrogenation.

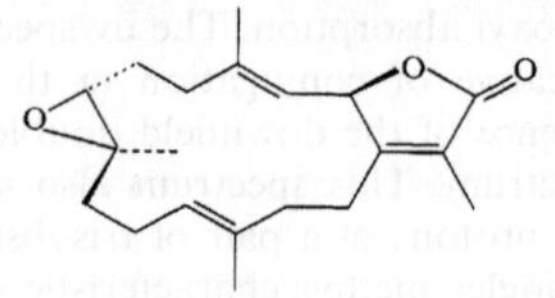

(117)

The authors expended considerable effort in futile attempts to interpret the mass spectra of **(117)** and its derivatives. They concluded that their failure was due to sarcophine's large number of functional groups and its many possible modes of fragmentation. (From their own work with cembranes, as well as results in the literature, the reviewers can corroborate that mass spectrometry provides only trivial bits of structural information. They feel, however, that an equally important reason for the general non-utility of mass spectometry is the cyclic character of the cembrane skeleton. Even though typical fragmentation processes may be occurring at individual functional groups, the cyclic skeleton should prevent observable loss of interpretable fragments.)

The structure and relative configuration obtained for sarcophine by X-ray diffraction are shown in **(117)**. The authors considered the use of circular dichroism for assignment of absolute configuration, but concluded that a suitable sector rule for unsaturated lactones was not yet available.

2-Episarcophine

KASHMAN, ZADOCK, and NEEMAN (*56*) reported two further cembranolides from *Sarcophytum glaucum*. Since the authors offered no names for these compounds, the reviewers suggest they be called 2-episarcophine and 2,8-bis-episarcophine.

The infrared and mass spectra of 2-episarcophine, ($C_{20}H_{28}O_3$) and sarcophine (**117**) were indistinguishable. Slight differences were observed in their pmr spectra, but the decoupling behavior of 2-episarcophine was identical to that of sarcophine. The CD curves of the two, however, differed markedly. The authors interpreted these differences as a reflection of an epimeric relationship at position 2, and offered structure (**118**).

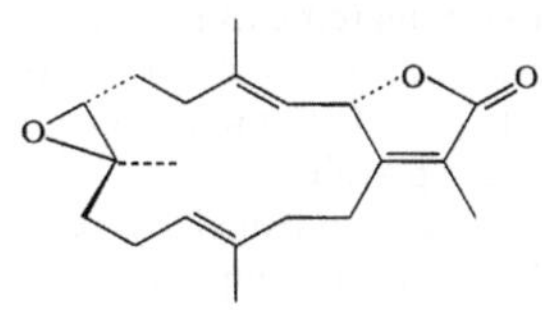

(**118**)

2,8-Bis-Episarcophine

The isolation of 2-episarcophine by chromatography was conducted in the absence of chloroform. However, when chloroform was employed no 2-episarcophine was obtained. Instead, an oil, also of composition $C_{20}H_{28}O_3$, was isolated. Kashman *et al.* (*56*) stated that the new isomer was indistinguishable from either sarcophine or 2-episarcophine by TLC, infrared and mass spectroscopy. They noted that the pmr spectrum of 2-episarcophine changes to that of the oily isomer while being run in $CDCl_3$, but reverts to the original after removal of the chloroform and solution of the residue in carbon tetrachloride. It was shown that phosgene, a normal trace contaminant of chloroform, was responsible for the peculiar behavior. As a possible explanation it was suggested that phosgene inverted the configuration of one of the epoxidic carbon atoms and that the oily isomer might be (**119**). The smooth reversibility of such a reaction would be remarkable, and since sarcophine itself displayed no such behavior, the authors recognized their position as "very delicate".

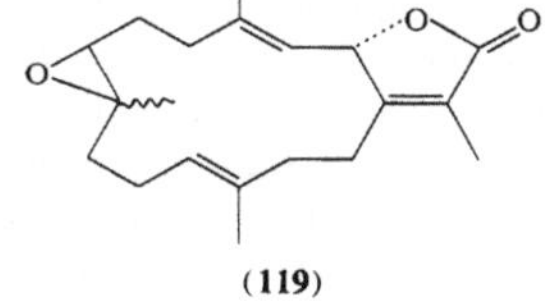

(**119**)

Eupalmerin Acetate

The cembranolide, eupalmerin acetate (**120**) was isolated from the gorgonian *Eunicea palmeri* collected near Miami, Florida. Its structure was established by chemical and spectral studies by REHM (*61*), and its absolute configuration was determined by EALICK, VAN DER HELM, and WEINHEIMER (*62*).

The ir spectrum of eupalmerin acetate, $C_{22}H_{32}O_5$, showed no hydroxyl absorption, but contained carbonyl absorptions for a γ-lactone and an acetate. The pmr spectrum confirmed the presence of one secondary acetate, and indicated a conjugated exo-methylene group in the lactone. In addition, there were signals for a methyl group on a trisubstituted double bond, an isolated secondary methyl group, and the methyl and hydrogen of a trisubstituted epoxide.

The epoxide function was confirmed by reaction with acidic silver periodate, thereby establishing that eupalmerin acetate had one carbocyclic ring.

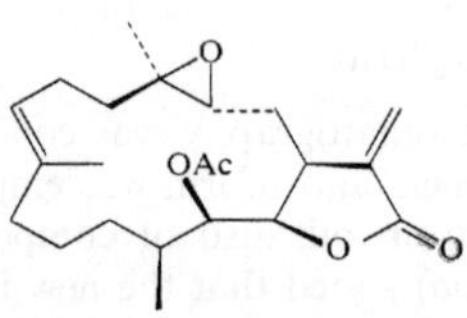

(120)

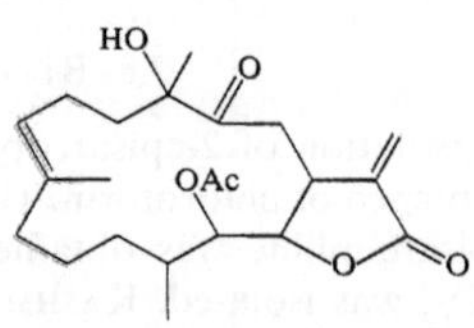

(121)

Double resonance studies showed that the two protons of the exo-methylene were coupled to the β-hydrogen of the lactone at position 1, and that this proton was also coupled to the hydrogen at the lactone alkyl terminus, position 14, confirming the size of the lactone ring. It was of interest that no coupling was evident between the γ-proton of the lactone and the proton of the vicinal secondary acetate which appeared as a doublet. This doublet collapsed to a singlet by irradiation in the methylene envelope at a frequency which also caused the doublet of the secondary methyl group to collapse to a singlet.

Hydrolytic reactions of eupalmerin acetate produced complex mixtures, apparently because of the reactivity of the epoxide function. Jónes oxidation transformed the epoxide to the tertiary ketol (**121**), in which the lactone proton at position 14 was now coupled to the carbinyl proton of the vicinal acetate. However, the signal of the latter again appeared as a doublet, indicating that a conformational change had eliminated the coupling observed in eupalmerin acetate. Since this coupling evidence was incompatible with a methylene group at position 12, the secondary methyl group could be assigned to this position.

The ketol (**121**) also exhibited a two proton doublet for the methylene *alpha* to the new carbonyl, and this was shown to be coupled to the β-proton of the lactone.

References, pp. 381—387

These experiments established the relationship of the secondary methyl, the acetate, the lactone and the epoxide functions. To characterize the balance of the molecule, the ketol was reduced with lithium aluminum hydride to the pentol (**122**) which was cleaved by periodate at the two vicinal diol functions. The initially formed unsaturated keto aldehyde was oxidized to the corresponding acid and converted to the methyl ester (**123**).

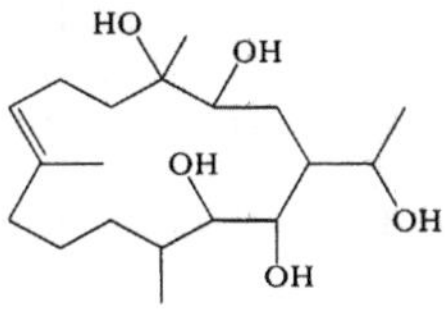

(**122**)

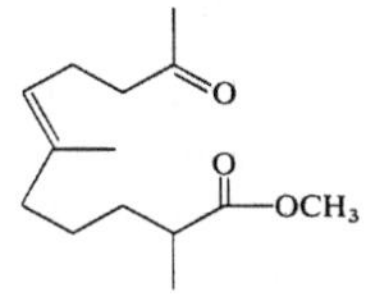

(**123**)

Racemization of the asymmetric α-methyl position had occurred in the initial keto aldehyde, preventing assessment of chirality at this center. However, the trisubstituted double bond could be assigned the *trans* geometry because of the high field (δ 1.60) resonance of the vinyl methyl group in (**123**). Ozonolysis of (**123**) afforded levulinic acid, and the methyl ester of 2-methyl-6-oxoheptanoic acid, thereby completing the assignment of structure (**120**) to eupalmerin acetate.

By an X-ray diffraction study, EALICK, VAN DER HELM, and WEINHEIMER (*62*) determined the absolute configuration shown in (**120**) for eupalmerin acetate. The same authors (*63*) also determined the structure of the interesting dibromo tetrahydropyran derivative (**124**) obtained from (**120**) by reaction with bromine. This compound resulted from transannular participation of the epoxide oxygen in the reaction at the double bond, and possessed the configurations expected for a concerted process.

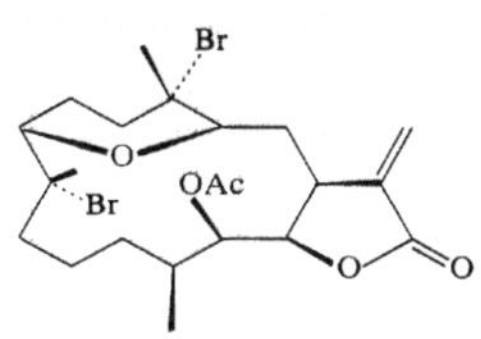

(**124**)

Lobolide

The cembranolide lobolide (**125**) was isolated from a *Lobophytum* species of soft coral from the Red Sea by KASHMAN and GROWEISS (*64*). The nature of its individual functional groups was deduced from the pmr and cmr spectra, which showed the presence of an α-methylenic γ-lactone, a trisubstituted epoxide, a primary acetate and two methyl-bearing trisubstituted *trans* double bonds. The relationship of the epoxy and lactone functions was established by decoupling of signals for the protons on carbons 1, 2, 3, and 14. Isolation of levulinaldehyde from an ozonolysis experiment, together with the demonstration of coupling between the allylic methylene at position 13 and the lactone hydrogen at position 14, secured the indicated locations of the two trisubstituted *trans* double bonds. Evidence for the position of the primary acetate moiety depended upon the greater lanthanide induced shift of the epoxy proton than of other signals in the deacetylated compound.

Lobolide, like sarcophine, is toxic to fish and is considered to play a role in the protection of the coral from predators.

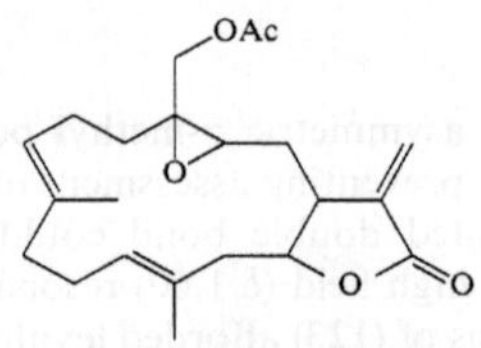

(125)

Eunicin

The cembranolide eunicin (**2**) was isolated by CIERESZKO *et al.* (*65*) from the Caribbean gorgonian *Eunicea mammosa* collected at Bimini. It was the first of several oxygen bridged cembranolides isolated from *E. mammosa* collected in different locales in the Caribbean. Unique among the Carribean gorgonians, this species elaborates a series of chemically similar cembranolides, each of which appears to be characteristic for the locale in which the gorgonian grows. Eunicin was found to possess mild *in vitro* antibacterial properties.

The structure of eunicin was determined by WEINHEIMER *et al.* (*66*) by chemical degradation, and also by HOSSAIN, NICHOLAS, and VAN DER HELM (*67*) by X-ray diffraction methods.

The ir spectrum of eunicin, (**2**), $C_{20}H_{30}O_4$, showed hydroxyl, and γ-lactone absorptions accounting for three of its four oxygens. The alcohol function was resistant to oxidation. The pmr spectrum showed characteristic signals for a conjugated lactone *exo*-methylene group, each hydrogen of which was coupled to the single proton at the endocyclic β-position. Also evident were a singlet methyl of a tertiary carbinol, a trisubstituted double bond bearing a methyl group, and an isolated secondary methyl. The fourth oxygen in the molecule was judged to exist as a cyclic disecondary ether because of a pair of downfield signals of one proton each.

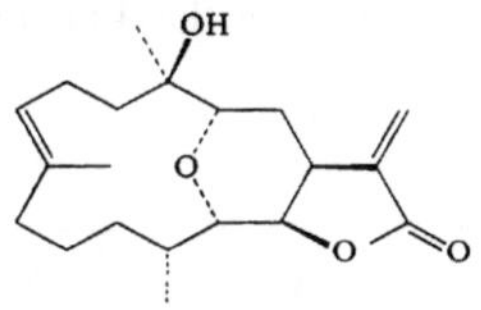

(2)

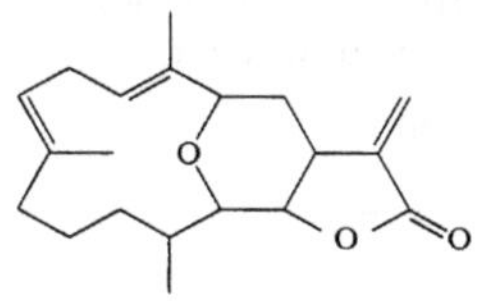

(126)

The tertiary alcohol function and one ether terminus were shown to be vicinal to each other because of the downfield shift experienced by the signal of the ether proton when eunicin was dehydrated with thionyl chloride-pyridine. The shift due to the newly allylic nature of the ether proton was evident in both the exocyclic and endocyclic (**126**) forms of the dehydration product.

The other ether terminus was shown by decoupling experiments with eunicin to be vicinal to the lactone proton at position 14. The fact that this ether proton showed no further coupling was incompatible with a methylene residue in the other vicinal position, thus placing the secondary methyl at position 12.

The position of the tertiary alcohol relative to the isolated double bond, and relative to position 1 of the cembrane skeleton was demonstrated by two series of oxidative degradations.

Reduction of the conjugated double bond of eunicin by sodium-butanol gave one epimeric form of dihydroeunicin which was hydroxylated at the remaining double bond by permanganate to form (**127**). Jones oxidation cleaved the diol function and formed a new γ-lactone (**128**) by cyclization with the tertiary hydroxyl, as indicated by the downfield shift of the singlet methyl at position 4. A similar shift occurred when eunicin was converted to its acetate.

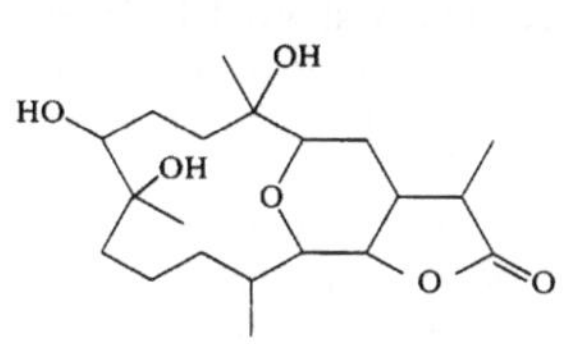

(127)

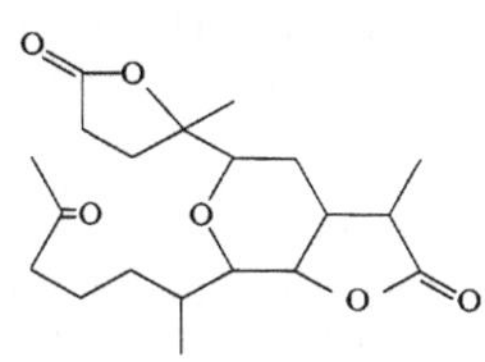

(128)

Catalytic hydrogenation of eunicin reduced the isolated double bond nonspecifically to a pair of saturated derivatives (**129**) epimeric at position 8, while concurrently isomerizing the exocyclic double bond of the lactone to the endocyclic position. Ozonolysis of (**129**) generated a carboxyl carbon from the β-position of the lactone to give the epimeric pair of aldehydro γ-lactones (**130**) in which the resonance of the 4-methyl was again shifted downfield.

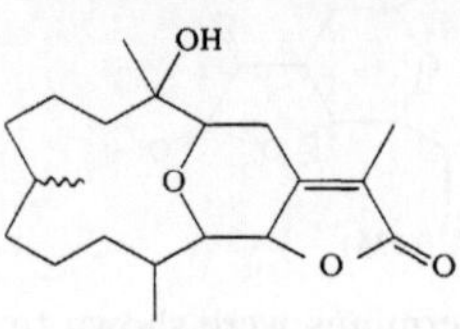

(**129**)

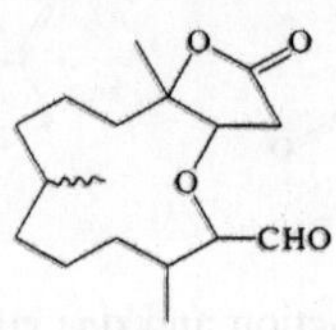

(**130**)

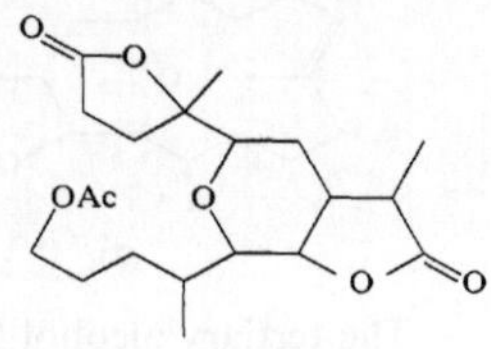

(**131**)

Baeyer-Villiger reaction of the methyl ketone (**128**) formed the acetate (**131**) in which the primary methylene signal appeared as a triplet, requiring simple methylene groups at positions 9 and 10. The remaining methylene group was required to be at position 11, thus completing assignment of structure (**2**) to eunicin.

In an accompanying paper, HOSSAIN *et al.* (*67*) reported the X-ray structure determination and absolute configuration of eunicin iodoacetate, the details of which are incorporated in structure (**2**).

Cueunicin and Cueunicin Acetate

The two ether bridged cembranolides, cueunicin (**132**) and cueunicin acetate (**133**) were isolated by GROSS (*68*) from the gorgonian *Eunicea mammosa* collected at Curacao. The structures of these compounds were deduced from chemical and spectral studies, which also led to proposals for their absolute configurations.

The unacetylated parent compound cueunicin, $C_{20}H_{30}O_4$, was a gum that could not be crystallized. Upon acetylation it formed the nicely crystalline cueunicin acetate which was the major cembranolide occurring in this gorgonian. The ir spectrum of cueunicin showed hydroxyl and γ-lactone absorptions, accounting for three of its four oxygens. The pmr spectrum indicated a conjugated α-methylenic unsaturation which also accounted for the uv maximum at 215 nm. The two methylenic protons were shown by double resonance to be coupled to a single β-proton in the lactone ring. The β-proton was also coupled to the proton at the lactone alkyl terminus, confirming the lactone ring size. Also evident in the pmr spectrum were a singlet methyl of a tertiary carbinol, a vinyl methyl of a trisubstituted double bond, and an isolated secondary methyl. The fourth oxygen in the molecule was judged to be present in the form of a cyclic disecondary ether on the basis of a pair of downfield signals, each corresponding to one proton. One of the ether protons, a double doublet, was coupled to the γ-proton of the lactone. The remaining vicinal coupling indicated that the secondary methyl group was located at position 12.

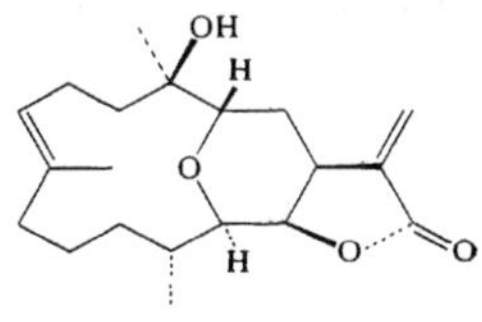

(132)

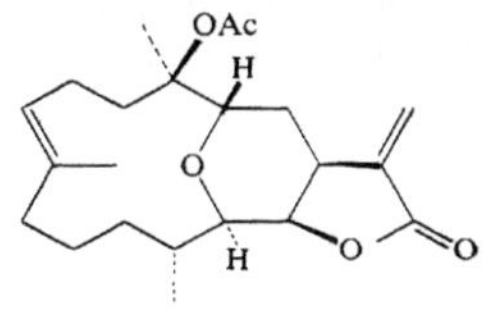

(133)

The location of the other ether terminus could not be determined from the spectra of cueunicin itself, but was established with the aid of the derivative (**134**). The latter compound was prepared from the single epoxide formed in the reaction of cueunicin with peracid. The epoxidic proton in this intermediate appeared as a double doublet due to a vicinal methylene group. Catalytic hydrogenation of the epoxide produced both the expected saturated α-methyl lactone and the butenolide (**134**) in which the conjugated double bond had migrated to the endocyclic position. Double resonance pmr studies on (**134**) showed that the newly allylic methylene at position 2 was coupled both to the α-methyl of the lactone and to the remaining ether proton. Since the vicinal methylene accounted for the multiplicity of the ether proton, the tertiary alcohol function was assigned to the other vicinal position. This assignment was consistent with the pronounced deshielding of the ether proton observed upon acetylation of the tertiary hydroxyl in several compounds in the series.

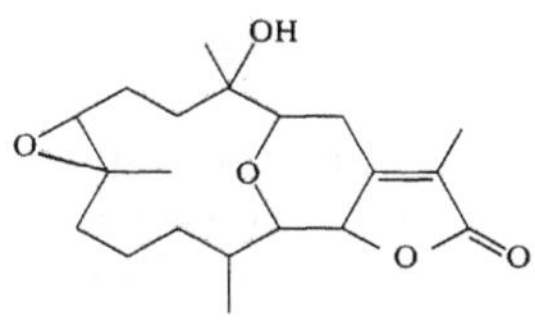

(134)

The structure elucidation was completed by the Lemieux oxidation of the α-methyl lactone (**135**) prepared by sodium borohydride reduction of cueunicin. The keto dilactone (**136**) analogous to that in the eunicin series was obtained. Cyclization of the new lactone at the tertiary hydroxyl was indicated by the downfield shift of the singlet methyl at position 4. High resolution mass spectrometry confirmed the identity of the two side chains situated as *alpha* substituents of its cyclic ether. The methyl ketone of (**136**) was converted by the Baeyer-Villiger reaction to the acetate (**137**) in which the primary methylene signal appeared as a triplet, requiring simple methylene groups at positions 9 and 10. The methylene group still unaccounted for was necessarily at position 11. The composition of the acetate-bearing side chain was confirmed by high resolution ms, thus completing the assignment of gross structures (**132**) and (**133**) to cueunicin and its acetate.

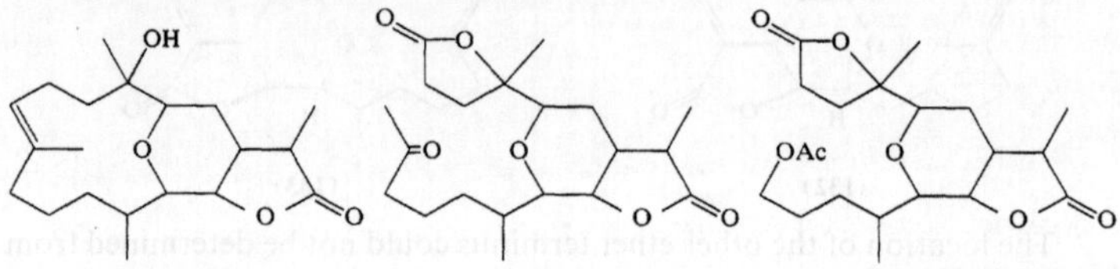

(**135**) (**136**) (**137**)

GROSS next evaluated the stereochemistry of the two cembranolides. The CD curve of the pyrazoline (**138**), obtained from cueunicin acetate and diazomethane, displayed a strongly negative Cotton effect at 324 nm which correlated in both magnitude and sign with the pyrazoline of eunicin. Since the lactone ring of eunicin is *cis*-fused and possesses the β-configuration, similar assignments could be made for cueunicin acetate.

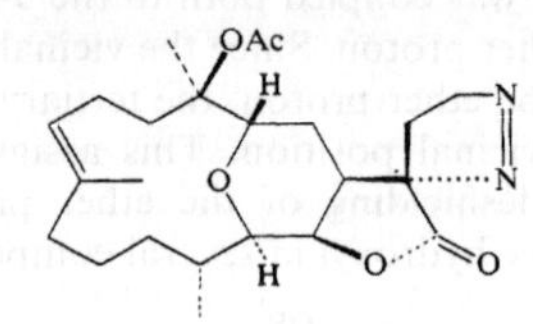

(**138**)

In the dihydro derivative (**139**) obtained from cueunicin acetate by sodium borohydride reduction, a strong nuclear Overhauser effect between the α-methyl group of the lactone and the ether proton at position 3 established that this ether proton also was β-oriented. Because of the strong deshielding influence on this ether proton caused by esterification of the vicinal tertiary hydroxyl, the hydroxyl group was also assigned the β-orientation.

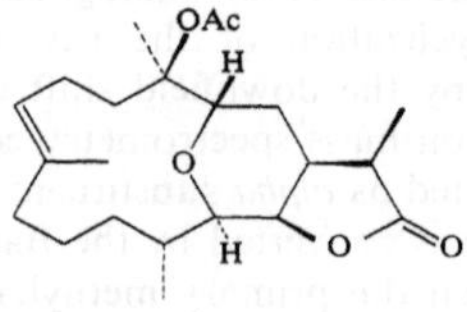

(**139**)

The absence of a nuclear Overhauser effect between the vinyl methyl and the vinyl proton in (**139**) showed the *trans* geometry of the tri-substituted double bond.

Lastly, comparison of coupling constants of eunicin and cueunicin led to assignment of the α-orientation to ether proton at position 13, and the α-orientation to the methyl group at 12. In eunicin, the *anti* disposition of the lactone proton (position 14) and the β-ether proton is reflected by a large *J*-value. Its small value in cueunicin was interpreted as a reflection of the opposite configuration (α) for the ether proton. Consistent with this difference from eunicin was the large $J_{12,13}$ in cueunicin, in contrast to its small value in eunicin, indicating that the secondary methyl group had the same orientation in each. These stereochemical postulates are incorporated in structures (**132**) and (**133**).

Jeunicin

The cembranolide, jeunicin (**140**) was isolated by VAN DER HELM *et al.* (*69*) from the gorgonian *Eunicea mammosa* collected at Jamaica. Its composition, $C_{20}H_{30}O_4$, indicated that it was isomeric with eunicin (**2**), present in *E. mammosa* from Bimini. It showed γ-lactone and hydroxyl absorptions in the ir, and its pmr features were generally similar to those of eunicin. However, formation of the 3-ketone by Jones oxidation indicated that one ether terminus was tertiary, and the hydroxyl secondary, in contrast to eunicin.

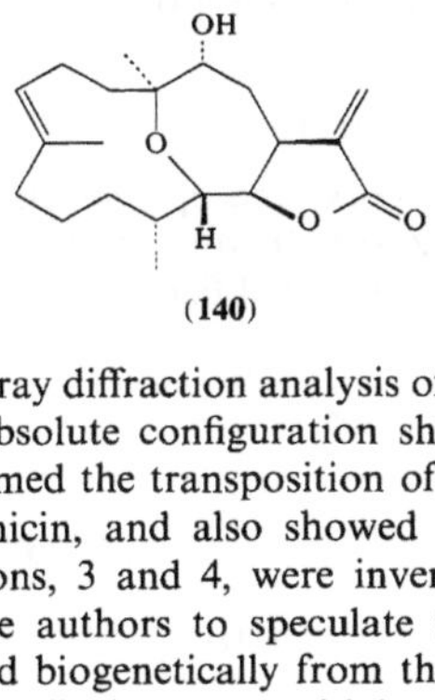

(**140**)

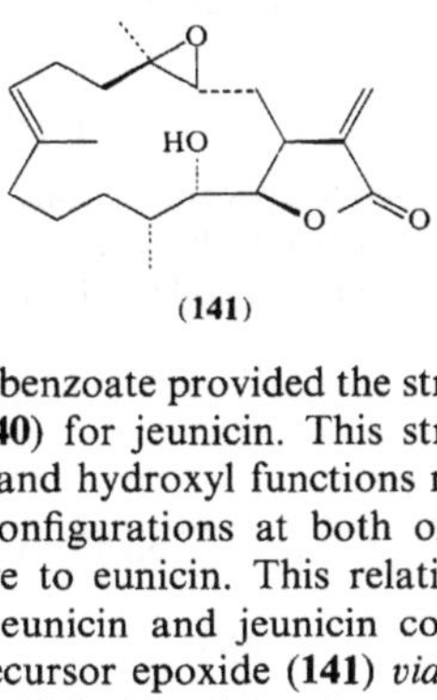

(**141**)

X-ray diffraction analysis of its *p*-iodobenzoate provided the structure and absolute configuration shown in (**140**) for jeunicin. This structure confirmed the transposition of the ether and hydroxyl functions relative to eunicin, and also showed that the configurations at both of these positions, 3 and 4, were inverted relative to eunicin. This relationship led the authors to speculate that both eunicin and jeunicin could be derived biogenetically from the same precursor epoxide (**141**) *via* trans-annular displacement, with inversion, by a 13-hydroxyl function.

It is of interest that Jamaican *E. mammosa* when initially collected in 1962 contained jeunicin as its only isolable cembranolide. Five years later, collections from the same locale contained mixtures of jeunicin and eunicin. Subsequently, the same stand of gorgonian was found to contain only eunicin.

13,14-Bis-Epijeunicin

The ether-bridged cembranolide, 13,14-bis-epijeunicin, $C_{20}H_{30}O_4$ (**142**), was isolated by WEINHEIMER, MATSON, VAN DER HELM and POLING (*70*) from the gorgonian *Eunicea mammosa* from Bimini. It accompanied eunicin (**2**) as a minor cembranoid in this organism, and was isolated as a result of guidance by cytotoxicity bioassays.

Its ir spectrum showed hydroxyl and γ-lactone carbonyl absorptions. The pmr spectrum showed characteristic signals for an exo-methylene lactone, and other signals indicating similarity to jeunicin (**140**).

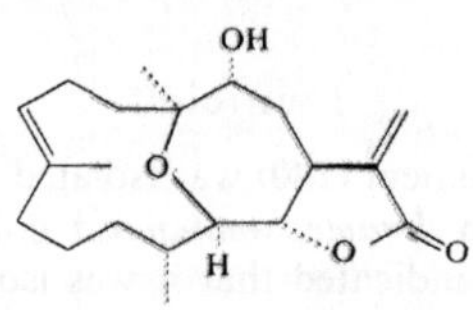

(**142**)

An X-ray diffraction study of this compound determined the structure and absolute configuration shown in (**142**). The compound is epimeric with jeunicin at one terminus of the seven-membered ether ring (position 13) and at the adjacent lactone alkyl terminus (position 14).

Peunicin and Epipeunicin

The two keto cembranolides, peunicin (**143**) and epipeunicin (**144**) have been isolated by CHANG and CIERESZKO (*71*) from the gorgonian *Eunicea mammosa* collected on the Caribbean shore of Panama.

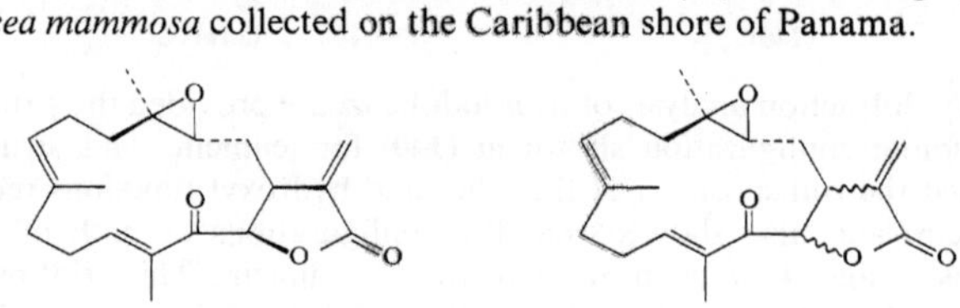

(**143**) (**144**)

Peunicin showed carbonyl absorptions in the infrared for a γ-lactone (1770 cm^{-1}) and for a conjugated ketone (1675 cm^{-1}), and a uv maximum at 241 nm. Its pmr spectrum exhibited the two downfield signals of a conjugated exo-methylene in the lactone, a three proton singlet for a methyl substituted oxirane, and the vinyl methyl and hydrogen signals of a trisubstituted double bond. The compound was unstable in chloroform.

The structure and absolute configuration of peunicin shown in (**143**) was determined by the X-ray diffraction method.

The spectra of epipeunicin were similar to those of peunicin, but its structure was not firmly established. Differences associated with the pmr signals of the lactone portion of the molecule suggested that it was epimeric with peunicin at either position 1 or position 14.

It should be noted that the use of the name peunicin for (**143**) corrupts the previously systematic usage of names based on eunicin for *Eunicea* cembranolides possessing an ethereal bridge. The names for such compounds were constructed from eunicin by adding one or two initial letters which were indicative of the locale in which the source gorgonian was collected, e. g., jeunicin (Jamaica); cueunicin (Curacao).

Pukalide

Pukalide (**145**), a furanocembranolide, was obtained by MISSAKIAN, BURRESON, and SCHEUER (*72*) from *Sinularia abrupta*, an Hawaiian soft coral. It was the first example of a furan moiety in a marine cembrane.

Interpretation of its ir and pmr spectra suggested the presence of a butenolide lactone, an isopropenyl group and a methyl substituted epoxide. The epoxide was assigned *trans* geometry because no nuclear Overhauser effect was observed between its methyl and hydrogen substituents. Each of these functions was affected in the catalytic reduction to hexahydropukalide (**146**), which was the only transformation product reported. The double bonds of the butenolide and isopropenyl groups were saturated, and the epoxide was cleaved to a tertiary methyl carbinol.

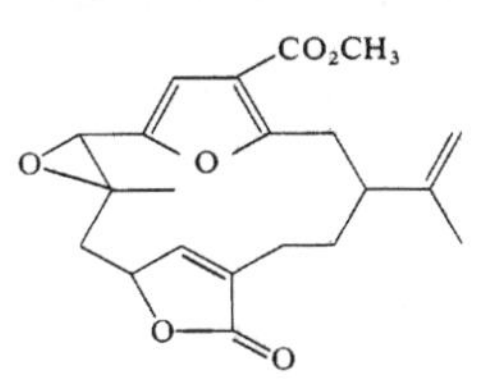

(**145**)

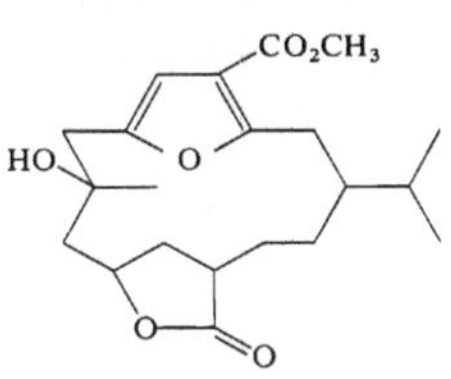

(**146**)

The furan ring was not affected by the reduction. The nature of this ring, and the presence of a β-carbomethoxy group and a β′-hydrogen were established with the aid of pmr and cmr correlations. Although the β-carbomethoxy group was assigned to skeletal position 4, its possible location at the alternate β-position of the furan, skeletal position 5, was not excluded.

Double resonance studies established the sequence of groupings from positions 1 through 12 in (**145**), accounting for all of the molecule except two methylenes which were necessarily placed at positions 13 and 14 to complete the required carbocycle.

Crassin Acetate

The cembranolide, crassin acetate (**147**), was isolated by CIERESZKO *et al.* (*65*) from the Caribbean gorgonian *Pseudoplexaura porosa* (= *Plexaura crassa*) in which it is present to the extent of 2% of the dry weight of the organism. It is found (*73*) in each of the four gorgonians of this genus which have been examined, irrespective of the locale of collection (cf. *Eunicea mammosa*). The compound has mild antibiotic (*65*) and antineoplastic (*73*) activities, and is toxic to fishes and juvenile forms of barnacles (cf. *73*).

The structure of crassin acetate was elucidated by WEINHEIMER, KARNS, SIFFORD, and CIERESZKO (*74*) by chemical methods, and its absolute configuration was determined by HOSSAIN and VAN DER HELM (*75*).

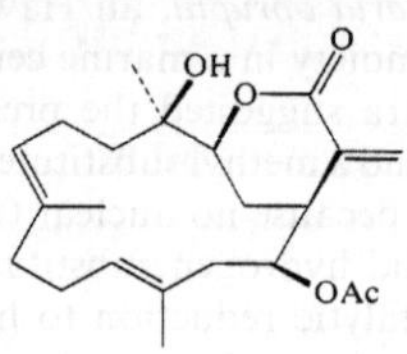

(**147**)

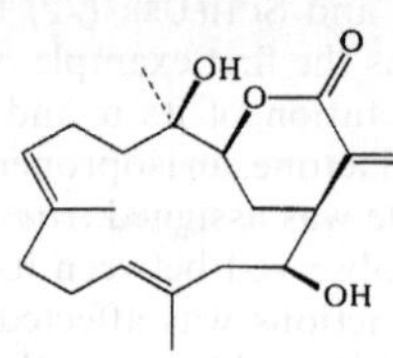

(**148**)

Crassin acetate had the composition $C_{22}H_{32}O_5$. The functional groups indicated by its ir and pmr spectra, in addition to the readily evident acetate, were an exomethylenic δ-lactone possessing one hydrogen in the β-position and one hydrogen at its alkyl oxygen terminus, a tertiary methyl carbinol and, from decoupling, two methyl-bearing trisubstituted double bonds. The structural elements and the molecular formula required a monocarbocyclic skeleton.

Hydrolysis of crassin acetate by aqueous base, followed by acidification, liberated the parent diterpene, crassin (**148**), which exhibited an upfield shift of the carbinyl proton of the secondary alcohol. Acetylation of crassin produced crassin acetate, demonstrating that relactonization after hydrolysis did not involve either the new secondary hydroxyl or the tertiary hydroxyl.

Dehydration of crassin acetate with phosphoryl chloride-pyridine led to a single product (**149**). Its structure was evident from the absence of the signal for the lactone proton, and the appearance of a new vinyl methyl in place of the carbinol methyl. This result demonstrated the vicinal relationship of the tertiary carbinol and the lactone alkyl terminus.

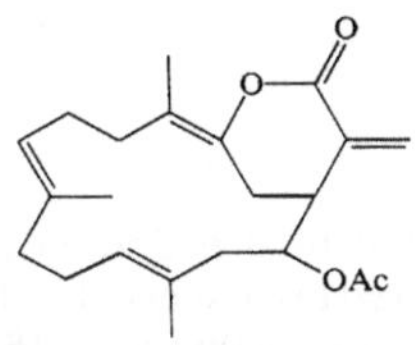

(149)

Controlled hydrogenation of crassin acetate gave predominantly one form of the saturated *alpha*-methyl lactone **(150)**. Alkaline hydrolysis of **(150)**, followed by acidification, formed the γ-lactone **(151)** resulting from relactonization to the secondary alcohol group liberated by deacetylation. That the new γ-lactone formed at the γ′-position rather than at the γ-position of the original lactone was indicated by the fact that the pmr signal of the lactone proton in crassin acetate appeared as a double doublet. Since the carbinol function occupied one vicinal position, this multiplicity required an unsubstituted methylene group at the γ-position in the original δ-lactone ring. As expected, the diol function in **(151)** reacted readily with periodic acid.

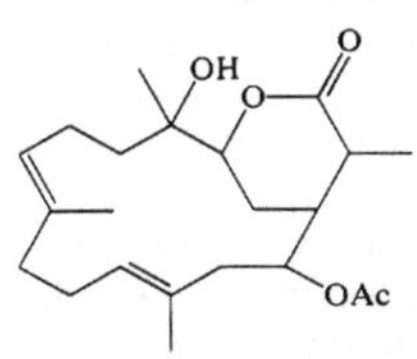

(150)

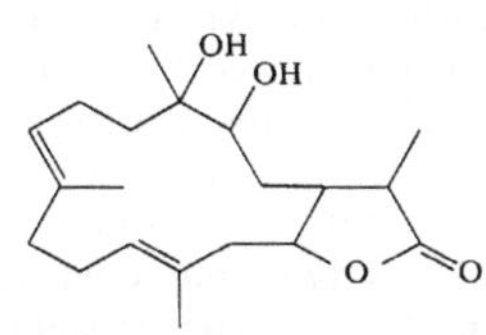

(151)

A related lactone was obtained by treatment of crassin acetate with methanolic sodium hydroxide. This reaction afforded compound **(152)**, in contrast to crassin which was formed with aqueous base. As with the aqueous reagent, deacetylation had occurred, but instead of simple relactionization, the methanolic reagent promoted translactonization to the deacetylated hydroxyl at position 14, accompanied by Michael addition of the liberated lactone hydroxyl to the conjugated exo-methylene function. Compound **(152)** was shown to have only one hydroxyl group which was resistant to oxidation.

The preceding evidence interrelated the oxygen functions in crassin acetate. The balance of the carbon skeleton was established by ozonolysis

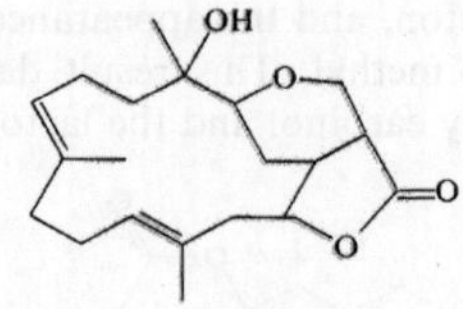

(152)

of **(153)** which was prepared by dehydration of dihydrocrassin acetate **(150)**. The ozonolysis produced two moles of levulinic acid and a ten-carbon keto lactonic acid isolated as the methyl ester **(154)**.

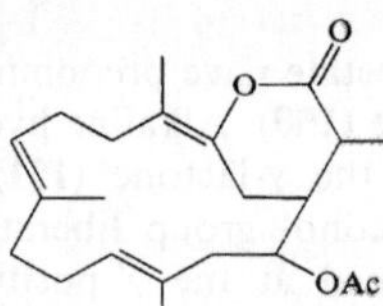

(153)

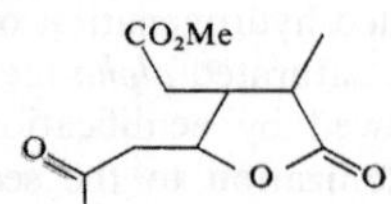

(154)

Compound **(154)** showed distinct carbonyl absorptions for the γ-lactone (1790 cm^{-1}), the ester (1740 cm^{-1}) and for the methyl ketone (1725 cm^{-1}) functions. Its pmr spectrum showed that each position of the lactone ring was monosubstituted, and that the substituents were methyl, carbomethoxymethyl, and acetonyl residues. The high resolution mass spectrum showed fragments corresponding to the latter two fragments, confirming the structure of **(154)**, and thereby establishing the complete structure of crassin-acetate.

An X-ray diffraction study of crassin *p*-iodobenzoate by HOSSAIN and VAN DER HELM (*15*) fully confirmed structure **(147)** for crassin acetate, and provided the stereochemical detail and absolute configuration shown therein.

Sinularin and Dihydrosinularin

The epoxy cembranolides, sinularin **(155)** and dihydrosinularin **(156)**, were isolated by WEINHEIMER, MATSON, HOSSAIN, and VAN DER HELM (*12*) from the soft coral *Sinularia flexibilis* collected on the Great Barrier Reef of Australia. These two compounds were the major constituents of the active fraction obtained from its extract in an isolation guided by a cytotoxicity bioassay. Accompanying them in minor amounts were sinulariolide **(157)** which was the major cembranolide of the Indonesian form of this species, and 11-epi-sinulariolide acetate **(165)**.

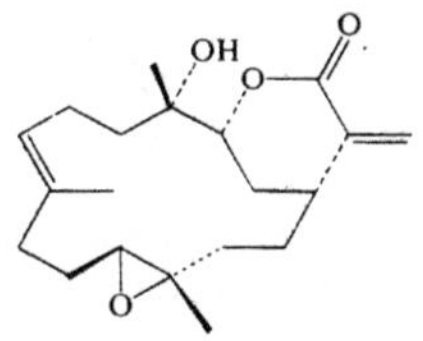

(155)

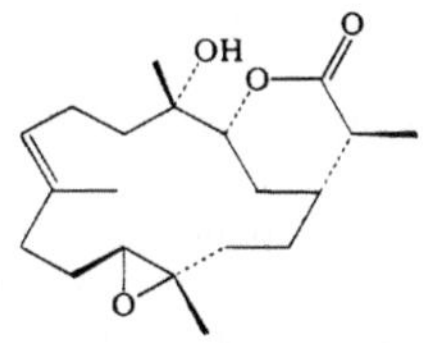

(156)

Sinularin, $C_{20}H_{30}O_4$, showed hydroxyl and lactone carbonyl absorptions in the infrared, and pmr signals for an α-methylenic lactone, a tertiary methyl carbinol, a methyl substituted epoxide, and the vinyl methyl and vinyl proton signals of a trisubstituted double bond. These structural elements required that the skeleton be monocarbocyclic.

Dihydrosinularin, $C_{20}H_{32}O_4$, showed similar spectral features, except for replacement of the exo-methylene group of the lactone by a secondary methyl group. The structures and absolute configurations shown in (**155**) and (**156**) were determined for each by the X-ray diffraction method.

These authors observed that the substitution pattern and relative stereochemistry of positions 1 through 9 of sinularin were identical to the corresponding portion of crassin acetate (**147**). However, because of their opposite absolute configurations, the two compounds are essentially enantiomeric in this portion of the molecule. This relationship prompted the proposal (*12*) of the configuration convention for cembranes described in the Introduction.

Sinulariolide

Tursch *et al.* (*76*) isolated sinulariolide (**157**), a cembranolide containing the unusual seven-membered ε-lactone, as the major cembranoid in the soft coral, *Sinularia flexibilis*, from Indonesian waters. Its complete structure was deduced from chemical transformations and degradations and was confirmed by an X-ray diffraction study.

Sinulariolide, $C_{20}H_{30}O_4$, showed hydroxyl and lactone carbonyl ($1720\ cm^{-1}$) absorptions in the ir. Its pmr and cmr spectra indicated an exo-methylene in the lactone, a methyl on a trisubstituted double bond, a secondary alcohol, and two methyls on carbon bearing oxygen. In the

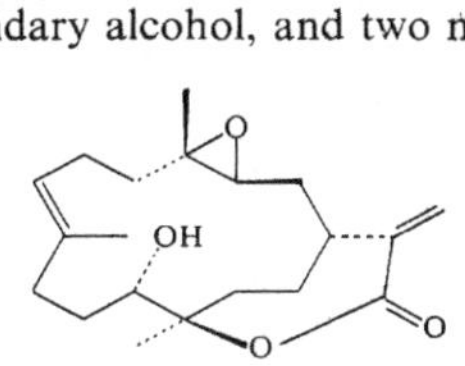

(157)

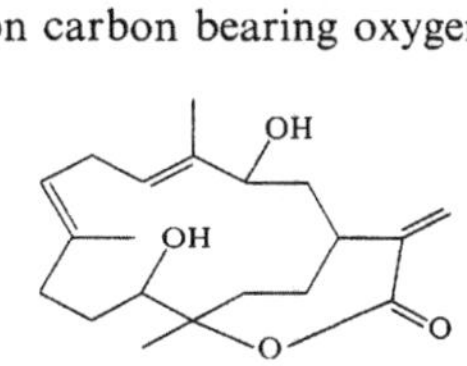

(158)

pmr spectrum one of the latter singlets was relatively deshielded, indicating it was at the lactone terminus, while the other exhibited a normal chemical shift for an epoxide methyl. Sinulariolide therefore contained one carbocyclic ring.

Since sinulariolide was sensitive to acids, most of the structure determination was based on reactions of (**158**), the allylic alcohol formed by isomerization of the epoxide with hydriodic acid. This compound was shown to be an allylic alcohol by selective oxidation with manganese dioxide to a conjugated ketone, thereby confirming the original function in sinulariolide as an epoxide. The doubly allylic methylene at position 6 of (**158**) was established by decoupling (270 MHz) from the two vicinal vinyl protons, demonstrating the relationship between the trisubstituted double bond and the original epoxide function.

On treatment with methanolic base, the lactone opened to form the methyl ester (**159**) which contained a new 1,2-diol function. Cleavage of the latter with periodate formed the expected keto aldehyde (**160**), which formed levulinaldehyde upon ozonolysis. These results related the hydroxyl function and the lactone alkyl terminus to the original trisubstituted double bond, and consequently to the epoxide.

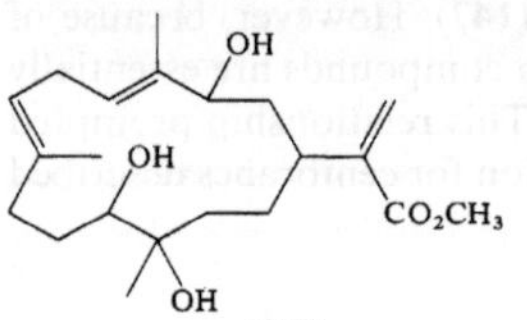

(**159**)

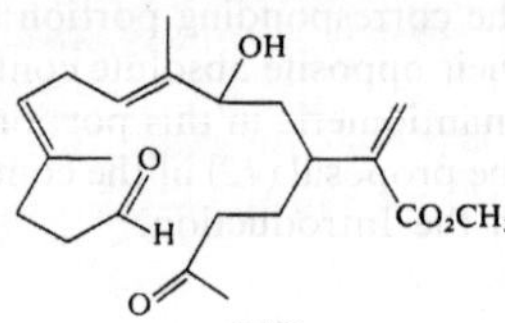

(**160**)

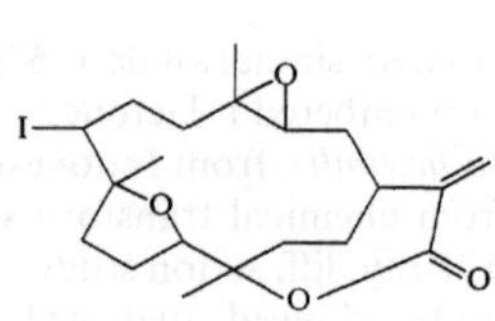

(**161**)

With N-iodosuccinimide, sinulariolide formed the iodo tetrahydrofuran (**161**) in which the iodine atom adventitiously deshielded the β-proton of the lactone (position 1). This permitted decoupling experiments which demonstrated the presence of a single methylene group between the β-position of the lactone and the epoxide function. The two methylene groups still unassigned were necessarily part of the seven membered lactone ring.

The proposed structure was confirmed by an X-ray diffraction study which also established the relative and absolute configuration shown in

(**157**) for sinulariolide. The full details of the latter study have recently been published by KARLSSON (*77*).

6-ξ-Hydroxysinulariolide

One of the minor constituents isolated by HÉRIN and TURSCH (*52*) from Indonesian *Sinularia flexibilis* was the 6-hydroxy derivative, $C_{20}H_{30}O_5$ (**162**), of sinulariolide. Since this compound was also encountered amongst the degradation products of sinulariolide (**157**) in their earlier study, it is likely that it is an artifact of air oxidation of the allylic site in sinulariolide.

The spectra of (**162**) showed close similarities to those of sinulariolide with the exception that an additional secondary hydroxyl group was indicated. This was confirmed by formation of a diacetate. The carbinyl proton of the new alcohol function was shown by decoupling to be vicinal to the vinyl proton of a trisubstituted double bond. This fact would locate the new hydroxyl at position 6 if indeed the compound otherwise possessed the sinulariolide structure.

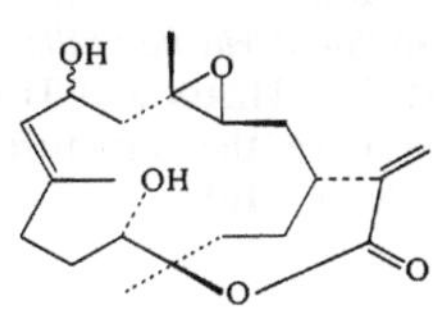

(**162**)

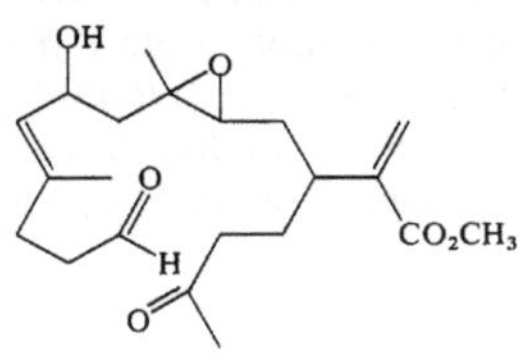

(**163**)

In a sequence of reactions analogous to those conducted with sinulariolide, the lactone of (**162**) was converted by treatment with base to the methyl ester [cf. (**159**)], which was cleaved with periodate to the keto aldehyde (**163**). Like (**160**), this compound also formed levulinaldehyde upon ozonolysis. These reactions demonstrated the similarity of the hydroxy derivative with sinulariolide and secured structure (**162**) for it.

11-Dehydrosinulariolide

Another minor cembranolide present in Indonesian *Sinularia flexibilis* was shown by HÉRIN and TURSCH (*52*) to be the keto derivative, 11-dehydrosinulariolide (**164**). Spectrally similar to sinulariolide, it showed no hydroxyl absorption in the ir, and its molecular ion ($C_{20}H_{28}O_4$) was two mass units smaller than that of sinulariolide. Its identity was established by synthesis of the identical compound from sinulariolide by oxidation with chromium trioxide in acetic acid.

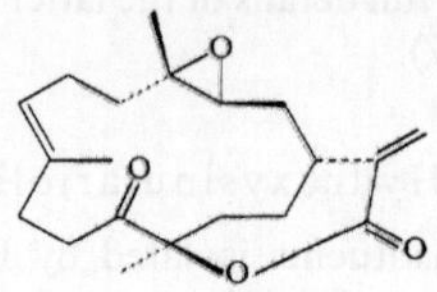

(164)

11-Epi-sinulariolide acetate

A fourth minor cembranolide reported by HÉRIN and TURSCH (*52*) from Indonesian *Sinularia flexibilis* was an unstable acetate of undetermined structure. It was recognized spectrally as an acetate, $C_{22}H_{32}O_5$, of a compound similar to sinulariolide, but it was not identical to the acetate prepared from sinulariolide. Structure determination of this compound was not completed, apparently because of its loss due to instability.

Recently, WEINHEIMER, MATSON, HOSSAIN, and VAN DER HELM (*78*) have isolated a stable acetate from Australian *Sinularia flexibilis*. Its ir and pmr spectra closely resemble those reported by HÉRIN and TURSCH. X-ray diffraction studies have shown it to have the structure and relative configuration of 11-epi-sinulariolide acetate (**165**).

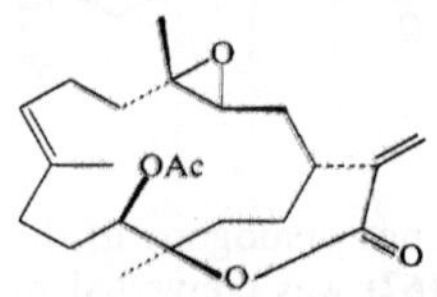

(165)

Other marine cembranoids

In a recent review article, TURSCH (*79*) mentioned four additional marine cembranoids that are under investigation in his laboratory and for which full reports have not yet appeared. The epoxycembrol, trocheliophorol (**166**), was isolated from *Sarcophyton trocheliophorum* collected in the Seychelles Islands, and its structure was deduced from chemical and spectral evidence. This compound was not present in the same species collected in Indonesia which, instead, afforded two other compounds. One of these was the dihydrofuran sarcophytoxide

(167), referred to by Tursch as a stereoisomer of deoxysarcophine **(113)**. The other was its positional isomer, isosarcophytoxide **(168)**. The structures of **(167)** and **(168)** were still provisional as was that of the highly oxygenated cembranolide, crassolide **(169)**, which was isolated from *Lobophytum crassum* collected in Indonesia.

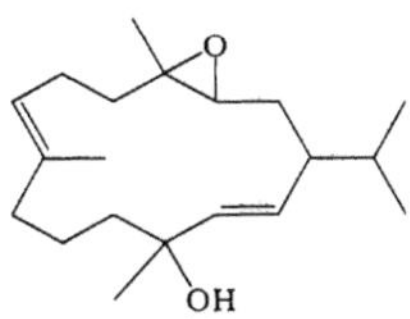

(166)

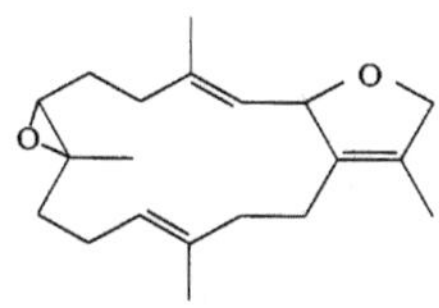

(167)

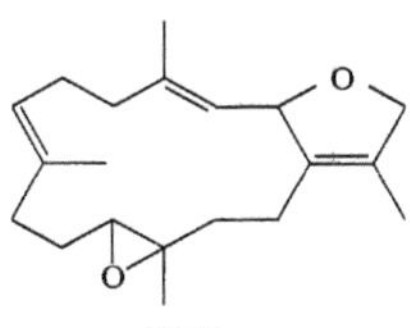

(168)

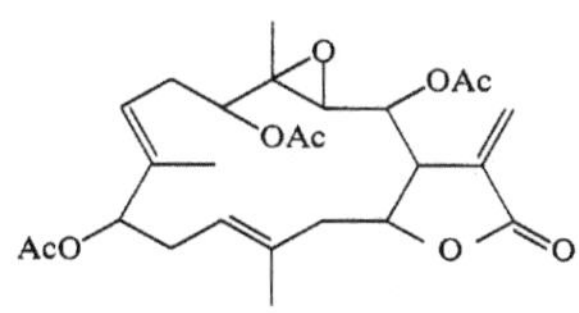

(169)

III. Synthesis of Cembranoids

A. (±)-Cembrane

Shortly after the structure of cembrene was established, Entwistle and Johnstone (*80*) reported a synthesis of the parent cycloalkane, cembrane **(176)**, which employed the acyloin condensation in the ring-forming step. The pimelate half-ester **(170)** was homologated to the methyl benzyl ester **(171)** by the Arndt-Eistert method and then hydrogenolyzed to the half-ester acid **(172)**. Electrolysis of this acid with the second half-ester acid **(173)** produced a mixture of diesters from which the mixed product **(174)** was isolated by preparative glc. Diester **(174)** was cyclized using the acyloin condensation to a mixture of ketols **(175)** which were reduced by the Clemmensen method to cembrane **(176)** which was identical with the hydrocarbon prepared by hydrogenation of cembrene.

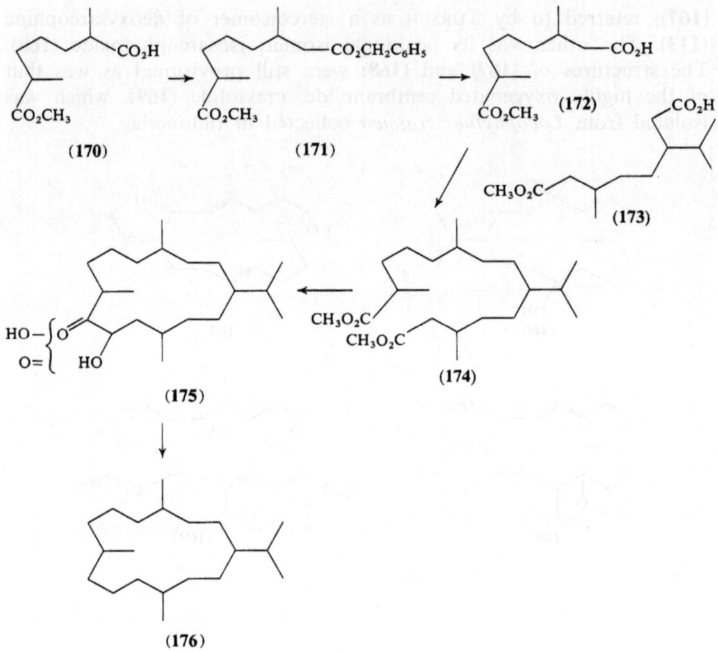

B. (±)-Cembrene

It was fitting that the first total synthesis of cembrene to appear in the literature was due to DAUBEN *et al.* (*81, 82*). For their method of cyclization, they chose the nickel carbonyl promoted coupling of allylic bromides which leads to the formation of cyclic 1,5-dienes. Since the conjugated diene system in cembrene was incompatible with the coupling reagent, their approach required a synthon for the diene system which could be converted to it following cyclization. Selected initially for this purpose was the 2-ene-4-one system which, upon conversion to the tertiary alcohol, could be dehydrated to cembrene.

Thus, the first objective of the synthesis was the enone **(179)**. The precursors **(177)** and **(178)**, each containing THP protected alcohol functions, were prepared and subjected to the ylid condensation which formed the desired enone **(179)** in 78% yield, with a *trans* selectivity of greater than 95% for the new disubstituted double bond. After conversion of the protected alcohol functions to the allylic bromides, however, it was found that the cyclization reaction failed, apparently because of reaction of the conjugated enone system with the nickel carbonyl catalyst.

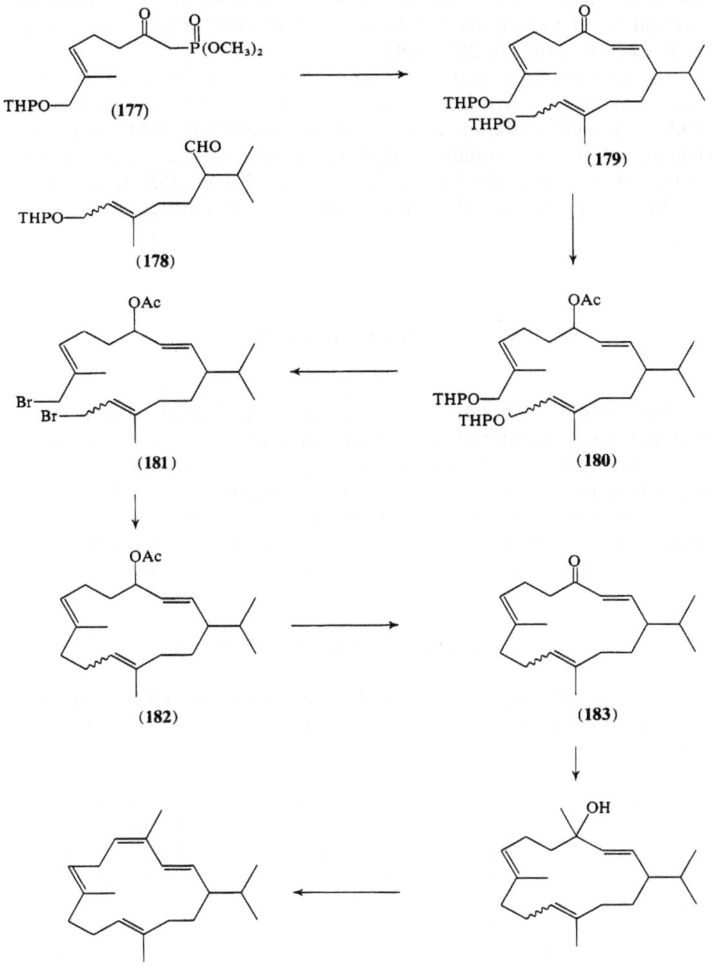

(177) (178) (179) (180) (181) (182) (183) (184) (185)

In an alternate approach, the enone system was converted to the allylic acetate by lithium aluminium hydride reduction followed by acetylation, forming the protected allylic ester (**180**). The THP blocking groups were removed selectively by mild acid hydrolysis, and the diol was converted to the dibromide (**181**) with phosphorus tribromide. Cyclization of this compound by nickel carbonyl formed a mixture of triene-acetates (**182**) in 25% yield.

The mixture of acetates was converted to the enones (**183**) with lithium aluminium hydride followed by Jones oxidation. Treatment of (**183**) with methyl lithium gave a mixture of alcohols (**184**) which was dehydrated by toluenesulfonic acid in benzene to form a mixture of hydrocarbons from which crystalline (±)-cembrene (**185**) was isolated by chromatography on silver nitrate impregnated silica gel.

C. Other Cembranoids

Two groups at Tohoku University have developed biogenetic-like syntheses for a number of cembranoid diterpenes. The late KITAHARA's group, now guided by KATO, has developed a cationic cyclization reaction which has led to the synthesis of (±)-mukulol, which has in turn been converted to (±)-cembrene and (±)-incensole. ITÔ, in another approach, has employed an anionic cyclization in the synthesis of (±)-nephthenol which has been transformed to (±)-cembrene-A.

(±)-Mukulol and (±)-Cembrene

KATO *et al.* (*83, 84*) found that all *trans* geranyl geranic acid chloride (**186**) cyclized to form the 14-membered ring chloro ketone (**187**) in 71% yield when treated with stannic chloride at −87° in dilute solution in methylene chloride. Hydrogenolysis of the chloro ketone with tributyltin hydride converted it quantitatively to the ketone (**188**). Reduction of the ketone with lithium aluminium hydride afforded, in addition to the saturated ketone which was the major product, a mixture of epimeric alcohols (**189**) and (**190**) (only one enantiomer depicted), in which the former, (±)-mukulol, predominated (28% yield).

Dehydration of (±)-mukulol at ice temperature with thionyl chloride-pyridine formed (±)-cembrene in 43% yield. (±)-Epimukulol also formed (±)-cembrene under these conditions.

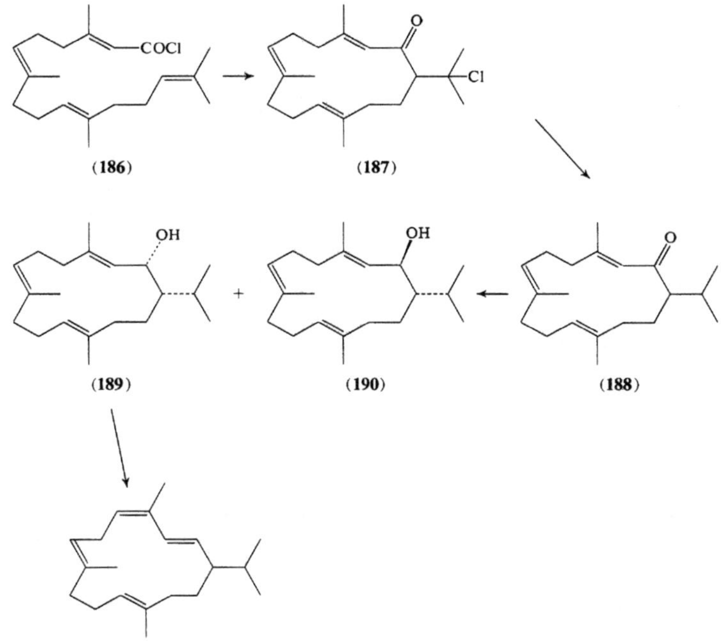

(186) (187) (189) (190) (188) (185)

(±)-Incensole

In the conversion of (±)-mukulol to (±)-incensole, KATO *et al.* (*85*) utilized the newly developed bromo etherification reagent, 2,4,4,6-tetrabromocyclohexadienone, to form the bromo tetrahydropyranyl ether **(191)*** in 54% yield. This ether was isomerized by boron trifluoride-etherate to the bromo tetrahydrofuran **(192)** in 46% yield, presumably *via* the intermediate conjugated 1,3-diene. Treatment of **(192)** with silver acetate, followed by lithium aluminium hydride reduction, afforded (±)-incensole **(193)** in 70% yield.

* The configuration of C-2 does not change during the transformation of (**189**) to (**191**), but its conformation does. The 2-hydroxyl group is directed away from the center of the ring in (**189**), and must instead be directed toward its center in order to form (**191**). This reorientation is accomplished by rotation (not shown) of C-2 about its bonds with C-1 and C-3, having the consequence of reversing the initial disposition of the substituents at C-2 with respect to the general plane of the ring. For such reasons, the designation of configuration by α and β in the cembranes is potentially ambigous, and must be used with care.

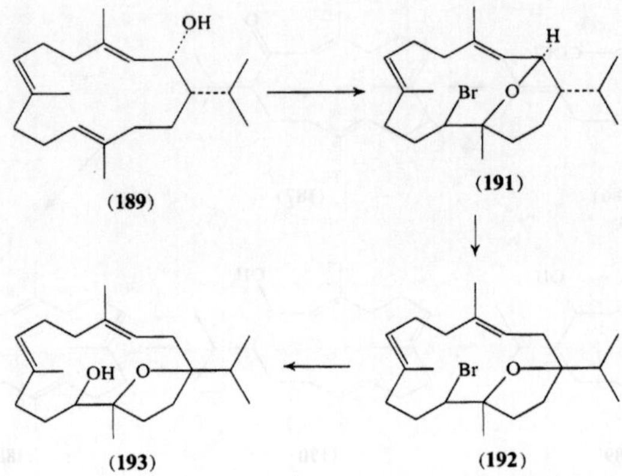

(189) (191) (192) (193)

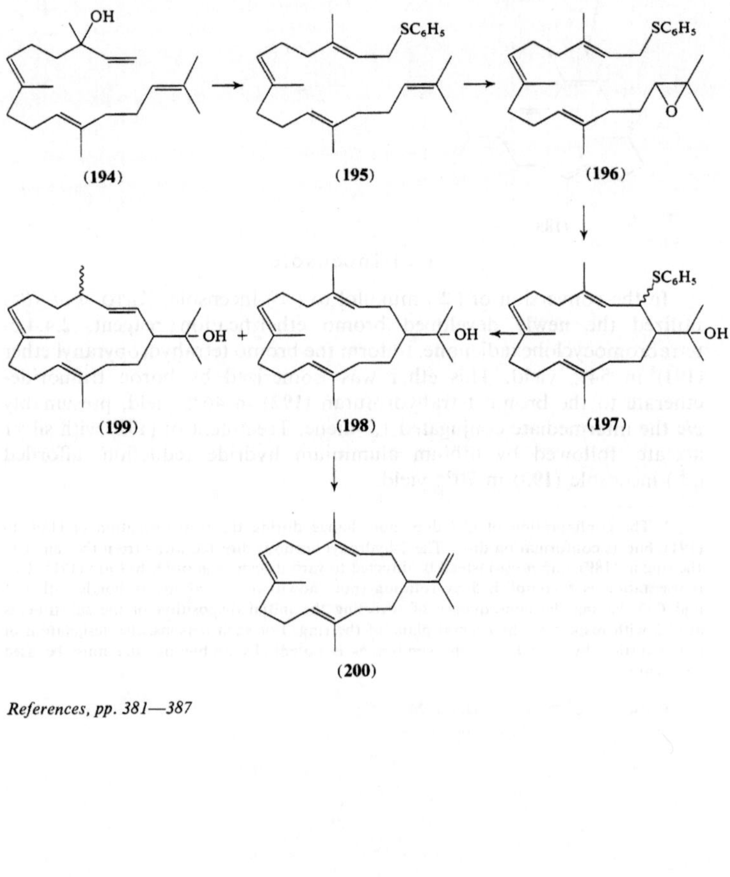

(194) (195) (196) (197) (198) (199) (200)

References, pp. 381—387

(±)-Nephthenol and (±)-Cembrene-A

Seeking greater levels of selectivity than they expected to be possible from cationic cyclization processes, KODAMA, MATSUKI and ITO (*86*) developed an anionic method which they applied to the synthesis of nephthenol.

All-*trans* geranyl geranyl phenyl thioether (**195**) was prepared in 73% yield from *trans, trans*-geranyl linalool (**194**) by treatment with phosphorus tribromide followed by sodium thiophenoxide. The thioether was converted to the terminal epoxide (**196**) in 42% yield by reaction with aqueous N-bromosuccinimide, followed by potassium carbonate. Cyclization of the epoxy thioether to (**197**) was achieved in 62% yield by treatment with butyl lithium in dilute THF at −78°. Desulfurization of (**197**) with lithium-ethylamine afforded (±)-nephthenol (**198**) in 30% yield. It was accompanied by 20% of the allylic isomerization product (**199**).

Dehydration of (±)-nephthenol with thionyl chloride-pyridine formed (±)-cembrene-A (**200**) as the sole product (>95%).

IV. Biosynthesis of Cembranoids

Although many authors have speculated upon the biosynthetic origin of cembranoids, we are aware of only one group which has described incorporation of labelled precursors into a specific natural cembranoid. RICE, PAPASTEPHANOU, and ANDERSON (*87*) of the University of Miami, using tissue slices of the gorgonian *Pseudoplexaura porosa,* have measured the incorporation of carbonate ^{14}C and acetate 2-^{14}C into crassin acetate (**147**). Although the net incorporation of $Na_2{}^{14}CO_3$ and $^{14}CH_3COOH$ was small, it was clearly established that incorporation had occurred and that acetate ^{14}C was incorporated better than carbonate ^{14}C. Less than half of the incorporated acetate was taken up by exchange of acetate in the cembranolide, as demonstrated by measurements on the acetic acid obtained by hydrolysis of the labelled crassin acetate.

The formation of labelled crassin acetate in this study demonstrated that the gorgonian or the gorgonian-zooxanthellae complex synthesized crassin acetate, thereby eliminating the possibility that this compound originated in the feeding or digestive processes.

More recently, ANDERSON (*88*) has found that the zooxanthellae, algal symbionts in the gorgonian, are involved in crassin acetate synthesis. Using a soluble cell-free system prepared from isolated zooxanthellae, mevalonic acid and geranyl geranyl pyrophosphate were incorporated into crassin acetate.

Table 1. *Cembranoids From Resinous Plants*

a) Hydrocarbons

Name[1] (Compound Number)	Structure (Molecular Formula)	Physical Data[2]	Source	Reference
Cembrene (**1**)†	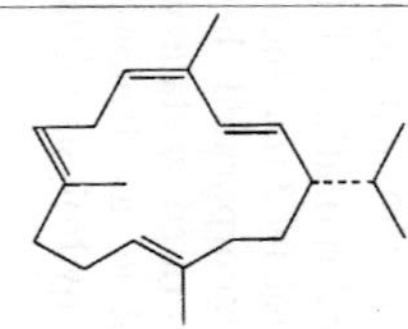 $C_{20}H_{32}$	mp 59—60° {α}$_D$ + 238°	*Picea obvata* *Pinus albicaulis* *Pinus sibirica* *Pinus thunberigii* *Larix sibirica* *Larix sukaczawii* *Larix czekanovskii* *Chamaecryparis obtusa* *Pseudotsuga menziesii*	(*27, 89*) (*1, 11, 14*) (*90, 91, 92*) (*2, 5, 6, 13*) (*90, 91*) (*91*) (*91*) (*93*) (*23*)
Cembrene-A (Neocembrene) (Neocembrene-A) (**16**)†	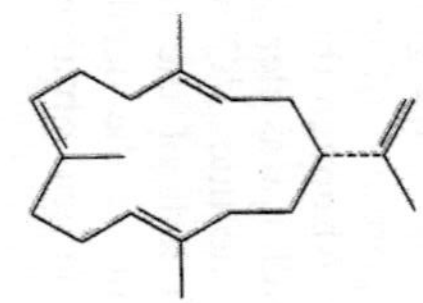 $C_{20}H_{32}$	bp 150—152° (0.8 mm) {α}$_D$ − 19.7°	*Commiphora mukol* *Picea obovata* *Pinus koraiensis*	(*15*) (*27*) (*16*)
Isocembrene (**21**)	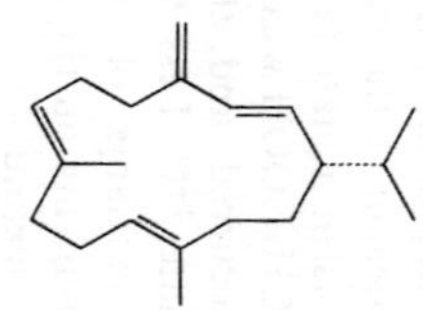 $C_{20}H_{32}$	{α}$_D$ + 60.3°	*Pinus sibirica*	(*17*)

[1] Compounds are presented in increasing order of C, H and O content. The symbol † indicates absolute configuration.
[2] Mp/bp and specific rotation, if available, from literature.

b) Oxygenated Compounds

Name (Compound Number)	Structure (Molecular formula)	Physical Data	Source	Reference
Mukulol **(22)**†	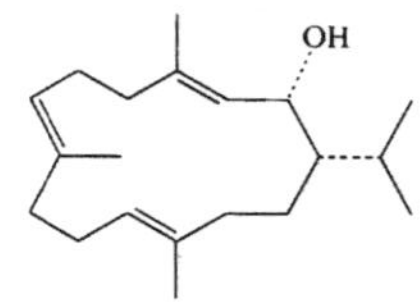 $C_{20}H_{34}O$	mp 37—38° $\{\alpha\}_D + 53°$	*Commiphora mukul*	(*15, 21*)
"Thunbergol" **(30)**	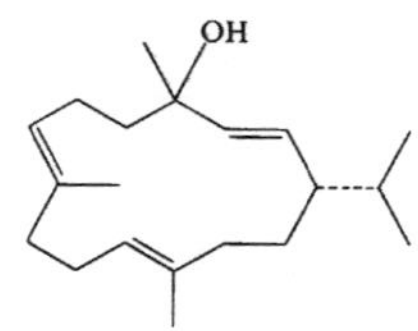 $C_{20}H_{34}O$	oil $\{\alpha\}_D + 74.4°$	*Pseudotsuga menziesii* *Pinus sibirica* *Picea obovata*	(*23*) (*24*) (*27*)
Isocembrol **(31)**	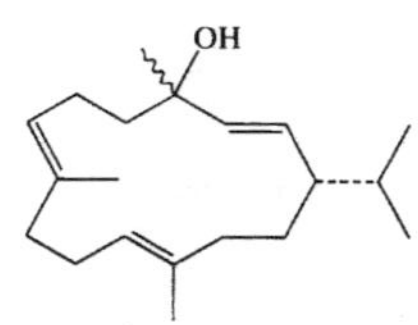 $C_{20}H_{34}O$	oil $\{\alpha\}_D + 80.1°$	*Pinus koraiensis* *Pinus sibirica* *Picea obovata*	(*26*) (*24, 25*) (*27*)

Table 1. b) (continued)

Name (Compound Number)	Structure (Molecular formula)	Physical Data	Source	Reference
4-Epiisocembrol **(32)**	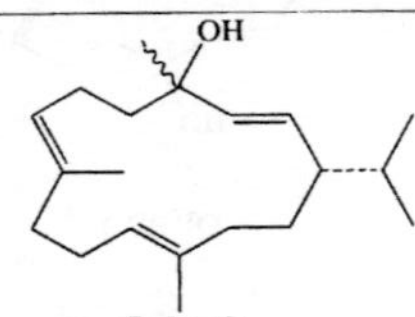 $C_{20}H_{34}O$	oil $\{\alpha\}_D + 110.5°$	*Pinus koraiensis* *Pinus sibirica*	*(28)* *(28)*
Cembrol **(33)**	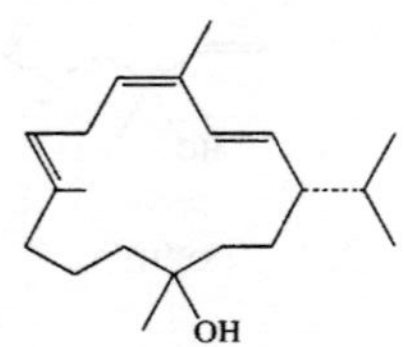 $C_{20}H_{34}O$	$\{\alpha\}_D + 59.6°$	*Pinus sibirica*	*(29)*
Incensole **(37)**	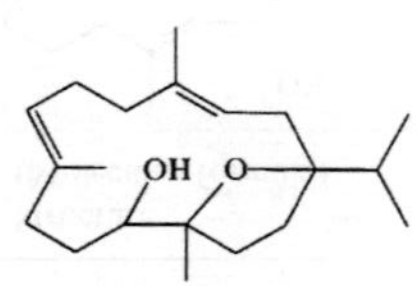 $C_{20}H_{34}O_2$	"low melting crystalline solid" $\{\alpha\}_D - 77.5°$	*Boswellia carteri*	*(31)*

Incensole oxide **(45)**	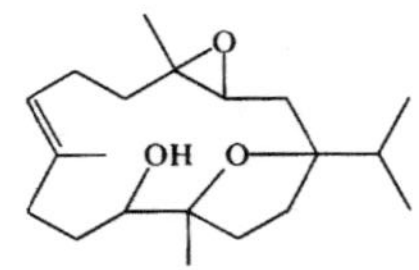 $C_{20}H_{34}O_3$	mp 164—165° $\{\alpha\}_D$ − 48°	*Boswellia carteri*	(*32*)
Isoincensole oxide **(51)**	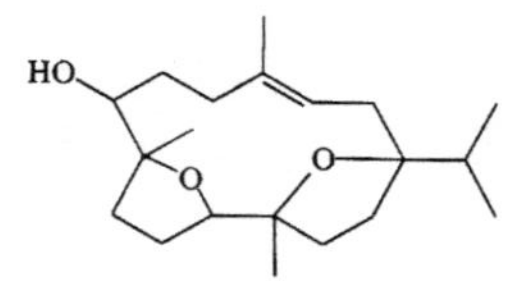 $C_{20}H_{34}O_3$	mp 84—85° $\{\alpha\}_D$ − 13.9°	*Boswellia carteri*	(*33, 34*)

Table 2. *Cembranoids From Tobacco (Nicotiana tabacum)*

Compound Number	Structure	Molecular Formula	Physical Data	Reference
(71)	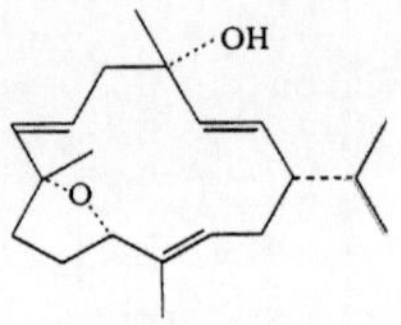	$C_{20}H_{32}O_2$	mp 95—96° {α}D + 86°	(*36*)
(74)	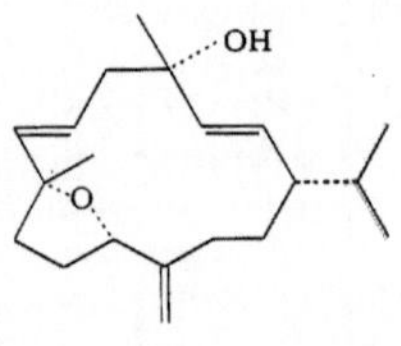	$C_{20}H_{32}O_2$	mp 109—110° {α}D + 77.4°	(*36*)
(75)	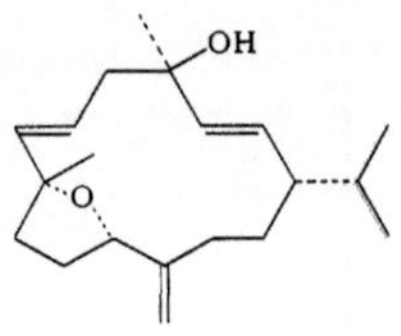	$C_{20}H_{32}O_2$	mp 108—109° {α}D + 72.5°	(*36*)
(78)	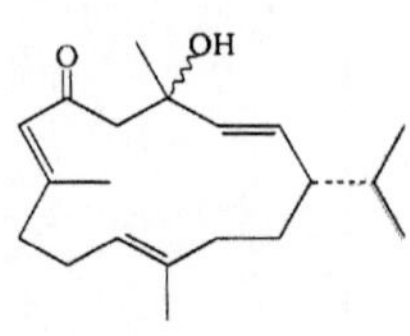	$C_{20}H_{32}O_2$	oil	(*37*)
(54)	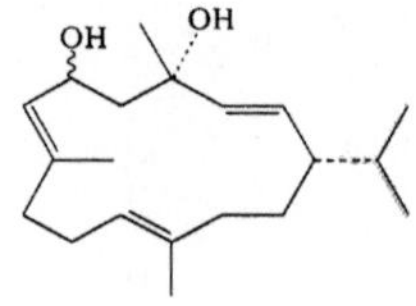	$C_{20}H_{34}O_2$	mp 65—66° {α}D + 281.6°	(*3*)

References, pp. 381—387

Table 2 (continued)

Compound Number	Structure	Molecular Formula	Physical Data	Reference
(55)	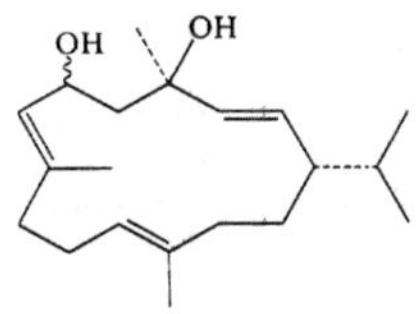	$C_{20}H_{34}O_2$	mp 127—127.5° {α}D + 162°	(*3*)
(66)	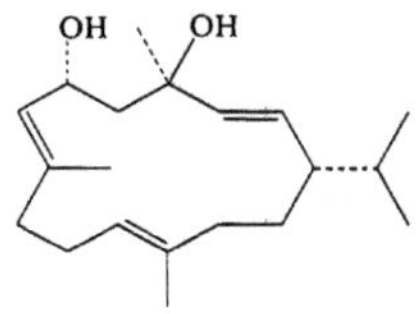	$C_{20}H_{34}O_2$	mp 123°	(*43*)
(68)	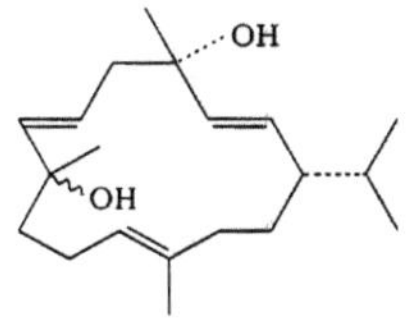	$C_{20}H_{34}O_2$	mp 118—120° {α}D + 100°	(*45*)
(69)	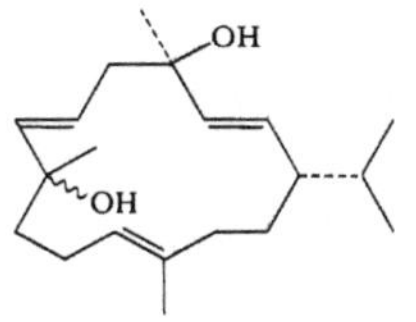	$C_{20}H_{34}O_2$	mp 150—152° {α}D + 40°	(*45*)
(77)	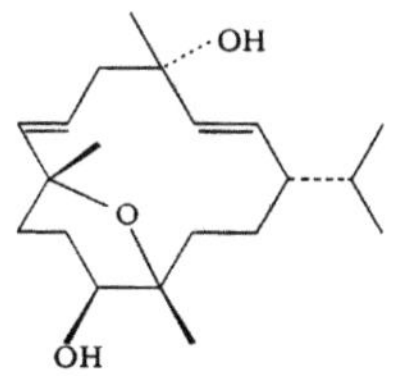	$C_{20}H_{34}O_3$	—	(*38*)

Table 3. *Cembranoids From Miscellaneous Plants*

Name (Compound Number)	Structure (Molecular formula)	Physical Data	Source	Reference
Ovatodiolide (**79**) †	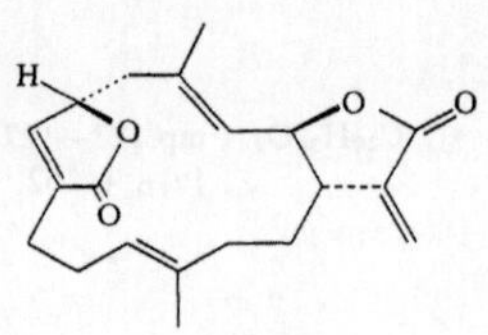 $C_{20}H_{24}O_4$	mp 148—150° 157° {α} + 21.8°	*Anisomeles ovata* *Anisomeles malabarica*	(*46*) (*49*)
Anisomelic Acid (**92**)	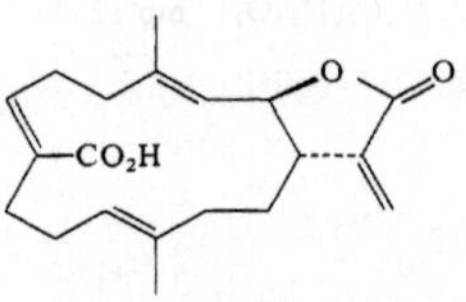 $C_{20}H_{26}O_4$	mp 155°	*Anisomeles malabarica*	(*49*)

Table 4. *Cembranoids From Insects*

Name Compound Number)	Structure (Molecular Formula)	Physical Data	Source	Reference
Cembrene-A (Neo-cembrene-A) (**16**) †	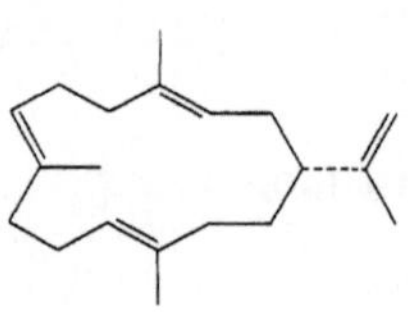 $C_{20}H_{32}$	bp 150—152° (0.8 mm)	*Nasutitermes exitosus*	(*51*)

References, pp. 381—387

Table 5. *Cembranoids From Marine Invertebrates*

Name (Compound Number)	Structure (Molecular formula)	Physical Data	Source	Reference
Peunicin (**143**) †	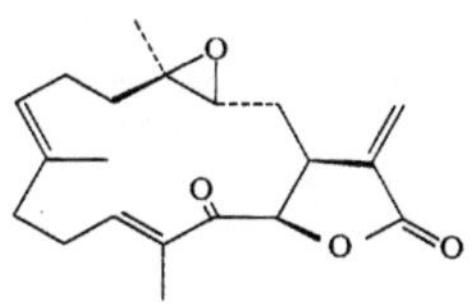 $C_{20}H_{26}O_4$	mp 177—178°	*Eunicea succinea*	(*71*)
Epipeunicin (**144**)	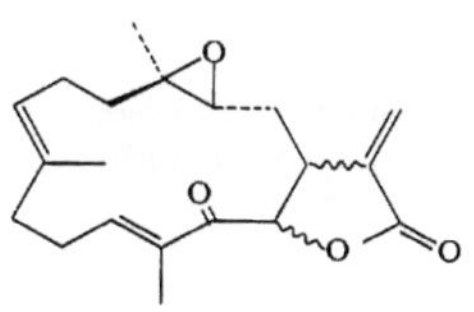 $C_{20}H_{26}O_4$	mp 145—146°	*Eunicea succinea*	(*71*)
Lobo-phytolide (**115**) †	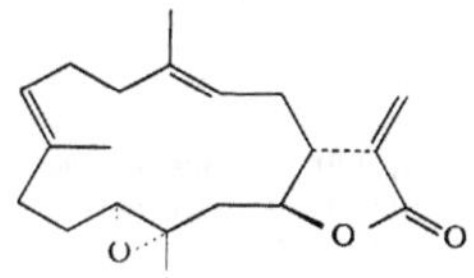 $C_{20}H_{28}O_3$	mp 137—138° {α}D + 7°	*Lobophytum cristagalli*	(*58*)
Sarcophine (**117**)	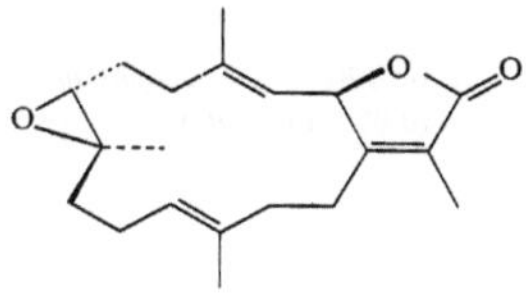 $C_{20}H_{28}O_3$	mp 133—134° {α}D + 92°	*Sarcophytum glaucum*	(*60*)
2-Epi-sarcophine (**118**)	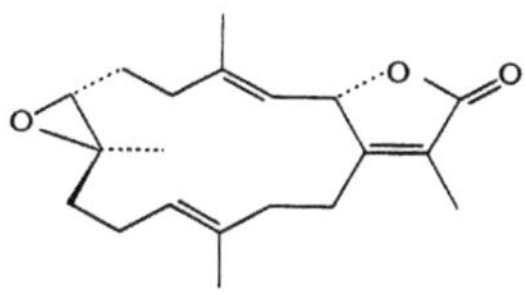 $C_{20}H_{28}O_3$	mp 70° {α}D − 16°	*Sarcophytum glaucum*	(*56*)

Table 5 (continued)

Name (Compound Number)	Structure (Molecular formula)	Physical Data	Source	Reference
2,8-Bis-epi-sarcophine (**119**)	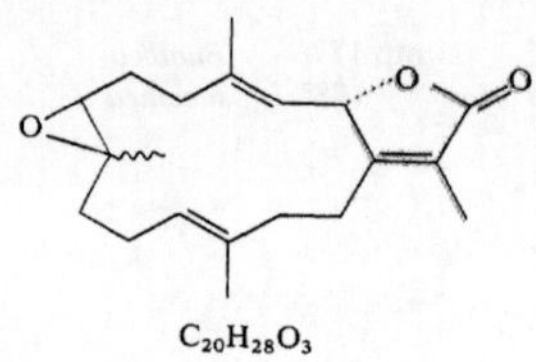 $C_{20}H_{28}O_3$	oil	*Sarcophytum glaucum*	(*56*)
11-dehydro-sinulariolide (**164**) †	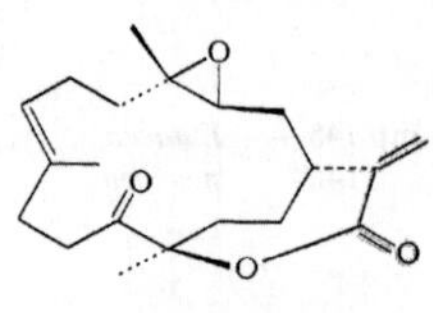 $C_{20}H_{28}O_4$	mp 120° {α}$_D$ + 87°	*Sinularia flexibilis*	(*52*)
Deoxo-sarcophine (**113**)	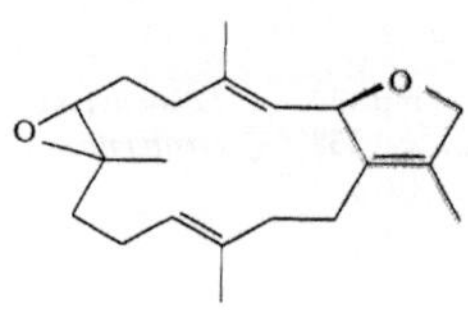 $C_{20}H_{30}O_2$	bp 120° (0.05 mm) {α}$_D$ + 40°	*Sarcophytum glaucum*	(*56*)
2-Epi-deoxo-sarcophine (**114**)	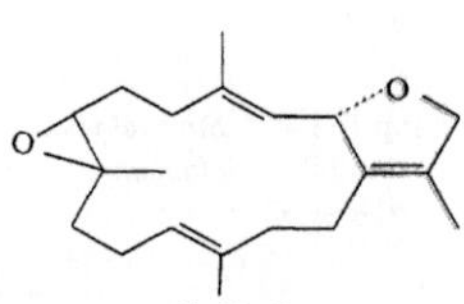 $C_{20}H_{30}O_2$	bp 120° (0.05 mm)	*Sarcophytum glaucum*	(*56*)
Sarco-phytoxide (**167**)	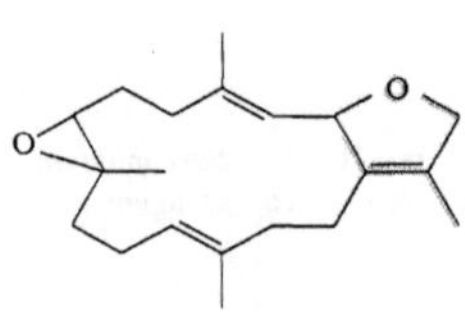 $C_{20}H_{30}O_2$	—	*Sarcophyton trocheliophorum*	(*79*)

References, pp. 381—387

Table 5 (continued)

Name (Compound Number)	Structure (Molecular formula)	Physical Data	Source	Reference
Iso-sarco-phytoxide (**168**)	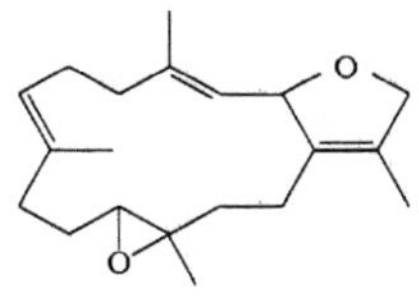 $C_{20}H_{30}O_2$	—	*Sarcophyton trocheliophorum*	(*79*)
Sinulariolide (**157**) †	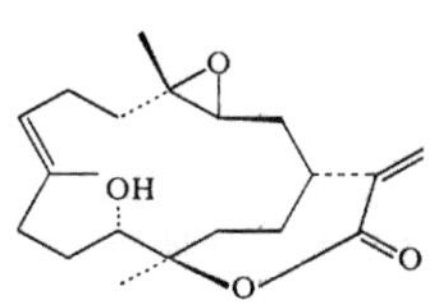 $C_{20}H_{30}O_4$	mp 170—173° {α}D + 76°	*Sinularia flexibilis*	(*76*)
Sinularin (**155**) †	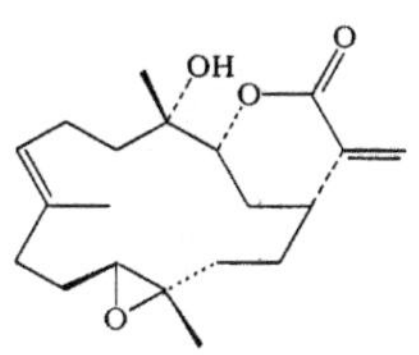 $C_{20}H_{30}O_4$	mp 150—152° {α}D − 127°	*Sinularia flexibilis*	(*12*)
Eunicin (**2**) †	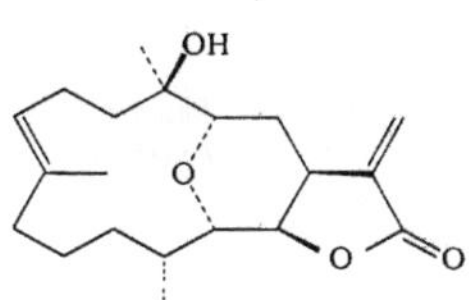 $C_{20}H_{30}O_4$	mp 155° {α}D − 89°	*Eunicea mammosa*	(*65*)
Cueunicin (**132**) †	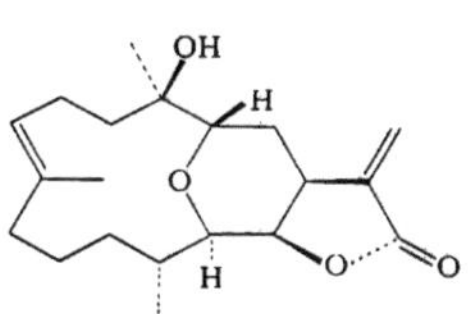 $C_{20}H_{30}O_4$	gum {α}D − 147°	*Eunicea mammosa*	(*68*)

Table 5 (continued)

Name (Compound Number)	Structure (Molecular formula)	Physical Data	Source	Reference
Jeunicin (**140**) †	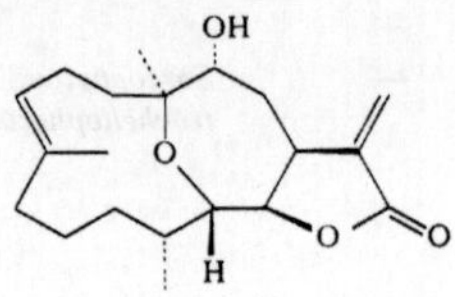 $C_{20}H_{30}O_4$	mp 139—141° {α}D + 12.8°	*Eunicea mammosa*	(*69*)
13,14-Bis-epijeunicin (**142**) †	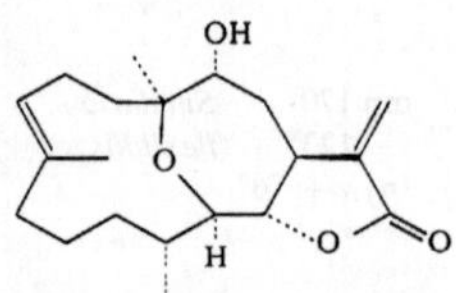 $C_{20}H_{30}O_4$	mp 147—147.5° {α}D − 31.1°	*Eunicea mammosa*	(*70*)
6-ξ-Hydroxy-sinulariolide (**162**)	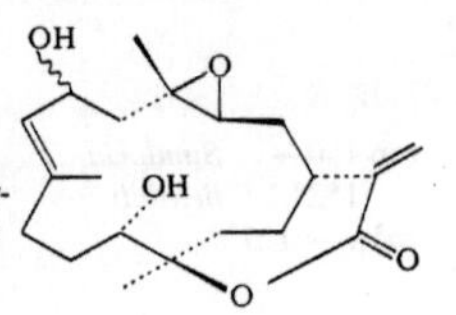 $C_{20}H_{30}O_5$	mp 192—194° {α}D + 54.5°	*Sinularia flexibilis*	(*52*)
Cembrene-A (**16**)	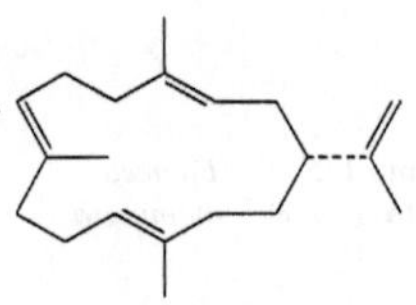 $C_{20}H_{32}$	bp 150—152° (0.8 mm) {α}D − 19.7°	*Sinularia flexibilis*	(*52*)
Flexibilene (**93**)	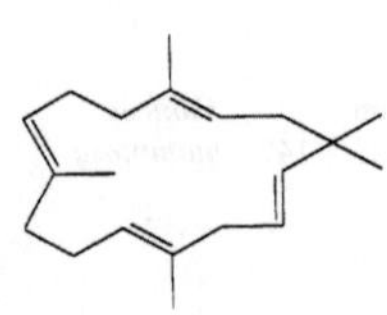 $C_{20}H_{32}$	—	*Sinularia flexibilis*	(*53*)

References, pp. 381—387

Table 5 (continued)

Name (Compound Number)	Structure (Molecular formula)	Physical Data	Source	Reference
10-Hydroxy-cembrene **(97)**	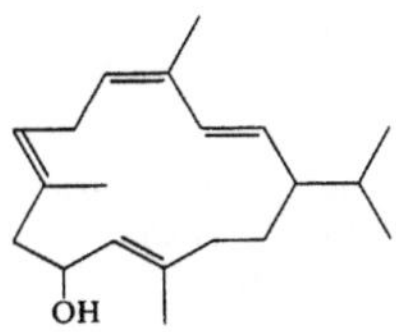 $C_{20}H_{32}O$	mp 143—145°	*Sarcophytum glaucum*	(*56*)
3,4,11,12-Diepoxy-cembrene-A **(105)**	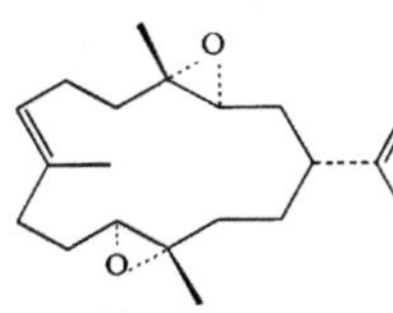 $C_{20}H_{32}O_2$	mp 66—68° {α}D + 63°	*Sinularia flexibilis*	(*52*)
Asperdiol **(112)** †	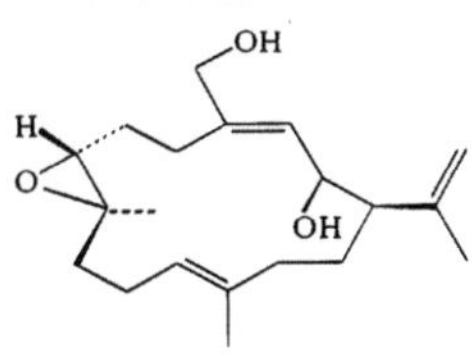 $C_{20}H_{32}O_3$	mp 109—110° {α}D − 87°	*Eunicea asperula* *Eunicea tourneforti*	(*57*) (*57*)
Dihydro-sinularin **(156)** †	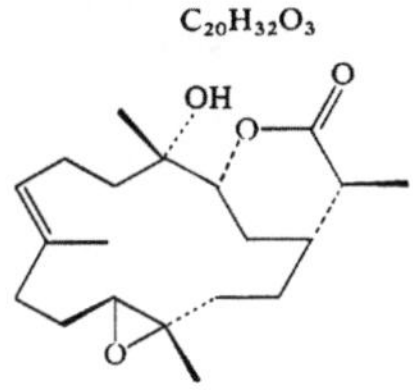 $C_{20}H_{32}O_4$	mp 110—112° {α}D − 45°	*Sinularia flexibilis*	(*12*)
Nephthenol **(95)**	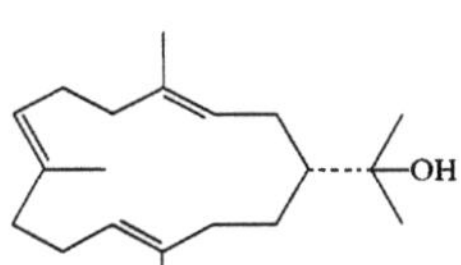 $C_{20}H_{34}O$	bp *ca.* 96° (0.03 mm) {α}D − 36°	*Nephthea sp.* *Lithophyton viridis*	(*54*) (*55*)

Table 5 (continued)

Name (Compound Number)	Structure (Molecular formula)	Physical Data	Source	Reference
2-Hydroxy-nephthenol (**98**)	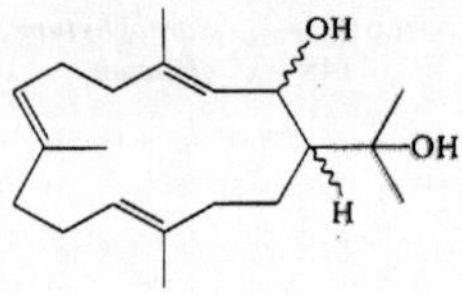 $C_{20}H_{34}O_2$	mp 98—99° {α}$_D$ − 104°	*Litophyton viridis*	(*55*)
Trochelio-phorol (**166**)	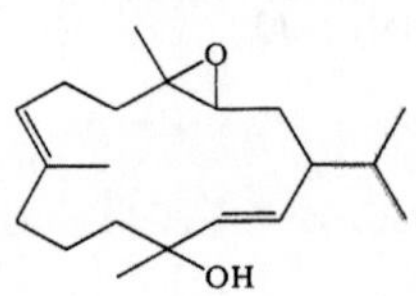 $C_{20}H_{34}O_2$	—	*Sarcophyton trocheliophorum*	(*79*)
2,7,11-cembra-triene-4,10-diol (**102**)	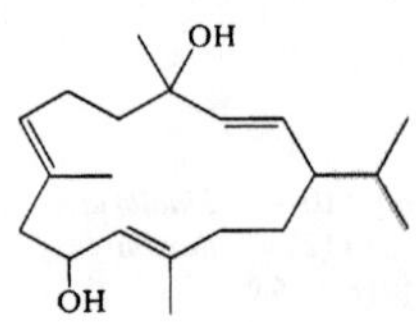 $C_{20}H_{34}O_2$	—	*Sarcophytum glaucum*	(*56*)
Pukalide (**145**)	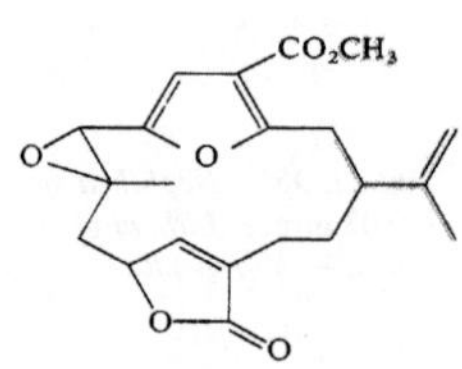 $C_{21}H_{24}O_6$	mp 204—206° {α}$_D$ + 44°	*Sinularia abrupta*	(*72*)

References, pp. 381—387

Table 5 (continued)

Name (Compound Number)	Structure (Molecular formula)	Physical Data	Source	Reference
Lobolide (**125**)	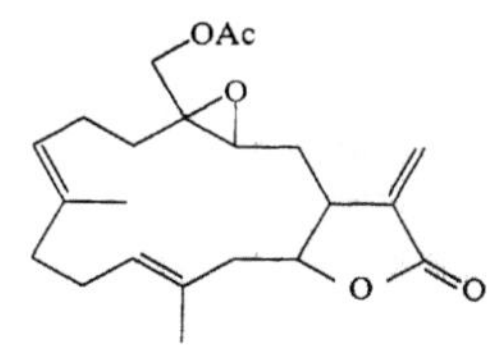 $C_{22}H_{32}O_5$	mp 114—115° $\{\alpha\}_D$ − 58°	*Lobophytum sp.*	*(64)*
Crassin Acetate (**147**) †	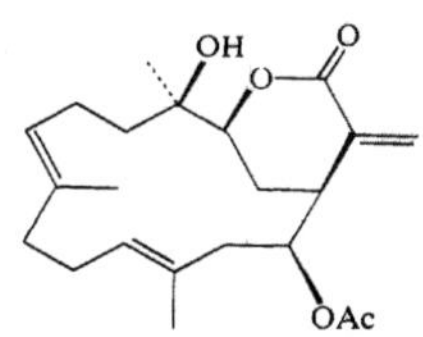 $C_{22}H_{32}O_5$	mp 138—140° $\{\alpha\}_D$ +70.4°	*Pseudoplexaura porosa* *Ps. wagenaari* *Ps. flagellosa* *Ps. crucis*	*(65, 73)* *(65, 73)* *(65, 73)* *(73)*
Eupalmerin Acetate (**120**) †	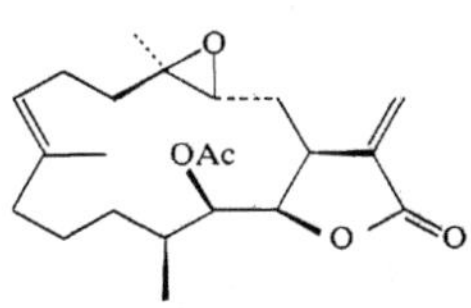 $C_{22}H_{32}O_5$	mp 157—159° $\{\alpha\}_D$ + 8°	*Eunicea palmeri*	*(61)*
Cueunicin acetate (**133**) †	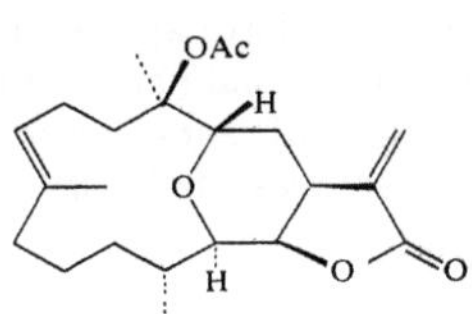 $C_{22}H_{32}O_5$	mp 141—142° $\{\alpha\}_D$ − 152°	*Eunicea mammosa*	*(68)*
11-Epi-sinulariolide acetate (**165**) †	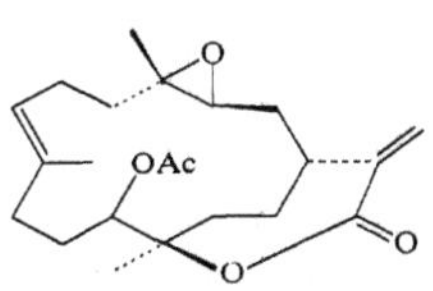 $C_{22}H_{32}O_5$	—	*Sinularia flexibilis*	*(78)*

Table 5 (continued)

Name (Compound Number)	Structure (Molecular formula	Physical Data	Source	Reference
Epoxy-nephthenol acetate (**103**)	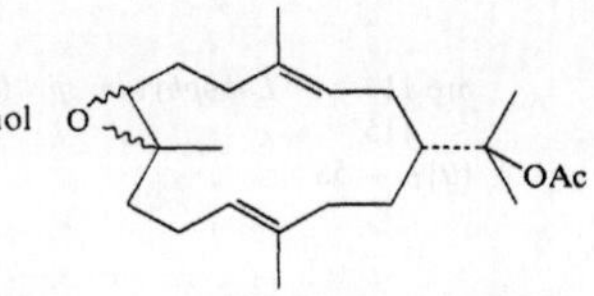 $C_{22}H_{36}O_3$	bp *ca.* 96° (0.03 mm) {α}$_D$ − 20.7°	*Nephthea sp.*	(*54*)
Crassolide (**169**)	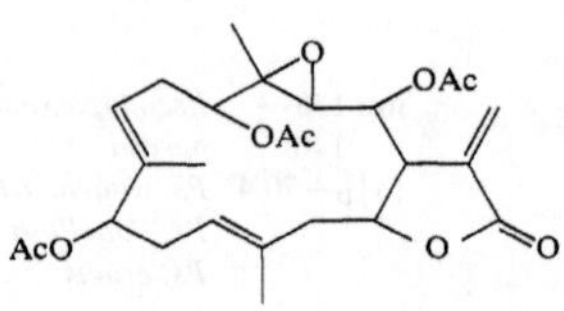 $C_{26}H_{34}O_9$	—	*Lobophytum crassum*	(*79*)

Addendum

In this section are reviewed articles which have appeared or become available to us subsequent to completion of the main body of this review. The topics include newly reported natural cembranes, the occurrence of previously reported cembranes in new sources, and further synthetic approaches to natural cembranes.

Natural Cembranes

3,15-Epoxycembra-7,11-dien-4-ol

JEFFERIES' group (*94*) at the University of Western Australia has isolated an epoxycembradienol (**201**), $C_{20}H_{34}O_2$, m. p. 114—5°, $[\alpha]_D$ −3.4°, from the resinous leaf coating of the plant *Eremophila georgei*. This compound is unusual in that both nuclear unsaturations have *cis*-geometry.

The absence of carbonyl absorption in (**201**), and the absence of hydroxyl absorption in its monoacetate, required that its second oxygen atom be ethereal. The presence of only one downfield methine proton indicated that the ether was of the secondary-tertiary class. A vicinal relationship between the tertiary methyl carbinol and the secondary ether terminus was demonstrated by the downfield shift of the newly allylic ether proton in both the *endo* (**202**) and *exo* forms of the dehydration product obtained from (**201**) with phosphoryl chloride and pyridine. Upon catalytic hydrogenation, each of the isomeric dehydration products sufferred partial hydrogenolysis to the oily saturated alcohol (**203**). The cembranoid skeleton of the latter was established by conversion to cembrane by a dehydration/hydrogenation sequence.

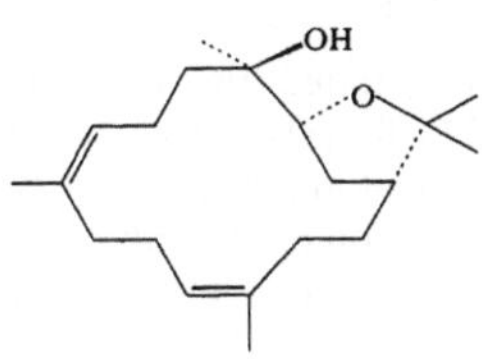

(**201**)

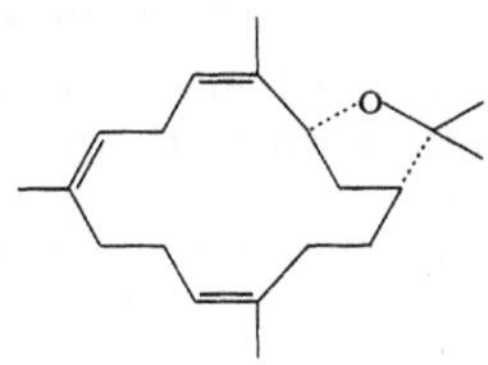

(**202**)

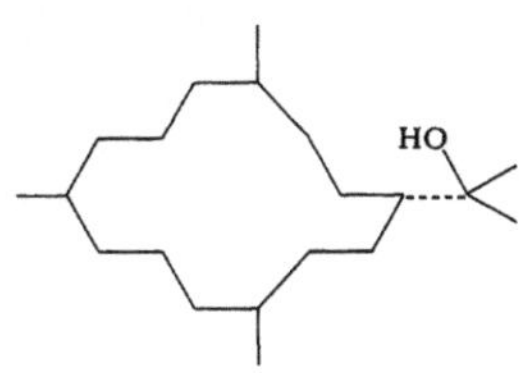

(**203**)

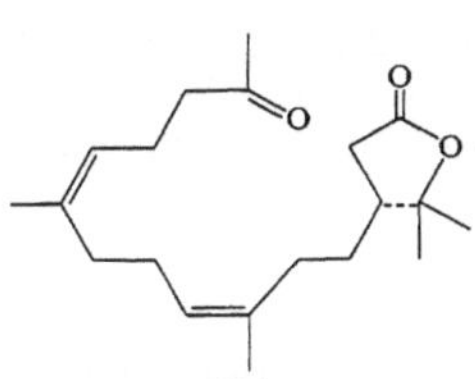

(**204**)

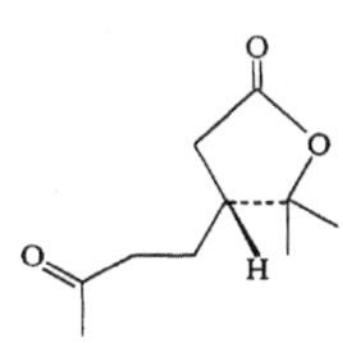

(**96**)

One of the two methyl substituted unsaturations was shown to be at position 7 by decoupling experiments involving the doubly allylic methylene group at position 6 in (**202**). The balance of the structure, and the absolute configuration at position 1, was secured by the formation of γ-lactone (**204**) from (**201**) on Jones oxidation, and the subsequent osmium tetroxide/periodate degradation of this lactone to form (−)-homoterpenyl methyl ketone (**96**). The authors concluded that the tetrahydrofuran ring was *cis*-fused to the carbocyclic ring because of the large coupling constants (11 and 13 Hz), indicative of anti-parallel geometry, between both H-1 and H-3, and one of the methylene protons at position 2. Their basis for assignment of *cis*-geometry to the two unsaturations was the relatively low field chemical shifts (δ 1.67) of the two vinyl methyl groups in the open chain lactone (**204**).

All of these assignments, as well as the configuration at the tertiary alcohol center, were confirmed by an X-ray crystallographic study by MASLEN *et al.* (*95*), which established the absolute configuration shown in (**201**) for the natural diterpene.

Isoovatodiolide

MANCHAND and BLOUNT (*96*) have isolated the new bis-lactone, isoovatodiolide (**205**), $C_{20}H_{24}O_4$, m. p. 127—8°, $[\alpha]_D$ −316°, which accompanies ovatodiolide (**79**) as a minor constituent in the leaves of *Anisomeles ovata*. It is epimeric with ovatodiolide at position 6. They also reported the occurrence of ovatodiolide in the leaves of *A. indica* from Trinidad.

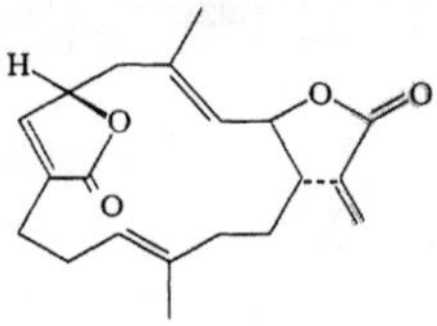

(**79**)

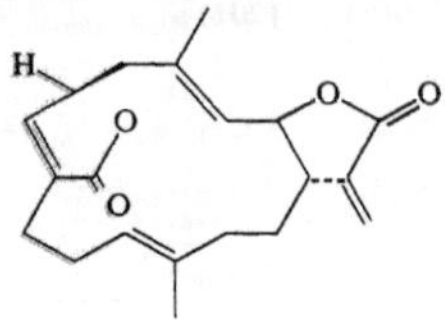

(**205**)

The structure and relative stereochemistry of both isoovatodiolide (**205**) and ovatodiolide (**79**) were determined by X-ray crystallography. Ovatodiolide was also shown to undergo facile acid catalyzed transannular ring closure to compounds possessing the carbobicyclic system of the gorgonian diterpene, eunicellin (**3**).

Cembrene-C

VANDERAH *et al.* (*97*) have reported the isolation of the optically inactive hydrocarbon, cembrene-C (**36**), $C_{20}H_{32}$, as an oil from a *Nephthea* soft coral from Enewetak. Previously, a compound of this structure had been encountered as an artifact in the fractionation of a pine oleoresin, and called pinacene (**36**). The new name cembrene-C was adopted because, like the name cembrene-A, it corresponded to the well established -A, -B, -C nomenclature for double bond isomers of the germacrenes, the sesquiterpene prenologs of the cembrenes.

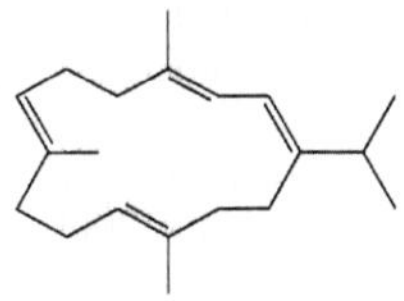

(**36**)

Cembrene-C occurred with cembrene-A in this soft coral, and was isolated by preparative glc. It is of interest that cembrene-C could not be recovered from column chromatography on silica gel, but was unaffected by Florisil. Unlike the hydrocarbon from other natural sources, both marine and terrestrial, the cembrene-A from this source was racemic.

The structure of cembrene-C was deduced from interpretation of its spectral characteristics and its conversion to cembrane by hydrogenation. The conjugated diene system was characterized by uv absorption at 252 nm, and by pmr absorptions for a pair of vicinal protons and a methyl group, all mutually coupled. The allylic chemical shift of the isopropyl methine proton required that the diene system be located as in structure (**36**). This assignment was confirmed by ozonolysis of cembrene-C to levulinic acid, 6-methyl-2,5-heptanedione and glyoxal. Although the geometry of the individual olefinic centers was not established, the failure of cembrene-C to react with several dienophiles indicated that the diene system was s-*trans*.

Nephthenol

Kashman *et al.* (*98*) have reported the isolation of nephthenol (**95**) from the two soft corals *Lithophyton arboreum* and *Lobophytum pauciflorum* from the Red Sea.

1,2-Dehydroepoxynephthenol

In their studies of Australian soft corals, Coll's group (*99*) at James Cook University in North Queensland has isolated the 1,2-dehydro derivative (**206**) of epoxynephthenol from a *Sarcophyton* species. This compound, $C_{20}H_{32}O_2$, m. p. 105—6°, $[\alpha]_D$ +61.8°, displayed hydroxyl and conjugated diene (252 nm) absorptions, as well as appropriate pmr and cmr absorptions for a methyl substituted oxirane and isolated olefin functions. The isopropylol grouping was regarded as allylic because of the downfield resonance of its two equivalent methyl groups (δ 1.36) compared to the corresponding signal (δ 1.20) in nephthenol (**95**). The two protons on the central atoms of the diene system appeared as an AB quartet, one half of which showed coupling to a low field (δ 1.78) methyl group.

Even on the assumption that the carbon skeleton was cembranoid, neither LIS nor decoupling experiments permitted the authors to distinguish between structure (**206**) and its positional isomer in which the oxirane and isolated olefin functions were transposed. The structure and relative configuration shown in (**206**) were finally determined by X-ray crystallography.

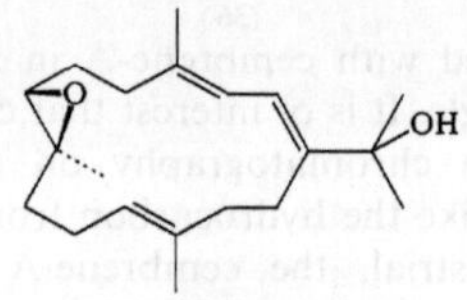

(**206**)

A Cembranolide

COLL's group (*100*) has isolated the simple, unsubstituted prototype of the γ-lactonic cembranolides from the Australian soft coral *Lobophytum michaelae*. The compound, $C_{20}H_{28}O_2$ (**207**), m. p. 101—2°, $[\alpha]_D$ +77.9°, was not given a trivial name.

The nature and size of the lactone ring was established by double resonance experiments involving the protons at the β- and γ-positions, and those of the α-methylenic group. Both the chemical shift (δ 5.35) of the γ-proton, and its apparent coupling with one of the three vinyl protons suggested it was allylic. The cembranoid character of the single carbocyclic ring, and the location of the methyl substituted unsaturations, as shown in (**207**), were deduced from the absence of pmr indications of a doubly allylic methylene group, and by the formation of two moles of levulinaldehyde upon ozonolysis.

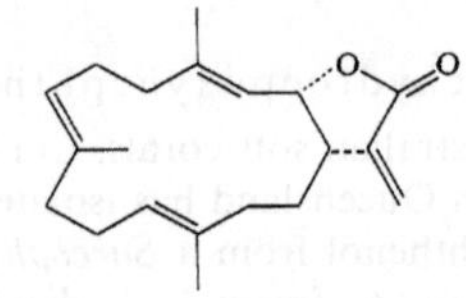

(**207**)

The relative configurations at positions 1 and 2 were based on the assignment of a *cis*-fusion of the lactone ring. This stereochemistry was deduced from the magnitudes of the vicinal coupling constant between the β- and γ-protons, and of the allylic coupling constants between the β-proton and the pair of α-methylene protons. These coupling constants in (**207**) more closely approximated those of the *cis*-fused cembranolide eunicin (**2**) than those of the *trans*-fused lobophytolide (**115**). The three nuclear unsaturations in (**207**) were assigned *trans* geometry because of the low field chemical shifts (*ca.* 15 ppm) of the three vinyl methyl carbon atoms.

11-Episinulariolide Acetate and 11-Dehydrosinulariolide

Kashman *et al.* (*98*) have isolated 11-episinulariolide acetate (**165**) and 11-dehydrosinulariolide (**164**) from *Sinularia notanda* from the Red Sea. The acetate was also present in minute quantities in *S. querciformis* from the same locale.

Sarcophine

Employing Uchida and Kuriyama's (*101*) newly proposed chirality rule for the Cotton effect of butenolides, Kashman (*102*) has been able to assign the absolute configuration at position 2 of sarcophine. On this basis, sarcophine possesses the absolute configuration shown in (**117′**), which is the enantiomer of (**117**).

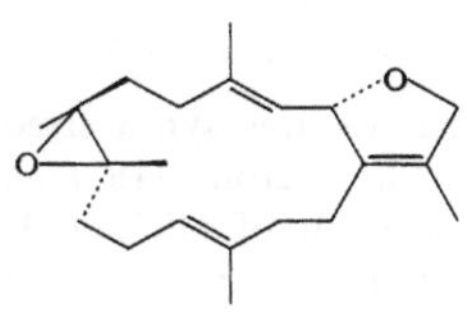

(**117′**)

Kashman *et al.* (*102*) have suggested names for two peculiar isomers of sarcophine. These two isomers, one oily, the other crystalline, were referred to earlier in this review as 2,8-bis-episarcophine (**119**) and 2-episarcophine (**118**), respectively. Kashman *et al.* now designate the oily isomer sarcophine-2, and the crystalline isomer sarcophine-3. It should be noted that the configurations shown in (**118**) and (**119**) may require modification in light of the above revision of the absolute configuration of sarcophine.

Crassolide

Some details of the Brussels group's structural studies of crassolide (**169**) were reported at a NATO conference by Braekman (*103*). The nmr spectra of crassolide showed the presence of three secondary acetates, an α-methylene γ-lactone, two methyl substituted double bonds and a methyl substituted oxirane. Decoupling experiments established the partial structure for that portion of the molecule extending from position 12 through position 1 to position 4. Ozonolysis experiments established the relative positions of the two nuclear double bonds.

Relevant to the location of the two remaining acetoxy residues was the fact that the triol obtained by potassium carbonate/methanol deacetylation of crassolide was not affected by periodate. The spectral characteristics of crassolide mono- and diepoxides permitted assignment of these acetoxy groups to positions 5 and 9. An interesting transposition of the γ-lactone and 2-acetoxy residues led to the formation of **(208)** during conversion of the monoepoxide to the diepoxide with *m*-chloroperbenzoic acid.

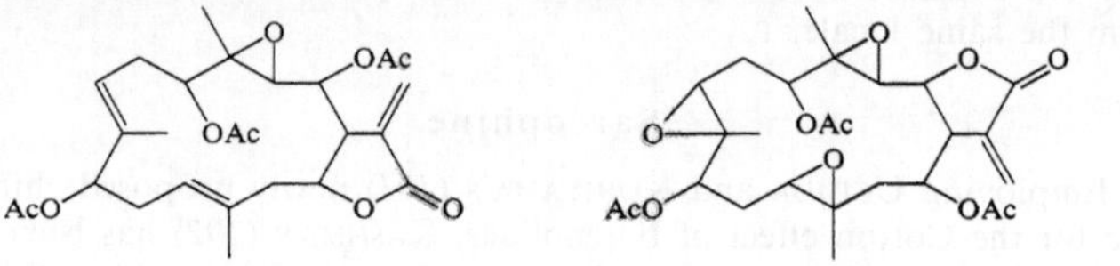

(169) **(208)**

Syntheses

Cembrene-A

KITAHARA *et al.* (*104*) have now synthesized (±)-cembrene-A **(200)** (neocembrene) from the chloroketone **(187)** employed earlier in their synthesis of mukulol. Treatment of **(187)** with LiBr/Li_2CO_3 in DMF gave a mixture of elimination products (86% yield) in which the non-conjugated isopropenyl isomer **(209)** predominated over the conjugated isopropylidene isomer in a 9 : 1 ratio. A 1 : 6 ratio was observed when LiCl was employed. Upon reduction of the ketone **(209)** with lithium aluminium hydride, the alcohol **(210)** and its epimer were obtained in 55 and 34% yields, respectively. Reduction of the acetate of **(210)** by lithium/ethylamine produced (±)-cembrene-A **(200)** in 51% yield, and reduction of the epimeric acetate gave (±)-cembrene-A in 35% yield. The synthetic (±)-cembrene-A possessed a level of termite trail-laying activity comparable with that of the natural trail pheromone. However, there remains a minor ambiguity concerning the identity of the natural pheromone since its rotation was not determined in the original investigation (*51*).

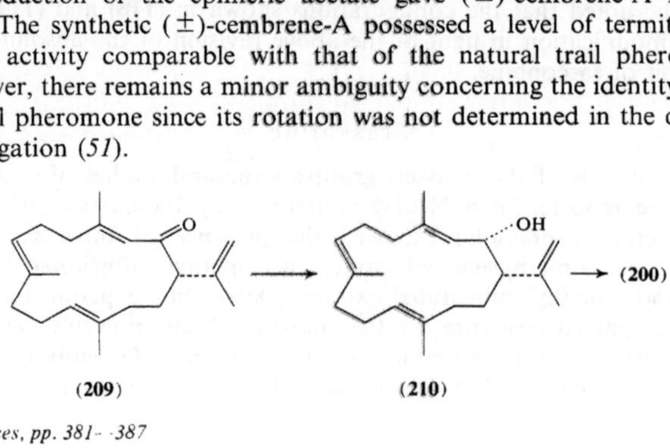

(209) **(210)**

Thunbergol and Isocembrol

Both thunbergol (cf. **30**) and isocembrol (cf. **31**) have recently been synthesized by stereospecific methods and appear to have precisely the same structure. However, direct comparison of the two compounds to establish their identity has not been reported.

RALDUGIN and PENTEGOVA, in a report (*105*) available to the reviewers only in abstract form, isolated isocembrol (**212**) from lithium aluminium hydride reduction of the cembrene epoxide (**211**). The epoxide apparently was employed as a mixture with the 7,8- and 11,12-monoepoxy derivatives of cembrene. The method of preparation of the epoxides was not indicated, nor was the basis for assignment of configuration to the 3,4-epoxide (**211**).

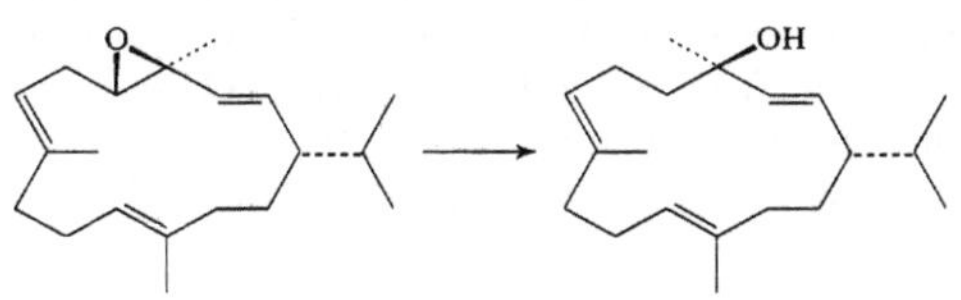

(**211**) (**212**)

Employing (±)-epi-mukulol (**190**) which they had synthesized previously, KATO *et al.* (*106*) have prepared (±) thunbergol (**212**) by a sequence which achieved the transposition of the elements of the allylic alcohol moiety. Both regio- and stereospecificity were achieved by epoxidation with t-butyl hydroperoxide in the presence of vanadyl acetylacetonate catalyst which preferentially introduced the 3,4-epoxide function *syn* to the 2-hydroxyl. The epoxy alcohol (**213**) produced in this manner was converted to (±) thunbergol (**212**) by reductive elimination of the corresponding epoxy mesylate with sodium in ammonia. The nmr spectrum of the synthetic thunbergol was completely superimposable on that of natural thunbergol. In a parallel sequence of reactions, these authors also converted (±)-mukulol to (±)-4-epi-thunbergol.

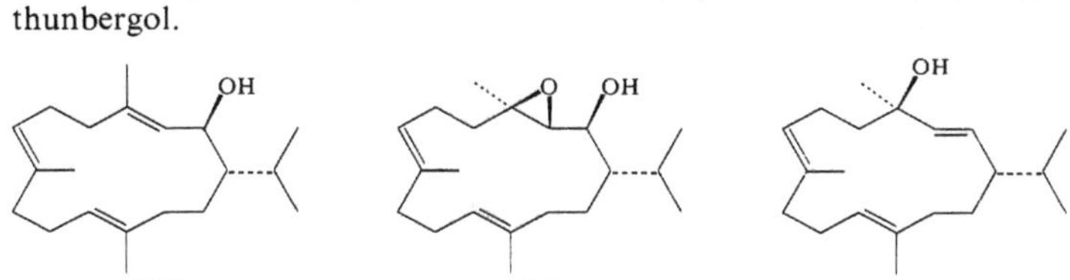

(**190**) (**213**) (**212**)

In their original investigation of thunbergol, KIMLAND and NORIN (*23*) had established *trans* geometry for its disubstituted double bond and had dehydrated thunbergol to cembrene, leaving only the configuration of the hydroxyl group in doubt. KATO *et al.* (*106*) suggest that this one remaining feature is definitely established by their synthesis. However, the reviewers feel that the assignment of the hydroxyl configuration must still be approached with care. This need is emphasized by the simple fact that despite strict fidelity in representing the stereochemical details of the transformation in each case, the thunbergol shown here (**212**) and that in the original paper have opposite configurations. In the paper by KATO *et al.*, (*106*), the compounds epimukulol, its *syn* epoxy derivative, and thunbergol are represented by the structures (**214**), (**215**) and (**216**). In order to re-orient the structure of thunbergol (**216**) for comparison with the style of structural notation used here, three operations are required. First, the hydroxyl group at position 4 must be re-oriented, as in (**217**). Next, the whole structure must be turned over, from top to bottom, giving (**218**). Finally, a partial turn in the counter-clockwise direction, and minor reorganization produces (**219**) as the equivalent of the thunbergol depicted by (**216**).

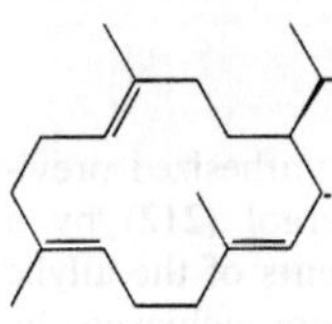

(**214**)

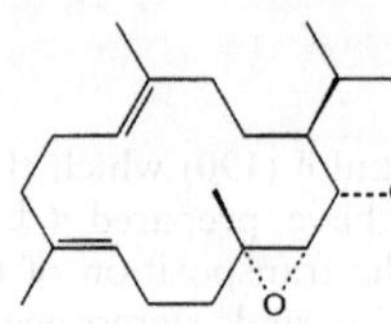

(**215**)

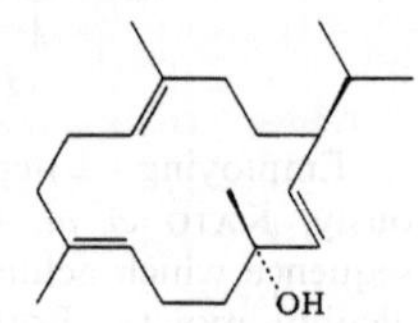

(**216**)

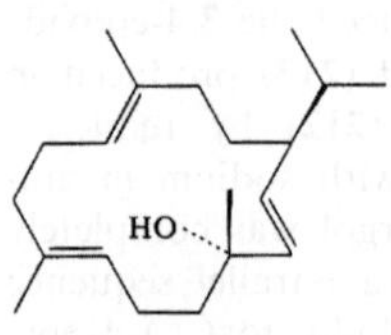

(**217**)

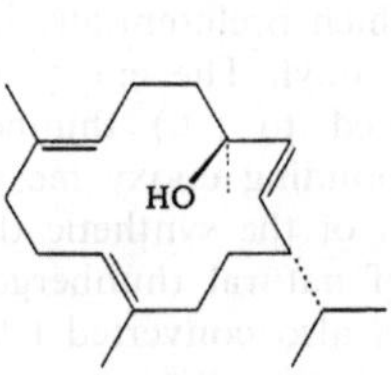

(**218**)

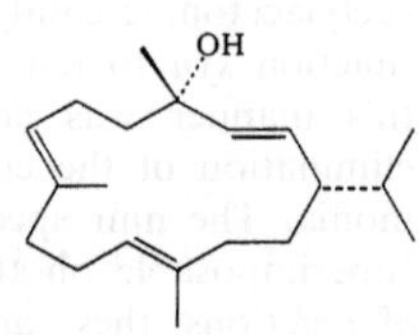

(**219**)

The discrepancy between (**212**) and (**219**) is due to conformational differences implicit in the two representations, (**190**) and (**214**), of epimukulol. The allylic alcohol portion of this molecule can adopt the two conformations (**220**) and (**221**) which differ only as rotamers about the 2,3-single bond. Each approximates the conformation required for the *syn* epoxidation reaction, but they differ in that the 2-hydroxyl group is *syn* to opposite faces of the 3,4-double bond. Consequently, opposite configurations would be established at position 4 in the epoxidation reaction, and these configurations would be retained in the alcohol formed in the final reductive elimination step. Structures (**190**) and (**214**) correspond with conformations (**220**) and (**221**), respectively, thus accounting for the differing configurations in the final alcohols.

Thus, assignment of the correct hydroxyl configuration for thunbergol must rest upon an interpretation of the conformational preference exhibited in the epoxidation reaction. It is obvious that the preference is extremely high, since only one of the two possible epoxyalcohols was observed experimentally. Dreiding models of epimukulol suggest that in conformation (**221**), the 4-methyl group can encounter significant transannular interactions which are absent in conformation (**220**). On the basis that the latter is the preferred conformation, we suggest that thunbergol is properly represented by structure (**212**).

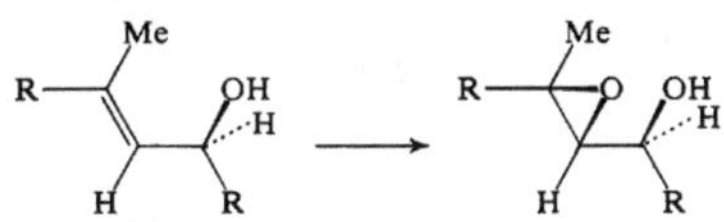

(**220**)

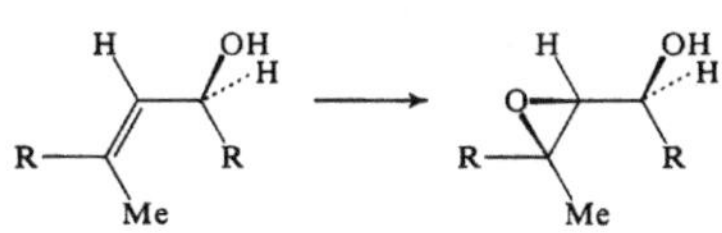

(**221**)

Isoincensole Oxide

In an extension of their work on the frankincense cembranes, KATO *et al.* (*107*) have reported the synthesis of (±)-isoincensole oxide (**224**) from (±)-incensole (**193′**). As in their elegant synthesis of the latter, the key transformation in the new study was the transannular bromo-etherification reaction using tetrabromocyclohexadienone. With this reagent, (±)-incensole was smoothly cyclized to a single stereoisomer of the bromo diether (**223**, X = Br). Replacement of the bromine using silver acetate gave predominantly the acetate (**223**, X = OAc) of retained configuration. Reduction of this acetate with LAH produced (±)-iso-incensole oxide (**224**), the solution spectra of which were identical to those of the natural cembranoid.

These authors also reported that, like mukulol, epi-mukulol (**190**) formed a bromotetrahydropyran (**222**). It is of interest that, upon BF_3 isomerization, it formed the *same* bromotetrahydrofuran (**192′**) that had been obtained from the mukulol-derived bromo-THP (**191′**).

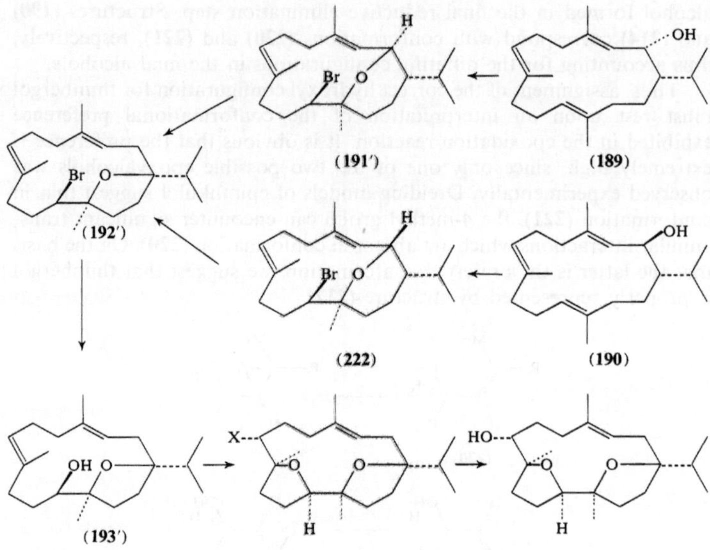

(191′) (189) (192′) (222) (190) (193′) (223) (224)

This observation supports the authors' earlier suggestion that the THP-THF isomerization occurs by an elimination-addition sequence involving the conjugated 1,3-diene intermediate. It also demonstrates that transannular attack by the 2-hydroxyl occurs at the same face of the 11,12-olefin in both mukulol and epi-mukulol, irrespective of the configuration of the hydroxyl group.

Kato *et al.* (*107*) considered the stereochemical features of the intermediates, and their implications for the relative configurations of incensole and isoincensole oxide. Despite the opportunity for double bond isomerization during the BF_3 promoted transformation, the absence of the relevant nuclear Overhauser effect confirmed that the Δ^3 olefin of THP (**191′**) remained *trans* in THF (**192′**), and also in incensole (**193′**).

The ether rings in both the THP (**191′**) and the THF (**192′**) were assigned *cis* geometry, the former because of the absence of certain transannular nuclear Overhauser effects expected in the *trans* isomer, the latter because models convincingly demonstrate that unless this ring is *cis* in incensole, it is sterically impossible to form the second THF ring of isoincensole oxide.

The stereochemical assignment of incensole (**193′**) was completed by a consideration of the configuration of the 11-hydroxyl. This was deduced from the presumed *trans* addition to the Δ^{11}-double bond during cyclization to the THP (**191′**), and the retention of configuration observed during replacement of the bromine by acetate.

The stereochemistry of isoincensole oxide (**224**) was similarly assigned on the presumption of *trans* addition in the formation of the second THF ring, and retention of configuration observed during replacement of the bromine atom. However, no evidence was presented concerning the configuration at position 8 relative to that at position 11. During cyclization, oxygen could add to either face of the Δ^{7}-double bond, leading either to the *cis*-THF (**223**) or the *trans*-THF epimeric to (**223**) at positions 7 and 8. Models suggest that less serious transannular steric interactions occur in the process leading to the cis isomer.

Related Cembrane Derivatives

RALDUGIN *et al.* (*108*) have studied the chromic acid oxidation of cembrene (*1*). Reaction occurred at the lone *cis* double bond, forming two epimeric 4,5-diols, one *cis*, the other *trans*.

A number of patents have issued to the two Tohoku groups (*109, 110, 111, 112, 113, 114*) which cover the syntheses of some of the natural cembranes described above, as well as a number of related compounds.

References

1. DAUBEN, W. G., W. E. THIESSEN, and P. R. RESNICK: Cembrene, A 14-Membered Ring Diterpene Hydrocarbon. J. Amer. Chem. Soc. **84,** 2015 (1962).

2. KOBAYASHI, H., and S. AKIYOSHI: Thunbergene, a Macrocyclic Diterpene. Bull. Chem. Soc. Japan **35,** 1044 (1962).

3. ROBERTS, D. L., and R. L. ROWLAND: Macrocyclic Diterpenes. α- and β-4,8,13-Duvatriene-1,3-diols from Tobacco. J. Organ. Chem. (U.S.A.) **27,** 3989 (1962).

4. WIENHAUS, H.: New Compounds in Turpentine Oil. Nord. Kemistmotet **1926,** 211 (1928). [Chem. Abstr. **24,** 1635 (1930).]

5. SEBE, Y.: Nippon Kogsku Kaishi **56,** 118 (1935).

6. AKIYOSHI, S.: Turpentine Oils of Japan and Her Neighboring Countries. Repts. Imp. Ind. Research Inst., Osaka, Japan **17,** No. 10, 1 (1937). [Chem. Abstr. **31,** 8172 (1937).]

7. HAAGEN-SMIT, A. J., T. H. WANG, and N. T. MIROV: Composition of Gum Turpentines of Pines. XIII. A Report on *Pinus albicaulis*. J. Amer. Pharm. Assoc. Sci. Ed. **40,** 557 (1951).

8. MIROV, N. T., and P. M. ILLOF: Composition of Gum Turpentines of Pines. XXIV. A Report on Two Asiatic Pines: *Pinus armandi* and *Pinus bungeana*. J. Am. Pharm. Assoc. **44,** 424 (1955).

9. ILLOF, P. M., and N. T. MIROV: Composition of Gum Turpentines of Pines. XXV. A Report on Two Pines: *Pinus koraiensis* from Korea and *P. peuce* from Macedonia. J. Am. Pharm. Assoc. **45,** 77 (1956).

10. MIROV, N. T.: Composition of Gum Turpentines of Pines. Technical Bulletin No. 1239. U. S. Dept. of Agriculture, 1961.
11. DAUBEN, W. G., W. E. THIESSEN, and P. R. RESNICK: Cembrene, A Fourteen-Membered Ring Diterpene Hydrocarbon. J. Organ. Chem. (U.S.A.) **30,** 1693 (1965).
12. WEINHEIMER, A. J., J. A. MATSON, M. B. HOSSAIN, and D. VAN DER HELM: Marine Anticancer Agents. Sinularin and Dihydrosinularin, New Cembranolides from the Soft Coral, *Sinularia flexibilis*. Tetrahedron Letters **1977,** 2973.
13. KOBAYASHI, H., and S. AKIYOSHI: Terpenoids. VI. The Structure of Thunbergene. Bull. Chem. Soc. Japan **36,** 823 (1963).
14. DREW, M. G. B., D. H. TEMPLETON, and A. ZALKIN: The Crystal and Molecular Structure of Cembrene. Acta Crystallogr. **B25,** 261 (1969).
15. PATIL, V. D., U. R. NAYAK, and S. DEV: Chemistry of *Ayurvedic* Crude Drugs — II. *Guggulu* (Resin from *Commiphora Mukul*)-2: Diterpenoid Constituents. Tetrahedron **29,** 341 (1973).
16. SHMIDT, E. N., N. K. KASHTANOVA, and V. A. PENTEGOVA: Neocembrene, a New Diterpenic Hydrocarbon from *Picea obovata* and *Pinus koraiensis*. Khim. Prir. Soedin. **6,** 694 (1970).
17. KASHTANOVA, N. K., A. I. LISINA, and V. A. PENTEGOVA: The New Diterpenes, Isocembrene and Isocembrol, from Oleoresin of *Pinus sibirica*. Khim. Prir. Soedin. **4,** 52 (1968).
18. PENTEGOVA, V. A., and N. K. KASHTANOVA: Diterpene hydrocarbons of the resin of *Pinus sibirica*. R. MAYR. Khim. Prir. Soedin. **1,** 223 (1965).
19. RÜCKER, G.: Über monocyclische Diterpene aus dem indischen Guggul-Harz *(Commiphora mukul)*. Arch. Pharm. **305,** 486 (1972).
20. RALDUGIN, V. A., O. B. SHELEPINA, I. P. SEKATSIS, A. I. REZVUKHIN, and V. A. PENTEGOVA: Configuration of the double bond and partial synthesis of allylcembrol. Khim. Prir. Soedin. **1976,** 108.
21. PRADSAD, R. S., and S. DEV: Chemistry of Ayurvedic Crude Drugs — IV. *Guggulu* (Resin from *Commiphora Mukul*)-4. Absolute Stereochemistry of Mukulol. Tetrahedron **32,** 1437 (1976).
22. ERDTMAN, H., B. KIMLAND, T. NORIN, and P. J. L. DANIELS: The Constituents of The "Pocket Resin" from Douglas Fir *Pseudotsuga menziesii* (Mirb.) Franco. Acta Chem. Scand. **22,** 938 (1968).
23. KIMLAND, B., and T. NORIN: Thunbergol, a New Macrocyclic Diterpene Alcohol. Acta Chem. Scand. **22,** 943 (1968).
24. KASHTANOVA, N. K., A. I. LISINA, and V. A. PENTEGOVA: New diterpenes, isocembrene and isocembrol, in the oleoresin of *Pinus sibirica*. Khim. Prir. Soedin. **4,** 52 (1968).
25. KASHTANOVA, N. K., A. I. LISINA, and V. A. PENTEGOVA: Composition of the neutral fraction from the Oleoresin *Pinus sibirica*. Khim. Prir. Soedin. **5,** 10 (1969).
26. RALDUGIN, V. A., and V. A. PENTEGOVA: New Diterpenoid components of the Oleoresin of *Pinus koraiensis*. Khim. Prir. Soedin. **1976,** 174.
27. SHMIDT, E. N., and V. A. PENTEGOVA: Chemical Composition of *Picea* Oleoresins. Diterpenoid from *Picea obovata* Oleoresin. Khim. Prir. Soedin. **6,** 769 (1970).
28. RALDUGIN, V. A., and V. A. PENTEGOVA: 4-Epiisocembrol, A New Diterpenoid from Oleoresins of *Pinus koraiensis* and *Pinus sibirica*. Khim. Prir. Soedin. **7,** 669 (1971). Chem. Abstr. **76,** 83509 r (1972).
29. LISINA, A. I., A. I. REZVUKHIN, and V. A. PENTEGOVA: Composition of the Neutral Part of the Resin *Pinus sibirica*. II. Oxygen Containing Compounds of the High Boiling Neutral Part of the Cedar Resin. Khim. Prir. Soedin. **1,** 250 (1965).
30. RALDUGIN, V. A., N. K. KASHTANOVA, and V. A. PENTEGOVA: α-, β-, and γ-Pinacenes of *Pinus koraiensis* and *P. sibirica*. Khim. Prir. Soedin. **7,** 604 (1971).

31. Corsano, S., and R. Nicoletti: The Structure of Incensole. Tetrahedron **23,** 1977 (1967).

32. Nicoletti, R., and M. L. Forcellese: The Structure of Incensole-Oxide. Tetrahedron **24,** 6519 (1968).

33. Forcellese, M. L., R. Nicoletti, and U. Petrossi: The Structure of Isoincensole-oxide. Tetrahedron **28,** 325 (1972).

34. Forcellese, M. L., R. Nicoletti, and C. Santarelli: The Revised Structure of Isoincensole-Oxide. Tetrahedron Letters **1973,** 3783.

35. Aasen, A. J., N. Junker, C. R. Enzell, J. Berg, and A. Pilotti: Tobacco Chemistry 36. Absolute Configuration of Tobacco Thunberganoids. Tetrahedron Letters **1976,** 2607.

36. Rowland, R. L., A. Rodgman, J. N. Schumacher, D. L. Roberts, L. C. Cook, and W. E. Walker, jr.: Macrocyclic Diterpene Hydroxy Ethers from Tobacco and Cigarette Smoke. J. Organ. Chem. (U.S.A.) **29,** 16 (1964).

37. Aasen, A. J., A. Pilotti, C. R. Enzell, J. Berg, and A. M. Pilotti: Tobacco Chemistry 31. (1*S*, 4*S*, 8*R*, 11*S*, 12*R*)-8,12-Epoxy-2E, 6E-thunbergadiene-4,11-diol, a New Constituent of Greek Tobacco. Acta Chem. Scand. **B 30,** 999 (1976).

38. Zane, A.: 4,8,13-Duvatriene-1-ol-3-one and 11-Isopropyl-4,8-dimethyl-3,7,12-pentadecatriene-2,14-dione Isomers from *Nicotiana tabacum.* Phytochemistry **12,** 731 (1973).

39. Reid, W. W.: The Action of Inhibitors on the Incorporation of $\{2^{14}C\}$-Mevalonate into the Triterpenes and Sterols of *Nicotiana tabacum.* Biochem. J. **100,** P13 (1966).

40. Chang, S. Y., and C. Grunwald: Duvatrienediol, Alkanes, and Fatty Acids in Cuticular Wax of Tobacco Leaves of Various Physiological Maturity. Phytochemistry **15,** 961 (1976).

41. Chang, S. Y., and C. Grunwald: Duvatrienediols in Cuticular Wax of Burley Tobacco Leaves, J. Lipid Res. **17,** 7 (1976).

42. Courtney, J. L., and S. McDonald: A New C20 α,β-Unsaturated Aldehyde (3,7,13-trimethyl-10-isopropyl-2,6,11,13-tetradecatetraen-1-al) (I) from Tobacco. Tetrahedron Letters **1967,** 459.

43. Springer, J. P., J. Clardy, R. H. Cox, H. G. Cutler, and R. J. Cole: The Structure of a New Type of Plant Growth Inhibitor Extracted from Immature Tobacco Leaves. Tetrahedron Letters **1975,** 2737.

44. Roberts, D. L., and W. A. Rhode: Isolation and Identification of Flavor Components of Burley Tobacco. Tobacco Science **16,** 107 (1972).

45. Rowland, R. L., and D. L. Roberts: Macrocyclic Diterpenes Isolated from Tobacco. α- and β-3,8,13-Duvatriene-1,5-diols. J. Organ. Chem. (U.S.A.) **28,** 1165 (1963).

46. An, H. D., R. Toubiana, and E. Lederer: Sur la Structure de l'Ovatodiolide, Nouveau Diterpene Isolé d'Anisomeles Ovata. Bull. soc. chim. France **1963,** 1192.

47. Immer, H., J. Polonsky, R. Toubiana, and H. D. An: Structure de l'Ovatodiolide Diterpene Macrocyclique Isolé d'Anisomeles Ovata. Tetrahedron **21,** 2117 (1965).

48. Toubiana, R., M.-J. Toubiana, A. T. McPhail, R. W. Miller, and K. Tori: Structures and Conformations of the Fourteen-Membered Ring Diterpene Ovatodiolide and Its Acid Cyclization Product: Nuclear Overhauser Effect Studies in Solution and X-ray Crystal Structure Analyses of Ovatodiolide and Ovatodiolic acid. J. Chem. Soc., Perkin II **1976,** 1881.

49. Purushothaman, K. K., R. B. Rao, and K. Kalyani: Ovatodiolide and Anisomelic Acid, Two Diterpenoid Lactones from *Anisomeles malabarica* R. Brit. Indian J. Chem. **13,** 1357 (1975).

50. Moore, B. P.: Isolation of the Scent-Trail Pheromone of an Australian Termite. Nature **211,** 746 (1966).

51. Birch, A. J., M. W. V. Brown, J. E. T. Corrie, and B. P. Moore: Neocembrene-A, a Termite Trail Pheromone. J. Chem. Soc. Perkin I. **1972,** 2654.

52. HÉRIN, M., and B. TURSCH: Chemical Studies of Marine Invertebrates. XXIV. Minor Cembrane Diterpenes from the Soft Coral *Sinularia flexibilis (Coelenterata, Octocorallia)*. Bull. soc. chim. Belges **85,** 707 (1976).
53. HÉRIN, M., M. COLIN, and B. TURSCH: Chemical Studies of Marine Invertebrates. XXV. Flexibilene, an Unprecedented Fifteen-membered Ring Diterpene Hydrocarbon from the Soft Coral *Sinularia flexibilis (Coelenterata, Octocorallia, Alcyonacea)*. Bull. soc. chim. Belges **85,** 801 (1976).
54. SCHMITZ, F. J., D. J. VANDERAH, and L. S. CIERESZKO: Marine Natural Products: Nephthenol and Epoxynephthenol Acetate, Cembrene Derivatives from a Soft Coral. J. Chem. Soc. Chem. Commun. **1974,** 407.
55. TURSCH, B., J. C. BRAEKMAN, and D. DALOZE: Chemical Studies of Marine Invertebrates — XIII. 2-Hydroxynephthenol, a Novel Cembrane Diterpene from the Soft Coral *Litophyton viridis (Coelenterata, Octocorallia, Alcyonacea)*. Bull. soc. chim. Belges **84,** 167 (1975).
56. KASHMAN, Y., E. ZADOCK, and I. NEEMAN: Some New Cembrane Derivatives of Marine Origin. Tetrahedron **30,** 3615 (1974).
57. WEINHEIMER, A. J., J. A. MATSON, D. VAN DER HELM, and M. POLING: Marine Anticancer Agents: Asperdiol, a Cembranoid from the Gorgonians. *Eunicea asperula* and *E. tourneforti*. Tetrahedron Letters **1977,** 1295.
58. TURSCH, B., J. C. BRAEKMAN, D. DALOZE, M. HÉRIN, and R. KARLSON: Chemical Studies of Marine Invertebrates. X. Lobophytolide, a New Cembrenolide Diterpene from the Soft Coral *Lobophytum cristagalli (Coelenterata, Octocorallia, Alcyonacea)*. Tetrahedron Letters **1974,** 3769.
59. KARLSSON, R.: Lobophytolide, a Cembranolide Diterpene. Acta Crystallogr. **B33,** 2032 (1977).
60. BERNSTEIN, J., U. SHMEULI, E. ZADOCK, Y. KASHMAN, and I. NEEMAN: Sarcophine, a New Epoxy Cembranolide from Marine Origin. Tetrahedron **30,** 2817 (1974).
61. REHM, S. J.: I. Eupalmerin Acetate: A New Marine Epoxy Cembranolide. II. *Eunicea palmeri* Bayer: An Ambivalent Species. Dissertation, University of Oklahoma, U.S.A., **1971.**
62. EALICK, S. E., D. VAN DER HELM, and A. J. WEINHEIMER: The Molecular Structures and Absolute Configurations of Eupalmarin Acetate and Eupalmerin Acetate Dibromide at Low Temperature. Acta Crystallogr. **B31,** 1618 (1975).
63. HELM, D. VAN DER, S. E. EALICK, and A. J. WEINHEIMER: 15(*R*)-Acetoxy-6(*S*),10(*S*)-dibromo-3a(*S*), 4,7,8,11,12,14,14(*S*), 15,15a(*R*)-decahydro-6,10,14-trimethyl-3-methylene-5(*R*), 9(*R*)-epoxycyclotetradeca{b}-2-furanone, $C_{22}H_{32}Br_2O_5$. Cryst. Struct. Commun. **3,** 167 (1974).
64. KASHMAN, Y., and A. GROWEISS: Lobolide: A New Epoxy Cembranolide from Marine Origin. Tetrahedron Letters **1977,** 1159.
65. CIERESZKO, L. S., D. S. SIFFORD, and A. J. WEINHEIMER: Chemistry of Coelenterates. I. Occurrence of Terpenoid Compounds in Gorgonians. Ann. New York Acad. Sci. **90,** 917 (1960).
66. WEINHEIMER, A. J., R. E. MIDDLEBROOK, J. O. BLEDSOLE, JR., W. E. MARSICO, and T. K. B. KARNS: Eunicin, an Oxa-bridged Cembranolide of Marine Origin. Chem. Commun. **1968,** 384.
67. HOSSAIN, M. B., A. F. NICHOLAS, and D. VAN DER HELM: The Molecular Structure of Eunicin Iodoacetate. Chem. Commun. **1968,** 385.
68. GROSS, R. A., JR.: I. Cueunicin and Cueunicin Acetate — Two New Marine Cembranolides. II. Structure and PMR Spectra of Natural Products. The Association of High Field Vinyl Methyl Signals with *Trans* Double Bond Geometry in Germacrene Derivatives. III. New Marine Diterpenoids. Dissertation, University of Oklahoma, U.S.A., **1974.**

69. Helm, D., van der, E. L. Enwall, A. J. Weinheimer, T. K. B. Karns, and L. S. Ciereszko: p-Iodobenzoate of Jeunicin. Acta Crystallogr. **B 32,** 1558 (1976).
70. Weinheimer, A. J., J. A. Matson, D. van der Helm, and M. Poling: Unpublished results.
71. Chang, C. Y.: Dissertation, University of Oklahoma, U.S.A., 1977 (with L. S. Ciereszko).
72. Missakian, M. G., B. J. Burreson, and P. J. Scheuer: Pukalide, a Furanocembranolide from the Soft Coral *Sinularia abrupta.* Tetrahedron **31,** 2513 (1975).
73. Weinheimer, A. J., and J. A. Matson: Crassin Acetate, the Principal Antineoplastic Agent in four Gorgonians of the *Pseudoplexaura* Genus. Lloydia **38,** 378 (1975).
74. Weinheimer, A. J., T. K. B. Karns, D. H. Sifford, and L. S. Ciereszko: Unpublished results.
75. Hossain, M. B., and D. van der Helm: The Crystal Structure of Crassin p-Iodobenzoate. Rec. trav. chim. Pays-Bas **88,** 1413 (1969).
76. Tursch, B., J. C. Braekman, D. Daloze, M. Hérin, R. Karlsson, and D. Losman: Chemical Studies of Marine Invertebrates — XI. Sinulariolide, a New Cembranolide Diterpene from the Soft Coral *Sinularia flexibilis (Coelenterata, Octocorallia, Alcyonacea).* Tetrahedron **31,** 129 (1975).
77. Karlsson, R.: The Structure and Absolute Configuration of Sinulariolide, a Cembranolide Diterpene. Acta Crystallogr. **B 33,** 2027 1977).
78. Weinheimer, A. J., J. A. Matson, M. B. Hossain, and D. van der Helm: Study in Progress.
79. Tursch, B.: Some Recent Developments in the Chemistry of Alcyonaceans. Pure and Appl. Chem. **48,** 1 (1976).
80. Entwistle, I. D., and R. A. W. Johnstone: Synthesis of 1,5,9-Trimethyl-12-isopropylcyclotetradecane (Cembrane). Chem. Commun. (London) **1966,** 136.
81. Dauben, W. G., G. H. Beasley, M. D. Broadhurst, B. Muller, D. J. Peppard, P. Pesnelle, and C. Suter: A Synthesis of Cembrene. A 14-Membered Ring Diterpene. J. Amer. Chem. Soc. **96,** 4724 (1974).
82. Dauben, W. G., G. H. Beasley, M. D. Broadhurst, B. Muller, D. J. Peppard, P. Pesnelle, and C. Suter: A Synthesis of (±)-Cembrene, a Fourteen-Membered Ring Diterpene. J. Amer. Chem. Soc. **97,** 4973 (1975).
83. Kato, T., T. Kobayashi, and Y. Kitahara: Cyclization of Polyenes XVI. Biogenetic Type Synthesis of Cembrene Type Compounds. Tetrahedron Letters **1975,** 3299.
84. Kato, T., T. Kobayashi, T. Kumagai, and Y. Kitahara: Cyclization of Polyenes. XVIII. Selective Synthesis of Cembrene. Synthetic Commun. **6,** 365 (1976).
85. Kato, T., C. C. Yen, T. Kobayashi, and Y. Kitahara: Cyclization of Polyenes. XXI. Synthesis of DL-incensole. Chem. Letter **1976,** 1191.
86. Kodama, M., Y. Matsuki, and S. Itô: Syntheses of Macrocyclic Terpenoids by Intramolecular Cyclization 1. (±)-Cembrene-A, a Termite Trail Pheromone, and (±)-Nephthenol. Tetrahedron Letters **1975,** 3065.
87. Rice, J. R., C. Papastephanou, and D. G. Anderson: Isolation, Localization and Biosynthesis of Crassin Acetate in *Pseudoplexaura porosa* (Houttuyn). Biol. Bull. **138,** 334 (1970).
88. Anderson, D. G.: Personal Communication.
89. Shibanova, G. I., and V. A. Pentegova: Extractives from *Picea obovata* bark. Khim. Prir. Soedin. **7,** 214 (1971).
90. Shmidt, E. N., A. I. Lisina, and V. A. Pentegova: The Neutral Part of the Oleoresin of Siberian Larch *(Larix sibirica).* Sintetich. Produkty iz Kanifoli Skipidara, Akad. Nauk Belorussk. SSR, Tsentr. Nauchn. Issled. i Proekktn. Inst. Lesokhim. Prom., Tr. Vses. Nauchn. Tekhn. Soveshch., Gorki **1963,** 326. Chem. Abst. **62,** 14732f (1965).

91. Shmidt, E. N., L. E. Chupakhina, and V. A. Pentegova: Diterpenoids of the Oleoresins of Three Species of the Genus *Larix: Larix sibirica, Larix sukaczawii,* and *Larix czekanovskii.* Izv. Sib. Otd. Akad. Nauk SSSR, Ser. Khim. Nauk **1975,** 173.
92. The Wealth of India (Raw Materials), Vol. II, pp. 313—315. Council of Sci. and Indust. Research, Delhi (1950).
93. Fujise, Y., I. Maruta, S. Ito, and T. Nozoe: Chemical Constituents of Essential Oil of *Chamaecyparis obtusa* (Sieb. et Zucc.). Chem. Pharm. Bull. (Tokyo) **12,** 991 (1964).
94. Ghisalberti, E. L., P. R. Jefferies, J. R. Knox, and P. N. Shepard: The Chemistry of *Eremophila* sp. — VII. An Epoxycembradienol from *Eremophila georgei.* Tetrahedron **33,** 3301 (1977).
95. Maslen, E. N., C. L. Raston, and A. H. White: Crystal Structure of an Epoxycembradienol, 3,15-Epoxy-4-hydroxycembra-7(**Z**),11(*Z*)-diene. Tetrahedron **33,** 3305 (1977).
96. Manchand, P. S., and J. F. Blount: Stereostructures of the Macrocyclic Diterpenoids Ovatodiolide and Isoovatodiolide. J. Organ. Chem. (U.S.A.) **42,** 3824 (1977).
97. Vanderah, D. J., N. Rutledge, F. J. Schmitz, and L. S. Ciereszko: Marine Natural Products: Cembrene-A and Cembrene-C from a Soft Coral, *Nephthea sp.* J. Organ. Chem. **43,** 1614 (1978).
98. Kashman, Y., M. Bodner, Y. Loya, and Y. Benayaha: Cembranolids from Marine Origin (Red Sea), Survey, and Isolation of a New Sinulariolide Derivative. Israel Journal of Chemistry **16,** 1 (1977).
99. Coll, J. C., G. B. Hawes, N. Liyanage, W. Oberhänsli, and R. J. Wells: Studies of Australian Soft Corals. I. A New Cembrenoid Diterpene from a *Sarcophyton* Species. Aust. J. Chem. **30,** 1305 (1977).
100. Coll, J. C., S. J. Mitchell, and G. J. Stokie: Studies of Australian Soft Corals. II. A Novel Cembrenoid Diterpene from *Lobophytum michaelae.* Austral. J. Chem. **30,** 1859 (1977).
101. Uchida, L., and K. Kuriyama: The π-π* Circular Dichroism of α,β-Unsaturated γ-Lactones. Tetrahedron Letters 3761 (1974).
102. Kashman, Y.: Cembrane Derivatives from *Sarcophytum glaucum.* In: Marine Natural Products Chemistry, pp. 17—21. (Faulkner, D. J., and W. H. Fenical, Eds.) New York: Plenum Press. 1977.
103. Braekman, J. C.: Recent Developments in Terpenoid and Steroid Chemistry of Alcyonacea. In: Marine Natural Products Chemistry, pp. 5—15. (Faulkner, D. J., and W. H. Fenical, Eds.) New York: Plenum Press. 1977.
104. Kitahara, Y., T. Kato, and T. Kobayashi: Cyclization of Polyenes XVII. Synthesis and Pheromone Activity of DL-Neocembrene. Chemistry Letters 219 (1976).
105. Raldugin, V. A., and V. A. Pentegova: Absolute Configuration of Isocembrol and Its Stereospecific Synthesis from Cembrene. Khim. Prir. Soedin. **1977,** 577.
106. Kato, T., M. Suzuki, M. Takahashi, and Y. Kitahara: Cyclization of polyenes. XXII. Synthesis and Stereochemistry of Thunbergol. Chem. Lett. **1977,** 465.
107. Kato, T., C. C. Yen, T. Uyehara, and Y. Kitahara: Cyclization of Polyenes. XXIII. Synthesis and Stereochemistry of Isoincensole Oxide. Chem. Lett. **1977,** 565.
108. Raldugin, V. A., L. Ya. Korokikh, A. I. Rezvukhin, and V. A. Pentegova: Oxidative Transformations of Cembrane Diterpenoids. III. Epoxy Cembrenes. Khim. Prir. Soedin. **1977,** 525.
109. Kodama, M., K. Shimada, and S. Itô: Syntheses of Macrocyclic Terpenoids by Intramolecular Cyclization. III. Model Reactions for the Synthesis of Cembrane and Thunbergol Derivatives. Tetrahedron Letters **1977,** 2763.
110. Itô, S., and M. Kodama: Cembrene-A. Japan. Kokai **77 31, 049.** (Cl. C07C13/02), 09 Mar 1977, Appl. 75/105, 782, 01 Sep 1975. [CA. **87,** 201818t 1977.]

111. KITAHARA, Y., T. KATO, and T. KOBAYASHI: 2-Isopropyl-5,9,13-trimethylcyclotetradecanone. Japan. Kokai **77 36, 648** (Cl. C07C49/27), 22 Mar 1977, Appl. 75/112, 385, 17 Sep. 1975. [CA **87,** 136062a, 1977.]
112. — — — 2-(1-Methyl-1-chloroethyl)-5,9,13-trimethyl-cyclotetradeca-5,9,13-trienone. Japan Kokai **77 36 647** (Cl. C07C49/27), 22 Mar 1977, Appl. 75/112, 384. 17 Sep 1975. [CA **87,** 136063b, 1977.]
113. — — — Unsaturated 14-membered cyclic compounds. Japan Kokai **77 39, 650** (Cl. C07C33/05), 28 Mar 1977, Appl. 75/114, 515, 22 Sep 1975. [CA **87,** 117974h, 1977.]
114. — — — Unsaturated 14-membered cyclic compounds. Japan Kokai **77 39, 651** (Cl. C07C33/05), 28 Mar 1977, Appl. 75/114, 516, 22 Sep 1975. [CA **87,** 117975i, 1977.]

(Received January 4, 1978)

(77) [illegible], Y., T. Kato, and [illegible]: 2-[illegible]-1,9-[illegible]cyclotetradecadiene. Japan. Kokai 77 36, 6-8 (Cl. C07C 49/27), 22 Mar. 1977, Appl. 75/112 [illegible], 19 Sep. 1975. [CA 87, [illegible] (1977)].

(78) [illegible] Methyl-[illegible]-[illegible]-cyclotetradecadiene [illegible]. Japan. Kokai 77 36,647 (Cl. C07C 49/27), 22 Mar. 1977, Appl. 75/112, [illegible] 19 Sep. 1975. [CA 87, [illegible] (1977)].

(79) [illegible] Unsaturated 14-membered cyclic compounds. Japan. Kokai 77 [illegible] (Cl. C07C [illegible]), 28 Mar. 1977, Appl. 75/114, 511, 22 Sep. 1975. [CA 87, [illegible] (1977)].

(80) [illegible] Unsaturated 14-membered cyclic compounds. Japan. Kokai 77 [illegible] (Cl. C07C [illegible]), 28 Mar. 1977, Appl. 75/114, [illegible], 22 Sep. 1975. [CA 87, [illegible] (1977)].

(Received January [illegible], 1978)

Namenverzeichnis. Author Index

Kursiv gedruckte Seitenzahlen beziehen sich auf Literaturverzeichnisse

Page numbers printed in *italics* refer to References

Sachverzeichnis. Subject Index

Von · By

A. SIEGEL, Wien

Fortschritte der Chemie organischer Naturstoffe

Progress in the Chemistry of Organic Natural Products

Volume 34: 63 figures. X, 620 pages. 1977. ISBN 3-211-81415-9.

Contents: C. R. ENZELL, I. WAHLBERG, and A. J. AASEN, Isoprenoids and Alkaloids of Tobacco – A. R. PINDER, The Chemistry of the Eremophilane and Related Sesquiterpenes – D. GROSS, Phytoalexine und verwandte Pflanzenstoffe – K. H. OVERTON and D. J. PICKEN, Studies in Secondary Metabolism with Plant Tissue Cultures – D. P. CHAKRABORTY, Carbazole Alkaloids – J. JACOB, Bürzeldrüsenlipide – W. VOELTER, Hypothalamus-Regulationshormone – Author Index – Subject Index.

Volume 35: VIII, 589 pages. 1978. ISBN 3-211-81460-4.

Contents: O. R. GOTTLIEB, Neolignans – K. HERRMANN, Hydroxyzimtsäuren und Hydroxybenzoesäuren enthaltende Naturstoffe in Pflanzen – G. PATTENDEN, Natural 4-Ylidenebutenolides and 4-Ylidenetetronic Acids – R. D. H. MURRAY, Naturally Occurring Plant Coumarins – G. OHLOFF, Recent Developments in the Field of Naturally-Occurring Aroma Components – Author Index – Subject Index.

Springer-Verlag Wien · New York

the EU authorized representative is:
Springer Nature Customer Service Center GmbH
Europaplatz 3, 69115 Heidelberg, Germany

MIX
Papier aus verantwortungsvollen Quellen
Paper from responsible sources
FSC® C105338

If you have any concerns about our products,
you can contact us on
ProductSafety@springernature.com

In case Publisher is established outside the EU,
the EU authorized representative is:
Springer Nature Customer Service Center GmbH
Europaplatz 3, 69115 Heidelberg, Germany

Printed by Libri Plureos GmbH
in Hamburg, Germany